D1484147

Die Grundlehren der mathematischen Wissenschaften

in Einzeldarstellungen
mit besonderer Berücksichtigung
der Anwendungsgebiete

Band 171

Ivan Singer

Best Approximation in Normed Linear Spaces by Elements of Linear Subspaces

Publishing House of the Academy
of the Socialist Republic of Romania, Bucharest 1970
Springer-Verlag New York • Heidelberg • Berlin 1970

Prof. Ivan Singer
Academy of the Socialist Republic of Romania
Institute of Mathematics, Bucharest

This monograph is a translation of the original Romanian version
"Cea mai bună aproximare în spații vectoriale normate prin elemente din subspații vectoriale"

Translated by Radu Georgescu

 Library of Congress Catalog Card Number 73–110407
Title No. 5154

CONTENTS

	Page
CONTENTS	5
PREFACE	9
PREFACE TO THE ENGLISH EDITION	11
INTRODUCTION	13

Chapter I

BEST APPROXIMATION IN NORMED LINEAR SPACES BY ELEMENTS OF ARBITRARY LINEAR SUBSPACES 17

§1. Characterizations of elements of best approximation 17

1.1. The first theorem of characterization of elements of best approximation in general normed linear spaces 18
1.2. Geometrical interpretation 24
1.3. Applications in the spaces $C(Q)$ 29
1.4. Applications in the spaces $C_R(Q)$ 33
1.5. Applications in the spaces $L^1(T, \nu)$ 45
1.6. Applications in the spaces $C^1(Q, \nu)$ and $C^1_R(Q, \nu)$ 55
1.7. Applications in the spaces $L^p(T, \nu)$ $(1 < p < \infty)$ and in inner product spaces 56
1.8. The second theorem of characterization of elements of best approximation in general normed linear spaces 58
1.9. Geometrical interpretation 67
1.10. Applications and geometrical interpretation in the spaces $C(Q)$ 69
1.11. Applications in linear subspaces of the spaces $C(Q)$ 75
1.12. Applications in the spaces $L^1(T, \nu)$ 83
1.13. Other characterizations of elements of best approximation in general normed linear spaces 87
1.14. Orthogonality in general normed linear spaces 91

§2. Existence of elements of best approximation 93

§3. Uniqueness of elements of best approximation 103

3.1. Uniqueness of elements of best approximation in general normed linear spaces 103
3.2. Applications in the spaces $C(Q)$ and $C_R(Q)$ 117

3.3. Applications in the spaces $L^1(T, \nu)$ and $L^1_R(T, \nu)$ 120
3.4. Applications in the spaces $C^1(Q, \nu)$ and $C^1_R(Q, \nu)$ 123

§4. **k-dimensional $\mathfrak{L}_G(x)$ sets** 125

4.1. Preliminaries 125
4.2. k-semi-Čebyšev linear subspaces in general normed linear spaces 126
4.3. Applications in the spaces $C(Q)$ and $C_R(Q)$ 131
4.4. Applications in the spaces $L^1(T, \nu)$, $L^1_R(T, \nu)$, $C^1(Q, \nu)$ and $C^1_R(Q, \nu)$ 133

§5. **Interpolative best approximation, best approximation by elements of linear manifolds and their equivalence to best approximation by elements of linear subspaces** 135

§6. **The operators π_G and the functionals e_G. Deviations. Elements of ε-approximation** 139

6.1. The operators π_G 140
6.2. The functionals e_G 147
6.3. The functionals e_{G_n} for increasing or decreasing sequences $\{G_n\}$ of closed linear subspaces 151
6.4. The deviation of a set from a linear subspace 156
6.5. Elements of ε-approximation 162

Chapter II

BEST APPROXIMATION IN NORMED LINEAR SPACES BY ELEMENTS OF LINEAR SUBSPACES OF FINITE DIMENSION.................................... 165

§1. **Characterizations of polynomials of best approximation** 166

1.1. Preliminary lemmas 166
1.2. Characterizations of polynomials of best approximation in general normed linear spaces 170
1.3. Applications in the spaces $C(Q)$, $C_R(Q)$ and $C_0(T)$ . . . 178
1.4. The conjugate space of the space $C_E(Q)$ and the extremal points of its unit cell 191
1.5. Applications in the spaces $C_E(Q)$ 201
1.6. Applications in the spaces $L^1(T, \nu)$, $L^1_R(T, \nu)$, $C^1(Q, \nu)$ and $C^1_R(Q, \nu)$ 203

§2. **Uniqueness of polynomials of best approximation** 206

2.1. Preliminary lemmas 206
2.2. Finite dimensional Čebyšev subspaces in general normed linear spaces 210
2.3. Applications in the spaces $C(Q)$, $C_R(Q)$, $C_0(T)$ and $L^\infty(T, \nu)$ 215
2.4. Applications in the spaces $C_E(Q)$ 225
2.5. Applications in the spaces $L^1(T, \nu)$, $L^1_R(T, \nu)$, $C^1(Q, \nu)$ and $C^1_R(Q, \nu)$ 226

§3. **Finite dimensional k-Čebyšev subspaces** 237

3.1. Finite dimensional k-Čebyšev subspaces in general normed linear spaces 238
3.2. Applications in the spaces $C(Q)$ and $C_R(Q)$ 240

§4. **Polynomial interpolative best approximation. Best approximation by elements of finite dimensional linear manifolds** 242

4.1. The case of general normed linear spaces 242
4.2. Applications in the spaces $C(Q)$ and $C_R(Q)$ 244

§5. **The operators π_G and the functionals e_G for linear subspaces G of finite dimension** 246

5.1. The operators π_G for linear subspaces G of finite dimension 246
5.2. The operators π_{G_n} for increasing sequences $\{G_n\}$ of linear subspaces of finite dimension 252
5.3. The functionals e_{G_n} for increasing sequences $\{G_n\}$ of linear subspaces of finite dimension 262

§6. **n-dimensional diameters. Best n-dimensional secants** 268

6.1. Preliminary lemmas 269
6.2. n-dimensional diameters 274
6.3. Best n-dimensional secants 282
6.4. Best n-dimensional $\mathfrak{V}$-secants. Čebyšev centers. Closest points to a set. Best n-nets. Best n-coverings 287

Chapter III

BEST APPROXIMATION IN NORMED LINEAR SPACES BY ELEMENTS OF CLOSED LINEAR SUBSPACES OF FINITE CODIMENSION 291

§1. **Best approximation by elements of factor-reflexive closed linear subspaces** 292

§2. **Best approximation by elements of closed linear subspaces of finite codimension** 295

2.1. Best approximation by elements of closed linear subspaces of finite codimension in general normed linear spaces . . . 295
2.2. Applications in the spaces $C_R(Q)$ 302
2.3. Applications in the spaces $L^1_R(T, \nu)$ 325

§3. **Best approximation in conjugate spaces by elements of weakly* closed linear subspaces of finite codimension** 333

3.1. Weakly* closed Čebyšev subspaces of finite codimension in general conjugate spaces 333
3.2. Applications in the spaces $L^1_R(T, \nu)^*$ 335
3.3. Applications in the spaces $C_R(Q)^*$, $L^\infty_R(T, \nu)^*$ and $((c_0)_R)^*$ 339

§4. **The operators π_G and the functionals e_G for closed linear subspaces G of finite codimension. Diameters of order n** 350

4.1. The operators π_G for closed linear subspaces G of finite codimension 350
4.2. The operators π_{G^n} for decreasing sequences $\{G^n\}$ of closed linear subspaces of finite codimension 353
4.3. The functionals e_{G^n} for decreasing sequences $\{G^n\}$ of closed linear subspaces of finite codimension 355
4.4. Diameters of order n 357

Appendix I

BEST APPROXIMATION IN NORMED LINEAR SPACES BY ELEMENTS OF NON-LINEAR SETS 359

§1. Best approximation by elements of convex sets 360
§2. The problem of convexity of Čebyšev sets 364
§3. Best approximation by elements of finite dimensional surfaces 371
§4. Best approximation by elements of arbitrary sets 374

Appendix II

BEST APPROXIMATION IN METRIC SPACES BY ELEMENTS OF ARBITRARY SETS 377

§1. Properties of the sets $\mathfrak{P}_G(x)$. A characterization of elements of best approximation 379

§2. Proximinal sets 382

§3. Properties of the mappings $\mathfrak{P}_G$ 386

§4. Properties of the mappings π_G and of the functionals e_G 390

BIBLIOGRAPHY 393

PREFACE

The existing monographs on approximation theory and on the constructive theory of functions (e.g. N. I. Ahiezer [1], J. R. Rice [192], A. F. Timan [250], I. P. Natanson [158] etc.), treat the problems of best approximation in classical style, with the methods and language of the theory of functions, the use of functional analysis being reduced to a few elementary results on best approximation in normed linear spaces and in Hilbert spaces. In contrast to these, the present monograph attempts to give a modern theory of best approximation, using in a consequent manner the methods of functional analysis.

From the vast field of best approximation, this monograph presents in more detail the results on best approximation in normed linear spaces by elements of linear subspaces, which today constitute a unified theory. The more general problems of best approximation are exposed, briefly, in Appendices I and II. A glance at the table of contents shows that together with results in general normed linear spaces there are given many applications of them in various concrete spaces.

In order to limit the size of the present monograph, we deliberately have omitted some problems related to those treated herein (e.g. applications to extremal problems of the theory of analytic functions, methods of computation of elements of best approximation, the problem of moments, connections with linear programming, etc.). Also, some results related to those presented here are mentioned without proof.

Being the first of this kind in the literature, the present monograph is based exclusively on papers published in mathematical journals; the references to these are given in the text after each result. The bibliography given at the end does not aim at being complete, it includes only papers which are effectively quoted in the text.

The present monograph is intended for a large circle of mathematicians. Firstly, it is addressed to specialists in approximation theory and the constructive theory of functions, offering to them the methods of functional analysis for the study of these classical domains of mathematics; the necessity and advantages of these methods are shown in the "Introduction". Secondly, it is addressed to those working in functional analysis, offering them an important field of applications. Also, taking into account that in the problems investigated there are combined the methods of functional analysis, geometry, general topology, measure theory and other mathematical disciplines, we hope that the present monograph will be useful to other categories of readers as well (e.g. to specialists in the geometry of convex bodies, etc.).

The reader is assumed to know the elements of functional analysis and the mathematical disciplines mentioned above, e.g. within the limits of university courses. However, in order to facilitate the reading and to make the book accessible to University students as well, we have indicated, in connection with the results used (of functional analysis, theory of measure, etc.), a reference to a treatise containing the proof of the respective result; when we have used results which are not to be found in monographs but only in papers published in journals, we have mentioned them in the form of lemmas, giving also their proof.

In conclusion, I wish to express my thanks to Miron Nicolescu, member of the Academy, for the invitation to write this monograph and for the constant interest shown during its elaboration. Also, I extend my thanks to my colleagues N. Dinculeanu and C. Foiaș for valuable discussions on the proofs of certain theorems and to V. Klee of the University of Washington in Seattle for some bibliographical indications.

Bucharest, July 1, 1966

THE AUTHOR

PREFACE TO THE ENGLISH EDITION

This is a translation of the original Romanian monograph, with a number of corrections of misprints and errors. I wish to express my thanks to those friends and colleagues, G. Godini (Bucharest), J. Blatter and G. Pantelidis (Bonn), A. Garkavi (Moscow), M. I. Kadec (Harkov), G. Alexits (Budapest) and others, who have called my attention to some of these corrections.

State College, Pennsylvania IVAN SINGER

November 29, 1968

INTRODUCTION

Starting from problems concerning certain mechanisms (e.g. the motion of the connecting-rod of a steam engine), P. L. Čebyšev was led to state, a century ago [32], the problem of finding, for a real continuous function $x(t)$ on a segment $[a, b]$, an algebraic polynomial $g_0(t) = \sum_{i=1}^{n} \alpha_i^{(0)} t^{i-1}$ of degree $\leqslant n-1$ such that the "deviation" of the polynomial $g_0(t)$ from the function $x(t)$ on the segment $[a, b]$ be the least possible among the deviations of all algebraic polynomials $g(t) = \sum_{i=1}^{n} \alpha_i t^{i-1}$ of degree $\leqslant n-1$, or, in other words, the problem of *best approximation* of the function $x(t)$ by algebraic polynomials $g(t)$ of degree $\leqslant n-1$.

In the mechanical problems considered, for the measuring of the deviation between $g(t)$ and $x(t)$ on the segment $[a, b]$, P. L. Čebyšev has found as being the most suitable the number $\max_{t \in [a,b]} |x(t) - g(t)|$; thus, the problem amounts to the minimization of this maximum when $g(t)$ runs over the set of all algebraic polynomials of degree $\leqslant n-1$. In order to include other important cases as well, the problem has been generalized by other mathematicians, the interval $[a, b]$ being replaced by a compact space Q, the real-valued functions by complex-valued functions or by functions with values in more general spaces, and the algebraic polynomials $g(t) = \sum_{i=1}^{n} \alpha_i t^{i-1}$ by the linear combinations $g(q) = \sum_{i=1}^{n} \alpha_i x_i(q)$ of a system of n continuous functions $x_1(q), \ldots, x_n(q)$ linearly independent on Q, or even by elements of a family $\{g(\alpha_1, \ldots, \alpha_n ; q)\}$ of continuous functions on Q, depending on n scalar parameters $\alpha_1, \ldots, \alpha_n$.

On the other hand, practical necessities have required the consideration of certain not necessarily continuous functions

$x(t)$, $g(t)$ and the measuring of their "deviation" on the segment $[a, b]$ by other numbers, for example by the number $\sqrt{\int_a^b |x(t) - g(t)|^2 dt}$ ("the mean-square deviation" used e.g. in the "method of least squares"), $\int_a^b |x(t) - g(t)| dt$, etc. Obviously, a polynomial $g_0(t)$ which is the solution of the minimum problem of Čebyšev is not, in general, a solution of the minimum problem obtained by taking one of the above integral deviations, and conversely. The results which have been obtained on best approximation in the sense of Čebyšev (for instance on the characterization or uniqueness of polynomials $g_0(t)$ "of best approximation") are completely different from those obtained for the case of integral deviations, apparently with no connection to the former. Also, the methods (of the theory of functions) by which these results have been obtained are different too, requiring different artifices for each definition of the deviation between $x(t)$ and $g(t)$.

At this stage there arose, in a natural way, the problem: what is the cause of the diversity of these results and methods? Are there general methods for obtaining all these results? Could all these results be particular cases of a more general theory? The answer is affirmative. Namely, the above functions $x(t)$ and $g(t)$ can be considered as "points" x, g of more general "spaces" (e.g. of the space of all continuous functions on $[a, b]$, or the space of all square summable functions on $[a, b]$, etc.) and the above deviation between these functions can be considered as the "distance" between the points x, g in the respective space. The problem of best approximation amounts then to the problem of finding, for a given point and a given set G in a metric space E, a point $g_0 \in G$ which should be nearest to x among all points of the set G, i.e. such that

$$\rho(x, g_0) = \inf_{g \in G} \rho(x, g),$$

where ρ denotes the distance in the metric space E. However, in order to obtain a satisfactory theory of best approximation, it is convenient to observe that the above "spaces" of functions have not only the property that for any two functions there is defined their deviation, the "distance" between them, but also the property that the "sum" of two functions from the space, i.e. the function $(x + y)(t) = x(t) + y(t)$, as well as the "product" of a function from the space with a scalar, i.e. the function $(\alpha x)(t) = \alpha x(t)$, belong to the same space. Therefore, we shall take as E not an arbitrary metric space, but a normed linear space; naturally, the distance in E is that induced by the norm, i.e.

$$\rho(x, y) = \|x-y\| \qquad (x, y \in E).$$

Thus, the problem of best approximation consists in finding, for a given element x and a given set G in a normed linear space E, an element $g_0 \in G$ such that

$$\|x - g_0\| = \inf_{g \in G} \|x - g\|.$$

Every $g_0 \in G$ with this property is called an *element of best approximation* of x (by elements of the set G). We shall denote by $\mathfrak{L}_G(x)$ the set of all elements of best approximation of x by elements of the set G, i.e.

$$\mathfrak{L}_G(x) = \{g_0 \in G \mid \|x - g_0\| = \inf_{g \in G} \|x - g\|\},$$

and for brevity, instead of "g_0 is an element of best approximation of x by elements of the set G", we shall write

$$g_0 \in \mathfrak{L}_G(x).$$

In the present monograph we shall deal with the case when G is a linear subspace of E. The cases when G is a more general set in a normed linear space E or in an arbitrary metric space will be considered, briefly, in Appendices I and II.

The study of the problem of best approximation within the general framework of normed linear spaces has, among others, the following advantages as compared with its study by means of the classical methods of the theory of functions : a) From the results on best approximation in normed linear spaces one obtains as particular cases both the well known classical results in various concrete functional spaces (the space of continuous functions, the space of integrable functions, etc.) and many new results in these spaces; various known results, which up to now seemed to have no connection with each other, are thus obtained in a unified way. b) Within the framework of normed linear spaces the problem of best approximation amounts to the problem of minimizing a distance, hence it is geometrized, and thus in its study one can use arguments based on geometric intuition (but finally rigorously analytic). c) Within a more general framework, the mutual connections of the phenomena become clearer and the arguments simpler, than in certain particular cases. The — otherwise ingenious — particular artifices, used in the classical theory to prove certain theorems on best approximation, are more complicated and more tedious than the arguments of the theory of normed linear spaces, which give general theorems on best approximation, containing the preceding ones as particular cases (see e.g. the theorems on characterization or uniqueness of polynomials of best approximation).

The remark that normed linear spaces constitute the natural framework for the study of the problem of best approximation was made in 1938 by M. Nicolescu [161] and, independently, by M. G. Krein [125], who have also obtained the first results in this direction. After a relatively stagnant period of about twenty years, new methods and results have been given in the papers [209]—[234], which have made possible the development of a general theory of best approximation in normed linear spaces. Thus, important results on best approximation by elements of linear subspaces, especially in concrete spaces, have been obtained by A. L. Garkavi ([66]—[74]), R. R. Phelps ([176]—[177 a]) and others. A substantial part of the present monograph is based on the papers [209]—[234]. Although in the first ones of these papers the results have been given only for real Banach spaces, in the papers [230], [232] it has been remarked that they remain valid, with the same proof, for general (real or complex) normed linear spaces; in the monograph these results are given in their general form, with reference to the first papers in which they appear.

CHAPTER I

BEST APPROXIMATION IN NORMED LINEAR SPACES BY ELEMENTS OF ARBITRARY LINEAR SUBSPACES

§ 1. CHARACTERIZATIONS OF ELEMENTS OF BEST APPROXIMATION

In the present paragraph we shall give characterizations of elements of best approximation and some consequences of these characterizations in arbitrary (complex or real) normed linear spaces, and we shall apply them to various concrete spaces. Since we have *)

$$\mathfrak{L}_G(x) = \begin{cases} x & \text{for } x \in G \\ \emptyset & \text{for } x \in \overline{G} \setminus G, \end{cases} \tag{1.1}$$

for any linear subspace G of a normed linear space E, it will be sufficient to characterize the elements of best approximation of the elements $x \in E \setminus \overline{G}$. In order to exclude the trivial case when such elements x do not exist, throughout the sequel by "linear subspace" $G \subset E$ we shall understand "proper linear subspace G which is not dense in E", that is, we shall assume, without special mention, that $\overline{G} \neq E$.

*) We recall (see the "Introduction") that by $\mathfrak{L}_G(x)$ we denote the set of all elements of best approximation of x (by means of the elements of G), i.e. the set $\{g_0 \in G \mid \|x - g_0\| = \inf_{g \in G} \|x - g\|\}$.

1.1. THE FIRST THEOREM OF CHARACTERIZATION OF ELEMENTS OF BEST APPROXIMATION IN GENERAL NORMED LINEAR SPACES

The following theorem, although obtained by a simple application of a corollary of the Hahn-Banach theorem, has useful applications in various concrete spaces, as will be shown in the subsequent sections.

THEOREM 1.1 ([212], theorem 1.1). *Let E be a normed linear space, G a linear subspace of E, $x \in E \setminus \overline{G}$ and $g_0 \in G$. We have $g_0 \in \mathfrak{L}_G(x)$ if and only if there exists an*) $f \in E^*$ with the following properties:*

$$\|f\| = 1, \tag{1.2}$$

$$f(g) = 0 \qquad (g \in G), \tag{1.3}$$

$$f(x - g_0) = \|x - g_0\|. \tag{1.4}$$

Proof. Assume that $g_0 \in \mathfrak{L}_G(x)$. Then, since $x \in E \setminus \overline{G}$, we have**) $\|x - g_0\| = \rho(x, G) > 0$. Consequently, by virtue of a well known ***) corollary of the Hahn-Banach theorem, there exists an $f_0 \in E^*$ such that $\|f_0\| = \frac{1}{\|x-g_0\|}$, $f_0(g) = 0$ $(g \in G)$ and $f_0(x) = 1$, whence the functional $f = \|x - g_0\| f_0 \in E^*$ satisfies (1.2), (1.3) and (1.4).

Conversely, assume that there exists an $f \in E^*$ satisfying (1.2), (1.3) and (1.4). Then for any $g \in G$ we have

$$\|x - g_0\| = |f(x - g_0)| = |f(x - g)| \leqslant \|f\| \, \|x - g\| = \|x - g\|,$$

whence $g_0 \in \mathfrak{L}_G(x)$, which completes the proof of theorem 1.1.

We observe that the sufficiency part of theorem 1.1, and under the natural assumption that $\overline{G} \neq E$, also the necessity part of this theorem, remain true for every $x \in E$ (since for $x \in \overline{G}$ and $g_0 \in \mathfrak{L}_G(x)$ we have, by (1.1), $x = g_0$). This remark remains valid also for the other results in general normed linear spaces, containing the assumption $x \in E \setminus \overline{G}$, which will be given in the sequel. On the other hand, as will be seen in the subsequent sections, *in applications in concrete normed linear spaces the hypothesis $x \in E \setminus \overline{G}$ will be essential* and cannot be replaced by the hypothesis $x \in E$. In fact, $x \in E \setminus \overline{G}$ implies $x - g_0 \neq 0$,

---

*) We denote by E^* the conjugate space of E, i.e. the space of all continuous linear functionals on E, endowed with the usual vector operations and with the norm $\|f\| = \sup_{\substack{x \in E \\ \|x\| \leqslant 1}} |f(x)|$.

**) We recall (see the "Introduction") that we denote by ρ the distance defined by the norm in E, i.e. $\rho(x, y) = \|x - y\|$ $(x, y \in E)$.

***) See e.g. N. Dunford and J. Schwartz [49], p. 64, lemma 12.

whence, by (1.4) and (1.2) it follows that f has a non-zero "maximal element" and in concrete normed linear spaces we shall make full use of this property of f; also, in certain constructions in concrete spaces we shall make direct use of the hypothesis $x \in E \setminus \bar{G}$.

We shall give now a number of equivalent variants of the conditions of theorem 1.1. For this purpose let us prove first

LEMMA 1.1 *Let E be a normed linear space, G a linear subspace of $E, x \in E \setminus \bar{G}$, $g_0 \in G$ and $f \in E^*$. Then*

a) *f satisfies (1.2) and (1.4) if and only if it satisfies (1.2) and*

$$\operatorname{Re} f(x - g_0) = \|x - g_0\|. \tag{1.5}$$

b) *f satisfies (1.3) if and only if*

$$\operatorname{Re} f(g) = 0 \qquad (g \in G). \tag{1.6}$$

c) *f satisfies (1.2), (1.3) and*

$$|f(x - g_0)| = \|x - g_0\|, \tag{1.7}$$

if and only if *) $f_1 = [\operatorname{sign} f(x - g_0)]f$ *satisfies (1.2), (1.3) and (1.4).*

d) *f satisfies (1.2), (1.3) and*

$$|\operatorname{Re} f(x - g_0)| = \|x - g_0\|, \tag{1.8}$$

if and only if either $f_1 = f$, or $f_2 = -f$ satisfies (1.2), (1.3) and (1.4).

Proof. a) Obviously $(1.4) \Rightarrow (1.5)$, hence $(1.2) \cap (1.4) \Rightarrow$ $\Rightarrow (1.2) \cap (1.5)$. Conversely, if f satisfies (1.2) and (1.5), we have

$$\|x - g_0\| = \operatorname{Re} f(x - g_0) \leqslant |f(x - g_0)| \leqslant \|x - g_0\|,$$

whence $\operatorname{Re} f(x - g_0) = |f(x - g_0)|$, and consequently $f(x - g_0)$ is real and positive. Taking into account (1.5), it follows that we have (1.4).

b) Obviously $(1.3) \Rightarrow (1.6)$. Conversely, if f satisfies (1.6), then, taking into account that $ig \in G$ and the well known **) relation $\operatorname{Im} f(g) = -\operatorname{Re} f(ig)$, we have

$$f(g) = \operatorname{Re} f(g) - i \operatorname{Re} f(ig) = 0 \qquad (g \in G).$$

*) We recall that for a complex number α, by definition

$$\operatorname{sign} \alpha = \begin{cases} e^{-i \arg \alpha} = \dfrac{\bar{\alpha}}{|\alpha|} & \text{for } \alpha \neq 0, \\ 0 & \text{for } \alpha = 0. \end{cases}$$

**) See e.g. N. Dunford and J. Schwartz [49], p. 64.

c) If $f_1 = [\operatorname{sign} f(x - g_0)]f$ satisfies (1.2), (1.3) and (1.4) then obviously $f = e^{i \arg f(x-g_0)} f_1$ satisfies (1.2), (1.3) and (1.7). Conversely, if f satisfies (1.2), (1.3) and (1.7), f_1 satisfies (1.2), (1.3) and

$$f_1(x - g_0) = |f(x - g_0)| = \|x - g_0\|.$$

d) If $f_1 = f$ or $f_2 = -f$ satisfies (1.2), (1.3) and (1.4), then, obviously, f satisfies (1.2), (1.3) and (1.8). Conversely, if f satisfies (1.2), (1.3) and (1.8), we have

$$\|x - g_0\| = |\operatorname{Re} f(x - g_0)| \leqslant |f(x - g_0)| \leqslant \|x - g_0\|,$$

whence $|\operatorname{Re} f(x - g_0)| = |f(x - g_0)|$. Consequently, $f(x - g_0)$ is real, whence either $f_1 = f$, or $f_2 = -f$ satisfies (1.2), (1.3) and (1.4), which completes the proof.

Using lemma 1.1, one obtains immediately the following corollary of theorem 1.1 :

COROLLARY 1.1. *Let E be a normed linear space, G a linear subspace of E, $x \in E \setminus \overline{G}$ and $g_0 \in G$. The following statements are equivalent* :

1° $g_0 \in \mathfrak{P}_G(x)$.
2° *There exists an $f \in E^*$ satisfying (1.2), (1.3) and (1.5).*
3° *There exists an $f \in E^*$ satisfying (1.2), (1.3) and (1.7).*
4° *There exists an $f \in E^*$ satisfying (1.2), (1.3) and (1.8).*
5° *There exists an $f \in E^*$ satisfying (1.2), (1.6) and (1.5).*
6° *There exists an $f \in E^*$ satisfying (1.2), (1.6) and (1.8).*

Of course, using lemma 1.1 one can also give other equivalent conditions, but those above admit significant geometrical interpretations, as will be shown in section 1.2.

For a linear subspace G of a normed linear space E we shall denote by $G^{\perp}$ the set

$$G^{\perp} = \{f \in E^* \mid f(g) = 0 \quad \text{for all} \quad g \in G\}. \tag{1.9}$$

If Γ is a linear subspace of a conjugate space E^* (where E is a normed linear space), we shall denote by $\Gamma_{\perp}$ the set

$$\Gamma_{\perp} = \{x \in E \mid \gamma(x) = 0 \quad \text{for all} \quad \gamma \in \Gamma\}, \tag{1.10}$$

and by $\|x\|_{\Gamma}$ the numbers

$$\|x\|_{\Gamma} = \sup_{\substack{\gamma \in \Gamma \\ \|\gamma\| \leqslant 1}} |\gamma(x)| \qquad (x \in E). \tag{1.11}$$

For any linear subspace Γ of E^* we have, obviously,

$$\|x\|_{\Gamma} \leqslant \|x\| \qquad (x \in E). \tag{1.12}$$

With these notations, we can state the following corollary of theorem 1.1 :

COROLLARY 1.2 ([228], theorems 1 and 4). a) *Let E be a normed linear space, G a linear subspace of E, $x \in E \setminus \overline{G}$ and $g_0 \in G$. We have $g_0 \in \mathfrak{P}_G(x)$ if and only if*

$$\|x - g_0\|_{G^\perp} = \|x - g_0\|. \tag{1.13}$$

b) *Let E^* be the conjugate space of a normed linear space E, let Γ be a $\sigma(E^*, E)$-closed linear subspace of E^*, and let $f \in E^* \setminus \Gamma = E^* \setminus \overline{\Gamma}$, $\gamma_0 \in \Gamma$. We have $\gamma_0 \in \mathfrak{P}_\Gamma(f)$ if and only if* *)

$$\|(f - \gamma_0)|_{\Gamma_\perp}\| = \|f - \gamma_0\|. \tag{1.14}$$

Proof. a) Assume that $g_0 \in \mathfrak{P}_G(x)$. Then, by virtue of theorem 1.1, there exists an $f \in E^*$ satisfying (1.2), (1.3) and (1.4). Consequently,

$$\|x - g_0\|_{G^\perp} = \sup_{\substack{f' \in G^\perp \\ \|f'\| \leqslant 1}} |f'(x - g_0)| \geqslant |f(x - g_0)| = \|x - g_0\|,$$

which, together with (1.12), gives (1.13).

Conversely, if we have (1.13), then

$$\|x - g_0\| = \|x - g_0\|_{G^\perp} = \|x - g\|_{G^\perp} \leqslant \|x - g\| \quad (g \in G),$$

whence $g_0 \in \mathfrak{P}_G(x)$.

b) Assume that $\gamma_0 \in \mathfrak{P}_\Gamma(f)$. Then, since Γ is $\sigma(E^*, E)$-closed, for any number c such that $0 < c < \|f - \gamma_0\|$ there exists, by virtue of a theorem of S. Banach **), an element $x'_c \in \Gamma_\perp$ such that

$$\|x'_c\| \leqslant \frac{1}{c}, \quad f(x'_c) = 1.$$

Consequently, the element $x_c \in \Gamma_\perp$, defined by

$$x_c = \frac{1}{\|x'_c\|} x'_c,$$

satisfies

$$\|x_c\| = 1, \quad f(x_c) = \frac{1}{\|x'_c\|} \geqslant c.$$

---

*) For $\varphi \in E^*$ we denote by $\varphi|_{\Gamma_\perp}$ the restriction of φ to the linear subspace $\Gamma_\perp$ of E.

**) See e.g. S. Banach [8], p. 122, theorem 1. Other simpler proofs of this theorem have been given in the papers [220] and [221].

Since c has been an arbitrary number with the property that $0<c<\|f-\gamma_0\|$, it follows that we have

$$\|(f-\gamma_0)|_{\Gamma_\perp}\| = \sup_{\substack{x\in\Gamma_\perp \\ \|x\|\leqslant 1}} |(f-\gamma_0)(x)| = \sup_{\substack{x\in\Gamma_\perp \\ \|x\|\leqslant 1}} |f(x)| \geqslant \|f-\gamma_0\|.$$

Since the converse inequality is obvious, it follows that we have (1.14).

Conversely, the sufficiency of condition (1.14) is a consequence of the sufficiency of the condition in part a) proved above, since, obviously, we have the inequalities

$$\|(f-\gamma_0)|_{\Gamma_\perp}\| \leqslant \|f-\gamma_0\|_{\Gamma^\perp} \leqslant \|f-\gamma_0\|.$$

This completes the proof of corollary 1.2.

We observe that in the above proof only part a) has been obtained, in fact, as a corollary of theorem 1.1, whereas part b) has been obtained directly from Banach's theorem. One can also prove a result of the type of theorem 1.1 for $\sigma(E^*, E)$-closed linear subspaces of E^* (namely, that we have $\gamma_0\in\mathfrak{P}_\Gamma(f)$ if and only if for any number c with $0<c<\|f-\gamma_0\|$ there exists an element $x_c\in E$ such that $\|x_c\|=1$, $\gamma(x_c)=0$ $(\gamma\in\Gamma)$ and $f(x_c)\geqslant c$), and from this result one can deduce then corollary 1.2 b); however, we shall not make use of this result in the following.

We mention that the "duality" relations

$$\inf_{g\in G}\|x-g\| = \max_{\substack{f\in G^\perp \\ \|f\|\leqslant 1}} |f(x)| \qquad (x\in E),$$

$$\min_{\gamma\in\Gamma}\|f-\gamma\| = \sup_{\substack{x\in\Gamma_\perp \\ \|x\|\leqslant 1}} |f(x)| \qquad (f\in E^*)$$

(where G is a linear subspace of E and Γ a $\sigma(E^*, E)$-closed linear subspace of E^*), which obviously imply the necessity parts of theorem 1.1 and corollary 1.2, have been given, essentially, by M. G. Krein [125] and rediscovered subsequently by a number of other authors (see e.g. J. Dieudonné [44], S. M. Nikolsky [162], W. W. Rogosinsky [196], S. Ya. Havinson [85], H. S. Shapiro [207], F. F. Bonsall [20]). But theorem 1.1 of characterization of elements of best approximation has been given and used for the first time in the paper [212]; this fact is explicitly mentioned e.g. in the paper of R. C. Buck [29], p. 31.

Let us consider now, more generally, *the problem of simultaneous characterization* of a set of elements of best approximation, i.e. the following problem: given $E, G\subset E$ and x as above, and a subset M of G, what are the necessary and sufficient conditions in order that *every* element $g\in M$ be an element of best approximation of x (by means of the elements of G) ?

The answer is given by the following result, which is an immediate consequence of theorem 1.1*) and of the fact that $\|x-g_1\| = \|x-g_2\|$ for all pairs $g_1, g_2 \in M \subset \mathfrak{P}_G(x)$:

COROLLARY 1.3 ([229], p. 509). *Let E be a normed linear space, G a linear subspace of E, $x \in E \setminus \overline{G}$ and $M \subset G$. We have $M \subset \mathfrak{P}_G(x)$ if and only if there exists an $f \in E^*$ satisfying (1.2), (1.3) and*

$$f(x - g_0) = \|x - g_0\| \qquad (g_0 \in M). \tag{1.15}$$

Naturally, one can also give other equivalent conditions, corresponding to those of corollary 1.1.

By virtue of a well-known corollary of the Hahn-Banach theorem**), for any element x' of a normed linear space E there exists an $f \in E^*$ such that

$$\|f\| = 1, \ f(x') = \|x'\|.$$

The elements $x' \in E$ for which there exists only one such $f \in E^*$ are called, following M. G. Krein [125], *normal elements* ***) and in this case we shall denote the respective functional $f \in E^*$ by $f_{x'}$. Using theorem 1.1, one obtains the following characterization of the elements of best approximation g_0 for which $x - g_0$ is a normal element:

COROLLARY 1.4. *Let E be a normed linear space, G a linear subspace of E, $x \in E \setminus \overline{G}$ and $g_0 \in G$ such that $x - g_0$ is normal. We have $g_0 \in \mathfrak{P}_G(x)$ if and only if*

$$f_{x-g_0}(g) = 0 \qquad (g \in G). \tag{1.16}$$

Proof. Assume that $g_0 \in \mathfrak{P}_G(x)$. Then, by virtue of theorem 1.1 there exists an $f \in E^*$ satisfying (1.2), (1.3) and (1.4). Since $x - g_0$ is normal, it follows from (1.2) and (1.4) that $f = f_{x-g_0}$. Consequently, by virtue of (1.3) we have (1.16).

Conversely, if we have (1.16), then the functional f_{x-g_0} satisfies (1.2), (1.3) and (1.4), whence by virtue of theorem 1.1 we have $g_0 \in \mathfrak{P}_G(x)$, which completes the proof.

Another characterization of elements of best approximation g_0 for which $x - g_0$ is a normal element has been given in [230], theorem 8, but corollary 1.4 above is more convenient for applications in concrete spaces.

---

*) Obviously the converse statement is also true, theorem 1.1 being even a particular case of corollary 1.3.

**) See e.g. N. Dunford and J. Schwartz [49], p. 65, corollary 14.

***) Some authors use for normal elements $x' \in E$, the term "smooth point" (V. Klee [111]: smooth point; G. Köthe [122]: Flachpunkt) of the cell $S(0, \|x'\|)$, since the normal elements $x' \in E$ are the points in which the cell $S(0, \|x'\|)$ has only one support hyperplane. In the paper [211] we have used the term "monoplane point" of the cell $S(0, \|x'\|)$. The elements $x' \in E$ which are not normal are called sometimes "conical points" of the cell $S(0, \|x'\|)$ (N. V. Efimov and S. B. Stečkin [50]).

1.2. GEOMETRICAL INTERPRETATION

We recall that a set V in a normed linear space E is called a *linear manifold* if it is of the form $V = x_0 + G = \{x_0 + g \mid g \in G\}$, where G is a linear subspace of E. A closed linear manifold $H \subset E$ is called a *hyperplane* *) if there exists no closed linear manifold $H_1 \subset E$ such that $H \subset H_1$ and $H \neq H_1 \neq E$. For any $f \in E^*$, $f \neq 0$ and any scalar α, the set $\{y \in E \mid f(y) = \alpha\}$ is a hyperplane and conversely, for any hyperplane $H \subset E$ there exist an $f \in E^*$ and α uniquely determined by H up to a common scalar multiple, such that **)

$$H = \{y \in E \mid f(y) = \alpha\}. \tag{1.17}$$

LEMMA 1.2 (G. Ascoli [7]). *Let E be a normed linear space, H a hyperplane (1.17) of E and $x \in E$. Then the distance of the point x to the hyperplane H is*

$$\rho(x, H) = \frac{|f(x) - \alpha|}{\|f\|}. \tag{1.18}$$

Proof. For any $y \in H$, we have

$$\|x - y\| \geqslant \frac{|f(x - y)|}{\|f\|} = \frac{|f(x) - \alpha|}{\|f\|},$$

whence $\rho(x, H) \geqslant \frac{|f(x) - \alpha|}{\|f\|}$. On the other hand, if $0 < \varepsilon < \|f\|$, there exists a $z \in E$ such that $|f(z)| > (\|f\| - \varepsilon)\|z\|$. Multiplying this relation by $\left|\frac{f(x) - \alpha}{f(z)}\right|$ and putting

$$y = x - \frac{f(x) - \alpha}{f(z)} z,$$

we obtain $|f(x) - \alpha| > (\|f\| - \varepsilon)\|x - y\|$, whence

$$\|x - y\| < \frac{|f(x) - \alpha|}{\|f\| - \varepsilon}.$$

Since $\varepsilon > 0$ was arbitrary and $y \in H$, it follows that we have $\rho(x, H) \leqslant \frac{|f(x) - \alpha|}{\|f\|}$, which, together with the opposite inequality shown above, completes the proof of lemma 1.2.

*) Sometimes it is also used the term "closed hyperplane".

**) See e.g. G. Ascoli [7]; see also N. Bourbaki [23], Chap. I, p. 26, theorem 1.

Let $x \in E$ and $r > 0$. We say that a set $A \subset E$ *supports* the cell $S(x, r) = \{y \in E \mid \|y - x\| \leqslant r\}$, or that A is a *support set* of the cell $S(x, r)$, if $\rho(A, S(x, r)) = 0$ and $A \cap \operatorname{Int} S(x, r) = \emptyset$.

LEMMA 1.3. *A set A in a normed linear space E supports the cell $S = S(x, r) \subset E$ if and only if we have*

$$\rho(x, A) = r. \tag{1.19}$$

Proof. Assume that $\rho(x, A) = \alpha \neq r$. If $\alpha < r$, let $\varepsilon > 0$ be such that $\alpha + \varepsilon < r$. Then, taking an $y \in A$ such that $\|x - y\| \leqslant \rho(x, A) + \varepsilon = \alpha + \varepsilon < r$, we have $y \in A \cap \operatorname{Int} S$, and consequently A does not support the cell S. If $\alpha > r$, let $\varepsilon > 0$ be such that $\alpha > r + \varepsilon$. Then, taking an $y \in A$ such that $\rho(y, S) \leqslant \leqslant \rho(A, S) + \varepsilon$, we have $\rho(A, S) \geqslant \rho(y, S) - \varepsilon = \|y - x\| - - r - \varepsilon \geqslant \alpha - r - \varepsilon > 0$ and consequently A does not support the cell S.

Assume now that $\rho(x, A) = r$ and let $\varepsilon > 0$ be arbitrary. Then there exists an $y \in A$ such that $\|x - y\| \leqslant r + \varepsilon$. Put

$$z = \frac{\varepsilon}{r + \varepsilon} x + \frac{r}{r + \varepsilon} y.$$

We have then $\|x - z\| = \dfrac{r}{r + \varepsilon} \| x - y\| < r$, whence $z \in S(x, r) = S$, and $\|y - z\| = \dfrac{\varepsilon}{r + \varepsilon} \|x - y\| < \varepsilon$. Since $\varepsilon > 0$ was arbitrary, it follows that $\rho(A, S) = 0$. On the other hand, if there exists an $y \in A \cap \operatorname{Int} S$, we have $\rho(x, A) \leqslant \|x - y\| < r$, which contradicts the hypothesis $\rho(x, A) = r$. Consequently $A \cap \operatorname{Int} S = \emptyset$, and therefore A supports the cell S, which completes the proof.

LEMMA 1.4. *Let E be a normed linear space, $x \in E$ and $r > 0$. For any $f \in E^*$ with $\|f\| = 1$ the hyperplane $H \subset E$ defined by*

$$H = \{y \in E \mid f(x - y) = r\} = \{y \in E \mid f(y) = f(x) - r\} \tag{1.20}$$

supports the cell $S = S(x, r)$, and conversely, for any support hyperplane H of the cell S there exists a unique $f \in E^*$ with $\|f\| = 1$ such that we have (1.20).

Proof. Let H be a hyperplane of the form (1.20) with $\|f\| = 1$. Then by virtue of lemma 1.2 we have

$$\rho(x, H) = |f(x) - [f(x) - r]| = r,$$

and consequently, by virtue of lemma 1.3, H supports the cell S.

Conversely, let $H = \{y \in E \mid f_1(y) = \alpha_1\}$ be a support hyperplane of the cell S. Putting $f_2 = \dfrac{f_1}{\|f_1\|}$, $\alpha_2 = \dfrac{\alpha_1}{\|f_1\|}$, we have

$\|f_2\| = 1$ and $H = \{y \in E \mid f_2(y) = \alpha_2\}$. Since H supports the cell S, we have, by virtue of lemma 1.3, $\rho(x, H) = r$, whence, by virtue of lemma 1.2,

$$|f_2(x) - \alpha_2| = r.$$

Put $f = e^{i \arg [f_2(x) - \alpha_2]} f_2$. We have then $\|f\| = 1$ and

$$f(x - y) = |f_2(x) - \alpha_2| = r \qquad (y \in H),$$

whence $H \subset \{y \in E \mid f(x - y) = r\}$. Since both sets in this inclusion are hyperplanes, they must coincide, whence H is of the form (1.20), with $\|f\| = 1$.

In order to show the uniqueness of f, assume that we have (1.20) and $H = \{y \in E \mid f'(x - y) = r\}$, with $\|f'\| = 1$, $f' \neq f$. Then

$$H \subset \{y \in E \mid (f - f')(y) = (f - f')(x)\}.$$

Since $f \neq f'$, both sets in this inclusion are hyperplanes and therefore they must coincide. Since x belongs to the second set, it follows that we also have $x \in H$, whence, by (1.20), $r = 0$, which contradicts the hypothesis that $r > 0$. Consequently, $f' = f$, which completes the proof of lemma 1.4.

We observe that if for an $f \in E^*$ the hyperplane defined by (1.20) supports the cell $S(x, r)$, then we have necessarily

$$\|f\| = 1,$$

since by (1.19) and (1.18) we have

$$r = \rho(x, H) = \frac{|f(x) - [f(x) - r]|}{\|f\|} = \frac{r}{\|f\|}.$$

We recall that a complex normed linear space E may be considered also as a real normed linear space $E_{(r)}$, since multiplication by real scalars is well defined. In this case the linear manifolds of E, considered as linear manifolds in $E_{(r)}$, are called *real linear manifolds*.

LEMMA 1.5. *Let E be a normed linear space, G a linear subspace of E, $x \in E \setminus \overline{G}$, $g_0 \in G$ and $f \in E^*$. Then*

a) *The following statements are equivalent:*

1° *f satisfies (1.2), (1.3) and (1.4).*

2° *f satisfies (1.2), (1.3) and (1.5).*

3° *f satisfies (1.2), (1.5) and (1.6).*

4° *The hyperplane*

$$H = \{y \in E \mid f(x - y) = \|x - g_0\|\} = \{y \in E \mid f(y) = f(x) - \|x - g_0\|\} \tag{1.21}$$

supports the cell $S(x, \|x - g_0\|)$ and passes) through G.*

*) I.e. $G \subset H$.

5° *The real hyperplane*

$$H_1 = \{y \in E \mid \operatorname{Re} f(x - y) = \|x - g_0\|\} \tag{1.22}$$

supports the cell $S(x, \|x - g_0\|)$ *and passes through* G.

b) *The following statements are equivalent:*

1° f *satisfies (1.2), (1.3) and (1.7).*

2° f *satisfies (1.2) and the hyperplane*

$$H_2 = \{y \in E \mid f(y) = 0\} \tag{1.23}$$

supports the cell $S(x, \|x - g_0\|)$ *and passes through* G.

c) *The following statements are equivalent:*

1° f *satisfies (1.2), (1.3) and (1.8).*

2° f *satisfies (1.2), (1.6) and (1.8).*

3° f *satisfies (1.2) and the real hyperplane*

$$H_3 = \{y \in E \mid \operatorname{Re} f(y) = 0\} \tag{1.24}$$

supports the cell $S(x, \|x - g_0\|)$ *and passes through* G.

Proof. a) 1°⇔2°⇔3° by virtue of lemma 1.1.

Assume now that we have 1°. Then by (1.2) and lemma 1.4 it follows that the hyperplane H defined by (1.21) supports the cell $S(x, \|x - g_0\|)$. On the other hand, if $g \in G$, we have, taking into account (1.3) and (1.4),

$$f(x - g) = f(x - g_0) = \|x - g_0\|,$$

whence $g \in H$. Thus, 1°⇒4°.

Conversely, assume that we have 4°. Since H supports the cell $S(x, \|x - g_0\|)$, by virtue of the remark which follows the proof of lemma 1.4, we have (1.2). On the other hand, by $G \subset H$ and (1.21) we have

$$f(g) = f(x) - \|x - g_0\| = \text{const.} \qquad (g \in G)$$

whence, taking into account that $0 \in G$ and $f(0) = 0$, we obtain (1.3), and for $g = g_0 \in G$ we obtain (1.4). Thus, 4°⇒1°.

The equivalence 3°⇔5° follows from the equivalence 1°⇔4° applied to E considered as a real space $E_{(r)}$, taking into account*)

$$\|f\|_E = \|\operatorname{Re} f\|_{E_{(r)}} \qquad (f \in E^*),$$

where the functional $\operatorname{Re} f \in (E_{(r)})^*$ is defined by $(\operatorname{Re} f)(x) = \operatorname{Re} f(x)$ $(x \in E_{(r)})$.

b) Assume that we have 1°. Then, for the hyperplane H_2 defined by (1.23) we have, taking into account lemma 1.2,

$$\rho(x, H_2) = |f(x)| = |f(x - g_0)| = \|x - g_0\|,$$

*) See e.g. M. M. Day [43], chap. II, § 1, proposition (10).

whence the hyperplane H_2 supports, by virtue of lemma 1.3, the cell $S(x, \|x - g_0\|)$. On the other hand, it follows from (1.3) that $G \subset H_2$. Thus, $1^\circ \Rightarrow 2^\circ$.

Conversely, assume that we have 2°. Then from (1.2), lemma 1.2 and lemma 1.3 we obtain

$$|f(x)| = \rho(x, H_2) = \|x - g_0\|,$$

whence, since $g_0 \in G \subset H_2$, it follows

$$|f(x - g_0)| = |f(x)| = \|x - g_0\|,$$

i.e. (1.7). Finally, (1.3) follows from $G \subset H_2$.

c) $1^\circ \Leftrightarrow 2^\circ$ by virtue of lemma 1.1 and the equivalence $2^\circ \Leftrightarrow 3^\circ$ is a consequence of part b) applied to E considered as a real space $E_{(r)}$. This completes the proof.

From the preceding it follows that theorem 1.1 and corollary 1.1 are equivalent to the following geometrical result:

THEOREM 1.2. *Let E be a normed linear space, G a linear subspace of E, $x \in E \setminus \overline{G}$ and $g_0 \in G$. The following statements are equivalent:*

1° $g_0 \in \mathfrak{P}_G(x)$.

2° *There exists a hyperplane H which supports the cell $S(x, \|x - g_0\|)$ and passes through G.*

3° *There exists a real hyperplane H which supports the cell $S(x, \|x - g_0\|)$ and passes through G.*

Proof. $1^\circ \Rightarrow 2^\circ$ and $1^\circ \Rightarrow 3^\circ$ by virtue of theorem 1.1 and lemma 1.5.

Assume that we have 2°. Then, since H supports the cell $S(x, \|x - g_0\|)$, there exists, by lemma 1.4, a unique $f \in E^*$ with $\|f\| = 1$, such that we have (1.21). Since, in addition, H passes through G, it follows by lemma 1.5 that f satisfies (1.2), (1.3) and (1.4), whence, by virtue of theorem 1.1, $g_0 \in \mathfrak{P}_G(x)$. Thus, $2^\circ \Rightarrow 1^\circ$.

Finally, applying the implication $2^\circ \Rightarrow 1^\circ$ to E considered as a real space $E_{(r)}$, we obtain the implication $3^\circ \Rightarrow 1^\circ$, which completes the proof.

We observe that the other equivalent conditions of corollary 1.1, respectively lemma 1.5, also lead to the conditions of theorem 1.2.

The geometrical interpretation of corollary 1.3 is similar.

We recall that a support hyperplane H of the cell S is called, following S. Mazur [145], *tangent hyperplane of $S(x, r)$ at the point $y_0 \in \operatorname{Fr} S(x, r)$*, if H is the only support hyperplane of $S(x, r)$ passing through y_0. Since the element $x' = x - g_0 \in E$ is normal*) if and only if the cell $S(x, \|x - g_0\|)$

*) See section 1.1.

admits a tangent hyperplane at the point g_0, corollary 1.4 is equivalent to the following geometrical result:

COROLLARY 1.5. *Let E be a normed linear space, G a linear subspace of E, $x \in E \setminus \overline{G}$ and $g_0 \in G$ such that $x - g_0$ is normal. We have $g_0 \in \mathfrak{L}_G(x)$ if and only if there exists a tangent hyperplane H of the cell $S(x, \|x - g_0\|)$ at the point g_0, passing through G.*

Naturally, theorem 1.2 and corollary 1.5 can be proved also directly, observing that $g_0 \in \mathfrak{L}_G(x)$ if and only if the linear subspace G supports the cell $S(x, \|x - g_0\|)$ and applying the well known separation theorem of S. Mazur *).

1.3. APPLICATIONS IN THE SPACES $C(Q)$

For a compact **) space Q we shall denote by $C(Q)$ the space of all numerical (complex or real) continuous functions on Q, endowed with the usual vectorial operations and with the norm $\|x\| = \max_{q \in Q} |x(q)|$.

THEOREM 1.3. *Let $E = C(Q)$ (Q compact), G a linear subspace of E, $x \in E \setminus \overline{G}$ and $g_0 \in G$. We have $g_0 \in \mathfrak{L}_G(x)$ if and only if there exists a Radon measure ***) μ (complex or real, according to the space E) on Q, with the following properties ****):*

$$|\mu| (Q) = 1, \tag{1.25}$$

$$\int_Q g(q) \, d\mu(q) = 0 \quad (g \in G), \tag{1.26}$$

$$\frac{d\mu}{d|\mu|} \in C(Q), \tag{1.27}$$

$$x(q) - g_0(q) = \left[\operatorname{sign} \frac{d\mu}{d|\mu|} (q) \right] \max_{t \in Q} |x(t) - g_0(t)| \quad (q \in S(\mu)), \tag{1.28}$$

where (1.27) is meant in the sense that $\frac{d\mu}{d|\mu|}$ can be made continuous on Q by changing its values on a set of $|\mu|$-measure zero, in (1.28) is taken this continuous function $\frac{d\mu}{d|\mu|}$ and $S(\mu)$ is the carrier of the measure μ.

*) See e.g. G. Köthe [122], p. 191, theorem (2).

**) In the sense of N. Bourbaki [25] (i.e.: bicompact Hausdorff).

***) For the notions and theorems of measure theory which will be used in the sequel, see N. Bourbaki [22], P. R. Halmos [82], N. Dunford and J. Schwartz [49], N. Dinculeanu [48], or the monographs in Romanian of N. Dinculeanu [46], [47].

****) For the definition of sign α (α complex) see section 1.1.

Proof. By virtue of theorem 1.1, we have $g_0 \in \mathfrak{L}_G(x)$ if and only if there exists a Radon measure μ on Q such that we have (1.25), (1.26) and

$$\int_Q [x(q) - g_0(q)]\, \mathrm{d}\mu(q) = \max_{q \in Q} |x(q) - g_0(q)|\,. \tag{1.29}$$

We shall now show that these conditions are equivalent to (1.25) — (1.28). Assume first that we have (1.25), (1.26) and (1.29). Then from (1.29), (1.25) and $x - g_0 \neq 0$ it follows that*)

$$\frac{\mathrm{d}\mu}{\mathrm{d}|\mu|}(q) = \frac{\overline{x(q) - g_0(q)}}{\max_{t \in Q} |x(t) - g_0(t)|} \qquad |\mu|\text{-a.e. on } Q. \tag{1.30}$$

Indeed, assume the contrary, i.e. that there exists a set $A \subset Q$ with $|\mu|(A) > 0$, such that

$$\frac{\mathrm{d}\mu}{\mathrm{d}|\mu|}(q) \neq \frac{\overline{x(q) - g_0(q)}}{\|x - g_0\|} \qquad |\mu|\text{-a.e. on A.}$$

Then

$$\mathrm{Re}\left(\frac{\mathrm{d}\mu}{\mathrm{d}|\mu|}(q)[x(q) - g_0(q)]\right) < \|x - g_0\| \qquad |\mu|\text{-a.e. on A,}$$

since otherwise, taking into account that we always have**)

$$\left|\frac{\mathrm{d}\mu}{\mathrm{d}|\mu|}(q)\right| = 1 \qquad |\mu|\text{-a.e. on } Q, \tag{1.31}$$

there would exist a set $A_1 \subset A$ with $|\mu|(A_1) > 0$ such that

$$\|x - g_0\| \leqslant \mathrm{Re}\left(\frac{\mathrm{d}\mu}{\mathrm{d}|\mu|}(q)\,[x(q) - g_0(q)]\right) \leqslant$$

$$\leqslant \left|\frac{\mathrm{d}\mu}{\mathrm{d}|\mu|}(q)[x(q) - g_0(q)]\right| \leqslant \|x - g_0\|$$

$$|\mu|\text{ - a.e. on A}_1,$$

whence $\frac{\mathrm{d}\mu}{\mathrm{d}|\mu|}(q)[x(q) - g_0(q)] =$ real and positive $|\mu|$-a.e.

---

*) We shall write briefly "a.e." for "almost everywhere"

**) By $\mu = \frac{\mathrm{d}\mu}{\mathrm{d}|\mu|}|\mu|$.

on A_1, hence $= \|x - g_0\|$ $|\mu|$-a.e. on A_1, and thus

$$\frac{d\mu}{d|\mu|}(q) = \frac{\|x - g_0\|}{x(q) - g_0(q)} = \frac{\overline{x(q) - g_0(q)}}{\|x - g_0\|} \qquad |\mu| \text{ - a.e. on } A_1,$$

contradicting the hypothesis. Consequently, we obtain

$$\operatorname{Re}\int_Q [x(q) - g_0(q)] d\mu(q) = \operatorname{Re}\int_Q [x(q) - g_0(q)] \frac{d\mu}{d|\mu|}(q) d|\mu|(q) =$$

$$= \int_Q \operatorname{Re}\left([x(q) - g_0(q)] \frac{d\mu}{d|\mu|}(q) \right) d|\mu|(q) < \int_Q \|x - g_0\| d|\mu|(q) =$$

$$= \|x - g_0\|,$$

which contradicts (1.29). This proves the relation (1.30).

Changing now the values of $\frac{d\mu}{d|\mu|}$ so as to have (1.30) everywhere on Q, we will have (1.27), whence, taking into acount (1.31), there follow the relations *)

$$\left| \frac{d\mu}{d|\mu|}(q) \right| = 1 \qquad (q \in S(\mu)), \tag{1.32}$$

and thus, by (1.30) (everywhere on Q), we obtain (1.28).

Conversely, assume that we have (1.25) — (1.28). Then by (1.28), (1.27), (1.31) and (1.25) it follows that

$$\int_Q [x(q) - g_0(q)] d\mu(q) = \|x - g_0\| \int_Q \left[\operatorname{sign} \frac{d\mu}{d|\mu|}(q) \right] d\mu(q) =$$

$$= \|x - g_0\| \int_Q \left[\operatorname{sign} \frac{d\mu}{d|\mu|}(q) \right] \frac{d\mu}{d|\mu|}(q) d|\mu|(q) =$$

$$= \|x - g_0\| \int_Q \left| \frac{d\mu}{d|\mu|}(q) \right| d|\mu|(q) = \|x - g_0\| |\mu|(Q) = \|x - g_0\|,$$

i.e. (1.29), which completes the proof.

Let us consider now the problem of simultaneous characterization of a set $M \subset G$ of elements of best approximation.

*) In fact, if there existed a $q_0 \in S(\mu)$ such that $\left| \frac{d\mu}{d|\mu|}(q_0) \right| \neq 1$, then, taking an open neighbourhood V of q_0 such that $\left| \frac{d\mu}{d|\mu|}(q) \right| \neq 1$ $(q \in V)$, we would have $|\mu|(V) > 0$ (by $q_0 \in S(\mu)$), which contradicts (1.31).

THEOREM 1.4. *Let $E = C(Q)$ (Q compact), G a linear subspace of E, $x \in E \setminus \overline{G}$ and $M \subset G$. We have $M \subset \mathfrak{L}_G(x)$ if and only if there exists a Radon measure μ on Q satisfying (1.25), (1.26), (1.27) and*

$$x(q) - g_0(q) = \left[\operatorname{sign} \frac{d\mu}{d|\mu|}(q)\right] \max_{t \in Q} |x(t) - g_0(t)| \quad (q \in S(\mu),\ g_0 \in M). \tag{1.33}$$

Proof. Assume that we have $M \subset \mathfrak{L}_G(x)$. Then there exists, by virtue of corollary 1.3, a Radon measure μ on Q satisfying (1.25), (1.26) and

$$\int_Q [x(q) - g_0(q)]\, d\mu(q) = \max_{q \in Q} |x(q) - g_0(q)| \quad (g_0 \in M). \tag{1.34}$$

It follows*) from (1.25), (1.34) and $x \in E \setminus \overline{G}$ that for every $g_0 \in M$ there exists a set of $|\mu|$-measure zero $N_{g_0} \subset Q$ such that we have

$$\frac{d\mu}{d|\mu|}(q) = \frac{\overline{x(q) - g_0(q)}}{\max_{t \in Q} |x(t) - g_0(t)|} \qquad (q \in Q \setminus N_{g_0}).$$

Consequently, for any pair $g_0, g_0' \in M$ we have

$$g_0(q) = g_0'(q) \qquad (q \in Q \setminus (N_{g_0} \cup N_{g_0'})),$$

hence

$$(g_0 - g_0')\,|\mu| = 0,$$

whence**)

$$g_0(q) = g_0'(q) \qquad (q \in S(\mu)).$$

Now, in the same way as at the end of the above proof of the necessity part of theorem 1.3, it follows that we have (1.27) and (1.33). This proves the necessity of the conditions of theorem 1.4.

On the other hand, the sufficiency of the conditions of theorem 1.4 is an immediate consequence of the sufficiency of the conditions of theorem 1.3, which completes the proof of theorem 1.4.

We mention that another simultaneous characterization of a set $M \subset G (\subset C(Q))$ of elements of best approximation, less convenient for applications, has been given by Z. S. Romanova ([198], theorem 1).

---

*) See the above proof of the relations (1.30).

**) See N. Bourbaki [22], Chap. III, p. 73, corollary of proposition 10, and p. 70, proposition 2.

1.4. APPLICATIONS IN THE SPACES $C_R(Q)$

For a compact space Q, we shall denote by $C_R(Q)$ the space of all continuous *real-valued* functions on Q, endowed with the usual vector operations and with the norm $\|x\| = \max_{q \in Q} |x(q)|$.

Naturally, the results of the preceding section are applicable, in particular, also in the spaces $C_R(Q)$. However, in these spaces one can obtain additional information, making use of certain special properties of the real Radon measures.

THEOREM 1.5 ([229], theorem 1). *Let $E = C_R(Q)$ (Q compact), G a linear subspace of $E, x \in E \setminus \overline{G}$ and $g_0 \in G$. We have $g_0 \in \mathfrak{P}_G(x)$ if and only if there exist two disjoint sets $Y_{g_0}^+$ and $Y_{g_0}^-$ closed in Q, and a real Radon measure μ on Q, with the following properties:*

$$|\mu|(Q) = 1, \tag{1.35}$$

$$\int_Q g(q) \mathrm{d}\mu(q) = 0 \qquad (g \in G), \tag{1.36}$$

μ is non-decreasing on) $Y_{g_0}^+$, non-increasing*

$$\text{on } Y_{g_0}^- \text{ and } Y_{g_0}^+ \cup Y_{g_0}^- \supset S(\mu) \text{ (the carrier of } \mu), \tag{1.37}$$

$$x(q) - g_0(q) = \begin{cases} \max_{t \in Q} |x(t) - g_0(t)| \text{ for } q \in Y_{g_0}^+ \\ -\max_{t \in Q} |x(t) - g_0(t)| \text{ for } q \in Y_{g_0}^- . \end{cases} \tag{1.38}$$

First proof. By virtue of theorem 1.3, we have $g_0 \in \mathfrak{P}_G(x)$ if and only if there exists a *real* Radon measure μ on Q, satisfying (1.25) — (1.28). We shall now show that these conditions are equivalent to (1.35) — (1.38).

Assume first that we have (1.25) — (1.28). Put**)

$$Y_{g_0}^+ = \{q \in Q \mid x(q) - g_0(q) = \max_{t \in Q} |x(t) - g_0(t)|\}, \tag{1.39}$$

$$Y_{g_0}^- = \{q \in Q \mid x(q) - g_0(q) = -\max_{t \in Q} |x(t) - g_0(t)|\}. \tag{1.40}$$

*) We write for brevity "on $Y_{g_0}^+$" instead of "on the σ-field of μ-measurable subsets of $Y_{g_0}^+$".

**) In fact here we should use the notations Y_{x,g_0}^+, Y_{x,g_0}^- or, in order to be in accordance with (1.47) and (1.48), the notations $Y_{x-g_0}^+$, $Y_{x-g_0}^-$. However, the notations $Y_{g_0}^+$, $Y_{g_0}^-$ are simpler and point out the fact that in the last part of the present section $x \in E$ will be fixed, while $g_0 \in M \subset \mathfrak{P}_G(x)$ will be variable.

Then $Y_{g_0}^+$ and $Y_{g_0}^-$ are disjoint and closed in Q and we have (1.38). Now let $q \in S(\mu)$ be arbitrary. Then, since for real $\alpha \neq 0$ we have sign $\alpha = \pm 1$, it follows by (1.32) and (1.28) that $q \in Y_{g_0}^+ \cup Y_{g_0}^-$. Thus, $S(\mu) \subset Y_{g_0}^+ \cup Y_{g_0}^-$. On the other hand, by (1.28) and (1.39) we have sign $\frac{d\mu}{d|\mu|}(q) = 1$ for $q \in Y_{g_0}^+$, whence, taking into account (1.32), we obtain

$$\frac{d\mu}{d|\mu|}(q) = 1 \qquad (q \in S(\mu) \cap Y_{g_0}^+),$$

whence

$$\mu(Y_{g_0}^+) = \int_{Y_{g_0}^+} d\mu(q) = \int_{Y_{g_0}^+} \frac{d\mu}{d|\mu|}(q)\, d|\mu|(q) = \int_{Y_{g_0}^+} d|\mu|(q) = |\mu|\,(Y_{g_0}^+)$$

which shows that μ is non-decreasing on $Y_{g_0}^+$. Similarly, μ is non-increasing on $Y_{g_0}^-$, hence we have (1.37). Thus, (1.25)—(1.28) implies (1.35) — (1.38).

Conversely, assume that there exist two disjoint sets $Y_{g_0}^+$ and $Y_{g_0}^-$ closed in Q and a real Radon measure μ on Q such that we have (1.35) — (1.38). Then, by (1.37), we have

$$\mu = \begin{cases} |\mu| \text{ on } Y_{g_0}^+ \\ -|\mu| \text{ on } Y_{g_0}^-, \end{cases}$$

whence, taking into account that $\mu = \frac{d\mu}{d|\mu|}|\mu|$, we obtain

$$\frac{d\mu}{d|\mu|} = \begin{cases} 1 \ |\mu|\text{-a.e. on } Y_{g_0}^+ \\ -1 \ |\mu|\text{-a.e. on } Y_{g_0}^-. \end{cases}$$

Since by virtue of the classical lemma of Uryson there exists a continuous real-valued function on Q, equal to 1 on $Y_{g_0}^+ \cap S(\mu)$ and to -1 on $Y_{g_0}^- \cap S(\mu)$, and since by (1.37) we have $Y_{g_0}^+ \cup Y_{g_0}^- \supset S(\mu)$, it follows that we have (1.27). Denoting now by $\frac{d\mu}{d|\mu|}$ the continuous function constructed in this way and taking into account (1.38) and again $Y_{g_0}^+ \cup Y_{g_0}^- \supset S(\mu)$, we obtain (1.28). Thus, (1.35) — (1.38) implies (1.25) — (1.28), which completes the proof.

Second proof ([229], p. 508). By virtue of theorem 1.1, we have $g_0 \in \mathfrak{L}_G(x)$ if and only if there exists a *real* Radon measure μ on Q such that we have (1.25), (1.26) and (1.29). We shall now show that these conditions are equivalent to (1.35) — (1.38).

Assume first that we have (1.25), (1.26) and (1.29). Define $Y^+_{g_0}$ by (1.39) and $Y^-_{g_0}$ by (1.40). Then $Y^+_{g_0}$ and $Y^-_{g_0}$ are disjoint and closed in Q and we have (1.38). On the other hand, (1.29) and $x - g_0 \neq 0$, together with (1.25), imply (1.37). In fact, if μ were not non-decreasing on $Y^+_{g_0}$, there would exist a set $A \subset Y^+_{g_0}$ with $|\mu|(A) > 0$, such that $\mu(A) < |\mu|(A)$, whence

$$\int_A [x(q) - g_0(q)]\mathrm{d}\mu(q) = \mu(A)\|x - g_0\| < |\mu|(A)\|x - g_0\| =$$

$$= \int_A \|x - g_0\| \mathrm{d}|\mu|(q),$$

whence, taking into account (1.25),

$$\int_Q [x(q) - g_0(q)]\mathrm{d}\mu(q) < \int_Q \|x - g_0\| \mathrm{d}|\mu|(q) = \|x - g_0\|,$$

which contradicts (1.29); similarly, $-\mu$ is non-decreasing on $Y^-_{g_0}$, whence μ is non-increasing on $Y^-_{g_0}$; finally, if there existed a $q_0 \in S(\mu)$ such that $q_0 \notin Y^+_{g_0} \cup Y^-_{g_0}$, whence $|x(q_0) - g_0(q_0)| < \|x - g_0\|$, then, taking an open neighbourhood U of q_0 such that $|x(q) - g_0(q)| < \|x - g_0\|$ $(q \in U)$, we would have $|\mu|(U) > 0$ (since $q_0 \in S(\mu)$) and

$$\int_U [x(q) - g_0(q)]\mathrm{d}\mu(q) \leqslant \int_U |x(q) - g_0(q)|\mathrm{d}|\mu|(q) < \int_U \|x - g_0\| \mathrm{d}|\mu|(q)$$

whence, by (1.25),

$$\int_Q [x(q) - g_0(q)]\mathrm{d}\mu(q) < \int_Q \|x - g_0\| \mathrm{d}|\mu|(q) = \|x - g_0\|,$$

which contradicts (1.29). Thus (1.25), (1.26) and (1.29) imply (1.35) — (1.38).

Conversely, assume that there exist two disjoint sets $Y^+_{g_0}$ and $Y^-_{g_0}$ closed in Q, and a real Radon measure μ on Q such that we have (1.35) — (1.38). Then, by (1.37), (1.38) and (1.35), we have

$$\int_Q [x(q) - g_0(q)]\mathrm{d}\mu(q) = \int_{S(\mu) \cap Y^+_{g_0}} \max_{t \in Q} |x(t) - g_0(t)|\mathrm{d}\mu(q) +$$

$$+\int_{S(\mu)\cap Y^-_{g_0}}(-\max_{t\in Q}|x(t)-g_0(t)|)\mathrm{d}\mu(q)=$$

$$=\int_Q \max_{t\in Q}|x(t)-g_0(t)|\mathrm{d}|\mu|(q)=\max_{t\in Q}|x(t)-g_0(t)|,$$

i.e. (1.29). Thus, (1.35)—(1.38) imply (1.25), (1.26) and (1.29) which completes the proof.

Consider now the problem of simultaneous characterization of a set $M\subset G$ of elements of best approximation. By virtue of corollary 1.3 we have $M\subset\mathfrak{L}_G(x)$ if and only if there exists a real Radon measure on Q satisfying (1.25), (1.26) and (1.34). Combining the necessity part of this result and the second proof above of the necessity part of theorem 1.5, it follows that if $M\subset\mathfrak{L}_G(x)$, then there exist a (common) real Radon measure μ on Q and, for every $g_0\in M$, two disjoint sets $Y^+_{g_0}$ and $Y^-_{g_0}$ closed in Q (namely those defined, for every $g_0\in M$, by (1.39) and (1.40) respectively), such that we have (1.35) — (1.38) for all $g_0\in M$. The following problem arises naturally : is it possible to find, in this case, instead of the sets $Y^+_{g_0}$ and $Y^-_{g_0}$, two disjoint closed sets Y^+ and Y^-, independent of $g_0\in M$, such that we have (1.35) — (1.38) for all $g_0\in M$, with $Y^+_{g_0}$, $Y^-_{g_0}$ replaced by Y^+, Y^-? The answer is affirmative, as shown by

THEOREM 1.6 ([229], theorem 2). *Let $E=C_R(Q)$ (Q compact), G a linear subspace of E, $x\in E\setminus\overline{G}$ and $M\subset G$. We have $M\subset\mathfrak{L}_G(x)$ if and only if there exist two disjoint sets Y^+ and Y^- closed in Q, and a real Radon measure μ on Q, with the following properties :*

$$|\mu|(Q)=1, \tag{1.41}$$

$$\int_Q g(q)\mathrm{d}\mu(q)=0 \qquad (g\in G), \tag{1.42}$$

μ is non-decreasing on Y^+, non-increasing on Y^- and $Y^+\cup Y^-\supset S(\mu)$ (the carrier of μ), (1.43)

$$x(q)-g_0(q)=\begin{cases}\max_{t\in Q}|x(t)-g_0(t)| & \text{for } q\in Y^+\\ -\max_{t\in Q}|x(t)-g_0(t)| & \text{for } q\in Y^-\end{cases}\qquad (g_0\in M). \tag{1.44}$$

Proof. Assume that we have $M\subset\mathfrak{L}_G(x)$. Then, by theorem 1.4, there exists a real Radon measure μ on Q satisfying (1.25) — (1.27) and (1.33). From (1.27) and (1.33) it follows that

$$g_0(q)=g_0'(q) \qquad (q\in S(\mu);\ g_0, g_0'\in M). \tag{1.45}$$

Consider an *arbitrary* element $g_0 \in M$ and put

$$Y^+ = Y_{g_0}^+ \cap S(\mu), \quad Y^- = Y_{g_0}^- \cap S(\mu), \tag{1.46}$$

where $Y_{g_0}^+$, $Y_{g_0}^-$, are the sets (1.39) and (1.40) respectively. Then Y^+ and Y^- are well defined, disjoint and closed in Q. Since $g_0 \in \mathfrak{L}_G(x)$, by virtue of the argument used in the first proof above of the necessity part of theorem 1.5 we have (1.37) and (1.38), whence, by (1.46) and (1.45), we obtain (1.43) and (1.44), which proves the necessity of the conditions of theorem 1.6.

On the other hand, the sufficiency of the conditions of theorem 1.6 is an immediate consequence of the sufficiency of the conditions of theorem 1.5, which completes the proof of theorem 1.6.

As in the case of theorem 1.5, we shall also give a direct proof of theorem 1.6 (i.e. which is not based on theorem 1.4, but on corollary 1.3). For this purpose we shall prove first some results on the maximal elements of continuous linear functionals on $C_R(Q)$ which are also of interest for other applications.

Let $f \in C_R(Q)^*$. Any element $x \in C_R(Q) \setminus \{0\}$ with the property

$$f(x) = \|f\| \; \|x\|$$

is called a *maximal element* of f; in general, f need not have any maximal element or it may have several linearly independent maximal elements. The Radon measure μ on Q corresponding to f, i.e. satisfying

$$f(x) = \int_Q x(q) \mathrm{d}\mu(q) \qquad (x \in C(Q)),$$

$$\|f\| = |\mu|\,(Q),$$

is called*) the *kernel* of f.

We recall that for every real Radon measure μ on Q there exists a μ-measurable subset A of Q such that μ is non-decreasing on A and non-increasing on $Q \setminus A$; the pair A, $Q \setminus A$ is called *a Hahn decomposition of Q with respect to* μ. More generally, if M is a μ-measurable subset of Q, by a *Hahn decomposition of M with respect to* μ, or briefly by *a Hahn decomposition of M*, we shall mean a pair of μ-measurable subsets B, $M \setminus B$ of M, with the property that μ is non-decreasing on B and non-increasing on $M \setminus B$. Now let M be the carrier $S(\mu)$ of the real Radon measure μ. Then, obviously, the trace on $S(\mu)$ of any Hahn decomposition of Q is a Hahn

*) See e.g. W. W. Rogosinski [197].

decomposition of $S(\mu)$, and conversely, any Hahn decomposition of $S(\mu)$ may be extended to a Hahn decomposition of Q. In general, neither the Hahn decomposition of Q nor that of $S(\mu)$ is necessarily unique. However, we have

LEMMA 1.6 ([225], theorem 5). *Let $f \in C_R(Q)^*$. Then*

a) *In order that f have a maximal element it is necessary and sufficient that the carrier $S(\mu)$ of the kernel μ of f admit a Hahn decomposition into two* **closed** *sets*

$$S(\mu)^+ \text{ and } S(\mu)^- = S(\mu) \setminus S(\mu)^+.$$

b) *In this case, the above decomposition is unique.*

Proof. Assume that f has a maximal element $x \in C_R(Q)$. Put

$$Y_x^+ = \{q \in Q \mid x(q) = \max_{t \in Q} |x(t)|\}, \tag{1.47}$$

$$Y_x^- = \{q \in Q \mid x(q) = -\max_{t \in Q} |x(t)|\}. \tag{1.48}$$

Then Y_x^+, Y_x^- are disjoint closed sets, at least one of them is non-void and μ is*) non-decreasing on Y_x^+, non-increasing on Y_x^-, and we have $Y_x^+ \cup Y_x^- \supset S(\mu)$. Consequently, the pair of sets

$$S(\mu)_x^+ = S(\mu) \cap Y_x^+, \quad S(\mu)_x^- = S(\mu) \cap Y_x^-$$

is a Hahn decomposition of $S(\mu)$ into two closed sets.

Conversely, assume that there exists a Hahn decomposition of $S(\mu)$ into two closed sets $S(\mu)^+$, $S(\mu)^-$. Since $S(\mu)^+$, $S(\mu)^-$ are then disjoint closed subsets of Q, there exists, by virtue of the classical lemma of Uryson, an $x \in C_R(Q)$ such that

$$x(q) = \begin{cases} \max_{t \in Q} |x(t)| & \text{for } q \in S(\mu)^+ \\ -\max_{t \in Q} |x(t)| & \text{for } q \in S(\mu)^-. \end{cases} \tag{1.49}$$

We have then

$$f(x) = \int_Q x(q)\, d\mu(q) = \int_{S(\mu)^+} \max_{t \in Q} |x(t)|\, d\mu(q) +$$

$$+ \int_{S(\mu)^-} (-\max_{t \in Q} |x(t)|)\, d\mu(q) = \int_Q \max_{t \in Q} |x(t)|\, d|\mu|(q) = \|f\|\, \|x\|,$$

whence x is a maximal element of f.

---

*) See above the second proof of the necessity part of theorem 1.5; for another proof see S. I. Zuhovitzky [268], theorem 7.

b) Assume that $S(\mu)$ admits two Hahn decompositions into closed sets, say $S(\mu)_1^+$, $S(\mu)_1^-$ and $S(x)_2^+$, $S(\mu)_2^-$. We have then*)

$$\mu(S(\mu)_1^+ \cap S(\mu)_2^+) = \mu(S(\mu)_1^+) = \mu(S(\mu)_2^+),$$

whence $\mu(U) = 0$, where we have put

$$U = S(\mu)_1^+ \setminus [S(\mu)_1^+ \cap S(\mu)_2^+].$$

On the other hand, $[Q \setminus S(\mu)] \cup U$ is an *open set*. In fact,

$$[Q \setminus S(\mu)] \cup U = [Q \setminus S(\mu)] \cup \{S(\mu) \setminus [S(\mu)_1^- \cup (S(\mu)_1^+ \cap S(\mu)_2^+)]\} =$$

$$= Q \setminus [S(\mu)_1^- \cup S(\mu)_1^+ \cap (S(\mu)_2^+)],$$

where $S(\mu)_1^-$, $S(\mu)_1^+$, $S(\mu)_2^-$ are closed by our hypothesis.

Since μ is non-decreasing on U (because $U \subset S(\mu)_1^+$), from $\mu(U) = 0$ it follows that $\mu(A) = 0$ for all subsets $A \subset U$. Consequently,

$$\mu(B) = 0 \text{ for all subsets } B \subset [Q \setminus S(\mu)] \cup U \tag{1.50}$$

(since $\mu(D) = 0$ for all subsets $D \subset Q \setminus S(\mu)$, by the definition of $S(\mu)$). Now, the fact that the set $[Q \setminus S(\mu)] \cup U$ is open and (1.50) imply, by the definition of $S(\mu)$,

$$[Q \setminus S(\mu)] \cup U \subset Q \setminus S(\mu),$$

whence

$$U \subset Q \setminus S(\mu).$$

This, together with $U \subset S(\mu)_1^+ \subset S(\mu)$, gives

$$U = \emptyset,$$

i.e. $S(\mu)_1^+ = S(\mu)_1^+ \cap S(\mu)_2^+$. Consequently,

$$S(\mu)_1^+ \subset S(\mu)_2^+.$$

Changing the roles of $S(\mu)_1^+$ and $S(\mu)_2^+$, we obtain

$$S(\mu)_2^+ \subset S(\mu)_1^+,$$

whence, finally,

$$S(\mu)_1^+ = S(\mu)_2^+$$

and

$$S(\mu)_1^- = S(\mu) \setminus S(\mu)_1^+ = S(\mu) \setminus S(\mu)_2^+ = S(\mu)_2^-,$$

which completes the proof of lemma 1.6.

*) See e.g. P. R. Halmos [82], § 29.

COROLLARY 1.6 ([225], corollary 2). *Let* $f \in C_R(Q)^*$ *have at least one maximal element, let* μ *be the kernel of* f, *let* $S(\mu)^+$, $S(\mu)^-$ *be (by lemma 1.6) the unique Hahn decomposition of* $S(\mu)$ *into two closed sets and let* $x \in C_R(Q)$. *The following statements are equivalent:*

1° x *is a maximal element of* f.

2° *We have*

$$S(\mu)^+ = S(\mu) \cap Y_x^+, \quad S(\mu)^- = S(\mu) \cap Y_x^-, \tag{1.51}$$

where Y_x^+, Y_x^- *are the sets defined by (1.47), (1.48).*

3° *We have (1.49).*

Proof. Assume that we have 1°. Then (see the first part of the above proof of lemma 1.6 a)) the pair of sets $S(\mu)_x^+ = S(\mu) \cap Y_x^+$, $S(\mu)_x^- = S(\mu) \cap Y_x^-$ constitutes a Hahn decomposition of $S(\mu)$ into two closed sets. By virtue of lemma 1.6 b) we may write

$$S(\mu)_x^+ = S(\mu)^+, \quad S(\mu)_x^- = S(\mu)^-,$$

whence 2° follows. Thus, 1°⇒2°.

Assume now that we have 2°. Then

$$S(\mu)^+ \subset Y_x^+, \quad S(\mu)^- \subset Y_x^-,$$

and thus, by (1.47) and (1.48), we have 3°. Consequently, 2°⇒3°.

Finally, the implication 3°⇒1° has been proved previously (see the end of the above proof of lemma 1.6 a)). This completes the proof of corollary 1.6.

We mention also the following *simultaneous proof of lemma 1.6 and corollary 1.6.* Assume that $f \in C_R(Q)^*$ has a maximal element $x \in C_R(Q)$. Defining then Y_x^+, Y_x^- by (1.47) and (1.48) respectively, it follows as in the first part of the above proof of lemma 1.6 a), that the pair of sets $S(\mu)_x^+ = S(\mu) \cap Y_x^+$, $S(\mu)_x^- = S(\mu) \cap Y_x^-$ constitutes a Hahn decomposition of $S(\mu)$ into two closed sets.

Assume now that $\{S(\mu)^+, S(\mu)^-\}$ is an arbitrary Hahn decomposition of $S(\mu)$ into closed sets. We claim that we have then (1.49). In fact, if there were $q_0 \in S(\mu)^+$ such that $x(q_0) < \max_{t \in Q} |x(t)|$, then by taking an open neighbourhood U of q_0 with $U \cap S(\mu)^- = \emptyset$ and $x(q) < \max_{t \in Q} |x(t)|$ $(q \in U)$, we would have

$$0 < |\mu|(U) = |\mu|(U \setminus S(\mu)) + |\mu|(S(\mu) \cap U) = |\mu|(S(\mu) \cap U) =$$

$$= \mu(S(\mu)^+ \cap U) - \mu(S(\mu)^- \cap U) = \mu(S(\mu)^+ \cap U),$$

whence $\int_{S(\mu)^+\cap U} x(q)\mathrm{d}\mu(q) < \max_{t\in Q}|x(t)|\mu(S(\mu)^+\cap U)$, whence

$$\int_Q x(q)\mathrm{d}\mu(q) = \int_{S(\mu)^+} x(q)\mathrm{d}\mu(q) - \int_{S(\mu)^-} x(q)\mathrm{d}\mu(q) <$$

$$< \max_{t\in Q}|x(t)|[\mu(S(\mu)^+) - \mu(S(\mu)^-)] = \|x\|\,\|f\|,$$

and thus x would not be a maximal element of f; similarly, there exists no $q_0\in S(\mu)^-$ such that $x(q_0) > -\max_{t\in Q}|x(t)|$, whence we have (1.49), which proves the implication 1°⇒3° of corollar y 1.6. At the same time, since $\{S(\mu)^+, S(\mu)^-\}$ has been an arbitrary Hahn decomposition of $S(\mu)$ into closed sets, it follows from (1.49) that this decomposition is unique, which proves the implication 3°⇒2° of corollary 1.6. Finally, the implication 2°⇒1° of corollary 1.6 is immediate (see the end of the above proof of lemma 1.6 a)), which completes the proof.

Now we can give the

Second proof of theorem 1.6 ([229], p. 510). Assume that we have $M\subset\mathfrak{L}_G(x)$. Then, as has been observed in the preceding, it follows from corollary 1.3 that there exists a real Radon measure μ on Q satisfying (1.25), (1.26) and (1.34). Take an *arbitrary* element $g_0\in M$ and define Y^+ and Y^- by (1.46). Then Y^+ and Y^- are well defined, disjoint and closed in Q, and the relations (1.34), (1.25) and $x - g_0 \neq 0$ imply*) (1.43). Finally, from (1.34) and corollary 1.6 it follows that we have (1.44), which proves the necessity of the conditions of theorem 1.6.

On the other hand, the sufficiency of the conditions of theorem 1.6 is an immediate consequence of the sufficiency of the conditions of theorem 1.5, which completes the proof of theorem 1.6.

We now shall give, for the particular case when M is a finite subset $M = \{g_0', g_1', \ldots, g_{k+1}'\}$ of $\mathfrak{L}_G(x)$ (k being an integer with $0 \leqslant k < \infty$), one more proof of the necessity part of theorem 1.6, which does not involve corollary 1.6. The method used in this proof (given, essentially, in [226], p. 166) is of interest for other applications too (see sections 1.5 and 1.6, the proofs of theorems 1.8 and 1.10).

*) See above the second proof of the necessity part of theorem 1.5.

Namely, let μ be a real Radon measure on Q satisfying (1.25), (1.26) and

$$\int_Q [x(q)-g_i'(q)]\, d\mu(q)=\max_{q\in Q} |x(q) - g_i'(q)| \quad (i = 0, 1, \ldots, k+1); \tag{1.52}$$

such a measure exists by virtue of corollary 1.3. Put

$$Y^+ = \left\{q\in S(\mu) \,\middle|\, x(q) - \frac{1}{k+2}\sum_{i=0}^{k+1} g_i'(q) = \max_{t\in Q} \left| x(t) - \frac{1}{k+2}\sum_{i=0}^{k+1} g_i'(t)\right|\right\}, \tag{1.53}$$

$$Y^- = \left\{q\in S(\mu) \,\middle|\, x(q) - \frac{1}{k+2}\sum_{i=0}^{k+1} g_i'(q) = -\max_{t\in Q} \left| x(t) - \frac{1}{k+2}\sum_{i=0}^{k+1} g_i'(t)\right|\right\}. \tag{1.54}$$

Then the sets Y^+, Y^- are closed in Q and, by $x\in E\setminus\overline{G}$, they are disjoint. On the other hand, by (1.52) and (1.53) we have

$$\frac{1}{k+2}\sum_{i=0}^{k+1} \|x-g_i'\| = \frac{1}{k+2}\sum_{i=0}^{k+1}\int_Q [x(q) - g_i'(q)]d\mu(q) =$$
$$= \int_Q\left[x(q) - \frac{1}{k+2}\sum_{i=0}^{k+1} g_i'(q)\right] d\mu(q) \leqslant \left\|x - \frac{1}{k+2}\sum_{i=0}^{k+1} g_i'\right\| \leqslant$$
$$\leqslant \frac{1}{k+2}\sum_{i=0}^{k+1}\|x - g_i'\|,$$

whence

$$\int_Q\left[x(q) - \frac{1}{k+2}\sum_{i=0}^{k+1} g_i'(q)\right] d\mu(q) = \max_{t\in Q}\left|x(t) - \frac{1}{k+2}\sum_{i=0}^{k+1} g_i'(t)\right|,$$

which, together with (1.25), (1.53), (1.54) and $x - \frac{1}{k+2}\sum_{i=0}^{k+1} g_i' \neq 0$, implies*) (1.43), even with

$$Y^+ \cup Y^- = S(\mu). \tag{1.55}$$

---

*) See the second proof above of the necessity part of theorem 1.5.

Finally, let us prove (1.44). From (1.52) and (1.25) it follows that

$$\max_{q\in Q} |x(q) - g_i'(q)| = \int_Q [x(q)-g_i'(q)]\mathrm{d}\mu(q) \leqslant \int_Q |x(q) -$$

$$- g_i'(q)|\mathrm{d}|\mu|(q) \leqslant \max_{q\in Q} |x(q) - g_i'(q)| \quad (i = 0,1,\ldots, k+1),$$

whence we obtain the corresponding equalities, whence*)

$$|x(q) - g_i'(q)| = \max_{t\in Q} |x(t) - g_i'(t)| \quad (q\in S(\mu);\, i = 0,1,\ldots,k+1), \tag{1.56}$$

and thus, taking into account $g_0', g_1', \ldots, g_{k+1}' \in \mathfrak{L}_G(x)$,

$$|x(q) - g_0'(q)| = |x(q) - g_1'(q)| = \ldots = |x(q)-g_{k+1}'(q)| \; (q\in S(\mu)). \tag{1.57}$$

Assume now that there exists a $q_0\in S(\mu)$ and integers i, j with $0 \leqslant i, j \leqslant k+1$ such that

$$g_i'(q_0) \neq g_j'(q_0).$$

Then we have, by virtue of (1.57) and taking into account that the functions $x - g_i'$, $x - g_j'$ are real-valued,

$$x(q_0) - g_i'(q_0) = - x(q_0) + g_j'(q_0),$$

whence

$$x(q_0) - \frac{g_i'(q_0) + g_j'(q_0)}{2} = 0,$$

whence, by putting $Y_{i,j}^{+} = \left\{q\in Q \,\middle|\, x(q) - \frac{g_i'(q) + g_j'(q)}{2} = \right.$

$$\left. = \max_{t\in Q} \left| x(t) - \frac{g_i'(t) + g_j'(t)}{2} \right| \right\}, \; Y_{i,j}^{-} = \left\{q\in Q \,\middle|\, x(q) - \right.$$

$$\left. - \frac{g_i'(q) + g_j'(q)}{2} = - \max_{t\in Q} \left| x(t) - \frac{g_i'(t) + g_j'(t)}{2} \right| \right\}$$ and taking into account that by the preceding we have $S(\mu) = Y_{i,j}^{+} \cup Y_{i,j}^{-}$, it follows that we have

$$\max_{t\in Q} \left| x(t) - \frac{g_i'(t) + g_j'(t)}{2} \right| = 0,$$

---

*) See N. Bourbaki [22], Chap. III, p. 72, proposition 9.

i.e. $x = \frac{g'_i + g'_j}{2} \in G$, which contradicts the hypothesis $x \in E \setminus \bar{G}$. Thus, we have

$$g'_0(q) = g'_1(q) = \ldots = g'_{k+1}(q) \quad (q \in S(\mu) = Y^+ \cup Y^-). \quad (1.58)$$

Now let $q \in Y^+$ be arbitrary. By (1.58), (1.53) and $x \in E \setminus \bar{G}$ we have then

$$x(q) - g'_i(q) = x(q) - \frac{1}{k+2} \sum_{i=0}^{k+1} g'_i(q) = \max_{t \in Q} \left| x(t) - \frac{1}{k+2} \sum_{i=0}^{k+1} g'_i(t) \right| > 0 \quad (i = 0, 1, \ldots, k+1),$$

whence, by (1.56), it follows that

$$x(q) - g'_i(q) = \max_{t \in Q} \left| x(t) - g'_i(t) \right| \quad (i = 0, 1, \ldots, k+1).$$

Similarly, for $q \in Y^-$ we obtain

$$x(q) - g'_i(q) = - \max_{t \in Q} \left| x(t) - g'_i(t) \right| \quad (i = 0, 1, \ldots, k+1).$$

Thus, for $M = \{g'_0, g'_1, \ldots, g'_{k+1}\}$ we also have (1.44), which completes the proof.

Finally, we observe that from theorem 1.5 we immediately obtain the following result of E. W. Cheney and A. A. Goldstein ([35], theorem 1) : *Let L be a real topological linear space, Q a compact subset of L and x a continuous real-valued function on Q. In order that a $z_0^* \in L^*$ minimize the expression $\Delta(z^*) = \max_{q \in Q} |x(q) - z^*(q)|$ it is necessary and sufficient that 0 belong o the closed convex hull of the set*

$$A = \{q \in Q \mid x(q) - z_0^*(q) = \Delta(z_0^*)\} \cup - \{q \in Q \mid x(q) - z_0^*(q) = - \Delta(z_0^*)\}. (1.59)$$

In fact, let $E = C_R(Q)$, let $u : L^* \to E$ be defined by $u(z^*) = z^*|_Q$ $(z^* \in L^*)$ and let $G = u(L^*) \subset E$, $g_0 = u(z_0^*) \in G$. We have $\Delta(z_0^*) = \min_{z^* \in L^*} \Delta(z^*)$ if and only if $g_0 \in \mathfrak{P}_G(x)$, which happens, by theorem 1.5, if and only if there exists a real Radon measure μ on Q with the properties (1.35) — (1.37), where $Y_{g_0}^+$, $Y_{g_0}^-$ are the sets (1.39) and (1.40) respectively. But in this case, for the measure μ_0 on A defined by

$$\int_A x(q)\,\mathrm{d}\mu_0(q) = \int_{Y_{g_0}^+} x(q)\,\mathrm{d}\mu(q) - \int_{Y_{g_0}^-} x(-q)\,\mathrm{d}\mu(q) \quad (x \in C_R(A))$$

we have $\mu_0 \geqslant 0$, $\|\mu_0\| = 1$ and

$$\int_A z^*(q)\mathrm{d}\mu_0(q) = \int_Q [u(z^*)](q)\mathrm{d}\mu(q) = 0 \qquad (z^* \in L^*),$$

and thus*) 0 belongs to the closed convex hull of the set A. Conversely, if this condition is satisfied, there exists a Radon measure μ_0 on A with $\mu_0 \geqslant 0$, $\|\mu_0\| = 1$ and $\int_A z^*(q)\mathrm{d}\mu_0(q) = 0 (z^* \in L^*)$. Then for the measure μ on Q defined by

$$\int_Q x(q)\,\mathrm{d}\mu(q) = \int_{Y_{g_0}^+} x(q)\,\mathrm{d}\mu_0(q) -$$

$$-\int_{-Y_{g_0}^- \setminus Y_{g_0}^+} x(-q)\,\mathrm{d}\mu_0(q) \quad (x \in C_R(Q))$$

we have (1.35) — (1.37), which completes the proof. For similar results on the characterization and existence of $z_0^* \in L^*$ see B. R. Kripke and R. T. Rockafellar [130], and E. W. Cheney and A. A. Goldstein [36].

1.5. APPLICATIONS IN THE SPACES $L^1(T, \nu)$

Let (T, ν) be a positive measure space**) and, for $1 \leqslant p < \infty$ (respectively $p = \infty$), let $L^p(T, \nu)$ be the space of all equivalence classes of (complex or real) functions of pth power ν-integrable (respectively ν-measurable and ν-essentially bounded on T), endowed with the usual vector operations and with the norm***) $\|x\| = \left[\int_T |x(t)|^p \mathrm{d}\nu(t)\right]^{\frac{1}{p}}$ (respectively $\|x\| =$

*) See N. Bourbaki [22], Chap. III, p. 87, proposition 7.

**) See e.g. P. R. Halmos [82], § 17 or N. Dunford and J. Schwartz [49], p. 126. For brevity we shall not specify the σ-field of subsets of T on which the measure ν is defined; this will cause no confusion.

***) For simplicity we shall use the same notation for a function in $\mathfrak{L}^p$ and for its equivalence class in L^p $(1 \leqslant p \leqslant \infty)$.

$= \operatorname{ess\,sup}_{t\in T} |x(t)|$). For a function x' on T we shall use the notation

$$Z(x') = \{t\in T \mid x'(t) = 0\}. \tag{1.60}$$

We shall now deduce from theorem 1.1 a number of characterizations of elements of best approximation in the spaces $L^1(T, \nu)$, collected in the following theorem:

THEOREM 1.7. *Let $E = L^1(T, \nu)$, where (T, ν) is a positive measure space, let G be a linear subspace of E, $x\in E\setminus \bar{G}$ and $g_0\in G$. The following statements are equivalent:*

1° $g_0\in\mathfrak{P}_G(x)$.

2° *There exists a countably additive and ν-absolutely continuous set function m defined on the sets of finite measure, such that*

$$\sup_{0<\nu(A)<\infty} \frac{|m(A)|}{\nu(A)} = 1, \tag{1.61}$$

$$\int_T g(t)\, \mathrm{d}m(t) = 0 \qquad (g\in G), \tag{1.62}$$

$$\int_T [x(t) - g_0(t)]\, \mathrm{d}m(t) = \int_T |x(t) - g_0(t)|\, \mathrm{d}\nu(t). \tag{1.63}$$

3° *We have*

$$\left|\int_{T\setminus Z(x-g_0)} g(t) \operatorname{sign}[x(t) - g_0(t)]\mathrm{d}\nu(t)\right| \leqslant \int_{Z(x-g_0)} |g(t)|\mathrm{d}\nu(t) \quad (g\in G). \tag{1.64}$$

4° *We have*

$$\left|\int_{T\setminus P_0} g(t)\operatorname{sign}\, [x(t)-g_0(t)]\, \mathrm{d}\nu(t)\right| \leqslant \int_{P_0} |g(t)|\mathrm{d}\nu(t) \quad (g\in G), \tag{1.65}$$

where

$$P_0 = Z(x - g_0)\setminus \bigcap_{g\in G} Z(g). \tag{1.66}$$

These statements are implied by each of the following statements, which are equivalent to each other:

5° *There exists a $\beta\in L^\infty(T, \nu)$ such that*

$$\operatorname{ess\,sup}_{t\in T} |\beta(t)| = 1, \tag{1.67}$$

$$\int_T g(t)\,\beta(t)\,\mathrm{d}\nu(t) = 0 \qquad (g \in G), \qquad (1.68)$$

$$\int_T [x(t) - g_0(t)]\,\beta(t)\,\mathrm{d}\nu(t) = \int_T |x(t) - g_0(t)|\,\mathrm{d}\nu(t). \qquad (1.69)$$

6° *There exists a* $\beta \in L^\infty(T,\nu)$ *satisfying (1.67), (1.68) and*

$$\beta(t)[x(t) - g_0(t)] = |x(t) - g_0(t)| \quad \nu\text{-}a.e. \text{ on } T. \qquad (1.70)$$

7° *There exists a* ν*-measurable function* α *on the set* $Z(x - g_0)$ *such that**)

$$|\alpha(t)| \leqslant 1 \qquad \nu\text{-}a.e. \text{ on } Z(x - g_0), \qquad (1.71)$$

$$\int_{Z(x-g_0)} g(t)\alpha(t)\mathrm{d}\nu(t) + \int_{T \setminus Z(x-g_0)} g(t)\operatorname{sign}[x(t) - g_0(t)]\,\mathrm{d}\nu(t) = 0 \quad (g \in G). \qquad (1.72)$$

In the particular case when the positive measure space (T,ν) *has the property that the dual* $L^1(T,\nu)^*$ *is canonically equivalent***) *to* $L^\infty(T,\nu)$ *(hence in particular****) *when* (T,ν) *is* σ*-finite or when* T *is locally compact and* ν *is a positive Radon measure on* T*), all the statements* 1°—7° *are equivalent.*

Proof. The equivalence 1° $\Leftrightarrow$ 2° follows from theorem 1.1 and the general form****) of continuous linear functionals on $L^1(T,\nu)$, while the implication 5° $\Rightarrow$ 1° and in the particular case when $L^1(T,\nu)^* \equiv L^\infty(T,\nu)$ canonically, the equivalence 1° $\Leftrightarrow$ 5°, are immediate consequences of theorem 1.1.

In order to show the equivalence 5° $\Leftrightarrow$ 6°, let us observe that $\beta \in L^\infty(T,\nu)$ satisfies (1.67) and (1.69) if and only if it satisfies (1.67) and (1.70). In fact, it is obvious that (1.67), (1.70) $\Rightarrow$ $\Rightarrow$ (1.67), (1.69). Conversely, assume that we have (1.67) and $\beta(t)[x(t) - g_0(t)] \neq |x(t) - g_0(t)|$ on a set $A \subset T$ with $\nu(A) > 0$. Then, by (1.67), we have

$$\operatorname{Re}(\beta(t)[x(t) - g_0(t)]) < |x(t) - g_0(t)| \quad \nu\text{-a.e. on } A,$$

*) For the definition of sign ζ (ζ complex) see section 1.1.

**) We shall use the term "equivalent" in the sense of S. Banach [8], i.e.: linearly isometric. The canonical equivalence $L^1(T,\nu)^* \equiv L^\infty(T,\nu)$ is given by the correspondence $f \to \beta$, where

$$f(y) = \int_T y(t)\beta(t)\mathrm{d}\nu(t) \qquad (y \in L^1(T,\nu));$$

by the symbol $L^1(T,\nu)^* \equiv L^\infty(T,\nu)$ we shall always understand this canonical isometry.

***) See N. Dunford and J. Schwartz [49], p. 289, theorem 5 and p. 387. An example due to T. Botts (given in the paper [206] of J. Schwartz; see also E. J. Mc Shane [146]) shows that the canonical equivalence $L^1(T,\nu)^* \equiv$ $\equiv L^\infty(T,\nu)$ may not hold if (T,ν) is not σ-finite.

****) See J. Schwartz [206], theorem 2.

since otherwise there would exist a set $A_1 \subset A$ with $\nu(A_1) > 0$ such that

$$|x(t)-g_0(t)| \leqslant \mathrm{Re}(\beta(t)[x(t)-g_0(t)]) \leqslant |\beta(t)[x(t)-g_0(t)]| \leqslant |x(t)-g_0(t)| \quad \nu\text{-a.e. on } A_1,$$

whence $\beta(t)[x(t)-g_0(t)] =$ real and positive ν-a.e. on A_1 and hence $= |x(t)-g_0(t)|$ ν-a.e. on A_1, which contradicts our hypothesis. Consequently, we obtain

$$\mathrm{Re}\int_T [x(t)-g_0(t)]\beta(t)\mathrm{d}\nu(t) = \int_T \mathrm{Re}([x(t)-g_0(t)]\beta(t))\,\mathrm{d}\nu(t) < $$

$$< \int_T |x(t)-g_0(t)|\mathrm{d}\nu(t),$$

which contradicts (1.69). Thus, $5° \Leftrightarrow 6°$.

Let us prove now the equivalence $6° \Leftrightarrow 7°$. If we have $6°$, then from (1.70) we get

$$\beta(t) = \mathrm{sign}[x(t)-g_0(t)] \qquad \nu\text{-a.e. on } T\setminus Z(x-g_0),$$

whence, by (1.67) and (1.68), the function $\alpha = \beta|_{Z(x-g_0)}$ satisfies (1.71) and (1.72). Conversely, if we have $7°$, then the function β on T defined by

$$\beta(t) = \begin{cases} \mathrm{sign}\,[x(t)-g_0(t)] \text{ for } t \in T\setminus Z(x-g_0) \\ \alpha(t) \text{ for } t \in Z(x-g_0), \end{cases}$$

satisfies (1.67), (1.68) and (1.70). Thus, $6° \Leftrightarrow 7°$.

We shall prove now the implication $3° \Rightarrow 1°$. Assume that we have $3°$ and let $g_1 \in G$ be arbitrary. Let us put

$$\beta_1(t) = \begin{cases} \mathrm{sign}\,[x(t)-g_0(t)] \text{ for } t \in T\setminus Z(x-g_0) \\ -\dfrac{\int_{T\setminus Z(x-g_0)} g_1(\tau)\,\mathrm{sign}\,[x(\tau)-g_0(\tau)]\,\mathrm{d}\nu(\tau)}{\int_{Z(x-g_0)} g_1(\tau)\,\mathrm{d}\nu(\tau)}\,\mathrm{sign}\,g_1(t) \\ \qquad\qquad \text{for } t \in Z(x-g_0), \end{cases}$$

with the convention $\frac{0}{0} = 0$. Then we have $\beta_1 \in L^\infty(T,\nu)$ and

$$\operatorname*{ess\,sup}_{t \in T} |\beta_1(t)| = 1,$$

$$\int_T g_1(t)\,\beta_1(t)\,\mathrm{d}\nu(t) = \int_{T\setminus Z(x-g_0)} g_1(t)\,\mathrm{sign}[x(t)-g_0(t)]\mathrm{d}\nu(t) -$$

$$-\int_{Z(x-g_0)} g_1(t) \frac{\int_{T\setminus Z(x-g_0)} g_1(\tau)\,\mathrm{sign}[x(\tau)-g_0(\tau)]\,\mathrm{d}\nu(\tau)}{\int_{Z(x-g_0)} |g_1(\tau)|\,\mathrm{d}\nu(\tau)} \mathrm{sign}\, g_1(t)\mathrm{d}\nu(t) = 0,$$

$$\int_T [x(t)-g_0(t)]\,\beta_1(t)\,\mathrm{d}\nu(t) = \int_{T\setminus Z(x-g_0)} |x(t)-g_0(t)|\,\mathrm{d}\nu(t) =$$

$$= \int_T |x(t)-g_0(t)|\,\mathrm{d}\nu(t).$$

Consequently, we have

$$\|x-g_0\| = \int_T |x(t)-g_0(t)|\,\mathrm{d}\nu(t) = \int_T [x(t)-g_0(t)]\,\beta_1(t)\,\mathrm{d}\nu(t) =$$

$$= \int_T [x(t)-g_0(t)-g_1(t)]\,\beta_1(t)\,\mathrm{d}\nu(t) \leqslant$$

$$\leqslant \int_T |x(t)-g_0(t)-g_1(t)|\,\mathrm{d}\nu(t) = \|x-g_0-g_1\|$$

(this also follows from the implication $5^\circ \Rightarrow 1^\circ$ above, applied to the linear subspace $[g_1]$ generated by g_1 and to the elements $x-g_0 \in E\setminus[g_1]$, $0 \in [g_1]$). Since g_1 was an arbitrary element of G, it follows that $g_0 \in \mathfrak{P}_G(x)$. Thus, $3^\circ \Rightarrow 1^\circ$.

Conversely, assume that we have 1° and let $g \in G$ be arbitrary. Put

$$T^g = [T\setminus Z(g_0)] \cup [T\setminus Z(g)] \cup [T\setminus Z(x)],$$
$$\nu^g = \nu|_{T^g}.$$

Then (T^g, ν^g) is a positive measure space and we have, by 1°,

$$\int_{T^g} |x(t)-g_0(t)|\,\mathrm{d}\nu^g(t) = \int_T |x(t)-g_0(t)|\,\mathrm{d}\nu(t) \leqslant$$

$$\leqslant \int_T |x(t)-g_0(t)-\lambda g(t)|\,\mathrm{d}\nu(t) = \int_{T^g} |x(t)-g_0(t)-\lambda g(t)|\,\mathrm{d}\nu^g(t)$$

for any scalar λ, whence $0 \in \mathfrak{P}_{[g|_{T^g}]}(x|_{T^g} - g_0|_{T^g})$ in $E^g = L^1(T^g, \nu^g)$. Consequently by virtue of theorem 1.1 there exists an $f_0 \in (E^g)^*$ such that

$$\|f_0\| = 1,$$
$$f_0(g) = 0,$$
$$f_0(x|_{T^g} - g_0|_{T^g}) = \int_{T^g} |x(t)-g_0(t)|\,\mathrm{d}\nu^g(t).$$

Since the sets $T\setminus Z(g_0)$, $T\setminus Z(g)$, $T\setminus Z(x)$, whence also their union T^g, are of σ-finite ν-measure *), (T^g, ν^g) is σ-finite and hence we have $(E^g)^* = L^1(T^g, \nu^g)^* \equiv L^\infty(T^g, \nu^g)$. Consequently, there exists a $\beta_0 \in L^\infty(T^g, \nu^g)$ with the following properties:

$$\operatorname*{ess\,sup}_{t \in T} |\beta_0(t)| = 1,$$

$$\int_{T^g} g(t)\beta_0(t)\,d\nu^g(t) = 0,$$

$$\int_{T^g} [x(t) - g_0(t)]\beta_0(t)\,d\nu^g(t) = \int_{T^g} |x(t) - g_0(t)|\,d\nu^g(t).$$

Putting

$$\beta(t) = \beta^g(t) = \begin{cases} \beta_0(t) & \text{for } t \in T^g \\ 0 & \text{for } t \in T\setminus T^g, \end{cases}$$

we obtain (1.67), (1.69) and

$$\int_T g(t)\beta(t)\,d\nu(t) = 0.$$

From (1.67) and (1.69) it follows that

$$\beta(t) = \operatorname{sign}[x(t) - g_0(t)] \quad \nu\text{-a.e. on } T\setminus Z(x - g_0),$$

whence

$$\int_{Z(x-g_0)} g(t)\beta(t)\,d\nu(t) + \int_{T\setminus Z(x-g_0)} g(t)\operatorname{sign}[x(t) - g_0(t)]\,d\nu(t) = 0,$$

whence, taking into account (1.67),

$$\left|\int_{T\setminus Z(x-g_0)} g(t)\operatorname{sign}[x(t) - g_0(t)]\,d\nu(t)\right| \leqslant \int_{Z(x-g_0)} |g(t)|\,d\nu(t).$$

Since $g \in G$ was arbitrary, it follows that we have (1.64). Thus, $1° \Rightarrow 3°$.

Finally, the equivalence $3° \Leftrightarrow 4°$ follows from the relations

$$\int_{Z(x-g_0)} |g(t)|\,d\nu(t) = \int_{P_0} |g(t)|\,d\nu(t) \qquad (g \in G),$$

$$\int_{T\setminus Z(x-g_0)} g(t)\operatorname{sign}[x(t) - g_0(t)]\,d\nu(t) = \int_{T\setminus P_0} g(t)\operatorname{sign}[x(t) - g_0(t)]\,d\nu(t) \qquad (g \in G),$$

which completes the proof of theorem 1.7.

---

*) See e.g. P. R. Halmos [82], §25, theorem 6.

In the particular case when $T \subset (-\infty, \infty)$, ν is the Lebesgue measure, G is a linear subspace of finite dimension and the scalars are real, the equivalence $1° \Leftrightarrow 6°$ is due, essentially, to M. G. Krein [125]. In the general case it has been given by S. Ya. Havinson [87] and the remark that the assumption $L^1(T, \nu)^* \equiv L^\infty(T, \nu)$ is necessary for the validity of this equivalence has been made by B. R. Kripke and T. J. Rivlin [128]. The equivalences $1° \Leftrightarrow 5°$ and $1° \Leftrightarrow 7°$ have been deduced from the general theorem 1.1 in the papers [212], p. 183 and [230], p. 337, theorem 6, respectively. The equivalence $1° \Leftrightarrow 4°$ has been given by B. R. Kripke and T. J. Rivlin ([128], corollary of theorem 1.6). In the particular case when $T = [0, 1]$, $\nu =$ the Lebesgue measure and the scalars are real, the equivalence $1° \Leftrightarrow 3°$ was given by V. N. Nikolsky ([169], p. 106, formula (1)) and in the general case by B. R. Kripke and T. J. Rivlin ([129], corollary 1.4); the above proofs of the equivalences $1° \Leftrightarrow 3° \Leftrightarrow 4°$ have been given, essentially, in the paper [232], p. 353—354.

COROLLARY 1.7. *Let $E = L^1(T,\nu)$, where (T,ν) is a positive measure space with the property that the dual $L^1(T,\nu)^*$ is canonically equivalent to $L^\infty(T,\nu)$, and let G be a linear subspace of E, $x \in E \setminus \overline{G}$ and $g_0 \in \mathfrak{L}_G(x)$. Then there exist a ν-measurable set $U_{g_0} \subset T$ with $\nu(U_{g_0}) > 0$ and a $\beta \in L^\infty(T,\nu)$ such that we have (1.67), (1.68) and*

$$|\beta(t)| = 1 \ \nu\text{-a.e. on } U_{g_0}, \tag{1.73}$$

$$g_0(t) = x(t) \ \nu\text{-a.e. on } T \setminus U_{g_0}. \tag{1.74}$$

Proof. By virtue of the implication $1° \Rightarrow 6°$ of theorem 1.7 there exists a $\beta \in L^\infty(T,\nu)$ satisfying (1.67), (1.68) and (1.70). Put

$$U_{g_0} = T \setminus Z(x - g_0). \tag{1.75}$$

Then by $x \in E \setminus \overline{G}$ we have $\nu(U_{g_0}) > 0$, and by the definition of U_{g_0} we have (1.74). Finally, by (1.75) and (1.70) we have (1.73), which completes the proof.

Let us now consider the case of a set $M \subset G$ of elements of best approximation. By virtue of corollary 1.3 we have $M \subset \mathfrak{L}_G(x)$ if and only if there exists a $\beta \in L^\infty(T,\nu)$ such that we have (1.67), (1.68) and (1.69) for all $g_0 \in M$. Combining the necessity part of this result with the above proof of corollary 1.7, it follows that if $M \subset \mathfrak{L}_G(x)$, then there exist a $\beta \in L^\infty(T,\nu)$ and, for each $g_0 \in M$, a ν-measurable set $U_{g_0} \subset T$ with $\nu(U_{g_0}) > 0$, such that we have (1.67), (1.68), (1.73) and (1.74) for all $g_0 \in M$. The following problem arises naturally: is it possible to find, in this case, instead of the sets U_{g_0}, a ν-measurable set $U \subset T$ with $\nu(U) > 0$, independent of $g_0 \in M$, such as to have (1.67), (1.68), (1.73) and (1.74) for all $g_0 \in M$, with the sets U_{g_0} replaced by U? We shall now show that in the particular case when M is a finite set the answer is affirmative and we shall prove for such sets M one more property, which will be used in §3. Namely, we have

THEOREM 1.8. *Let $E = L^1(T,\nu)$, where (T,ν) is a positive measure space with the property that the dual $L^1(T,\nu)^*$ is canonically equivalent to $L^\infty(T,\nu)$, and let G be a linear subspace of E, $x \in E \setminus \overline{G}$ and $M = \{g'_0, g'_1, \ldots, g'_{k+1}\} \subset \mathfrak{P}_G(x)$, where k is an integer with $0 \leqslant k < \infty$. Then:*

a) *There exist a ν-measurable set $U \subset T$ with $\nu(U) > 0$ and a $\beta \in L^\infty(T,\nu)$ such that we have (1.67), (1.68) and*

$$|\beta(t)| = 1 \quad \nu\text{-a.e. on } U, \tag{1.76}$$

$$g'_0(t) = g'_1(t) = \ldots = g'_{k+1}(t) = x(t) \quad \nu\text{-a.e. on } T \setminus U. \tag{1.77}$$

b) *If at least two of the elements $g'_0, g'_1, \ldots, g'_{k+1}$ are distinct then there exists a ν-measurable subset U_0 of the above set U, with $\nu(U_0) > 0$, such that we have*

$$\sum_{i=1}^{k+1} |g'_i(t) - g'_0(t)| \neq 0 \quad \nu\text{-a.e. on } U_0, \tag{1.78}$$

$$\beta(t) = \pm \operatorname{sign}[g'_i(t) - g'_0(t)] \quad \nu\text{-a.e. on } T \setminus Z(g'_i - g'_0) \quad (i = 1, \ldots, k+1), \tag{1.79}$$

$$g'_0(t) = g'_1(t) = \ldots = g'_{k+1}(t) \quad \nu\text{-a.e. on } T \setminus U_0. \tag{1.80}$$

Proof. a) By virtue of corollary 1.3 and the assumption $L^1(T,\nu)^* \equiv L^\infty(T,\nu)$, there exists a $\beta \in L^\infty(T,\nu)$ satisfying (1.67), (1.68) and

$$\int_T [x(t) - g'_i(t)]\,\beta(t)\,d\nu(t) = \int_T |x(t) - g'_i(t)|\,d\nu(t) \quad (i = 0, 1, \ldots, k+1). \tag{1.81}$$

Put

$$U = T \setminus Z\left(x - \frac{1}{k+2}\sum_{i=0}^{k+1} g'_i\right). \tag{1.82}$$

By $x \in E \setminus \bar{G}$ we have then $\nu(U) > 0$. Furthermore, by (1.81) and (1.67) we have

$$\int_T \left| x(t) - \frac{1}{k+2} \sum_{i=0}^{k+1} g_i'(t) \right| \mathrm{d}\nu(t) \leqslant \frac{1}{k+2} \sum_{i=0}^{k+1} \int_T |x(t) - g_i'(t)| \mathrm{d}\nu(t) =$$

$$= \int_T \left[x(t) - \frac{1}{k+2} \sum_{i=0}^{k+1} g_i'(t) \right] \beta(t) \mathrm{d}\nu(t) \leqslant$$

$$\leqslant \int_T \left| x(t) - \frac{1}{k+2} \sum_{i=0}^{k+1} g_i'(t) \right| \mathrm{d}\nu(t), \tag{1.83}$$

and hence the equalities, which imply, on the one hand (taking into account (1.82)), that we have (1.76) and, on the other hand,

$$\left| x(t) - \frac{1}{k+2} \sum_{i=0}^{k+1} g_i'(t) \right| = \frac{1}{k+2} \sum_{i=0}^{k+1} |x(t) - g_i'(t)| \ \nu\text{-a.e. on } T, \tag{1.84}$$

and from (1.82) and (1.84) it follows that we also have (1.77).

b) Assume that at least two of the elements $g_0', g_1', \ldots, g_{k+1}'$ are distinct and put

$$U_0 = T \setminus \bigcap_{i=1}^{k+1} Z(g_i' - g_0') = \bigcup_{i=1}^{k+1} [T \setminus Z(g_i' - g_0')]. \tag{1.85}$$

By our hypothesis we have then $\nu(U_0) > 0$, and by (1.77) we have $T \setminus U \subset T \setminus U_0$, whence $U_0 \subset U$. Also, by the definition (1.85) of the set U_0, we have (1.78) and (1.80).

Finally, from (1.81) and (1.67) it follows (see the proof of the equivalence $5° \Leftrightarrow 6°$ of theorem 1.7) that we have

$$\beta(t)[x(t) - g_i'(t)] = |x(t) - g_i'(t)| \ (i = 0,1,\ldots,k+1) \ \nu\text{-a.e. on } T,$$

whence

$$\beta(t)[g_i'(t) - g_0'(t)] = |x(t) - g_0'(t)| - |x(t) - g_i'(t)| = \text{real}$$
$$(i = 1,\ldots,k+1) \nu\text{-a.e. on } T.$$

Consequently, taking into account that by (1.76) and $U_0 \subset U$ we have $|\beta(t)| = 1$ ν-a.e. on U_0, whence also on each $T \setminus Z(g_i' - g_0')$ (by (1.85)), it follows that we have (1.79), which completes the proof of theorem 1.8.

In the particular case when $g_0' = g_1' = \ldots = g_{k+1}'$, theorem 1.8 a) reduces to corollary 1.7. On the other hand, in the particular case when the scalars are real, theorem 1.8 b) is an obvious consequence of theorem 1.8 a); however, it presents some interest for applications in the case when the scalars are complex (see §3).

Finally, let us consider the problem of characterization of the elements of best approximation g_0 for which $x - g_0$ is a normal element. In this case we have the following result, in which the assumption $L^1(T,\nu)^* \equiv L^\infty(T,\nu)$ is no more involved:

THEOREM 1.9. *Let $E = L^1(T,\nu)$, where (T,ν) is a positive measure space, let G be a linear subspace of E, $x \in E \setminus \overline{G}$ and $g_0 \in G$ such that $x - g_0$ is normal. We have $g_0 \in \mathfrak{L}_G(x)$ if and only if*

$$\int_T g(t) \operatorname{sign}[x(t) - g_0(t)] d\nu(t) = 0 \qquad (g \in G). \tag{1.86}$$

Proof. For the normal element $x - g_0$ we have, obviously*),

$$f_{x-g_0}(y) = \int_T y(t) \operatorname{sign}[x(t) - g_0(t)] \, d\nu(t) \qquad (y \in L^1(T, \nu)), \tag{1.87}$$

and theorem 1.9 follows now from corollary 1.4.

We observe that if $x' \in L^1(T,\nu)$ is normal, then

$$\nu[Z(x')] = 0, \tag{1.88}$$

where $Z(x') = \{t \in T \mid x'(t) = 0\}$, since otherwise there would exist two distinct functionals $f_1, f_2 \in L^1(T,\nu)^*$ of the form

$$f_i(y) = \int_T y(t) \, \beta_i(t) \, d\nu(t) \qquad (y \in L^1(T,\nu)\,;\, i = 1,2)$$

(where $\beta_i \in L^\infty(T,\nu)$), such that $\|f_i\| = 1$, $f_i(x') = \|x'\|$ $(i = 1,2)$. *If $L^1(T,\nu)^*$ is canonically equivalent to $L^\infty(T,\nu)$*, the converse is also true, i.e. *(1.88) characterizes the normal elements* $x' \in L^1(T, \nu)$, since the relations $\int_T x'(t)\beta_i(t) d\nu(t) = 1 = \operatorname{ess\,sup}_{t \in T} |\beta_i(t)|$ $(i = 1, 2)$ imply $\beta_i(t) = \operatorname{sign} x'(t)$ ν-a.e. on $T \setminus Z(x')$ $(i = 1, 2)$, whence $\beta_1(t) = \beta_2(t)$ ν-a.e. on $T \setminus Z(x')$. Applying this remark to $x' = x - g_0$, we see that in the particular case when $L^1(T,\nu)^* \equiv L^\infty(T, \nu)$ canonically, theorem 1.9 also follows from theorem 1.7 (equivalence $1° \Leftrightarrow 7°$).

In the particular case when $T \subset (-\infty,\infty)$, ν is the Lebesgue measure, G is a subspace of finite dimension and the scalars are real, the characterization (1.88) of the normal elements and theorem 1.9 have been given, essentially, by M.G. Krein [125]. For $T = (-\infty, \infty)$, $\nu =$ the Lebesgue measure, $G =$ the subspace consisting of the entire functions $g_\sigma \in L^1(T, \nu)$

*) For the definition of f_{x-g_0} see section 1.1.

of degree $\leqslant \sigma$ and the scalars = the real numbers, theorem 1.9 can be found in A. F. Timan [250], p. 95, section 2.12.6 and in I. I. Ibragimov [92], p. 258, theorem 6.6.4.

1.6. APPLICATIONS IN THE SPACES $C^1(Q,\nu)$ AND $C^1_R(Q,\nu)$

Let Q be a compact space and ν a positive Radon measure on Q. We shall denote by $C^1(Q,\nu)$ the linear subspace of $L^1(Q,\nu)$ consisting of the equivalence classes of the (complex or real) continuous functions on Q, endowed with the usual vector operations and with the norm $\|x\| = \int_Q |x(q)|\,d\nu(q)$. In the sequel we shall consider only the case when the carrier $S(\nu) = Q$; in this case each equivalence class of $C^1(Q,\nu)$ contains only one continuous function.

Since the spaces $C^1(Q, \nu)$ are dense linear subspaces of the corresponding $L^1(Q,\nu)$ spaces, whence $C^1(Q,\nu)^* \equiv L^1(Q,\nu)^*$ canonically, the results of the foregoing section are applicable, in particular, also in $C^1(Q,\nu)$ spaces. However, by using certain properties of continuous functions, one can obtain in these spaces additional results. Namely, let us observe that *if $T = Q$ (compact) and ν is a positive Radon measure on Q with the carrier $S(\nu) = Q$, then for $x \in C^1(Q,\nu)$ and $G \subset C^1(Q,\nu)$ the sets $U \subset Q$ and $U_0 \subset U$ of theorem 1.8 may be taken open, and the function β of that theorem may be taken continuous on U.* In fact, the first statement follows from formulas (1.82) and (1.85) and the second statement follows from the fact that by (1.67), (1.82) and (1.83) we have

$$\beta(t) = \operatorname{sign}\Big[x(t) - \frac{1}{k+2}\sum_{i=0}^{k+1} g_i'(t)\Big] \qquad \nu\text{-a.e. on } U,$$

and from the fact that by (1.82) the function $\operatorname{sign}\Big(x - \frac{1}{k+2}\sum_{i=0}^{k+1} g_i'\Big)$ is continuous on U.

In the particular case when the scalars are real and the functions considered take real values, we shall denote the space $C^1(Q,\nu)$ by $C^1_R(Q,\nu)$.

THEOREM 1.10. *Let $E = C^1_R(Q,\nu)$, where Q is a compact space and ν is a positive Radon measure on Q with the carrier $S(\nu) = Q$, and let G be a linear subspace of E, $x \in E \setminus \overline{G}$ and $M = \{g_0', g_1', \ldots, g_{k+1}'\} \subset \mathfrak{P}_G(x)$, where k is an integer with $0 \leqslant k < \infty$. Then there exist two disjoint sets U_1 and U_2 open in Q,* *with* $U_1 \cup U_2 \neq \emptyset$, *and a real* ν*-measurable function* α *defined on* $Q \setminus (U_1 \cup U_2)$, *with the following properties:*

$$|\alpha(q)| \leqslant 1 \qquad \nu\text{-}a.e. \text{ on } Q \setminus (U_1 \cup U_2), \tag{1.89}$$

$$\int_{Q \setminus (U_1 \cup U_2)} g(q)\, \alpha(q) d\nu(q) + \int_{U_1} g(q)\, d\nu(q) - \int_{U_2} g(q)\, d\nu(q) = 0 \quad (g \in G), \tag{1.90}$$

$$g_0'(q) = g_1'(q) = \ldots = g_{k+1}'(q) = x(q) \quad (q \in Q \setminus (U_1 \cup U_2)). \tag{1.91}$$

The proof is similar to that of theorem 1.8 a), putting

$$U_1 = \{q \in Q \,|\, x(q) - \frac{1}{k+2} \sum_{i=0}^{k+1} g_i'(q) > 0\}, \tag{1.92}$$

$$U_2 = \{q \in Q \,|\, x(q) - \frac{1}{k+2} \sum_{i=0}^{k+1} g_i'(q) < 0\}, \tag{1.93}$$

$$\alpha = \beta|_{Q \setminus (U_1 \cup U_2)}, \tag{1.94}$$

and taking into account that by (1.83), (1.92) and (1.93) we have now

$$\beta(q) = \begin{cases} 1 & \nu\text{-a.e. on } U_1 \\ -1 & \nu\text{-a.e. on } U_2. \end{cases} \tag{1.95}$$

We observe that theorem 1.10 may be even deduced from theorem 1.8 a), by putting $U_1 = U \cap \{q \in Q \,|\, \beta(q) = 1\}$, $U_2 = U \cap \{q \in Q \,|\, \beta(q) = -1\}$ and $\alpha = \beta|_{Q \setminus (U_1 \cup U_2)}$.

In the above it has not been assumed that there exists an i with $1 \leqslant i \leqslant k+1$, such that $g_i' \neq g_0'$. In the particular case when $g_0' = g_1' = \ldots = g_{k+1}'$, theorem 1.10 gives a necessary condition in order to have $g_0 \in \mathfrak{P}_G(x)$.

1.7. APPLICATIONS IN THE SPACES $L^p(T,\nu)$ $(1 < p < \infty)$ AND IN INNER PRODUCT SPACES

THEOREM 1.11. *Let* $E = L^p(T,\nu)$ $(1 < p < \infty)$, *where* (T,ν) *is a positive measure space, let* G *be a linear subspace of* $E, x \in E \setminus \overline{G}$ *and* $g_0 \in G$. *We have* $g_0 \in \mathfrak{P}_G(x)$ *if and only if*

$$\int_T g(t)\,|x(t) - g_0(t)|^{p-1} \operatorname{sign}[x(t) - g_0(t)] d\nu(t) = 0 \quad (g \in G). \tag{1.96}$$

Proof. In the spaces $L^p(T, \nu)$ $(1 < p < \infty)$ every element x' is*) normal. But for the normal element $x - g_0$ we have, obviously,

$$f_{x-g_0}(y) = \int_T y(t) \frac{|x(t) - g_0(t)|^{p-1}}{\|x - g_0\|^{p-1}} \operatorname{sign}[x(t)-g_0(t)] \, d\nu(t) \quad (y \in L^p(T, \nu)) \tag{1.97}$$

and theorem 1.11 follows now from corollary 1.4.

In the particular case when T is a segment $[a, b] \subset (-\infty, \infty)$, ν is the Lebesgue measure, G is a subspace of finite dimension (respectively, an arbitrary linear subspace) and the scalars are real, theorem 1.11 has been given e.g. by A. F. Timan [250], p. 74, section 2.8.25 (respectively by V. N. Nikolsky [169], p. 103). In the particular case when $T = (-\infty, \infty)$, ν is the Lebesgue measure, G is the subspace consisting of the entire functions $g_\sigma \in L^p(T, \nu)$ of degree $\leqslant \sigma$ and the scalars are real, theorem 1.11 can be found in A. F. Timan [250], p. 95, section 2.12.6 and in I. I. Ibragimov [92], p. 258, theorem 6.6.4.

In the particular case when $p = 2$ the condition (1.96) may be written in the form

$$\int_T g(t)\left[\overline{x(t) - g_0(t)}\right] d\nu(t) = 0 \qquad (g \in G),$$

i.e.

$$(g, x - g_0) = 0 \qquad (g \in G). \tag{1.98}$$

By the same argument as that used in the above proof of theorem 1.11, one can prove (see [212], p. 183) the following well known theorem of characterization of elements of best approximation in abstract inner product spaces.

THEOREM 1.11′. *Let* $E = \mathcal{H} =$ *an inner product space***), *G a linear subspace of* E, $x \in E \setminus \overline{G}$ *and* $g_0 \in G$. *We have* $g_0 \in \mathfrak{P}_G(x)$ *if and only if we have (1.98).*

The necessity part of theorem 1.11 is given e.g. in N. I. Ahiezer [1], chap. I, § 13.

---

*) In fact, this follows from the strict convexity of $(L^p)^* \equiv L^q \left(\frac{1}{p} + \frac{1}{q} = 1\right)$.

**) By *inner product space* we mean here, as well as in the sequel, a normed linear space endowed with a scalar product such that $\|x\| = \sqrt{(x, x)}$ (*separated prehilbertian space* in the sense of N. Bourbaki [23], Chap. V, p. 128, definition 3); we do not assume the completeness of the space.

1.8. THE SECOND THEOREM OF CHARACTERIZATION OF ELEMENTS OF BEST APPROXIMATION IN GENERAL NORMED LINEAR SPACES

We recall that in a (real *or complex*) linear space L any set of the form $\{\lambda x + (1-\lambda)y \mid 0 \leqslant \lambda \leqslant 1\}$, where $x, y \in L$, is called a *segment*; the points $\lambda x + (1-\lambda)y$ with $0 < \lambda < 1$ are called *interior points* of the segment. A set $A \subset L$ is called *convex* if together with any two points x, y it contains the whole segment generated by them, that is, if the relations $x, y \in A$ and $0 \leqslant \lambda \leqslant 1$ imply $\lambda x + (1-\lambda)y \in A$.

A set $\mathfrak{M}$ in a (real or complex) topological linear space L is called*) an *extremal subset* of a closed convex set A if: a) $\mathfrak{M}$ is a closed convex subset of A and b) together with every interior point of a segment in A it contains the whole segment, i.e. the relations $x, y \in A$, $\lambda x + (1-\lambda)y \in \mathfrak{M}$ and $0 < \lambda < 1$ imply $x, y \in \mathfrak{M}$. An extremal subset of A consisting of a single point (i.e. a point of A which is not an interior point of any segment in A) is called an *extremal point* of A. We shall denote by $\mathcal{E}(A)$ the set of all extremal points of A.

LEMMA 1.7 ([230], lemma 1). *Let $\mathfrak{M}$ be an extremal subset of a closed convex set A in a topological linear space L. Then*

$$\mathcal{E}(\mathfrak{M}) = \mathcal{E}(A) \cap \mathfrak{M}. \tag{1.99}$$

Proof. The inclusion $\mathcal{E}(A) \cap \mathfrak{M} \subset \mathcal{E}(\mathfrak{M})$ is obvious. On the other hand, since $\mathfrak{M}$ is an extremal subset of A, every extremal point of $\mathfrak{M}$ is an extremal point of A, whence $\mathcal{E}(\mathfrak{M}) \subset \mathcal{E}(A) \cap \mathfrak{M}$. Consequently, we have (1.99), which completes the proof.

In the sequel we shall consider the particular case when L is the conjugate space of a normed linear space E, endowed with the weak topology $\sigma(E^*, E)$, and A is the unit cell

$$S_{E^*} = \{f \in E^* \mid \|f\| \leqslant 1\}.$$

In this case we shall say, for brevity, that *$\mathfrak{M}$ is an extremal subset of the cell S_{E^*} endowed with $\sigma(E^*, E)$*. Since S_{E^*} is compact for $\sigma(E^*, E)$, so is any extremal subset $\mathfrak{M}$ of the cell S_{E^*} endowed with $\sigma(E^*, E)$.

LEMMA 1.8 ([218], p. 118, first part of theorem 1). *Let E be a normed linear space and let $\mathfrak{F}$ be a non-void convex subset of the set $\{x \in E \mid \|x\| = 1\}$. Then the set*

$$\mathfrak{M} = \mathfrak{M}_{\mathfrak{F}} = \bigcap_{x \in \mathfrak{F}} \{f \in E^* \mid \|f\| = 1,\ f(x) = 1\} \tag{1.100}$$

is a non-void extremal subset of the cell S_{E^*} endowed with $\sigma(E^*, E)$.

*) See e. g. M. M. Day [43], Chap. V, §1; in N. Dunford and J. Schwartz ([49], p. 439) these sets are called closed extremal subsets.

Proof. The set $\mathfrak{M}$ is non-void by virtue of a well known theorem of S. Mazur*). Also, the set $\mathfrak{M}$ is convex and closed for $\sigma(E^*, E)$, since we have

$$\mathfrak{M} = \bigcap_{x \in \mathfrak{F}} \{f \in E^* \mid f(x) = 1\} \cap S_{E^*}.$$

Finally, assume that for an $f \in \mathfrak{M}$ and a λ with $0 < \lambda < 1$ we have $f = \lambda f_1 + (1 - \lambda) f_2$, where $f_1, f_2 \in S_{E^*}$. Then, since $f \in \mathfrak{M}$, we have

$$1 = f(x) = \lambda f_1(x) + (1 - \lambda) f_2(x) \qquad (x \in \mathfrak{F}),$$

whence, by

$$0 < \lambda < 1, \ |f_1(x)| \leqslant \|x\|, \ |f_2(x)| \leqslant \|x\| \qquad (x \in \mathfrak{F})$$

we obtain

$$f_1(x) = f_2(x) = \|x\| \qquad (x \in \mathfrak{F}),$$

i.e. $f_1, f_2 \in \mathfrak{M}$, which completes the proof.

In this general form we shall use lemma 1.8 in section 1.10 (see the proof of lemma 1.10). In the present section we shall use only the following immediate consequence of lemma 1.8:

COROLLARY 1.8 ([218], p. 119, section 3, paragraph d)). *Let E be a normed linear space and $x' \in E$, $x' \neq 0$. Then the set*

$$\mathfrak{M} = \mathfrak{M}_{x'} = \{f \in E^* \mid \|f\| = 1, \ f(x') = \|x'\|\} \tag{1.101}$$

is a non-void extremal subset of the cell S_{E^*} endowed with $\sigma(E^*, E)$.

Using theorem 1.1, lemma 1.7 and corollary 1.8, we now shall prove the following property of the elements of best approximation in normed linear spaces (whence we shall deduce then a second theorem of characterization of elements of best approximation):

THEOREM 1.12. *Let E be a normed linear space, G a linear subspace of E, $x \in E \setminus \overline{G}$ and $g_0 \in \mathfrak{P}_G(x)$. Then there exists an $f_0 \in E^*$ with the following properties:*

$$f_0 \in \mathcal{E}(S_{E^*}), \tag{1.102}$$

$$\operatorname{Re} f_0(g_0) \geqslant 0, \tag{1.103}$$

$$f_0(x - g_0) = \|x - g_0\|. \tag{1.104}$$

Theorem 1.12 has been given, for a particular class of normed linear spaces, called "Choquet spaces", in [230], theorem 3. G. Choquet has shown [37] that the theorem remains valid in arbitrary normed linear spaces. Subsequently, in [232], it has

*) See e.g. G. Köthe [122], p. 191, theorem 2.

been shown that even the proof given in [230] for Choquet spaces can be adapted, by changing its final part, so as to give theorem 1.12. We now shall give both proofs of theorem 1.12.

First proof ([230], p. 334 and [232], p. 346). By virtue of corollary 1.8 above, the set

$$\mathfrak{M}_{x, g_0} = \{f \in E^* \mid \|f\| = 1,\ f(x - g_0) = \|x - g_0\|\} \tag{1.105}$$

is a non-void extremal subset of the cell S_{E^*} endowed with $\sigma(E^*, E)$. Consequently, by virtue of the Krein-Milman theorem*), the set $\mathcal{E}(\mathfrak{M}_{x, g_0})$ is non-void and thus, taking into account lemma 1.7, there exists an $f_0 \in E^*$ satisfying (1.102) and (1.104).

Assume now that for all $f_0 \in E^*$ satisfying (1.102) and (1.104) we have

$$\operatorname{Re} f_0(g_0) < 0. \tag{1.106}$$

We claim that *in this case* the set

$$\mathfrak{N} = \{f \in \mathfrak{M}_{x, g_0} \mid \operatorname{Re} f(g_0) = 0\} \tag{1.107}$$

is a non-void extremal subset of the set $\mathfrak{M}_{x, g_0}$ endowed with $\sigma(E^*, E)$, whence also of the cell S_{E^*} endowed with $\sigma(E^*, E)$. In fact, since $g_0 \in \mathfrak{P}_G(x)$, by virtue of theorem 1.1 the set $\mathfrak{N}$ is non-void. Also $\mathfrak{N}$ is convex and $\sigma(E^*, E)$-closed, since

$$\mathfrak{N} = \mathfrak{M}_{x, g_0} \cap \{f \in E^* \mid \operatorname{Re} f(g_0) = 0\}.$$

Finally, assume that for an $f \in \mathfrak{N}$ and a λ with $0 < \lambda < 1$ we have $f = \lambda f_1 + (1 - \lambda) f_2$, where $f_1, f_2 \in \mathfrak{M}_{x, g_0}$. Then, by (1.107),

$$\operatorname{Re}\,[\lambda f_1(g_0) + (1 - \lambda) f_2(g_0)] = 0,$$

whence $\lambda \operatorname{Re} f_1(g_0) = (\lambda - 1) \operatorname{Re} f_2(g_0)$. Since by our hypothesis we have (1.106) for all $f_0 \in \mathcal{E}(\mathfrak{M}_{x, g_0}) = \mathcal{E}(S_{E^*}) \cap \mathfrak{M}_{x, g_0}$ and since the mapping $f \to f(g_0)$ of E^* into the field of scalars is linear and continuous for $\sigma(E^*, E)$, by virtue of the Krein-Milman theorem applied to $\mathfrak{M}_{x, g_0}$ we have $\operatorname{Re} f(g_0) \leqslant 0$ for all $f \in \mathfrak{M}_{x, g_0}$. Applying this to f_1 and f_2 and taking into account $\lambda > 0$, $\lambda - 1 < 0$, we obtain $\operatorname{Re} f_1(g_0) = \operatorname{Re} f_2(g_0) = 0$, i.e. $f_1, f_2 \in \mathfrak{N}$, which proves that $\mathfrak{N}$ is an extremal subset of the set $\mathfrak{M}_{x, g_0}$ endowed with $\sigma(E^*, E)$, whence also of the cell S_{E^*} endowed with $\sigma(E^*, E)$.

Consequently, by virtue of the Krein-Milman theorem applied to $\mathfrak{N}$, there exists an $f_0 \in \mathcal{E}(\mathfrak{N}) = \mathcal{E}(S_{E^*}) \cap \mathfrak{N}$, which contradicts the hypothesis that we have (1.106) for all $f_0 \in E^*$ satisfying (1.102) and (1.104). This completes the proof.

*) See e.g. N. Dunford and J. Schwartz [49], p. 440, theorem 4.

Second proof (G. Choquet [37]). Since $g_0 \in \mathfrak{P}_G(x)$, the set $\mathfrak{M}_{x,g_0}$ defined by (1.105) contains, by virtue of theorem 1.1, an element f_1 such that

$$f_1(g_0) = 0. \tag{1.108}$$

Let ψ be the mapping $f \to f(g_0)$ of E^* into the field of scalars K. This mapping is linear and continuous for $\sigma(E^*, E)$, whence the set $\psi(\mathfrak{M}_{x,g_0})$ is a compact convex subset of K containing, by (1.108), the point 0. In the case when K is the complex plane, let D be the closed half-plane

$$D = \{\zeta \in K \mid \operatorname{Re} \zeta \geqslant 0\}, \tag{1.109}$$

and let $d \subset D$ be a support line of $\psi(\mathfrak{M}_{x,g_0})$ parallel to the boundary of D. Then d, whence also D, contains an extremal point ζ_0 of $\psi(\mathfrak{M}_{x,g_0})$. In the case when K is the real line, the half-line D defined by (1.109) also contains an extremal point ζ_0 of $\psi(\mathfrak{M}_{x,g_0})$. Since in both cases $\mathfrak{M}_{x,g_0}$ is a convex $\sigma(E^*, E)$-compact subset of E^*, there exists, by virtue of a result of E. Bishop and K. de Leeuw ([18], lemma 4.4) an $f_0 \in \mathcal{E}(\mathfrak{M}_{x,g_0})$ such that*) $\psi(f_0) = \zeta_0$. Then the element f_0 has the properties (1.102), (1.103) and (1.104), which completes the proof.

We observe that in the second proof above, the half-plane D defined by (1.109) can be replaced by any other half-plane

$$\{\zeta \in K \mid \operatorname{Re} \beta\zeta \geqslant 0\}$$

(where $|\beta| = 1$), without changing the rest of the proof. Consequently, if $g_0 \in \mathfrak{P}_G(x)$, then for any scalar β of modulus $|\beta| = 1$ there exists an $f_0 \in E^*$ satisfying (1.102), (1.104) and

$$\operatorname{Re} \beta f_0(g_0) \geqslant 0. \tag{1.110}$$

Another proof of theorem 1.12 has been given in [232], p. 348. Since it makes use of theorem 1.1 of Chap. II, we shall present it in Chap. II, §1, section 1.1.

All these proofs of theorem 1.12 make use of the necessity part of theorem 1.1. We observe that conversely, even from the particular case $G = \{0\}$ of theorem 1.12 one can deduce the necessity part of theorem 1.1, as follows**): Let $x \in E$, $g_0 \in G$ be arbitrary. Then in the quotient space E/G, we have $\dot{g}_0 = \dot{0} \in \mathfrak{P}_{\{\dot{0}\}}(\dot{x})$ for the 0-dimensional linear subspace $\{\dot{0}\}$,

*) We mention, briefly, a proof of this result (following E. Bishop and K. de Leeuw [18]): by virtue of the Krein-Milman theorem there exists an $f_0 \in \mathcal{E}(\mathfrak{M}_{x,g_0} \cap \psi^{-1}(\zeta_0))$, and, taking into account $\zeta_0 \in \mathcal{E}[\psi(\mathfrak{M}_{x,g_0})]$, it is easy to verify that $f_0 \in \mathcal{E}(\mathfrak{M}_{x,g_0})$.

**) A similar argument, for corollary 1.11 instead of theorem 1.12, has been given by V. N. Nikolsky ([168], p. 337).

whence by virtue of theorem 1.12 there exists a $\varphi \in (E/G)^*$ such that

$$\varphi \in \mathcal{E}(S_{(E/G)^*}),$$

$$\varphi(\dot{x} - \dot{g}_0) = \| \dot{x} - \dot{g}_0 \| = \| \dot{x} \| = \inf_{g \in G} \| x - g \|.$$

Consequently, if $g_0 \in \mathfrak{P}_G(x)$, then for the functional $f \in E^*$ defined by

$$f(x) = \varphi(\dot{x}) \qquad (x \in \dot{x},\ \dot{x} \in E/G),$$

we have (1.2), (1.3) and (1.4), which concludes the proof.

We now can give the second theorem of characterization of elements of best approximation.

THEOREM 1.13. *Let E be a normed linear space, G a linear subspace of E, $x \in E \setminus \bar{G}$ and $g_0 \in G$. We have $g_0 \in \mathfrak{P}_G(x)$ if and only if for every $g \in G$ there exists an $f^g \in E^*$ such that*

$$f^g \in \mathcal{E}(S_{E^*}), \tag{1.111}$$

$$\operatorname{Re} f^g(g_0 - g) \geqslant 0, \tag{1.112}$$

$$f^g(x - g_0) = \| x - g_0 \|. \tag{1.113}$$

Proof ([230], p. 335). Let $g_0 \in \mathfrak{P}_G(x)$ and let g be an arbitrary element of G. Then $g_0 - g \in \mathfrak{P}_G(x - g)$, whence, by virtue of theorem 1.12 and of the relation $x - g - (g_0 - g) = x - g_0$, there exists an $f^g \in E^*$ satisfying (1.111), (1.112) and (1.113).

Conversely, assume that for every $g \in G$ there exists an $f^g \in E^*$ such that we have (1.111), (1.112) and (1.113). Then for every $g \in G$ we have

$$\|x - g_0\| = \operatorname{Re} f^g(x - g_0) \leqslant \operatorname{Re} f^g(x - g) \leqslant | f^g(x - g) | \leqslant$$
$$\leqslant \|f^g\| \, \|x - g\| = \| x - g \|,$$

whence $g_0 \in \mathfrak{P}_G(x)$, which completes the proof.

We observe that in the particular case when the element $x - g_0$ is normal, corollary 1.4 follows again from theorem 1.13. In fact, in this case the conditions $\|f\| = 1$ and (1.113) are equivalent to *)

$$f^g = f_{x - g_0} \qquad (g \in G),$$

*) We observe that if $x' \in E$ is normal, the subset $\mathfrak{M}$ of S_{E^*} defined by (1.101) consists of a single element, namely $f_{x'}$ whence we have, by corollary 1.8, $f_{x'} \in \mathcal{E}(S_{E^*})$.

and therefore the conditions (1.111), (1.112) and (1.113) are equivalent to

$$\operatorname{Re} f_{x-g_0}(g_0 - g) \geqslant 0 \qquad (g \in G).$$

Writing this for $g_1 = g_0 - g$, we obtain the equivalent condition

$$\operatorname{Re} f_{x-g_0}(g) \geqslant 0 \qquad (g \in G),$$

whence, taking into account $-g \in G$ $(g \in G)$, we obtain the equivalent condition

$$\operatorname{Re} f_{x-g_0}(g_0) = 0 \qquad (g \in G),$$

which, by virtue of lemma 1.1 b), is equivalent to the condition (1.16) of corollary 1.4.

Just as in the case of theorem 1.1, one can give several equivalent variants of the conditions of theorem 1.13, namely, we have

COROLLARY 1.9. *Let E be a normed linear space, G a linear subspace of E, $x \in E \setminus \overline{G}$ and $g_0 \in G$. The following statements are equivalent :*

1° $g_0 \in \mathfrak{P}_G(x)$.

2° *For every $g \in G$ there exists an $f^g \in E^*$ satisfying (1.111), (1.113) and*

$$\operatorname{Re} [\beta f^g(g_0 - g)] \geqslant 0, \tag{1.114}$$

where β is a scalar such that $|\beta| = 1$.

3° *For every $g \in G$ there exists an $f^g \in E^*$ satisfying (1.111), (1.112) and*

$$\operatorname{Re} f^g(x - g_0) = \|x - g_0\| . \tag{1.115}$$

4° *For every $g \in G$ there exists an $f^g \in E^*$ satisfying (1.111) and*

$$|f^g(x - g_0)| = \|x - g_0\| , \tag{1.116}$$

$$|f^g(x - g_0)| \leqslant f^g(x - g)| . \tag{1.117}$$

5° *For every $g \in G$ there exists an $f^g \in E^*$ satisfying (1.111), (1.116) and*

$$\operatorname{Re} [f^g(g_0 - g) \overline{f^g(x - g_0)}] \geqslant 0. \tag{1.118}$$

6° *For every $g \in G$ there exists an $f^g \in E^*$ satisfying (1.111), (1.116) and*

$$\operatorname{Re} [f^g(g) \overline{f^g(x - g_0)}] \geqslant 0. \tag{1.119}$$

Proof. Assume that we have 1°, whence the condition of theorem 1.13 is satisfied. Then, by (1.113) we have (1.116). On the other hand, by (1.113) and (1.112) we have

$$|f^g(x-g_0)| = \operatorname{Re} f^g(x-g_0) \leqslant \operatorname{Re} f^g(x-g) \leqslant |f^g(x-g)|, \tag{1.120}$$

whence (1.117). Thus, 1°⇒4°.

Conversely, if we have 4°, then for every $g \in G$ we have

$$\|x-g_0\| = |f^g(x-g_0)| \leqslant |f^g(x-g)| \leqslant \|x-g\|, \tag{1.121}$$

whence $g_0 \in \mathfrak{P}_G(x)$. Thus, 4°⇒1°.

The implication 1°⇒2° (with $\beta = 1$) follows immediately from theorem 1.13.

Assume now that we have 2°. Put

$$\varphi^g = f^{(1-\bar{\beta})g_0 + \bar{\beta}g} \qquad (g \in G). \tag{1.122}$$

We have then, taking into account 2°,

$$\varphi^g \in \mathcal{E}(S_{E^*}),$$

$$\operatorname{Re} \varphi^g(g_0 - g) = \operatorname{Re}\left\{\frac{1}{\bar{\beta}} \varphi^g[\bar{\beta}(g_0-g)]\right\} =$$

$$= \operatorname{Re}\left\{\frac{1}{\bar{\beta}} \bar{\beta}\, \beta f^{(1-\bar{\beta})g_0+\bar{\beta}g}[\bar{\beta}(g_0-g)]\right\} =$$

$$= \operatorname{Re}\{\beta f^{(1-\bar{\beta})g_0+\bar{\beta}g}[g_0-(1-\bar{\beta})g_0-\bar{\beta}g]\} \geqslant 0,$$

$$\varphi^g(x-g_0) = f^{(1-\bar{\beta})g_0+\bar{\beta}g}(x-g_0) = \|x-g_0\|,$$

whence for every $g \in G$ the functional φ^g defined by (1.122) satisfies (1.111), (1.112) and (1.113), and therefore, by theorem 1.13, we have $g_0 \in \mathfrak{P}_G(x)$. Thus, 2° ⇒ 1°.

The equivalence 1°⇔3° follows immediately from theorem 1.13 and lemma 1.1 a), and the implication 1°⇒5° follows immediately from theorem 1.13.

Assume now that we have 5°. Put

$$\psi^g = [\operatorname{sign} f^g(x-g_0)] f^g \qquad (g \in G). \tag{1.123}$$

We have then, taking into account 5°,

$$\psi^g \in \mathcal{E}(S_{E^*}),$$

$$\operatorname{Re} \psi^g(g_0 - g) = \operatorname{Re}\left\{\frac{\overline{f^g(x-g_0)}}{|f^g(x-g_0)|} f^g(g_0-g)\right\} \geqslant 0,$$

$$\psi^g(x-g_0) = |f^g(x-g_0)| = \|x-g_0\|,$$

whence, for every $g \in G$ the functional ψ^g defined by (1.123) satisfies (1.111), (1.112) and (1.113) and therefore, by theorem 1.13, we have $g_0 \in \mathfrak{P}_G(x)$. Thus $5° \Rightarrow 1°$.

Now again assume that we have 5°. Put

$$\varphi^g = f^{g_0 - g} \qquad (g \in G). \tag{1.124}$$

We have then, taking into account 5°,

$$\varphi^g \in \mathcal{E}(S_{E^*}),$$

$$\operatorname{Re}\,[\varphi^g(g)\,\overline{\varphi^g(x - g_0)}] = \operatorname{Re}\,[f^{g_0-g}(g)\,\overline{f^{g_0-g}(x - g_0)}] =$$

$$= \operatorname{Re}\,\{f^{g_0-g}\,[g_0 - (g_0 - g)]\,\overline{f^{g_0-g}(x - g_0)}\} \geqslant 0,$$

$$|\varphi^g(x - g_0)| = |f^{g_0-g}(x - g_0)| = \|x - g_0\|,$$

whence for every $g \in G$ the functional φ^g defined by (1.124) satisfies (1.111), (1.116) and (1.119). Thus, $5° \Rightarrow 6°$.

Finally, the implication $6° \Rightarrow 5°$ is shown in a similar way, by putting

$$\psi^g = f^{g_0 - g}, \tag{1.125}$$

which completes the proof of corollary 1.9.

We observe that the implication $1° \Rightarrow 2°$ of corollary 1.9 above, even with an arbitrary β of modulus $|\beta| = 1$, follows from the remark made after theorem 1.12 (see formula (1.10)), with the same argument as that used in the proof of the necessity part of theorem 1.13.

The equivalence $1° \Leftrightarrow 4°$ of corollary 1.9 has been given in [230], theorem 5, and the equivalence $1° \Leftrightarrow 6°$ in the paper [234], §1.

In both the theorem 1.13 and corollary 1.9, the condition (1.111) presents interest only as a necessary condition for $g_0 \in \mathfrak{P}_G(x)$, since in the sufficiency parts it may be replaced by $\|f^g\| = 1$. Using this remark we shall now prove

COROLLARY 1.10. *Let E be a normed linear space, G a linear subspace of E, $x \in E \setminus \overline{G}$ and $g_0 \in \mathfrak{P}_G(x)$. We have then*

$$\|x - g_0\| = \max_{f \in \mathcal{E}(\mathfrak{M}_{x,g_0})} |f(x) - f(g_0)| = \min_{g \in G} \max_{f \in \mathcal{E}(\mathfrak{M}_{x,g_0})} |f(x) - f(g)|, \tag{1.126}$$

where $\mathfrak{M}_{x,g_0}$ is the set defined by (1.105).

Proof. For every $y \in E$, let us denote by $\tilde{y}$ the continuous bounded function on*) $\mathcal{S}(\mathfrak{M}_{x,g_0})$ defined by

$$\tilde{y}(f) = f(y) \qquad (f \in \mathcal{S}(\mathfrak{M}_{x,g_0})), \tag{1.127}$$

and let E_1 be the space of all functions $\tilde{y}$ $(y \in E)$, endowed with the usual vector operations and with the norm**)

$$\|\tilde{y}\|_{E_1} = \max_{f \in \mathcal{S}(\mathfrak{M}_{x,g_0})} |\tilde{y}(f)| \tag{1.128}$$

(that is, $E_1 =$ the image of E in the space of all continuous and bounded functions on $\mathcal{S}(\mathfrak{M}_{x,g_0})$, under the mapping $y \to \tilde{y}$).

For every $\tilde{g}$ $(g \in G)$ define $\varphi^{\tilde{g}} \in E_1^*$ by

$$\varphi^{\tilde{g}}(\tilde{y}) = \tilde{y}(f^g) = f^g(y) \qquad (\tilde{y} \in E_1), \tag{1.129}$$

where $f^g \in E^*$ is a fixed functional with the properties (1.111) — — (1.113). Then, by (1.111) and (1.113) we have $f^g \in \mathcal{S}(\mathfrak{M}_{x,g_0})$, whence, by (1.129) and (1.128) , we obtain

$$|\varphi^{\tilde{g}}(\tilde{y}) = |\tilde{y}(f^g)| \leqslant \|\tilde{y}\| \qquad (\tilde{y} \in E_1),$$

hence

$$\|\varphi^{\tilde{g}}\| \leqslant 1. \tag{1.130}$$

From (1.112), (1.130) and (1.113) we obtain, taking into account (1.129) and (1.128),

$$\operatorname{Re} \varphi^{\tilde{g}}(\tilde{g}_0 - \tilde{g}) = \operatorname{Re} f^g(g_0 - g) \geqslant 0, \tag{1.131}$$

$$\|\tilde{x} - \tilde{g}_0\|_{E_1} \geqslant |\varphi^{\tilde{g}}(\tilde{x} - \tilde{g}_0)| = f^g(x - g_0) = \|x - g_0\| \geqslant \|\tilde{x} - \tilde{g}_0\|_{E_1},$$

hence

$$\varphi^{\tilde{g}}(\tilde{x} - \tilde{g}_0) = f^g(x - g_0) = \|\tilde{x} - \tilde{g}_0\|_{E_1}, \tag{1.132}$$

whence, taking into account (1.130),

$$\|\varphi^{\tilde{g}}\| = 1. \tag{1.133}$$

Consequently, applying theorem 1.13 and the remark made before corollary 1.10 to the linear subspace $G_1 = \{\tilde{g} \mid g \in G\}$ of

---

*) We recall that we consider the set $\mathcal{S}(\mathfrak{M}_{x,g_0})$ endowed with the weak topology $\sigma(E^*, E)$; in general, it is not closed in E^*, hence it is not compact.

**) We have seen in the first proof above of theorem 1.12 that this maximum is attained for an $f_0 \in \mathcal{S}(\mathfrak{M}_{x,g_0})$. Two functions $y_1, y_2 \in E_1$ with $\tilde{y}_1|_{\mathcal{S}(\mathfrak{M}_{x,g_0})} = \tilde{y}_2|_{\mathcal{S}(\mathfrak{M}_{x,g_0})}$ are considered identical.

E_1, it follows that we have $\tilde{g}_0 \in \mathfrak{L}_{G_1}(\tilde{x})$. But this means, taking into account (1.111), (1.113) and (1.128), that we have

$$\|x - g_0\| = \max_{f \in \mathcal{E}(\mathfrak{M}_{x,g_0})} |f(x) - f(g_0)| = \|\tilde{x} - \tilde{g}_0\|_{E_1} =$$

$$= \min_{\tilde{g} \in G_1} \|\tilde{x} - \tilde{g}\|_{E_1} = \min_{g \in G} \max_{f \in \mathcal{E}(\mathfrak{M}_{x,g_0})} |f(x) - f(g)|,$$

i.e. (1.126), which completes the proof of corollary 1.10.

In the particular case when E is a real Banach space, corollary 1.10 has been given, with a different proof, by A. L. Garkavi ([62], theorem 2); in the general case, it has been given in the paper [233], theorem 4. We shall give one more proof of corollary 1.10 in Chap. II, section 1.1.

1.9. GEOMETRICAL INTERPRETATION

In a normed linear space E a real hyperplane $H = \{y \in E \mid \operatorname{Re} f(y) = c\}$, where $f \in S_{E^*}$, is called ([232], p. 348) an *extremal hyperplane*, if we have*) $f \in \mathcal{E}(S_{E^*})$. For instance, if $H = \{y \in E \mid \operatorname{Re} f(y) = 1\}$ (where $f \in S_{E^*}$) is a real hyperplane tangent**) to the unit cell $S_E = \{x \in E \mid \|x\| \leqslant 1\}$ at a point $x_0 \in \operatorname{Fr} S_E$, then H is extremal ([232], proposition 1). Indeed, if $f \notin \mathcal{E}(S_{E^*})$, there exist $f_1, f_2 \in S_{E^*}$, $f_1 \neq f_2 \neq f$, such that $f = \dfrac{1}{2}(f_1 + f_2)$. Then from $\|f_1\| = \|f_2\| = 1 = \|x_0\|$ and $\operatorname{Re} \dfrac{f_1 + f_2}{2}(x_0) = 1$ it follows that we have $\operatorname{Re} f_1(x_0) = \operatorname{Re} f_2(x_0) = 1$. Consequently, the real hyperplanes $H_j = \{y \in E \mid \operatorname{Re} f_j(y) = 1\}$ $(j = 1, 2)$, $H_1 \neq H_2 \neq H$, support the unit cell S_E and pass through x_0, whence H is not tangent to S_E at x_0, which completes the proof. We observe that the converse statement is not true : there exist, in general, extremal support hyperplanes H of S_E which are not tangent to S_E at any point x_0 of the set $H \cap S_E$, as shown by the example $S_E =$ a convex lens in the plane.

A real hyperplane $H = \{y \in E \mid \operatorname{Re} f(y) = c\}$ in a linear space E is said to *separate* the set $A \subset E$ from the set $B \subset E$, if A is situated in one of the real half-spaces $\{y \in E \mid \operatorname{Re} f(y) \geqslant c\}$, $\{y \in E \mid \operatorname{Re} f(y) \leqslant c\}$, and B is situated in the other real half-space. In the sequel the particular case will occur when A is a point in E and B a cell in E.

---

*) For the notations S_{E^*} and $\mathcal{E}(S_{E^*})$ see the preceding section.

**) See section 1.2, the final part.

LEMMA 1.9 ([232], proposition 2). *Let E be a normed linear space, G a linear subspace of E, $x \in E \setminus \overline{G}$, $g_0 \in G$ and $f_0 \in S_{E^*}$. The following statements are equivalent:*

1° f_0 satisfies (1.102), (1.103) and (1.104).

2° The real support hyperplane $H_0 = \{y \in E \mid \operatorname{Re} f_0(x - y) = \|x - g_0\|\}$ of the cell $S(x, \|x - g_0\|)$ is extremal, passes through g_0 and separates 0 from $S(x, \|x - g_0\|)$.

Proof. By definition H_0 is extremal if and only if we have (1.102). If H_0 passes through g_0, we have, taking into account $\|f_0\| = 1$,

$$\|x - g_0\| = \operatorname{Re} f_0(x - g_0) \leqslant |f_0(x - g_0)| \leqslant \|x - g_0\|,$$

whence it follows that $f_0(x - g_0)$ is real and positive and therefore we have (1.104). Conversely, if we have (1.104), then obviously H_0 passes through g_0.

On the other hand, since by $\|f_0\| = 1$ we have

$$\|x - g_0\| \geqslant \|x - y\| \geqslant \operatorname{Re} f_0(x - y) \quad (y \in S(x, \|x - g_0\|)),$$

it follows that H_0 separates 0 from $S(x, \|x - g_0\|)$ if and only if

$$0 = \operatorname{Re} f_0(0) \leqslant \operatorname{Re} f_0(x) - \|x - g_0\|. \tag{1.134}$$

Assuming that we have (1.104) (or, what amounts to the same thing as we have seen, assuming that H_0 passes through g_0), the relation (1.134) holds if and only if we have

$$0 \leqslant \operatorname{Re} f_0(x) - \operatorname{Re} f_0(x - g_0),$$

i.e. (1.103), which concludes the proof of lemma 1.9.

Since by lemma 1.4 every real support hyperplane H of $S(x, \|x - g_0\|)$ is of the form $H = \{y \in E \mid \operatorname{Re} f(x - y) = \|x - g_0\|\}$, where $f \in S_{E^*}$, it follows that theorem 1.12 is equivalent to the following geometrical result:

THEOREM 1.14 ([232], theorem 5). *Let E be a normed linear space, G a linear subspace of E, $x \in E \setminus \overline{G}$* and *$g_0 \in \mathfrak{P}_G(x)$. Then there exists a real extremal hyperplane H_0 which supports the cell $S(x, \|x - g_0\|)$, passes through g_0 and separates 0 from $S(x, \|x - g_0\|)$.*

Similarly, theorem 1.13 is equivalent to the following geometrical result:

THEOREM 1.15 ([232], theorem 6). *Let E be a normed linear space, G a linear subspace of E, $x \in E \setminus \overline{G}$ and $g_0 \in G$. We have $g_0 \in \mathfrak{P}_G(x)$ if and only if for every $g \in G$ there exists a real extremal hyperplane H_g which supports the cell $S(x, \|x - g_0\|)$, passes through g_0 and separates g from $S(x, \|x - g_0\|)$.*

1.10. APPLICATIONS AND GEOMETRICAL INTERPRETATION IN THE SPACES $C(Q)$

Let Q be a compact space and let $E = C(Q)$. Then, as it is well known*), an $f \in E^*$ is an extremal point of S_{E^*} if and only if there exist a $q \in Q$ and a scalar α with $|\alpha| = 1$ such that

$$f(y) = \alpha y(q) \qquad (y \in C(Q)). \tag{1.135}$$

Using this result, we shall now deduce from theorem 1.13 the following classical theorem of characterization of elements of best approximation, due to A. N. Kolmogorov ([118], theorem 1)**) which includes as particular cases results of U. Barbuti ([9], theorems 2 and 3), L. Tonelli ([253], p. 69) and P. Kirchberger ([107], p. 138):

THEOREM 1.16. *Let $E = C(Q)$ (compact), G a linear subspace of E, $x \in E \setminus G$ and $g_0 \in G$. We have $g_0 \in \mathfrak{L}_G(x)$ if and only if for every $g \in G$ there exists a $q = q^g \in Q$ such that*

$$\operatorname{Re} \overline{[x(q) - g_0(q)]} g(q) \geqslant 0, \tag{1.136}$$

$$|x(q) - g_0(q)| = \max_{t \in Q} |x(t) - g_0(t)|. \tag{1.137}$$

Proof ([230], p. 336). By virtue of theorem 1.13 and of the general form (1.135) of the extremal points of S_{E^*}, we have $g_0 \in \mathfrak{L}_G(x)$ if and only if for every $g \in G$ there exist a $q = q^g \in Q$ and a scalar $\alpha = \alpha^g$ with $|\alpha| = 1$, such that

$$\operatorname{Re} \{\alpha[g_0(q) - g(q)]\} \geqslant 0,$$

$$\alpha[x(q) - g_0(q)] = \max_{t \in Q} |x(t) - g_0(t)|.$$

Since the only scalar α satisfying $|\alpha| = 1$ and $\operatorname{Im}\{\alpha[x(q) - g_0(q)]\} = 0$ is

$$\alpha = \operatorname{sign} [x(q) - g_0(q)] = \frac{\overline{x(q) - g_0(q)}}{|x(q) - g_0(q)|},$$

it follows that we have $g_0 \in \mathfrak{L}_G(x)$ if and only if for every $g \in G$ there exists a $q = q^g \in Q$ satisfying (1.137) and

$$\operatorname{Re} \{\overline{[x(q) - g_0(q)]} [g_0(q) - g(q)]\} \geqslant 0.$$

---

*) See e.g. N. Dunford and J. Schwartz [49], p. 441, lemma 6. In the sequel we shall give two more proofs of this result (see the remark made after the proof of lemma 1.11, and Chap. II, lemma 1.7).

**) As a matter of fact, Kolmogorov stated this theorem only for a finite dimensional G but his proof, given in [118] is obviously valid for an arbitrary linear subspace G of $C(Q)$.

Putting
$$q_1 = q_1^g = q^{g_0 - g} \qquad (g \in G),$$
it follows that we have $g_0 \in \mathfrak{L}_G(x)$ if and only if for every $g \in G$ there exists a $q_1 = q_1^g \in Q$ satisfying (1.136) and (1.137), which completes the proof.

Theorem 1.16 follows also from corollary 1.9, equivalence $1° \Leftrightarrow 6°$, taking into account the form (1.135) of the functionals $f \in \mathcal{E}(S_{C(Q)^*})$ and the fact that if an $f \in E^*$ satisfies (1.111), (1.116) and (1.119), then every βf, where $|\beta| = 1$, satisfies these conditions.

We observe that theorem 1.16 remains obviously valid also for every normed linear space E which is a linear subspace of a space $C(Q)$ (Q compact). Since for every normed linear space E there exists a $\sigma(E^*, E)$-compact set $\Gamma \subset S_{E^*}$ (e.g. one can take $\Gamma = S_{E^*}$) such that the natural mapping $y \to \tilde{y}$ of E into $C(\Gamma)$, defined by
$$\tilde{y}(f) = f(y) \qquad (f \in \Gamma,\ y \in E),$$
is an isometry, it follows that we have

Corollary 1.11. *Let E be a normed linear space, Γ a $\sigma(E^*, E)$-closed subset of S_{E^*} with the property that for every $x \in E \setminus \{0\}$ there exists an $f \in \Gamma$ such that $|f(x)| = \|x\|$, let G be a linear subspace of E, $x \in E \setminus \bar{G}$ and $g_0 \in G$. We have $g_0 \in \mathfrak{L}_G(x)$ if and only if for every $g \in G$ there exists an $f^g \in \Gamma$ satisfying (1.116) and (1.119).*

The result of corollary 1.11 has been given by V. N. Nikolsky ([168], [169]), with a direct proof, and the remark that this result is an immediate corollary of Kolmogorov's theorem by embedding E into $C(\Gamma)$, has been made in the paper [230] p. 339.

The non-trivial part of corollary 1.11, namely the necessity part, is also an immediate consequence of the implication $1° \Rightarrow 6°$ of corollary 1.9 *). Indeed, it has been shown in the paper [210] that the *natural mapping $y \to \tilde{y}$ of E into $C(\Gamma)$, where Γ is a $\sigma(E^*, E)$-closed subset of the unit cell S_{E^*}, is an isometry if and only if the circled hull $\tau(\Gamma) = \{\beta f \mid f \in \Gamma,\ |\beta| = 1\}$ of the set Γ contains the set $\mathcal{E}(S_{E^*})$* ([210], theorem 1). Consequently, if an $f^g \in E^*$ satisfies condition 6° of corollary 1.9, then by (1.111) we have $f^g \in \tau(\Gamma)$, whence there exists a scalar β with $|\beta| = 1$ such that $\bar{\beta} f^g \in \Gamma$; but obviously, $\bar{\beta} f^g$ satisfies also (1.116), (1.119), hence the condition of corollary 1.11, which proves our statement above.

*) This is not unexpected, since by virtue of the above corollary 1.9 $\Rightarrow$ theorem 1.16 $\Rightarrow$ corollary 1.11.

The converse of this statement is not true, that is, the implication $1°\Rightarrow 6°$ of corollary 1.9 does not follow from the necessity part of corollary 1.11. Indeed, taking into account the result of [210] mentioned above, the condition of corollary 1.11 implies only the following property of the elements $g_0 \in \mathfrak{P}_G(x)$: for every $g \in G$ there exists an $f^g \in \Gamma = w^*[\mathcal{E}(S_{E^*})]$ (= the closure of $\mathcal{E}(S_{E^*})$ in the weak topology $\sigma(E^*, E)$) such that we have (1.116) and (1.119). But this property is weaker than condition 6° of corollary 1.9, since *there exist normed linear spaces E for which the set $\mathcal{E}(S_{E^*})$ is not closed in the weak topology $\sigma(E^*, E)$.*

In fact, let e.g.

$$E = \{y \in C(Q) \mid y(q_0) = 0\},$$

where Q is a compact space containing at least one non-isolated point and where $q_0 \in Q$ is such a point. Then the null functional $f_0 = 0$ is not in $\mathcal{E}(S_{E^*})$, but we will show that it is in $w^*[\mathcal{E}(S_{E^*})]$. Let $V = V_{y_1, \ldots, y_n; \varepsilon}(0)$ be an arbitrary neighbourhood of $f_0 = 0$ for $\sigma(E^*, E)$. Since $y_1, \ldots, y_n \in E$, $\varepsilon > 0$ and q_0 is non-isolated, there exists a neighbourhood $U \neq \{q_0\}$ of q_0 such that

$$|y_k(q)| = |y_k(q) - y_k(q_0)| < \varepsilon \quad (q \in U;\ k = 1, \ldots, n). \tag{1.138}$$

Take an arbitrary $q_1 \in U \setminus \{q_0\}$ and let $f \in E^*$ be defined by

$$f(y) = y(q_1) \qquad (y \in E).$$

Since $q_1 \neq q_0$, we have then $\|f\| = 1$, and by $q_1 \in U$ and (1.138) we have $f \in V$; therefore it will be sufficient to show that $f \in \mathcal{E}(S_{E^*})$. Assume the contrary, i.e. $f \notin \mathcal{E}(S_{E^*})$. Then there exist $f_1, f_2 \in S_{E^*}$ with $f_1 \neq f \neq f_2$, such that $f = \frac{f_1 + f_2}{2}$.

By the Hahn-Banach theorem then there also exist two Radon measures μ_1, μ_2 on Q with $\|\mu_1\| = \|\mu_2\| = 1$, such that

$$f_j(y) = \int_Q y(q)\, d\mu_j(q) \qquad (y \in E,\ j = 1, 2),$$

whence for $\mu = \frac{\mu_1 + \mu_2}{2}$ we have $\|\mu\| = 1$ and

$$y(q_1) = f(y) = \int_Q y(q) d\mu(q) \qquad (y \in E). \tag{1.139}$$

Let us define $h_1, h_2, h \in C(Q)^*$ by

$$h_j(y) = \int_Q y(q) d\mu_j(q),\ h(y) = \int_Q y(q) d\mu(q) \qquad (y \in C(Q),\ j = 1, 2).$$

If there existed a neighbourhood U_1 of q_1 such that $U_1 \not\ni q_0$ and $|\mu|(U_1) < 1$, then, taking (by Uryson's lemma) a $y_1 \in E$ with $y_1(q_1) = 1$, $y_1(q) = 0$ $(q \in Q \setminus U_1)$ and $\|y_1\| = 1$, one would obtain

$$1 = y_1(q_1) = f(y_1) = \int_Q y_1(q)\, \mathrm{d}\mu(q) = \left|\int_{U_1} y_1(q) \mathrm{d}\mu(q)\right| \leqslant \|y_1\|\, |\mu|(U_1) < 1,$$

which is absurd. Consequently, since $\|\mu\| = 1$, we have $S(\mu) = \{q_1\}$, whence, taking into account (1.139),

$$h(y) = \int_Q y(q) \mathrm{d}\mu(q) = y(q_1) \qquad (y \in C(Q)),$$

and thus, by virtue of the general form (1.135) of the elements of $\mathcal{E}(S_{C(Q)^*})$, we have $h \in \mathcal{E}(S_{C(Q)^*})$. However, by the above we have $h_1, h_2 \in S_{C(Q)^*}$, $h_1 \neq h \neq h_2$ and $h = \frac{h_1 + h_2}{2}$, whence $h \notin \mathcal{E}(S_{C(Q)^*})$, and this contradiction completes the proof.

For the space $E = C(Q)$ we have $w^*[\mathcal{E}(S_{C(Q)^*})] = \mathcal{E}(S_{C(Q)^*})$ ([230], p. 331), hence in this space corollary 1.11 gives the same result as corollary 1.9, namely theorem 1.16. In fact, the mapping $q \to \varepsilon_q$, where

$$\varepsilon_q(y) = y(q) \qquad (y \in C(Q)),$$

is a homeomorphism *) of Q into a subset $\hat{Q}$ of $\mathcal{E}(S_{C(Q)^*})$, where $C(Q)^*$ is endowed with the weak topology $\sigma(C(Q)^*, C(Q))$. On the other hand, denoting by K the scalar field, the mapping $\psi : (\alpha, f) \to \alpha f$ of $K \times C(Q)^*$ into $C(Q)^*$ is continuous when $C(Q)^*$ is endowed with the weak topology $\sigma(C(Q)^*, C(Q))$. Since we have

$$\mathcal{E}(S_{C(Q)^*}) = \psi\, [\{\alpha \in K \mid |\alpha| = 1\} \times \hat{Q}],$$

and since the set $\{\alpha \in K \mid |\alpha| = 1\} \times \hat{Q}$ is compact, it follows that $\mathcal{E}(S_{C(Q)^*})$ is compact for $\sigma(C(Q)^*, C(Q))$, whence closed for this topology, which completes the proof.

Executing also in the reverse order the argument of the proof of theorem 1.16, given at the beginning of the present section, we see that this theorem of Kolmogorov is in fact equivalent to the particular case $E = C(Q)$ of theorem 1.13. Consequently, theorem 1.15 of the preceding section gives a geometrical interpretation to Kolmogorov's theorem as well. However, by using certain special properties of the space $C(Q)$, we now will show that in this case one can obtain more detailed geometrical information than that of theorem 1.15.

*) See e.g. N. Dunford and J. Schwartz [49], p. 441, lemma 7.

We recall that a subset $\mathfrak{F}$ of a cell $S(x, r)$ in a normed linear space E is called *) a *face* of the cell $S(x, r)$ if $\mathfrak{F}$ is a maximal convex subset of $\mathrm{Fr}\, S(x, r)$, that is, if $\mathfrak{F} \subset \mathrm{Fr}\, S(x, r)$ is convex and there does not exist a convex set $\mathfrak{F}_1 \neq \mathfrak{F}$ such that $\mathfrak{F} \subset \mathfrak{F}_1 \subset \mathrm{Fr}\, S(x, r)$.

LEMMA 1.10 ([232], proposition 3 **)). *In the space $E = C(Q)$ a real support hyperplane*

$$H_0 = \{y \in E \mid \mathrm{Re}\, f_0(x - y) = r\}$$

(where $f_0 \in S_{E^*}$) of the cell $S(x, r)$ is extremal if and only if the set

$$\mathfrak{F}_0 = H_0 \cap S(x, r)$$

is a face of $S(x, r)$.

Proof. Assume that H_0 is extremal, that is, $f_0 \in \mathcal{E}(S_{E^*})$. Then, as has been observed at the beginning of this section, there exist a $q_0 \in Q$ and a scalar α_0 with $|\alpha_0| = 1$, such that

$$f_0(y) = \alpha_0 y(q_0) \qquad (y \in E).$$

Consequently for $y \in \mathfrak{F}_0$ we have

$$r = \mathrm{Re}\, f_0(x - y) = \mathrm{Re}\, \alpha_0[x(q_0) - y(q_0)] \leqslant |\alpha_0 x(q_0) - - y(q_0)]| \leqslant \|x - y\| \leqslant r.$$

whence $\alpha_0[x(q_0) - y(q_0)]$ is real and positive and we have

$$\mathfrak{F}_0 = \{y \in E \mid \alpha_0[x(q_0) - y(q_0)] = r, \ \|x - y\| = r\}. \tag{1.140}$$

Assume now that $\mathfrak{F}_0$ is not a face of $S(x, r)$. Then there exists a convex subset $\mathfrak{F}_1$ of $\mathrm{Fr}\, S(x, r)$ such that $\mathfrak{F}_0 \subset \mathfrak{F}_1$, $\mathfrak{F}_0 \neq \mathfrak{F}_1$. Since $\mathfrak{F}_1$ is a convex subset of $\mathrm{Fr}\, S(x, r)$, the set

$$\mathfrak{M} = \bigcap_{y \in \mathfrak{F}_1} \{f \in S_{E^*} \mid f(x - y) = r, \ \|x - y\| = r\} \tag{1.141}$$

is by virtue of lemma 1.8 a non-void extremal subset of S_{E^*} endowed with $\sigma(E^*, E)$. Hence, by virtue of the Krein-Milman theorem, there exists an $f_1 \in \mathcal{E}(\mathfrak{M}) = \mathcal{E}(S_{E^*}) \cap \mathfrak{M}$. Since f_1 must be of the form

$$f_1(y) = \alpha_1 y(q_1) \qquad (y \in E),$$

where $q_1 \in Q, |\alpha_1| = 1$, it follows from $f_1 \in \mathfrak{M}$ and (1.141) that

$$\mathfrak{F}_1 \subset \{y \in E \mid \alpha_1[x(q_1) - y(q_1)] = r, \ \|x - y\| = r\}. \tag{1.142}$$

---

*) See e.g. [211]. Sometimes a different terminology is also used, e.g. S. Eilenberg [54] and R. F. Arens and J. L. Kelley [5] use the term "maximal convex subset of the surface of the cell", while K. Tatarkiewicz [242] uses the term "set of planeity of the cell".

**) For the case of real scalars the necessity part of lemma 1.10 has been given in [211], p. 107, lemma 1.3.

But this is impossible. For assume first that we have $q_0 \neq q_1$. Then, by the classical lemma of Uryson, there exists a (real) $z \in E$ such that

$$z(q_0) = r, z(q_1) < r, \ 0 \leqslant z(q) \leqslant r \quad (q \in Q),$$

whence, putting

$$y(q) = x(q) - \frac{1}{\alpha_0} z(q) \qquad (q \in Q),$$

we obtain $y \in \mathfrak{F}_0$, $y \notin \mathfrak{F}_1$ (since $|\alpha_0| = |\alpha_1| = 1$), which contradicts the hypothesis $\mathfrak{F}_0 \subset \mathfrak{F}_1$. Now assume that we have $q_0 = q_1$. Then, by $\mathfrak{F}_0 \subset \mathfrak{F}_1$, (1.140) and (1.141) we have $\alpha_0 = \alpha_1$, which contradicts the hypothesis $\mathfrak{F}_0 \neq \mathfrak{F}_1$. Thus we have proved that if H_0 is extremal, $\mathfrak{F}_0$ is a face of $S(x, r)$.

Conversely, assume that $\mathfrak{F}_0$ is a face of $S(x, r)$. Then, since $\mathfrak{F}_0$ is a convex subset of Fr $S(x, r)$, the set

$$\mathfrak{M}_1 = \bigcap_{y \in \mathfrak{F}_0} \{f \in S_{E^*} | f(x-y) = r, \|x - y\| = r\} \tag{1.143}$$

is, by virtue of lemma 1.8, a non-void extremal subset of S_{E^*} endowed with $\sigma(E^*, E)$. If $\mathfrak{M}_1$ contains more than one element, then, by virtue of the Krein-Milman theorem, there exist at least two distinct elements $f_1, f_2 \in \mathcal{E}(\mathfrak{M}_1) = \mathcal{E}(S_{E^*}) \cap \mathfrak{M}_1$. By f_1, $f_2 \in \mathfrak{M}_1$ and (1.143) and taking into account that $\mathfrak{F}_0$ is a maximal convex subset of Fr $S(x, r)$, it follows that

$$\{y \in E | f_1(x - y) = r, \|x-y\| = r\} = \{y \in E | \ f_2(x - y) = \\ = r, \|x - y\| = r\}. \tag{1.144}$$

But $f_1, f_2 \in \mathcal{E}(S_{E^*})$, $f_1 \neq f_2$, hence the equality (1.144) is impossible, as we have seen in the above proof of the necessity part. Consequently, the set $\mathfrak{M}_1$ defined by (1.143) contains exactly one element, say f_1. Since

$$\mathfrak{F}_0 = H_0 \cap S(x, r) = \{y \in E | f_0(x - y) = r, \ \|x - y\| = r\},$$

it follows that $f_0 = f_1 \in \mathcal{E}(S_{E^*})$, that is, H_0 is extremal, which completes the proof of lemma 1.10.

We observe that by the above argument one obtains, in particular, also a new proof of the following theorem due to S. Eilenberg ([54], theorem 5.3*)) : *A set* $\mathfrak{F}_0 \subset$ Fr $S(x, r) \subset E = C(Q)$ *(Q compact) is a face of* $S(x, r)$ *if and only if there exist a* $q_0 \in Q$ *and a scalar* α_0 *with* $|\alpha_0| = 1$, *such that*

$$\mathfrak{F}_0 = \{y \in E | \alpha_0[x(q_0) - y(q_0)] = r, \ \|x - y\| = r\}.$$

---

*) Actually, Eilenberg has proved this result only for real scalars and for the unit cell of $C(Q)$, but his proof, given in [54], can be adopted to the case of complex scalars and of an arbitrary cell $S(x, r)$.

Using this theorem of S. Eilenberg, one can give a shorter proof of the necessity part of lemma 1.10 above, but in this case the proof of the sufficiency part becomes longer (when proving the impossibility of (1.144)).

From lemma 1.10 and the equivalence, observed above, of theorems 1.15 and 1.16 in $E = C(Q)$, it follows that theorem 1.16, i.e. Kolmogorov's theorem of characterization of the elements of best approximation in $C(Q)$, is equivalent to the following geometrical result:

THEOREM 1.17 ([232], theorem 7). *Let $E = C(Q)$ (Q compact), G a linear subspace of E, $x \in E \setminus \overline{G}$ and $g_0 \in G$. We have $g_0 \in \mathfrak{L}_G(x)$ if and only if for every $g \in G$ there exists a real hyperplane H_g which supports the cell $S(x, \|x - g_0\|)$ in a face $\mathfrak{F}_g$ containing g_0 and separates g from $S(x, \|x - g_0\|)$.*

Of course, the equivalence of Kolmogorov's theorem to theorem 1.7 can be also proved directly (i.e. without the use of extremal hyperplanes and of lemma 1.10, but applying only Eilenberg's theorem mentioned above and a result corresponding to the particular case $E = C(Q)$ of lemma 1.8), but even in this case it is necessary to repeat certain arguments of the above proof of lemma 1.10 (impossibility of the equality (1.144) etc.).

1.11. APPLICATIONS IN LINEAR SUBSPACES OF THE SPACES $C(Q)$

As has been observed in the preceding section, theorem 1.16 remains valid for every normed linear space E which is a linear subspace of a $C(Q)$ space (Q compact) and this theorem of characterization of elements of best approximation, combined with the natural embedding $E \subset C(\Gamma)$ ($E =$ an arbitrary normed linear space), has led to corollary 1.11, which has been shown to be a strictly weaker result than corollary 1.9 (whence also than theorem 1.13).

In the present section we shall deduce from theorem 1.13 a more precise theorem of characterization of elements of best approximation in normed linear spaces E which are linear subspaces of a $C(Q)$ space (Q compact), by using effectively the hypothesis $G \oplus [x] \subset E$. Taking into account the canonical embedding $E \subset C(\Gamma)$ it will follow that this theorem is even equivalent to theorem 1.13 (hence also to corollary 1.9).

In order to apply theorem 1.13 (or corollary 1.9) to a linear subspace E of $C(Q)$ (Q compact), it is necessary to know the general form of the extremal points of S_{E^*}. But if E is such a space and $f_0 \in \mathcal{E}(S_{E^*})$, then, as it is well known,*) there exist a $q_0 \in Q$ and a scalar α with $|\alpha| = 1$ such that

$$f_0(y) = \alpha y(q_0) \qquad (y \in E). \tag{1.145}$$

However, the converse of this statement is not true, i.e. in general not every $f_0 \in E^*$ of the form (1.145) is in $\mathcal{E}(S_{E^*})$, as may be seen from the example $E = \{y \in C(Q) \,|\, y(q_0) = = \frac{1}{2} [y(q_1) + y(q_2)]\}$, where $q_j \in Q$ $(j = 0, 1, 2)$, $q_0 \neq q_1 \neq q_2$. Below we shall give a characterization of the extremal points of S_{E^*}, showing which of the functionals $f_0 \in E^*$ of the form (1.145) are in $\mathcal{E}(S_{E^*})$.

Let Q be a compact space and E a linear subspace of the space $C(Q)$. Then E defines in Q a relation of equivalence : two points $q_1, q_2 \in Q$ are said to be *E-equivalent* if we have $y(q_1) = y(q_2)$ for all $y \in E$. The equivalence class of a point $q_0 \in Q$ is then the set

$$A_{q_0}(E) = \{q \in Q \,|\, y(q) = y(q_0) \quad (y \in E)\}. \tag{1.146}$$

We shall denote by $\mathfrak{M}^1(Q)$ the set of all Radon measures on Q and by $\mathfrak{M}^1_+(Q)$ the set of all positive Radon measures on Q. For $q_0 \in Q$ we shall denote

$$\mathfrak{M}_{q_0}(E) = \{\mu \in \mathfrak{M}^1_+(Q) \,|\, \int_Q y(q)\, \mathrm{d}\mu(q) = y(q_0) \; (y \in E),\ \mu\,(Q) = 1\}. \tag{1.147}$$

The set $\mathfrak{M}_{q_0}(E)$ is always *non-void*, since it contains at least the measure ε_{q_0} (the evaluation at q_0). For E-equivalent points $q_1, q_2 \in Q$ we have obviously $\mathfrak{M}_{q_1}(E) = \mathfrak{M}_{q_2}(E)$.

We recall **) that the *Choquet boundary* of E is the set

$$\gamma(E) = \{q \in Q \,|\, \mu[A_q(E)] = 1 \quad (\mu \in \mathfrak{M}_q(E))\}. \tag{1.148}$$

In the particular case when the subspace E *separates* the points of Q (i.e. for every $q_0 \in Q$ we have $A_{q_0}(E) = \{q_0\}$), it is obvious that we have $q_0 \in \gamma(E)$ if and only if $\mathfrak{M}_{q_0}(E) = \{\varepsilon_{q_0}\}$.

---

*) See e.g. N. Dunford and J. Schwartz [49], p. 441, lemma 6. Moreover, this follows also from the form (1.135) of the extremal points of $S_{C(Q)^*}$, combined with lemma 1.2 of Chap. II (on the extremal extension of extremal functionals).

**) In the particular case when E contains the constant functions on Q (and thus the condition $\mu(Q) = 1$ of (1.147) may be omitted, since it is automatically satisfied), the Choquet boundary has been introduced by E. Bishop and K. de Leeuw [18]; in the case of an arbitrary linear subspace E of $C(Q)$ (and even in a more general case), it has been introduced by H. Bauer [10]. Taking into account the condition $\mu(Q) = 1$ in (1.147), the Choquet boundary $\gamma(E)$ is nothing else than the set of all points $q \in Q$ with the property that every measure $\mu \in \mathfrak{M}_q(E)$ is concentrated (N. Bourbaki [22], Chap. V, p. 53, definition 4) on the set $A_q(E)$.

This being said, we are now able to determine the general form of the extremal points of the unit cell S_{E^*}.

LEMMA 1.11 ([234], theorem 1). *Let Q be a compact space, E a linear subspace of the space $C(Q)$ and $\gamma(E)$ the Choquet boundary of E. In order that a functional $f_0 \in E^*$ be in $\mathcal{E}(S_{E^*})$ it is necessary and, if $E \ni 1$, sufficient, that there exist a $q_0 \in \gamma(E)$ and a scalar α with $|\alpha| = 1$, such that we have(1.145).*

Proof. Let $f_0 \in \mathcal{E}(S_{E^*})$ be arbitrary. Then there exist, as has been observed above, a $q_0 \in Q$ and a scalar α with $|\alpha| = 1$ such that we have (1.145). We shall show that for any such q_0 we have $q_0 \in \gamma(E)$.

Let $\mu \in \mathfrak{M}_{q_0}(E)$ be arbitrary, i.e. a positive Radon measure on Q with $\mu(Q) = 1$, satisfying

$$\int_Q y(q)\mathrm{d}\mu(q) = y(q_0) \qquad (y \in E), \qquad (1.149)$$

and let A be an arbitrary Borel set in Q, such that $0 < \mu(A) < 1$. Define $f_1, f_2 \in E^*$ by

$$f_1(y) = \frac{1}{\mu(A)}\int_A y(q)\mathrm{d}\mu(q) \qquad (y \in E), \qquad (1.150)$$

$$f_2(y) = \frac{1}{\mu(Q\setminus A)}\int_{Q\setminus A} y(q)\mathrm{d}\mu(q) \qquad (y \in E). \qquad (1.151)$$

Then we have, taking into account (1.145) and (1.149),

$$\bar{\alpha}f_0 = \mu(A)f_1 + [1 - \mu(A)]f_2,$$

whence, since $\bar{\alpha}f_0 \in \mathcal{E}(S_{E^*})$ (because $f_0 \in \mathcal{E}(S_{E^*})$ and $|\alpha| = 1$), it follows that we must have $\bar{\alpha}f_0 = f_1$, and thus

$$\int_Q y(q)\mathrm{d}\mu(q) = \frac{1}{\mu(A)}\int_A y(q)\mathrm{d}\mu(q) \qquad (y \in E) \qquad (1.152)$$

for every Borel set $A \subset Q$ with $\mu(A) > 0$.

Now let $q_1 \in S(\mu)$ be arbitrary. Then for every neighbourhood A of q_1 we have $\mu(A) > 0$, whence taking, for arbitrary $y \in E$ and $\varepsilon > 0$, an open neighbourhood $A = A_{y;\varepsilon}(q_1)$ of q_1 such that $|y(q) - y(q_1)| < \varepsilon \ (q \in A)$, we obtain

$$|y(q_1) - \frac{1}{\mu(A)}\int_A y(q)\mathrm{d}\mu(q)| = \frac{1}{\mu(A)}|\int_A [y(q_1) - y(q)]\mathrm{d}\mu(q)| < \varepsilon$$

$$(y \in E).$$

Since $\varepsilon > 0$ and $q_1 \in S(\mu)$ have been arbitrary, it follows, taking into account (1.152) and (1.149), that we have

$$y(q_1) = \int_Q y(q) \mathrm{d}\mu(q) = y(q_0) \qquad (y \in E,\ q_1 \in S(\mu)), \tag{1.153}$$

hence

$$S(\mu) \subset A_{q_0}(E),$$

whence, by $\mu(Q) = 1$, we obtain

$$\mu[A_{q_0}(E)] = 1.$$

Since μ was an arbitrary measure in $\mathfrak{M}_{J_0}(E)$, this proves that we have $q_0 \in \gamma(E)$.

Conversely, we now will show that if $E \ni 1$ (where 1 denotes the function $\equiv 1$ on Q), then every $f_0 \in E^*$ of the form (1.145), where $|\alpha| = 1$ and $q_0 \in \gamma(E)$, is in $\mathcal{E}(S_{E^*})$. Assume the contrary, i.e. that for such an f_0 we have $f_0 \notin \mathcal{E}(S_{E^*})$. Then for

$$f = \bar{\alpha} f_0 \tag{1.154}$$

we have also $f \notin \mathcal{E}(S_{E^*})$. Since by (1.154) and (1.145) f belongs to the set

$$M = \{\psi \in E^* \mid \|\psi\| = 1, \quad \psi(1) = 1\}, \tag{1.155}$$

and since by corollary 1.8 this set is an extremal subset of S_{E^*}, it follows that we have $f \notin \mathcal{E}(M)$, whence there exist $f_1, f_2 \in M$ with $f_1 \neq f \neq f_2$, such that

$$f = \bar{\alpha} f_0 = \frac{f_1 + f_2}{2}. \tag{1.156}$$

But by virtue of the Hahn-Banach theorem there exist $\mu_1, \mu_2 \in \mathfrak{M}^1(Q)$ with $\|\mu_j\| = \|f_j\|$ $(j = 1, 2)$ such that

$$f_j(y) = \int_Q y(q) \mathrm{d}\mu_j(q) \qquad (y \in E;\ j = 1, 2). \tag{1.157}$$

Since $f_1, f_2 \in M$, we have

$$|\mu_j|(Q) = \|\mu_j\| = \|f_j\| = 1 = f_j(1) = \mu_j(Q) \quad (j = 1, 2),$$

whence $\mu_1, \mu_2 \in \mathfrak{M}^1_+(Q)$. In addition, we have

$$\mu[A_{q_0}(E)] < 1, \tag{1.158}$$

since otherwise we would have

$$f_1(y) = \int_Q y(q) \mathrm{d}\mu_1(q) = y(q_0) = (\bar{\alpha} f_0)(y) = f(y) \qquad (y \in E),$$

whence $f_1 = f$, which contradicts the hypothesis $f_1 \neq f$. Put

$$\mu = \frac{\mu_1 + \mu_2}{2}. \tag{1.159}$$

Then we have $\mu \in \mathfrak{M}_+^1(Q)$ and

$$\int_Q y(q) d\mu(q) = \frac{f_1(y) + f_2(y)}{2} = (\bar{\alpha} f_0)(y) = y(q_0) \qquad (y \in E),$$

whence, since $E \ni 1$, we obtain in particular $\mu(Q) = \int_Q 1 \, d\mu(q) = 1$, and thus $\mu \in M_{q_0}(E)$. On the other hand, by (1.159) and (1.158) we have

$$\mu_1[A_{q_0}(E)] < 1. \tag{1.160}$$

Consequently, we have $q_0 \notin \gamma(E)$, in contradiction with the hypothesis $q_0 \in \gamma(E)$, which completes the proof of lemma 1.11.

We observe that the above argument yields also a simple proof of the result stated at the beginning of section 1.10, that an $f \in C(Q)^*$ belongs to the set $\mathcal{E}(S_{C(Q)^*})$ if and only if it is of the form (1.135). In fact, if the functional

$$f(y) = \int_Q y(q) d\mu(q) \qquad (y \in C(Q)) \tag{1.161}$$

is in $\mathcal{E}(S_{C(Q)^*})$, then the functional

$$|f|(y) = \int_Q y(q) d|\mu|(q) \qquad (y \in C(Q)) \tag{1.162}$$

is also in $\mathcal{E}(S_{C(Q)^*})$, since the relations

$$|\mu| = \frac{\mu_1 + \mu_2}{2}, \quad \mu_1 \neq |\mu| \neq \mu_2, \|\mu_1\| = \|\mu_2\| = 1$$

imply

$$\mu = \beta |\mu| = \frac{\beta\mu_1 + \beta\mu_2}{2}, \quad \beta\mu_1 \neq \mu \neq \beta\mu_2, \quad \|\beta\mu_1\| = \|\beta\mu_2\| = 1,$$

where we have denoted *) $\beta = \frac{d\mu}{d|\mu|}$. Consequently, by virtue

*) This argument is no longer valid if we replace $C(Q)$ by a linear subspace $E \subset C(Q)$.

of the argument of the above proof of the necessity part of lemma 1.11, we have

$$y(q_1) = \int_Q y(q)\mathrm{d}|\mu|(q) \qquad (y \in C(Q),\ q_1 \in S(|\mu|) = S(\mu)),$$

whence every $y \in C(Q)$ is constant on $S(\mu)$, whence $S(\mu)$ reduces to a single point q, and thus we have *) (1.135). Conversely, if the functional (1.161) is of the form (1.135), then $\mu = \varepsilon_q$. Consequently, if $f \notin \mathcal{E}(S_{C(Q)^*})$, then by virtue of the argument of the above proof of the sufficiency part of lemma 1.11, we obtain, taking into account $A_q(E) = \{q\}$,

$$1 = \varepsilon_q(\{q\}) = \mu[A_q(E)] < 1,$$

and this contradiction completes the proof.

In the sufficiency part of lemma 1.11 *the condition $E \ni 1$ is essential.* In fact, take e.g.

$$E = \{y \in C(Q) \,|\, y(q_0) = 0\}. \tag{1.163}$$

Then for every $\mu \in \mathfrak{M}_{q_0}(E)$ we have $\mu \in \mathfrak{M}_+^1(Q)$, $\|\mu\| = 1$ and

$$\int_Q y(q)\mathrm{d}\mu(q) = y(q_0) = 0 \qquad (y \in E),$$

whence $\mu = \varepsilon_{q_0}$. Since $A_{q_0}(E) = \{q_0\}$, it follows that we have $\mu[A_{q_0}(E)] = \varepsilon_{q_0}(\{q_0\}) = 1$ and consequently $q_0 \in \gamma(E)$. Nevertheless, the functional $f_0 \in E^*$ defined by (1.145) is $\equiv 0$, whence it is not in $\mathcal{E}(S_{E^*})$.

We now will show an important case when the condition $E \ni 1$ of lemma 1.11 may be omitted. Namely, we have

LEMMA 1.12 ([234], theorem 2). *Let E be a normed linear space, $Q = S_{E^*}$ endowed with the weak topology $\sigma(E^*, E)$, and $\widetilde{E}$ = the image of E in $C(S_{E^*})$ under the natural isometry $y \to \widetilde{y}$, i.e. the set of all functions $\widetilde{y} \in C(S_{E^*})$ of the form*

$$\widetilde{y}(f) = f(y) \qquad (f \in S_{E^*}), \tag{1.164}$$

where $y \in E$. Then we have

$$\mathcal{E}(S_{E^*}) = \gamma(\widetilde{E}). \tag{1.165}$$

Proof. The inclusion $\mathcal{E}(S_{E^*}) \subset \gamma(\widetilde{E})$ is a consequence of the necessity part of lemma 1.11. Indeed, let $f_0 \in \mathcal{E}(S_{E^*})$ and let $\widetilde{f}_0 \in \widetilde{E}^*$ be the image of f_0 under the natural isometry $E^* \to \widetilde{E}^*$, i.e.

$$\widetilde{f}_0(\widetilde{y}) = f_0(y) \qquad (y \in E). \tag{1.166}$$

*) See e.g. N. Bourbaki [22], Chap. III, p. 74, proposition 12.

Then $\tilde{f}_0 \in \mathcal{E}(S_{\tilde{E}^*})$, whence by virtue of the necessity part of lemma 1.11 there exist an $f_1 \in \gamma(\tilde{E})$ and a scalar α with $|\alpha| = 1$ such that

$$\tilde{f}_0(\tilde{y}) = \alpha \tilde{y}(f_1) \qquad (\tilde{y} \in \tilde{E}). \tag{1.167}$$

From (1.166), (1.167) and (1.164) it follows that we have

$$f_0(y) = \tilde{f}_0(\tilde{y}) = \alpha \tilde{y}(f_1) = \alpha f_1(y) \qquad (y \in E),$$

whence

$$f_0 = \alpha f_1. \tag{1.168}$$

Now let $\mu \in \mathfrak{M}_{f_0}(\tilde{E})$ be arbitrary, that is, $\mu \in \mathfrak{M}^1_+(S_{E^*})$, $\mu(S_{E^*}) = 1$ and

$$\int_{S_{E^*}} \tilde{y}(f) \mathrm{d}\mu(f) = \tilde{y}(f_0) = \tilde{y}(\alpha f_1) \qquad (\tilde{y} \in \tilde{E}). \tag{1.169}$$

Let ν be the image of the positive measure μ by*) the continuous mapping $f \to \bar{\alpha} f$ of S_{E^*} onto S_{E^*}, whence

$$\nu(A) = \mu(\alpha A)$$

for every Borel set $A \subset S_{E^*}$. Then we have $\nu \in \mathfrak{M}^1_+(S_{E^*})$, $\nu(S_{E^*}) = 1$ and, taking into account (1.164) and (1.169),

$$\int_{S_{E^*}} \tilde{y}(f) \mathrm{d}\nu(f) = \int_{S_{E^*}} \tilde{y}(\bar{\alpha} f) \mathrm{d}\mu(f) = \bar{\alpha} \int_{S_{E^*}} \tilde{y}(f) \mathrm{d}\mu(f) =$$

$$= \bar{\alpha} \tilde{y}(\alpha f_1) = \tilde{y}(f_1) \quad (\tilde{y} \in \tilde{E}),$$

that is, $\nu \in \mathfrak{M}_{f_1}(\tilde{E})$. Since $f_1 \in \gamma(\tilde{E})$ and since $\tilde{E}$ separates the points of S_{E^*}, it follows that we have

$$\mu[A_{f_0}(\tilde{E})] = \mu(\{f_0\}) = \nu(\{\bar{\alpha} f_1\}) = \nu(\{f_1\}) = \nu[A_{f_1}(\tilde{E})] = 1,$$

whence $f_0 \in \gamma(\tilde{E})$ and thus $\mathcal{E}(S_{E^*}) \subset \gamma(\tilde{E})$.

In order to prove the opposite inclusion, let $f_0 \in S_{E^*} \setminus \mathcal{E}(S_{E^*})$ be arbitrary. Then there exist $f_1, f_2 \in S_{E^*}$ with $f_1 \neq f_0 \neq f_2$, such that

$$f_0 = \frac{f_1 + f_2}{2}. \tag{1.170}$$

Define $\mu \in \mathfrak{M}^1(S_{E^*})$ by

$$\mu = \frac{\varepsilon_{f_1} + \varepsilon_{f_2}}{2}. \tag{1.171}$$

*) See N. Bourbaki [22], Chap. V, p. 73, definition 1.

Then we have $\mu \in \mathfrak{M}_+^1(S_{E^*})$, $\mu(S_{E^*}) = 1$ and

$$\int_{S_{E^*}} \tilde{y}(f)\,\mathrm{d}\mu(f) = \frac{\tilde{y}(f_1) + \tilde{y}(f_2)}{2} = \tilde{y}\left(\frac{f_1+f_2}{2}\right) = \tilde{y}(f_0) \quad (\tilde{y} \in \tilde{E}),$$

whence $\mu \in \mathfrak{M}_{f_0}(\tilde{E})$. On the other hand, since $\tilde{E}$ separates the points of S_{E^*} and since $f_1 \neq f_0 \neq f_2$, we have

$$\mu[A_{f_0}(\tilde{E})] = \mu(\{f_0\}) = 0,$$

whence $f_0 \notin \gamma(\tilde{E})$, which completes the proof of lemma 1.12.

Now let us return to the problem of best approximation in the spaces $E \subset C(Q)$ and to its applications to the problem of best approximation in arbitrary normed linear spaces. From theorem 1.13 (or corollary 1.9), the necessity part of lemma 1.11 and the sufficiency part of theorem 1.16, it follows that we have

THEOREM 1.18 ([234], theorem 3). *Let Q be a compact space, E a linear subspace of the space $C(Q)$, $\gamma(E)$ the Choquet boundary of E, G a linear subspace of E and $x \in E \setminus \overline{G}$. We have $g_0 \in \mathfrak{L}_G(x)$ if and only if for every $g \in G$ there exists a $q = q^g \in \gamma(E)$ such that we have (1.136) and (1.137).*

Conversely, from theorem 1.18 and lemma 1.12 it follows immediately theorem 1.13 (whence also corollary 1.9) for general normed linear spaces.

The question naturally arises, whether it would not have been simpler to follow the converse way, namely to give first a direct proof of theorem 1.18 and then to deduce theorem 1.13 and corollary 1.9 for general normed linear spaces with the aid of lemma 1.12. This problem also presents interest because it *arises in a similar manner for the results of Chap. II*, where the extremal points of S_{E^*} play an important role. Namely, from these results one can obtain *), by means of lemma 1.11, corresponding results for best approximation in the spaces $E \subset C(Q)$ (which can be formulated by replacing in the respective statements the extremal points of S_{E^*} by points of the Choquet boundary $\gamma(E)$) and the problem arises, whether it is not more convenient to give first a direct proof of these latter results in $E \subset C(Q)$ and then to deduce from them, with the aid of lemma 1.12, the general results of Chap. II (which can be formulated by replacing in the respective statements in $E \subset C(Q)$ the points of the Choquet boundary $\gamma(E)$ by the extremal points of S_{E^*}). The answer is *negative*. Indeed, even the proof that for $E \subset C(Q)$ the Choquet boundary $\gamma(E)$ is a "boundary" for E (i.e. that

*) For the explicit statements of such results see e.g. [234], theorems 4 and 5.

for every $x \in E$ there exists a $q \in \gamma(E)$ such that $|x(q)| = \|x\|$), necessitates either an application *) of the Krein-Milman theorem on the existence of extremal points, or an argument **) of application of Zorn's lemma, which is essentially equivalent to that used in the proof of the Krein-Milman theorem; on the other hand, the equivalent corresponding to this result in general normed linear spaces (namely, the existence, for every $x \in E$, of an $f \in \mathcal{E}(S_{E^*})$ such that $|f(x)| = \|x\|$), which is nothing else but the particular case $G = \{0\}$ of theorem 1.12, is proved, taking into account corollary 1.8, by a direct application of the Krein-Milman theorem, and the general results of Chap. II are also obtained with the aid of the Krein-Milman theorem.

Finally, we mention that the basic results of Chap. II (theorem 1.1 of Chap. II, § 1, theorem 2.1 of Chap. II, § 2, etc.) have been obtained in the papers [211], [212] (which have emphasized for the first time the role of extremal points of the unit cell of the conjugate space in problems of best approximation in arbitrary normed linear spaces), i.e. three years before the introduction of the notion of Choquet boundary by E. Bishop and K. de Leeuw [18].

1.12. APPLICATIONS IN THE SPACES $L^1(T,\nu)$

In order to be able to apply theorem 1.13 in the spaces $E = L^1(T, \nu)$, we need the general form of the extremal points of S_{E^*}. This is given by

LEMMA 1.13 ([211], lemma 1.4 and [230], lemma 3). *Let $E = L^1(T, \nu)$, where (T, ν) is a positive measure space with the property that the dual $L^1(T, \nu)^*$ is canonically equivalent to $L^\infty(T, \nu)$, and let $f \in E^*$. We have $f \in \mathcal{E}(S_{E^*})$ if and only if there exists a $\beta \in L^\infty(T, \nu)$ such that*

$$|\beta(t)| \equiv 1 \qquad \nu\text{-a. e. on } T, \tag{1.172}$$

$$f(x) = \int_T x(t)\beta(t)\mathrm{d}\nu(t) \qquad (x \in E). \tag{1.173}$$

Proof. Let $f \in S_{E^*}$. Then there exists, by virtue of the canonical equivalence $L^1(T, \nu)^* \equiv L^\infty(T, \nu)$, a $\beta \in L^\infty(T, \nu)$ such that we have (1.173) and

$$|\beta(t)| \leqslant 1 \qquad \nu\text{-a. e. on } T.$$

---

*) See E. Bishop and K. de Leeuw [18], the proof of lemma 6.1.

**) See H. Bauer [10], the proof of theorem 2.

Assume that β does not satisfy (1.172). Then there exists a measurable subset $A \subset T$ with $\nu(A) > 0$, such that

$$|\beta(t)| < 1 \qquad \nu\text{-a. e. on } A. \tag{1.174}$$

Put

$$\beta_1(t) = \overline{\text{sign } \beta(t)} = e^{i \arg \beta(t)} \qquad (t \in T),$$
$$\beta_2(t) = (2|\beta(t)| - 1)\, e^{i \arg \beta(t)} \qquad (t \in T).$$

Then for

$$f_j(x) = \int_T x(t)\beta_j(t) d\nu(t) \qquad (x \in E, j = 1, 2)$$

we will have $f_1, f_2 \in S_{E^*}, f = \frac{1}{2}(f_1 + f_2)$ and by (1.174), $f_1 \neq f_2$. Consequently $f \notin \mathcal{E}(S_{E^*})$.

Conversely, assume that $f \in S_{E^*} \setminus \mathcal{E}(S_{E^*})$. Then there exist $f_1, f_2 \in S_{E^*}, f_1 \neq f_2$, such that

$$f = \frac{1}{2}(f_1 + f_2), \tag{1.175}$$

whence also $\beta, \beta_1, \beta_2 \in L^\infty(T, \nu)$ uniquely determined by f, f_1 and f_2 respectively, and a measurable subset $A \subset T$ with $\nu(A) > 0$, such that

$$|\beta(t)|, |\beta_j(t)| \leqslant 1 \ \nu\text{-a. e. on } T, \qquad \beta_1(t) \neq \beta_2(t) \qquad \nu\text{-a. e. on } A, \tag{1.176}$$

$$f(x) = \int_T x(t)\beta(t) d\nu(t), \ f_j(x) = \int_T x(t)\, \beta_j(t) d\nu(t) \qquad (x \in E, j = 1, 2). \tag{1.177}$$

Then, from (1.175) and (1.177) it follows that we have

$$\beta(t) = \frac{1}{2}[\beta_1(t) + \beta_2(t)] \qquad \nu\text{-a. e. on } T,$$

whence, by (1.176),

$$|\beta(t)| < 1 \qquad \nu\text{-a. e. on } A$$

and thus condition (1.172) is not satisfied, which completes the proof of lemma 1.13.

We now will deduce from theorem 1.13 several characterizations of elements of best approximation in the spaces $L^1(T, \nu)$, collected in the following theorem :

THEOREM 1.19. *Let $E = L^1(T, \nu)$, where (T, ν) is a positive measure space, let G be a linear subspace of E, $x \in E \setminus \bar{G}$ and $g_0 \in G$. The following statements are equivalent:*

1° $g_0 \in \mathfrak{P}_G(x)$.

2° *For every $g \in G$ there exists a $\beta = \beta^g \in L^\infty(T, \nu)$ such that*

$$|\beta(t)| \equiv 1 \qquad \nu\text{-a. e. on } T, \tag{1.178}$$

$$\operatorname{Re} \int_T [g_0(t) - g(t)] \beta(t) \, d\nu(t) \geqslant 0, \tag{1.179}$$

$$\int_T [x(t) - g_0(t)] \beta(t) \, d\nu(t) = \int_T |x(t) - g_0(t)| \, d\nu(t). \tag{1.180}$$

3° *For every $g \in G$ there exists on the set $Z(x - g_0)$ (defined by (1.60)) a ν-measurable function $\alpha = \alpha^g$ such that*

$$|\alpha(t)| \equiv 1 \qquad \nu\text{-a. e. on } Z(x - g_0), \tag{1.181}$$

$$\begin{aligned} \int_T |x(t) - g_0(t)| d\nu(t) \leqslant \operatorname{Re} \int_{Z(x-g_0)} [x(t) - g(t)] \alpha(t) \, d\nu(t) + \\ + \operatorname{Re} \int_{T \setminus Z(x-g_0)} [x(t) - g(t)] \operatorname{sign} [x(t) - g_0(t)] \, d\nu(t). \end{aligned} \tag{1.182}$$

4° *For every $g \in G$ there exists on the set $Z(x - g_0)$ a ν-measurable function $\alpha = \alpha^g$ satisfying* (1.181) *and*

$$\begin{aligned} \operatorname{Re} \int_{T \setminus Z(x-g_0)} [g_0(t) - g(t)] \operatorname{sign} [x(t) - g_0(t)] \, d\nu(t) \geqslant \\ \geqslant - \operatorname{Re} \int_{Z(x-g_0)} [g_0(t) - g(t)] \alpha(t) \, d\nu(t). \end{aligned} \tag{1.183}$$

5° *We have*

$$\begin{aligned} \operatorname{Re} \int_{T \setminus Z(x-g_0)} [g_0(t) - g(t)] \operatorname{sign} [x(t) - g_0(t)] \, d\nu(t) \geqslant \\ \geqslant - \int_{Z(x-g_0)} |g_0(t) - g(t)| \, d\nu(t) \qquad (g \in G). \end{aligned} \tag{1.184}$$

6° *We have*

$$\begin{aligned} \operatorname{Re} \int_{T \setminus Z(x-g_0)} g(t) \operatorname{sign} [x(t) - g_0(t)] \, d\nu(t) \geqslant \\ \geqslant - \int_{Z(x-g_0)} |g(t)| \, d\nu(t) \qquad (g \in G). \end{aligned} \tag{1.185}$$

Proof. If we have 2°, then for every $g \in G$ the continuous linear functional $f = f^g \in E^*$, defined by

$$f(x) = \int_T x(t)\,\beta(t)\,\mathrm{d}\nu(t) \qquad (x \in E = L^1(T, \nu)),$$

satisfies (1.111), (1.112) and (1.113), whence, by virtue of theorem 1.13, we obtain $g_0 \in \mathfrak{L}_G(x)$. Thus, 2°⇒1°.

Furthermore, if we have 2°, then, by (1.178) and (1.180),

$$\beta(t) = \operatorname{sign}\,[x(t) - g_0(t)] \quad \nu\text{-}a.\ e.\ on\ T \setminus Z\,(x - g_0),$$

and consequently, by virtue of (1.178) and (1.179)*), the function $\alpha = \beta|_{Z(x-g_0)}$ on $Z(x - g_0)$ satisfies (1.181) and (1.182). Thus, 2°⇒3°.

Conversely, if we have (1.181) and (1.182), then the function β on T defined by

$$\beta(t) = \begin{cases} \alpha(t) & \text{for} \quad t \in Z(x - g_0) \\ \operatorname{sign}\,[x(t) - g_0(t)] & \text{for} \quad t \in T \setminus Z(x - g_0), \end{cases}$$

satisfies (1.178), (1.179) and (1.180). Thus 3°⇒2°.

The proof of the equivalence 2°⇔4° is similar to the above proof of the equivalence 2°⇔3°, defining the same functions α and β respectively and writing (1.179) in the equivalent form

$$\operatorname{Re}\int_{T\setminus Z(x-g_0)} [g_0(t) - g(t)]\beta(t)\mathrm{d}\nu(t) + \\ + \operatorname{Re}\int_{Z(x-g_0)} [g_0(t) - g(t)]\,\beta(t)\,\mathrm{d}\nu(t) \geqslant 0.$$

The implication 4°⇒5° is obvious. Conversely, the implication 5°⇒4° is obtained by putting

$$\alpha(t) = \begin{cases} \operatorname{sign}\,[g_0(t) - g(t)] & \text{for} \quad t \in Z(x - g_0) \setminus Z(g_0 - g) \\ 1 & \text{for} \quad t \in Z(x - g_0) \cap Z(g_0 - g). \end{cases}$$

Finally, the equivalence 5°⇔6° is obvious and the implication 1°⇒6° is an immediate consequence of the implication 1°⇒3° of theorem 1.7. This concludes the proof of theorem 1.19.

---

*) In fact, we have only to write (1.179) in the equivalent form

$$\operatorname{Re}\int_T [x(t) - g_0(t)]\,\beta(t)\mathrm{d}\nu(t) \leqslant \operatorname{Re}\int_T [x(t) - g(t)]\,\beta(t)\,\mathrm{d}\nu(t).$$

In the particular case when $L^1(T, \nu)$ is separable and $L^1(T, \nu)^* \equiv L^\infty(T, \nu)$ canonically, the equivalences 1°⇔2°⇔3° of theorem 1.19 have been given in [230], theorem 7. In the particular case when $T = [0, 1]$, ν is the Lebesgue measure and the scalars are real, the equivalences 1°⇔4°⇔5°⇔6° (but only with $|\alpha(t)| \leqslant 1$ instead of $|\alpha(t)| \equiv 1$ in 4°) have been given by V. N. Nikolsky ([169], § 9).

Of course, the conditions of theorem 1.19 are equivalent to the conditions 1°—4° of theorem 1.7. We observe that the equivalence of condition 6° of theorem 1.19 to condition 3° of theorem 1.7 can be easily proved directly as follows: condition 3° of theorem 1.7 evidently implies condition 6° of theorem 1.19. Conversely, assume that condition 6° of theorem 1.19 is satisfied. Applying this condition for the elements $\alpha g \in G$, where

$$\alpha = -\operatorname{sign} \int_{T \setminus Z(x-g_0)} g(t) \operatorname{sign}[x(t) - g_0(t)] \, d\nu(t),$$

we obtain

$$-\left| \int_{T \setminus Z(x-g_0)} g(t) \operatorname{sign}[x(t) - g_0(t)] d\nu(t) \right| \geqslant$$

$$\geqslant - \int_{Z(x-g_0)} |g(t)| \, d\nu(t) \qquad (g \in G),$$

whence condition 3° of theorem 1.7 is satisfied, which completes the proof.

1.13. OTHER CHARACTERIZATIONS OF ELEMENTS OF BEST APPROXIMATION IN GENERAL NORMED LINEAR SPACES

Let E be a normed linear space, G a linear subspace of E and $x \in E$. Since the condition $g_0 \in \mathfrak{L}_G(x)$ amounts, by definition, to the condition that the element $g_0 \in G$ should minimize the continuous functional Φ defined on G by

$$\Phi(g) = \Phi_{G,x}(g) = \|x - g\| \qquad (g \in G), \tag{1.186}$$

or, in other words, that the element $0 \in G$ should minimize the continuous function

$$\Psi(g) = \Phi(g_0 + g) = \|x - g_0 - g\| \qquad (g \in G), \tag{1.187}$$

or, what is equivalent, that for every $g \in G$ the scalar $\alpha = 0$ should minimize the continuous function of scalar variable

$$\chi_g(\alpha) = \Psi(\alpha g) = \|x - g_0 - \alpha g\| \qquad (\alpha = \text{scalar}), \tag{1.188}$$

it arises naturally the problem of obtaining characterizations of the elements of best approximation $g_0 \in \mathfrak{P}_G(x)$ with the aid of the methods of differential calculus in normed linear spaces *). The main difficulty is that in general the norm in E is not Gâteaux differentiable at each non-zero point in E. Nevertheless, it is known **) that the limits

$$\tau(x, y) = \lim_{t \to 0+} \frac{\|x + ty\| - \|x\|}{t} \qquad (x, y \in E) \quad (1.189)$$

always exist and we shall use them to give characterizations of the elements of best approximation.

THEOREM 1.20 ([233], theorem 1). *Let E be a normed linear space, G a linear subspace of E, $x \in E \setminus \overline{G}$ and $g_0 \in G$. The following statements are equivalent:*

1° $g_0 \in \mathfrak{P}_G(x)$.

2° *We have*

$$\tau(x - g_0, g) \geqslant 0 \qquad (g \in G). \qquad (1.190)$$

3° *We have*

$$-\tau(x - g_0, -g) \leqslant 0 \leqslant \tau(x - g_0, g) \qquad (g \in G). \quad (1.191)$$

4° *For every $g \in G$ there exists an $f^g \in E^*$ such that*

$$\|f^g\| = 1, \qquad (1.192)$$

$$\operatorname{Re} f^g(g) = 0, \qquad (1.193)$$

$$f^g(x - g_0) = \|x - g_0\|. \qquad (1.194)$$

Proof. If we have 1°, then

$$\|x - g_0 + tg\| \geqslant \|x - g_0\| \qquad (t > 0),$$

whence we infer 2°. Thus, 1°⇒2°.

If we have 2°, then applying (1.190) to $-g \in G$ instead of g we obtain

$$\tau(x - g_0, -g) \geqslant 0 \qquad (g \in G),$$

whence 3°. Thus, 2°⇒3°.

---

*) Even the function χ_g of the scalar variable α is obtained by composing the mapping $\alpha \to x - g_0 - \alpha g$ of the scalar field into E with the functional $x \to \|x\|$ on E.

**) See e.g. N. Dunford and J. Schwartz [49], p. 445, lemma 1.

The equivalence 3°⇔4° is a consequence of a classical theorem of G. Ascoli-S. Mazur *).

Finally, assume that we have 4°. Then for every $g \in G$ we have

$$\|x - g_0\| = f^{g_0-g}(x - g_0) = \operatorname{Re} f^{g_0-g}(x - g_0) = \operatorname{Re} f^{g_0-g}(x-g) \leqslant$$
$$\leqslant |f^{g_0-g}(x - g)| \leqslant \| f^{g_0-g}\| \, \|x - g\| = \|x - g\|,$$

i.e. $g_0 \in \mathfrak{P}_G(x)$. Thus, 4°⇒1°, which completes the proof.

We observe that one can also give the following direct proof of the implication 2°⇒1°: if we have 2°, then, since $\dfrac{\|x - g_0 + tg\| - \|x - g_0\|}{t}$ is for every $g \in G$ a non-decreasing **) function of the positive real variable t, we obtain (for $t = 1$)

$$\|x - g_0 + g\| \geqslant \|x - g_0\| \qquad (g \in G), \tag{1.195}$$

whence $g_0 \in \mathfrak{P}_G(x)$, which completes the proof. In the particular case of real scalars, the equivalence 1°⇔3° can be obtained also from the results of R. C. James [99] on orthogonality in normed linear spaces (see section 1.14).

By virtue of theorem 1.13, the statement 1° of theorem 1.20 is equivalent to the following:

5° *For every $g \in G$ there exists an $f^g \in E^*$ satisfying (1.111)—(1.113).*

We observe that, actually, the implication 1°⇒5° has been proved in the above via 1°⇒4°⇒5° (in the first proof of theorem 1.12, in order to show that the set $\mathfrak{M}$ defined by (1.107) is non-void, we have referred to theorem 1.1, according to which in 4° one may take a common $f^g = f$ for all $g \in G$, but what is effectively used is only the statement 4° above).

By the Ascoli-Mazur theorem mentioned above, the norm is Gâteaux differentiable at $x - g_0$, i.e. we have

$$-\tau(x - g_0, -y) = \tau(x - g_0, y) \quad (y \in E), \tag{1.196}$$

if and only if there exists a unique $f \in E^*$ satisfying

$$\|f\| = 1, \ f(x - g_0) = \|x - g_0\|, \tag{1.197}$$

that is, if and only if $x - g_0$ is a normal ***) element. Thus we obtain

---

*) We recall this theorem (see e.g. N. Dunford and J. Schwartz [49], p. 447, theorem 5): Let E be a normed linear space and let $x \in E$ and $f \in E^*$ be such that $f(x) = \|x\|$, $\|f\| = 1$; then we have $-\tau(x, -y) \leqslant \operatorname{Re} f(y) \leqslant \tau(x, y)$ $(y \in E)$. Conversely, if $-\tau(x, -y) \leqslant \alpha \leqslant \tau(x, y)$, then there exists an $f \in E^*$ such that $f(x) = \|x\|$, $\|f\| = 1$ and $\operatorname{Re} f(y) = \alpha$.

**) See e.g. G. Köthe [122], p. 351, proposition (1).

***) See section 1.1.

COROLLARY 1.12 ([233], corollary 1). *Let E be a normed linear space, G a linear subspace of E, $x \in E \setminus \overline{G}$ and $g_0 \in G$ such that the norm is Gâteaux differentiable at $x - g_0$. The following statements are equivalent :*

1° $g_0 \in \mathfrak{P}_G(x)$.

2° *We have*

$$\tau(x - g_0, g) = 0 \qquad (g \in G). \tag{1.198}$$

3° *For the (unique) functional $f \in E^*$ satisfying (1.197) we have*

$$\operatorname{Re} f(g) = 0 \qquad (g \in G). \tag{1.199}$$

By virtue of lemma 1.1 b), in (1.199) $\operatorname{Re} f(g)$ can be replaced by $f(g)$; thus we find again corollary 1.4.

Finally, we mention that *the continuous functional Φ on G defined by (1.186) cannot have any local minimum different from* $\inf_{g \in G} \Phi(g)$. For, assume the contrary, that there exist an element $g' \in G$ and a neighbourhood $U(g')$ of g' such that

$$\Phi(g') \leqslant \Phi(g) \qquad (g \in U(g')), \tag{1.200}$$

$$\inf_{g \in G} \Phi(g) < \Phi(g'). \tag{1.201}$$

Let $\varepsilon > 0$ be arbitrary such that

$$\inf_{g \in G} \Phi(g) + \varepsilon < \Phi(g')$$

and take $g'' \in G$ satisfying

$$\Phi(g'') < \inf_{g \in G} \Phi(g) + \varepsilon.$$

Then we have

$$\Phi(g'') < \Phi(g'). \tag{1.202}$$

Consider the elements $g_\lambda \in G$ $(0 \leqslant \lambda \leqslant 1)$ defined by

$$g_\lambda = \lambda g' + (1 - \lambda) g'' \qquad (0 \leqslant \lambda \leqslant 1). \tag{1.203}$$

Since for $\lambda = 1$ we have $g_1 = g'$ and since the mapping $\lambda \to g_\lambda$ of $[0,1]$ into G is continuous, it follows that there exists a $\lambda_0 \in (0, 1)$ such that $g_\lambda \in U(g')$ $(\lambda_0 < \lambda \leqslant 1)$, whence, by (1.200),

$$\Phi(g') \leqslant \Phi(g_\lambda) \qquad (\lambda_0 < \lambda \leqslant 1). \tag{1.204}$$

On the other hand, taking into account (1.202), we have

$$\Phi(g_\lambda) = \|x - g_\lambda\| = \|x - \lambda g' - (1 - \lambda)g''\| =$$

$$= \|\lambda(x - g') + (1 - \lambda)(x - g'')\| \leqslant \lambda\,\Phi(g') + (1 - \lambda)\,\Phi(g'') <$$

$$< \lambda\,\Phi(g') + (1 - \lambda)\,\Phi(g') = \Phi(g')\ (0 \leqslant \lambda \leqslant 1),$$

contradicting (1.204), which completes the proof.

In the particular case when the scalars are real and the linear subspace G is of finite dimension, the above remark has been made by S. I. Zuhovitzky ([267], p. 139).

1.14. ORTHOGONALITY IN GENERAL NORMED LINEAR SPACES

We recall the notion of orthogonality introduced by G. Birkhoff [17]: An element x of a normed linear space E is said to be *orthogonal**) to an element $y \in E$, and we write $x \perp y$, if we have

$$\|x + \alpha y\| \geqslant \|x\| \text{ for every scalar } \alpha. \tag{1.205}$$

This is an extension of the usual notion of orthogonality since *in an inner product space* $\mathcal{H}$ *we have* $x \perp y$ *if and only if* $(x, y) = 0$. Indeed, if $(x, y) \neq 0$, then for $\alpha = -\dfrac{(x, y)}{(y, y)}$ we have

$$\|x + \alpha y\|^2 = \left(x - \frac{(x, y)}{(y, y)}y,\ x - \frac{(x, y)}{(y, y)}y\right) = (x, x) - 2\,\frac{|(x, y)|^2}{(y, y)} +$$

$$+ \frac{|(x, y)|^2}{(y, y)^2}(y, y) = (x, x) - \frac{|x, y)|^2}{(y, y)} < (x, x) = \|x\|^2,$$

whence x is not orthogonal to y, while if $(x, y) = 0$, then for every scalar α we have

$$\|x + \alpha y\|^2 = (x + \alpha y, x + \alpha y) = \|x\|^2 + |\alpha|^2\|y\|^2 \geqslant \|x\|^2,$$

whence $x \perp y$.

An element x of a normed linear space E is said to be orthogonal to a set $M \subset E$ and we write $x \perp M$, if we have

$$x \perp y \qquad (y \in M). \tag{1.206}$$

The relationship between orthogonality and best approximation is given by

---

*) Some authors use a different terminology, e.g. N. Bourbaki ([23], Chap. V, p. 144, exercise 14) says that y is *quasi-normal* to x if we have (1.205).

LEMMA 1.14. *Let E be a normed linear space, G a linear subspace of E, $x \in E \setminus \overline{G}$ and $g_0 \in G$. We have $g_0 \in \mathfrak{P}_G(x)$ if and only if*

$$x - g_0 \perp G. \qquad (1.207)$$

Proof. By the definition of orthogonality, condition (1.207) means that

$$\|x - g_0 + \alpha g\| \geqslant \|x - g_0\| \qquad (g \in G, \ \alpha = \text{scalar}),$$

and this is obviously equivalent to $g_0 \in \mathfrak{P}_G(x)$.

Lemma 1.14 allows the application of known results on orthogonality in normed linear spaces to problems of best approximation in such spaces. Thus, from lemma 1.14 and the observation above on orthogonality in Hilbert spaces it follows theorem 1.11′ of section 1.7. Likewise, R. C. James [99] has proved that in a real normed linear space E we have $x \perp \alpha x + y$ if and only if

$$-\tau(x, -y) \leqslant \alpha\|x\| \leqslant \tau(x, y),$$

where $\tau(x, y)$ is the limit (1.189). Applying this result for $x - g_0$ instead of x and for $\alpha = 0$, $y = g \in G$, and taking into account lemma 1.14, we find again the equivalence 1°⇔3° of theorem 1.20 (the real case).

We mention that there are also other notions of orthogonality in normed linear spaces which are extensions of the usual notion of orthogonality in Hilbert spaces. Thus, for instance, *x is orthogonal to y*

a) *in the sense of B. D. Roberts* [195], if

$$\|x + \alpha y\| = \|x - \alpha y\| \text{ for every scalar } \alpha; \qquad (1.208)$$

b) *in the isosceles sense* (R. C. James [98]), if

$$\|x + y\| = \|x - y\|; \qquad (1.209)$$

c) *in the pythagorean sense* (R. C. James [98]), if

$$\|x - y\|^2 = \|x\|^2 + \|y\|^2 \qquad (1.210)$$

d) *in the sense of* [214], if

$$\left\| \frac{x}{\|x\|} + \frac{y}{\|y\|} \right\| = \left\| \frac{x}{\|x\|} - \frac{y}{\|y\|} \right\|. \qquad (1.211)$$

Similarly, x is said to be orthogonal to a set $M \subset E$ in one of the above senses, if we have (1.206) in the respective sense.

By means of these definitions of orthogonality one can introduce new notions of "best approximation". Namely, a

$g_0 \in G$ may be called *an element of best approximation* of an element $x \in E$), if we have $x - g_0 \perp G$ in one of the senses a), b), c), d), above. For these new notions of best approximation one can raise then the same problems of characterization, existence, uniqueness, etc. as for the usual best approximation (corresponding to the orthogonality in the sense of Birkhoff [17], i.e. to the orthogonality (1.205)). We leave to the reader the study of these problems.

§2. EXISTENCE OF ELEMENTS OF BEST APPROXIMATION

Formula (1.1) of §1 shows that if E is a normed linear space and G a linear subspace of E, then for every $x \in G$ the set $\mathfrak{L}_G(x)$ is non-void, and in the case when the subspace G is not closed, then for every $x \in \overline{G} \setminus G$ the set $\mathfrak{L}_G(x)$ is void; furthermore, for the elements x of $E \setminus \overline{G}$ the set $\mathfrak{L}_G(x)$ may be non-void or void. The linear subspaces $G \subset E$ which have the property *) that $\mathfrak{L}_G(x)$ is non-void for every $x \in E \setminus G$ (or, what is equivalent, for every $x \in E$) are called *proximinal linear subspaces* **) of E. In the present paragraph we shall give various characterizations of proximinal linear subspaces of normed linear spaces, as well as applications of these characterizations (e.g. to the problem of existence of proximinal and non-proximinal linear subspaces).

LEMMA 2.1. *Let E be a normed linear space and H a hyperplane in E, passing through 0. H is proximinal if and only if there exists an element $z \in E \setminus \{0\}$ such that*

$$0 \in \mathfrak{L}_H(z) \tag{2.1}$$

(*i.e. such that* $z \perp H$).

Proof. Assume that H is proximinal and take arbitrary $x \in E \setminus H$, $y_0 \in \mathfrak{L}_H(x)$. Then for $z = x - y_0 \neq 0$ we have (2.1).

Conversely, assume that there exists a $z \in E \setminus \{0\}$ such that we have (2.1) and let $x \in E \setminus H$ be arbitrary. Take $f \in E^*$ such that ***)

---

*) The other extreme case, when $\mathfrak{L}_G(x) = \emptyset$ for every $x \in E \setminus G$, is also possible (see the remark made after the proof of corollary 2.4).

**) For the sets $G \subset E$ having the property that for every $x \in E$ the set $\mathfrak{L}_G(x)$ is non-void, the term "*proximinal*" set (a combination of the words "proximity" and "minimal") has been proposed by Raymond Killgrove (see R. R. Phelps [174], p. 790). Some authors use the term *distance set* (V. Klee [108]), or *existence set* (N. V. Efimov and S. B. Stečkin [50]).

***) See §1, section 1.2.

$$H = \{y \in E \mid f(y) = 0\},$$

and put

$$y_0 = x - \frac{f(x)}{f(z)} z$$

(we have $f(z) \neq 0$, since otherwise we would have $z \in H$, whence $0 \in \mathfrak{L}_H(z) = \{z\}$, whence $z = 0$, which contradicts the hypothesis). We have then

$$f(y_0) = 0,$$

whence $y_0 \in H$. Also, since $\frac{f(z)}{f(x)}(y - y_0) \in H$ for every $y \in H$ (we have $f(x) \neq 0$ by the hypothesis $x \in E \setminus H$), it follows from (2.1) that we have

$$\|x - y_0\| = \left|\frac{f(x)}{f(z)}\right| \|z\| \leqslant \left|\frac{f(x)}{f(z)}\right| \left\|z - \frac{f(z)}{f(x)}(y - y_0)\right\| = \|x - y\| \qquad (y \in H),$$

whence $y_0 \in \mathfrak{L}_H(x)$. Since $x \in E \setminus H$ has been arbitrary, it follows that H is proximinal, which completes the proof.

We shall say, following [228], p. 359, that a linear subspace Γ of the conjugate space E^* has *the property* $(\mathcal{S}_*)$, if for every $x \in E$ there exists an element $y \in E$ such that *)

$$\gamma(y) = \gamma(x) \qquad (\gamma \in \Gamma), \tag{2.2}$$

$$\|y\| = \|x\|_\Gamma . \tag{2.3}$$

We shall deduce now from theorem 1.1 and lemma 2.1 various characterizations of proximinal linear subspaces of normed linear spaces, collected in

THEOREM 2.1. *Let E be a normed linear space and G a linear subspace of E. The following statements are equivalent:*

1° *G is proximinal.*

2° *G is closed and in every linear subspace $F_x \subset E$ $(x \in E \setminus G)$ of the form*

$$F_x = G \oplus [x] \tag{2.4}$$

there exists an element $z \in F_x \setminus \{0\}$ such that

$$0 \in \mathfrak{L}_G(z) \tag{2.5}$$

(i.e. such that $z \perp G$).

---

*) For the notation $\|x\|_\Gamma$ see §1, formula (1.11).

3. *G is closed and every functional* $\varphi\in(F_x)^*(x\in E\setminus G)$ *with the property*

$$G = \{y\in F_x \mid \varphi(y) = 0\} \tag{2.6}$$

has at least one maximal element *).

4° *G is closed and* **) $G^\perp$ *has the property* $(\mathcal{E}_*)$.

These statements are implied by — and in the case when the quotient space E/G is reflexive, equivalent to — the following statement:

5° *G is closed and for every* $\Phi\in(G^\perp)^*$ *there exists an element* $y\in E$ *such that*

$$f(y) = \Phi(f) \qquad (f\in G^\perp), \tag{2.7}$$

$$\|y\| = \|\Phi\|. \tag{2.8}$$

Proof. Assume that we have 1°. Then by § 1, formula (1.1), G is closed. On the other hand, again by 1°, G is proximinal also in every subspace $F_x\subset E$ $(x\in E\setminus G)$ of the form (2.4). Since G is a hyperplane in each F_x, it follows from the necessity part of lemma 2.1 that we have 2°. Thus 1°$\Rightarrow$2°.

Conversely, assume that we have 2°. Since G is a hyperplane in every subspace $F_x\subset E(x\in E\setminus G)$ of the form (2.4), it follows from the sufficiency part of lemma 2.1 that G is proximinal in every F_x. Since $E = \bigcup_{x\in E\setminus G} F_x$, we infer that G is proximinal also in E. Thus, 2°$\Rightarrow$1°.

Assume again that we have 2°. Then for $z\in F_x\setminus\{0\}$ satisfying (2.5) there exists, by virtue of theorem 1.1, a $\psi\in(F_x)^*$ such that

$$\|\psi\| = 1 \tag{2.9}$$

$$\psi(g) = 0 \qquad (g\in G), \tag{2.10}$$

$$\psi(z) = \|z\|. \tag{2.11}$$

Let now $\varphi\in(F_x)^*\setminus\{0\}$ be arbitrary with the property (2.6). Then there exists, taking into account (2.10), a scalar $\lambda \neq 0$ such that $\varphi = \lambda\psi$. Consequently, by (2.11) and (2.9), we have

$$\varphi(\bar{\lambda}z) = (\lambda\psi)(\bar{\lambda}z) = |\lambda|^2\psi(z) = |\lambda|^2\|z\| = \|\lambda\psi\|\,\|\bar{\lambda}z\| = \|\varphi\|\|\bar{\lambda}z\|,$$

i.e. $\bar{\lambda}z$ is a maximal element of φ. Thus, 2°$\Rightarrow$3°.

---

*) I.e. (see §1, section 1.4) an element $z\in F_x\setminus\{0\}$ such that $\varphi(z) = \|\varphi\|\,\|z\|$.

**) See §1, formula (1.9).

Conversely, assume that we have 3°. Let $\varphi \in (F_x)^*$ be arbitrary with the property (2.6) and let $z \in F_x$ be a maximal element of φ, i.e. $z \neq 0$, $\varphi(z) = \|\varphi\| \, \|z\|$. Then for $\psi \in (F_x)^*$ defined by

$$\psi = \frac{\varphi}{\|\varphi\|}$$

we have (2.9), (2.10) and (2.11), whence, by virtue of theorem 1.1, it follows that $0 \in \mathfrak{P}_{F_x}(z)$. Thus, $3° \Rightarrow 2°$.

Assume now that we have 1°. Then, as we have seen above, G is closed. Furthermore, by corollary 1.2 a), for every $x \in E$ there exists an element $g_0 \in G$ such that

$$\|x\|_{G^\perp} = \|x - g_0\|_{G^\perp} = \|x - g_0\|. \tag{2.12}$$

Put

$$y = x - g_0.$$

Then, by the definition (1.9) of $G^\perp$ we have

$$f(x) = f(x - g_0) = f(y) \quad (f \in G^\perp),$$

and by (2.12) we have

$$\|x\|_{G^\perp} = \|y\|,$$

whence $G^\perp$ has the property $(\mathscr{E}_*)$. Thus, $1° \Rightarrow 4°$.

Conversely, assume that we have 4° and let $x \in E \setminus G$ be arbitrary. Then there exists, by 4°, an element $y \in E$ such that

$$f(y) = f(x) \qquad (f \in G^\perp), \tag{2.13}$$

$$\|y\| = \|x\|_{G^\perp}. \tag{2.14}$$

Put

$$g_0 = x - y.$$

Then (2.13) and (2.14) become

$$g_0 \in (G^\perp)_\perp = \overline{G} = G,$$

$$\|x - g_0\| = \|x\|_{G^\perp} = \|x - g_0\|_{G^\perp},$$

whence, by virtue of corollary 1.2 a) of § 1, it follows that $g_0 \in \mathfrak{P}_G(x)$. Since $x \in E \setminus G$ has been arbitrary, it follows that G is proximinal. Thus, $4° \Rightarrow 1°$.

The implication $5° \Rightarrow 4°$ is obvious, considering for every $x \in E$ the functional $\Phi \in (G^\perp)^*$ defined by

$$\Phi(f) = f(x) \qquad (f \in G^\perp)$$

and taking into account

$$\|\Phi\| = \sup_{\substack{f \in G^{\perp} \\ \|f\| \leq 1}} | \Phi(f) | = \sup_{\substack{f \in G^{\perp} \\ \|f\| \leq 1}} |f(x)| = \|x\|_{\Gamma}.$$

Finally, assume that we have 1° and that the quotient space E/G is reflexive. Since $(E/G)^*$ is equivalent *) to $G^{\perp}$ by the mapping $\varphi \to f$, where

$$f(x) = \varphi(\dot{x}) \qquad (x \in E),$$

the conjugate space $(G^{\perp})^*$ is equivalent to $(E/G)^{**}$ by the mapping $\Phi \to \Psi$, where

$$\Psi(\varphi) = \Phi(f) \qquad (\varphi \in (E/G)^*).$$

But, by the reflexivity of E/G, for every $\Psi \in (E/G)^{**}$ there exists an $\dot{y} \in E/G$ such that

$$\Psi(\varphi) = \varphi(\dot{y}) \qquad (\varphi \in (E/G)^*),$$

$$\|\Psi\| = \|\dot{y}\|,$$

and by the proximinality of G, for every $\dot{y} \in E/G$ there exists (by virtue of the implication $1° \Rightarrow 2°$ proved above) an $y \in \dot{y}$ such that

$$\|y\| = \inf_{g \in G} \|y - g\| = \|\dot{y}\|.$$

Consequently, for every $\Phi \in (G^{\perp})^*$ there exists an element $y \in E$ such that

$$\Phi(f) = \Psi(\varphi) = \varphi(\dot{y}) = f(y) \quad (f \in G^{\perp}),$$

$$\|\Phi\| = \|\Psi\| = \|\dot{y}\| = \|y\|.$$

Thus, $1° \Rightarrow 5°$ when E/G is reflexive, which completes the proof of theorem 2.1.

The equivalences $1° \Leftrightarrow 3° \Leftrightarrow 4°$ of theorem 2.1 have been given in [228], theorems 2 and 2′. In the case when E/G is reflexive, the equivalence $1° \Leftrightarrow 5°$ has been given by A. L. Garkavi ([66], theorem 1).

We shall now obtain from theorem 2.1, as corollaries, various results on proximinal linear subspaces of normed linear spaces.

COROLLARY 2.1 (V. Klee [108]). *Let E be a normed linear space and let G be a linear subspace of E with the property that the unit cell $S_G = \{g \in G \mid \|g\| \leq 1\}$ is strictly sequentially compact* **) *for the weak topology $\sigma(E, E^*)$. Then G is proximinal.*

---

*) See e.g. M. M. Day [43], Chap. II, §1, lemma 1.

**) I.e. from every sequence $\{g_n\} \subset S_G$ one can extract a subsequence which is $\sigma(E, E^*)$-convergent *to an element of* S_G.

Proof. It follows from the hypothesis that G, whence also every subspace $F_x(x \in E \setminus G)$ of the form (2.4), is closed and that the unit cell $S_{F_x} = \{y \in F_x \mid \|y\| \leqslant 1\}$ is strictly sequentially compact for $\sigma(E, E^*)$, whence also for $\sigma(F_x, (F_x)^*)$. Now let $\varphi \in (F_x)^*$ be arbitrary with the property (2.6) and let $\{y_n\} \subset S_{F_x}$ be a sequence such that

$$|\varphi(y_n)| \geqslant \|\varphi\| - \frac{1}{n} \qquad (n = 1, 2, \ldots). \qquad (2.15)$$

Since S_{F_x} is strictly sequentially compact for $\sigma(F_x, (F_x)^*)$, let $\{y_{n_k}\}$ be a subsequence of $\{y_n\}$ converging to a $y \in S_{F_x}$ for $\sigma(F_x, (F_x)^*)$. From (2.15) we obtain then $|\varphi(y)| \geqslant \|\varphi\|$, whence, since $y \in S_{F_x}$,

$$|\varphi(y)| = \|\varphi\|,$$

which shows that the element $z \in F_x$ defined by

$$z = [\operatorname{sign} \varphi(y)]\, y$$

is a maximal element for φ. Consequently, by virtue of the implication $3° \Rightarrow 1°$ of theorem 2.1, G is proximinal, which completes the proof.

We mention that the initial proof of V. Klee [108] consists in a dual argument to that used below in the proof of theorem 2.3.

In particular, the condition of corollary 2.1 is satisfied if G is a subspace of finite dimension, since in this case S_G is even compact for the strong topology of E. Consequently, we obtain the following well known result (see e.g. N. I. Ahiezer [1], Chap. I, § 8; for other proofs of this result and for more general theorems see Appendix II):

COROLLARY 2.2. *Let E be a normed linear space and let G be a linear subspace of finite dimension of E. Then G is proximinal.*

In particular, it follows that every normed linear space E has at least one proximinal linear subspace; this also follows from lemma 2.1 or theorem 2.1, which show that *every normed linear space has at least one proximinal hyperplane.*

It arises naturally the problem of characterizing normed linear spaces E with the property that *all closed linear subspaces of E are proximinal.* Such a characterization is given by

COROLLARY 2.3. *In order that all closed linear subspaces of a normed linear space E be proximinal it is necessary and sufficient that the restriction of each $f \in E^*$ to every closed linear subspace of E have a maximal element.*

Proof. Assume that all closed linear subspaces of E are proximinal. Let $f \in E^*$ be arbitrary and let G' be an arbitrary closed linear subspace of E. If $f|_{G'} = 0$, then every $g \in S_{G'} \setminus \{0\}$ is a maximal element for $f|_{G'}$. If $f|_{G'} \neq 0$, let $x \in G'$ be such that $f(x) \neq 0$ and let

$$G = \{y \in G' \,|\, f|_{G'}(y) = 0\}. \tag{2.16}$$

Then

$$G' = F_x = G \oplus [x], \tag{2.17}$$

and since G is by hypothesis proximinal, the functional $f|_{G'} = f|_{F_x} \in (F_x)^*$ has, by virtue of the implication $1° \Rightarrow 3°$ of theorem 2.1, a maximal element $z \in F_x = G'$.

Conversely, assume that the restriction of each $f \in E^*$ to every closed linear subspace of E has a maximal element. Let G be an arbitrary closed linear subspace of E and let $x \in E \setminus G$, $\varphi \in (F_x)^*$ be arbitrary. By the Hahn-Banach theorem we have then $\varphi = f|_{F_x}$ for a suitable $f \in E^*$, whence, since F_x is closed, $\varphi = f|_{F_x}$ has, by virtue of the hypothesis, a maximal element. Consequently, by the implication $3° \Rightarrow 1°$ of theorem 2.1, G is proximinal, which completes the proof.

In the particular case when E is a Banach space, corollary 2.3 has been given by V. N. Nikolsky ([170], p. 121, thorem *)). However, in this case we have, taking into account characterizations of reflexive Banach spaces given by R. C. James and W. F. Eberlein, the following more complete result:

COROLLARY 2.4. *Let E be a Banach space. The following statements are equivalent:*

1° *All closed linear subspaces of E are proximinal.*

2° *All separable closed linear subspaces of E are proximinal.*

3° *All closed linear subspaces of E, of a certain fixed finite codimension* **) *m, where $1 \leq m \leq \dim E - 1$, are proximinal.*

4° *Every $f \in E^*$ has a maximal element.*

5° *E is reflexive.*

Proof. The implications $1° \Rightarrow 2°$ and $1° \Rightarrow 3°$ are obvious. The equivalence $4° \Leftrightarrow 5°$ is a characterization of reflexive Banach spaces given by R. C. James ([101], theorem 5).

Assume now that we have 3° and let $G_1 \subset E$ be a closed linear subspace of codimension $m-1$ of E and let $f \in G_1^*$ be arbitrary. Then the closed linear subspace G of E defined by

$$G = \{y \in G_1 \,|\, f(y) = 0\}$$

is a hyperplane in G_1, whence of codimension m in E, and thus by 3° it is proximinal in E, whence also in G_1. Consequently,

---

*) We observe that actually the proof given by V. N. Nikolsky in [170] does not make use of the completeness of E, whence it is also valid for an arbitrary normed linear space E.

**) We recall that by definition $\operatorname{codim} G = \dim E/G$.

by virtue of the implication 1°⇒3° of theorem 2.1, f has a maximal element. Since $f \in G_1^*$ was arbitrary, by virtue of implication 4°⇒5° the Banach space G_1 is reflexive. Since $\dim E/G_1 = m - 1 < \infty$, it follows that E is also reflexive. Thus 3°⇒5°.

On the other hand, assume that we have 2° and let E_1 be an arbitrary separable closed linear subspace of E. Then every closed linear subspace G of E_1 is a separable closed linear subspace of E, whence by 2°, G is proximinal and thus E_1 satisfies condition 1°. Consequently, by virtue of the implication 1°⇒5° proved above, E_1 is reflexive. Since E_1 was an arbitrary separable closed linear subspace of E, it follows by the theorem of W. F. Eberlein *) that E is reflexive. Thus, 2°⇒5°.

Finally, if we have 5°, then the condition of corollary 2.3 is satisfied, whence all closed linear subspaces of E are proximinal. Thus, 5°⇒1°, which completes the proof.

The implication 1°⇒5° (and, essentially, even the implication 2°⇒5°) of corollary 2.4 has been given by R. C. James (see R. R. Phelps [176], § 4). The implication 5°⇒1° (whence, in particular, also the implications 5°⇒2°, 5°⇒3°) of corollary 2.4 has been given, essentially, by M. M. Day ([42], p. 316, lemma); it is also an immediate consequence **) of corollary 2.1.

The above results permit us to give examples not only of non-proximinal closed linear subspaces, but even of *closed linear subspaces $G \subset E$ which are "very" non-proximinal, i.e. having the property that $\mathfrak{L}_G(x)$ is void for every $x \in E \setminus G$.* For, let E be an arbitrary non-reflexive Banach space. Then, by virtue of the implication 4°⇒5° of corollary 2.4, there exists an $f \in E^*$ which has no maximal element. Let G be the hyperplane

$$G = \{y \in E \mid f(y) = 0\}.$$

We claim that $\mathfrak{L}_G(x) = \emptyset$ for every $x \in E \setminus G$. Indeed, if for an $x \in E \setminus G$ we have $\mathfrak{L}_G(x) \neq \emptyset$, then for an arbitrary $y_0 \in \mathfrak{L}_G(x)$ and for $z = x - y_0 \neq 0$ we have $0 \in \mathfrak{L}_G(z)$, as has been observed also in the proof of the necessity part of lemma 2.1. Consequently, by virtue of the sufficiency part of the same lemma, G is proximinal, whence, on the basis of the implication 1°⇒3° of theorem 2.1, it follows that f has a maximal element,

*) See e.g. N. Dunford and J. Schwartz [49], p. 430, theorem 1.

**) We observe that an immediate consequence of corollary 2.1 is even the following result which is more general than the implication 5°⇒1° of corollary 2.4:

COROLLARY 2.1′. *Let E be a normed linear space and let G be a linear subspace of E with the property that G is a reflexive Banach space. Then G is proximinal.*

contradicting the assumption. This completes the proof of our statement.

We also observe that *a linear subspace G of a normed linear space E has the property that $\mathfrak{P}_G(x)$ is void for every $x \in E \setminus G$ if and only if there exists no $z \in E \setminus \{0\}$ such that*

$$z \perp G. \tag{2.18}$$

For, if a $z \in E \setminus \{0\}$ satisfies (2.18), then $0 \in \mathfrak{P}_G(z)$, whence $\mathfrak{P}_G(z) \neq \emptyset$. Conversely, if $x \in E \setminus G$, $\mathfrak{P}_G(x) \neq \emptyset$, $g_0 \in \mathfrak{P}_G(x)$, then by lemma 1.14 we have $x - g_0 \perp G$, whence the element $z = x - g_0 \neq 0$ satisfies (2.18).

Corollary 2.1 and, in particular, corollary 2.2 yield classes of proximinal linear subspaces of normed linear spaces. We now shall give, in conjugate spaces, other classes of proximinal linear subspaces.

COROLLARY 2.5. *Let E^* be the conjugate space of a normed linear space E and let Γ be a $\sigma(E^*, E)$-closed linear subspace of E^*. Then Γ is proximinal.*

Proof. Since Γ is $\sigma(E^*, E)$-closed, it is also strongly closed. Now let $f \in E^*$ be arbitrary. Then, by virtue of the Hahn-Banach theorem, there exists an $h \in E^*$ such that

$$h(x) = f(x) \qquad (x \in \Gamma_\perp), \tag{2.19}$$

$$\|h\| = \|f|_{\Gamma_\perp}\| \,. \tag{2.20}$$

Since Γ is $\sigma(E^*, E)$-closed, we have $\Gamma = (\Gamma_\perp)^\perp$, whence

$$\Gamma^\perp = [(\Gamma_\perp)^\perp]^\perp = (\Gamma_\perp)^{\perp\perp},$$

and consequently $\Gamma_\perp$ is dense in $\Gamma^\perp$ for the weak topology $\sigma(E^{**}, E^*)$. Since the mappings $\Phi \to \Phi(h)$ and $\Phi \to \Phi(f)$ of E^{**} into the field of scalars are continuous for this topology, it follows from (2.19) that we have

$$\Phi(h) = \Phi(f) \qquad (\Phi \in \Gamma^\perp), \tag{2.21}$$

which, together with (2.20) and $\|f|_{\Gamma_\perp}\| \leqslant \|f\|_{\Gamma^\perp} = \|h\|_{\Gamma^\perp} \leqslant \|h\|$, shows that $\Gamma^\perp$ has the property $(\mathscr{E}_*)$. Consequently, by virtue of implication $4° \Rightarrow 1°$ of theorem 2.1, Γ is proximinal, which completes the proof.

In the particular case when E is reflexive or separable, corollary 2.5 is a consequence of a result of V. Klee [108], namely of theorem 2.3 below. For an arbitrary normed linear space E, corollary 2.5 has been stated by R. A. Hirschfeld ([88], theorem 2), with a proof holding when E is reflexive or separable. A proof for the general case, based on the $\sigma(E^*, E)$-

compactness of S_Γ, has been given by R. R. Phelps ([176], § 1, p. 239). The above proof of corollary 2.5 has been given in [228], p. 361—362.

The proof of R. R. Phelps, mentioned above, shows that actually we also have the following result (which is more general than corollary 2.5, since if E is a non-complete normed linear space, the fact that a linear subspace Γ of E^* is $\sigma(E^*, E)$-closed implies the $\sigma(E^*, E)$-compactness of S_Γ, but the converse implication is not true, as shown by an example of J. Dieudonné [44], § 23):

THEOREM 2.2 *Let E^* be the conjugate space of a normed linear space E and let Γ be a linear subspace of E^* having the unit cell $S_\Gamma = \Gamma \cap S_{E^*}$ compact for $\sigma(E^*, E)$. Then Γ is proximinal.*

Proof. Let $f \in E^* \setminus \Gamma$ be arbitrary and put

$$S_n = S\left(f, \rho(f, \Gamma) + \frac{1}{n}\right) \qquad (n = 1, 2, \ldots). \qquad (2.22)$$

Then $\{S_n \cap \Gamma\}$ is a decreasing sequence of non-void $\sigma(E^*, E)$-compact sets, whence there exists a $\gamma_0 \in \bigcap_{n=1}^{\infty} (S_n \cap \Gamma)$. Evidently $\gamma_0 \in \mathfrak{P}_\Gamma(f)$, which completes the proof.

We also mention the following result of V. Klee [108], which yields one more class of proximinal linear subspaces of conjugate spaces:

THEOREM 2.3. *Let E^* be the conjugate space of a normed linear space E and let Γ be a linear subspace of E^*, having the unit cell $S_\Gamma = \Gamma \cap S_{E^*}$ strictly sequentially compact for $\sigma(E^*, E)$. Then Γ is proximinal.*

Proof. Let $f \in E^* \setminus \Gamma$ be arbitrary. Put

$$d = \rho(f, \Gamma). \qquad (2.23)$$

Then there exist a sequence $\{\gamma_n\} \subset \Gamma$ and an element $\gamma_0 \in \Gamma$ such that

$$\lim_{n \to \infty} \|f - \gamma_n\| = d, \qquad (2.24)$$

$$\gamma_0(x) = \lim_{n \to \infty} \gamma_n(x) \qquad (x \in E), \qquad (2.25)$$

whence

$$|f(x) - \gamma_0(x)| = \lim_{n \to \infty} |f(x) - \gamma_n(x)| \leqslant \overline{\lim_{n \to \infty}} \|f - \gamma_n\| \, \|x\| =$$

$$= d\|x\| \qquad (x \in E).$$

Consequently, $\|f - \gamma_0\| \leqslant d$, whence $\gamma_0 \in \mathfrak{P}_\Gamma(f)$, which completes the proof.

From examples of J. Dieudonné—L. Schwartz ([45], §11) and A. Grothendieck ([78], §4) it follows that between theorems 2.2 and 2.3 there is no relation of implication.

We shall not obtain theorems 2.2 and 2.3 above as corollaries of theorem 2.1, but conversely, we observe that from theorems 2.2 and 2.3 and the implication $1° \Rightarrow 3°$ (or even $1° \Rightarrow 4°$) of theorem 2.1, we obtain for instance the following result:

COROLLARY 2.6. *Let E^* be the conjugate space of a normed linear space E and let $\Phi \in E^{**}$ be a functional with the property that the unit cell S_H of the hyperplane*

$$H = \{f \in E^* \mid \Phi(f) = 0\} \tag{2.26}$$

is compact or strictly sequentially compact for the weak topology $\sigma(E^*, E)$. Then Φ has a maximal element.

§ 3. UNIQUENESS OF ELEMENTS OF BEST APPROXIMATION

We shall call a linear subspace G of a normed linear space E a *semi-Čebyšev subspace* if for every $x \in E$ the set $\mathfrak{L}_G(x)$ contains at most one element. An example of such subspaces is that of the (non-proximinal) subspaces G with the property that the set $\mathfrak{L}_G(x)$ is void for all $x \in E \setminus G$, considered in §2. G is called a *Čebyšev subspace* *) if it is simultaneously proximinal and semi-Čebyšev, i.e. if for every $x \in E$ the set $\mathfrak{L}_G(x)$ contains exactly one element. We shall give in the present paragraph various characterizations of semi-Čebyšev subspaces of normed linear spaces as well as applications of these characterizations (e.g. to the problem of existence of semi-Čebyšev and non-semi-Čebyšev subspaces). Combining the results of the present paragraph with those of §2 one can obtain corresponding characterizations of Čebyšev subspaces and applications of these characterizations.

3.1. UNIQUENESS OF ELEMENTS OF BEST APPROXIMATION IN GENERAL NORMED LINEAR SPACES

We shall give first necessary and sufficient conditions in order that an element $g_0 \in G$ be *the unique* element of best approximation of an $x \in E \setminus \overline{G}$.

*) The term "Čebyšev subspace" has been proposed by N. V. Efimov and S. B. Stečkin [50]. In his paper [176], R. P. Phelps has used for these subspaces the term *Haar subspace*, but later [177] he too, has adopted the term Čebyšev subspace.

THEOREM 3.1. *Let E be a normed linear space, G a linear subspace of E, $x \in E \setminus \overline{G}$ and $g_0 \in G$. The following statements are equivalent:*

1° $\mathfrak{P}_G(x) = \{g_0\}$.

2° *$g_0 \in \mathfrak{P}_G(x)$ and there do not exist $g \in G \setminus \{g_0\}$ and $f \in E^*$ such that*

$$\|f\| = 1, \tag{3.1}$$

$$f(g) = f(g_0), \tag{3.2}$$

$$f(x-g) = \|x-g\|. \tag{3.3}$$

3° *$g_0 \in \mathfrak{P}_G(x)$ and there do not exist $g \in G \setminus \{g_0\}$ and $f \in E^*$ with the properties (3.2), (3.3) and*

$$f \in \mathcal{E}(S_{E^*}). \tag{3.4}$$

4° *$g_0 \in \mathfrak{P}_G(x)$ and there do not exist $g \in G \setminus \{g_0\}$ and $f \in E^*$ with the properties (3.2), (3.4) and*

$$|f(x-g)| = \|x-g\|. \tag{3.5}$$

5° *$g_0 \in \mathfrak{P}_G(x)$ and there do not exist $g \in G \setminus \{g_0\}$ and $f \in E^*$ with the properties (3.4), (3.5) and*

$$\operatorname{Re}\,[f(g-g_0)\overline{f(x-g)}] \geqslant 0. \tag{3.6}$$

Proof. Assume that we have 1° and that there exist $g \in G \setminus \{g_0\}$ and $f \in E^*$ satisfying (3.1), (3.2) and (3.3). Then

$$\|x-g_0\| \geqslant |f(x-g_0)| = |f(x-g)+f(g-g_0)| =$$
$$= |f(x-g)| = \|x-g\|,$$

whence, by $g_0 \in \mathfrak{P}_G(x)$, it follows that we have $g \in \mathfrak{P}_G(x)$, which contradicts 1°. Thus, 1°$\Rightarrow$2°.

The implications 2°$\Rightarrow$3°$\Rightarrow$4° are obvious.

Now assume that we have 4°. Then for every $g \in G \setminus \{g_0\}$ and $f \in E^*$ with the properties (3.4) and (3.5) we have $f(g) \neq f(g_0)$. Consequently, for any such $g \in G \setminus \{g_0\}$ and $f \in E^*$ we have

$$\|x-g_0\|^2 \geqslant |f(x-g_0)|^2 = |f(x-g)+f(g-g_0)|^2 =$$
$$= \|x-g\|^2 + |f(g-g_0)|^2 + 2\operatorname{Re}[f(g-g_0)\overline{f(x-g)}] >$$
$$> \|x-g\|^2 + 2\operatorname{Re}\,[f(g-g_0)\overline{f(x-g)}],$$

whence, taking into account that $g_0 \in \mathfrak{P}_G(x)$, it follows that for any such $g \in G \setminus \{g_0\}$ and $f \in E^*$ we have $\operatorname{Re}\,[f(g-g_0)\overline{f(x-g)}] < 0$. Thus, 4°$\Rightarrow$5°.

Finally, assume that we have 5° and let $g \in G \setminus \{g_0\}$ be arbitrary. Then by corollary 1.9 (implication 1°$\Rightarrow$5°) it follows that $g \notin \mathfrak{P}_G(x)$. Thus, 5°$\Rightarrow$1°, which concludes the proof.

A result similar to the equivalences $1° \Leftrightarrow 4° \Leftrightarrow 5°$ of theorem 3.1 but weaker than them (namely, the condition $f \in \mathcal{E}(S_{E^*})$ being replaced by the condition $f \in \Gamma$, where $\Gamma \subset S_{E^*}$ is as in corollary 1.11), has been given by V. N. Nikolsky ([169], § 3).

We shall say, following [228], p. 359, that a linear subspace of the conjugate space E^* has *the property* $(\mathfrak{U}_*)$, if for every $x \in E$ there exists at most one element $y \in E$ such that

$$\gamma(y) = \gamma(x) \qquad (\gamma \in \Gamma), \tag{3.7}$$

$$\|y\| = \|x\|_\Gamma. \tag{3.8}$$

THEOREM 3.2. *Let E be a normed linear space and G a linear subspace of E. The following statements are equivalent:*

1° *G is a semi-Čebyšev subspace.*

2° *There do not exist $f \in E^*$ and $x_1, x_2 \in E$ with $x_1 - x_2 \in G \setminus \{0\}$, such that*

$$\|f\| = 1, \tag{3.9}$$

$$f(g) = 0 \qquad (g \in G), \tag{3.10}$$

$$f(x_1) = \|x_1\|, f(x_2) = \|x_2\|. \tag{3.11}$$

3° *There do not exist $f \in E^*$, $x \in E$ and $g_0 \in G \setminus \{0\}$ with the properties (3.9), (3.10) and*

$$f(x) = \|x\| = \|x - g_0\|. \tag{3.12}$$

The above statements are implied by — and if G is closed they are equivalent to — the following statements, which are equivalent to each other:

4° *$G^\perp$ has the property $(\mathfrak{U}_*)$.*

5° *For every $\Phi \in (G^\perp)^*$ there exists at most one element $y \in E$ such that*

$$f(y) = \Phi(f) \qquad (f \in G^\perp). \tag{3.13}$$

$$\|y\| = \|\Phi\|. \tag{3.14}$$

Proof. Assume that there exist $f \in E^*$ and $x_1, x_2 \in E$ with $x_1 - x_2 \in G \setminus \{0\}$, such that we have (3.9), (3.10) and (3.11). Put

$$g_0 = x_1 - x_2.$$

Then we have

$$f(x_1 - g_0) = f(x_2) = \|x_2\| = \|x_1 - g_0\|, \tag{3.15}$$

$$f(x_1) = \|x_1\|, \tag{3.16}$$

whence also $x_1 \in E \setminus \overline{G}$ (otherwise from (3.10), (3.16) and (3.15) one would obtain $0 = f(x_1) = \|x_1\|$, $0 = f(x_1 - g_0) = \|x_2\|$, which contradicts $x_2 - x_1 \neq 0$). Consequently, by virtue of theorem 1.1 we have $g_0 \in \mathfrak{L}_G(x_1)$ and $0 \in \mathfrak{L}_G(x_1)$. Since $g_0 \neq 0$, it follows that G is not a semi-Čebyšev subspace. Thus, $1° \Rightarrow 2°$.

Now assume that there exist $f \in E^*$, $x \in E$ and $g_0 \in G \setminus \{0\}$ with the properties (3.9), (3.10) and (3.12). Put

$$x_1 = x, \; x_2 = x - g_0.$$

Then we have (3.9), (3.10) and (3.11) and also $x_1 - x_2 = g_0 \in G \setminus \{0\}$. Thus, $2° \Rightarrow 3°$.

Assume now that G is not a semi-Čebyšev subspace. Then for a suitable $y \in E \setminus \overline{G}$ there exist $g_1, g_2 \in \mathfrak{L}_G(x)$, $g_1 \neq g_2$. Consequently, putting

$$x = y - g_1, \quad g_0 = g_2 - g_1,$$

we have $x \in E \setminus \overline{G}$, $g_0 \in G \setminus \{0\}$ and 0, $g_0 \in \mathfrak{L}_G(x)$. By virtue of the simultaneous characterization of a set of elements of best approximation given in § 1, corollary 1.3, there exists then an $f \in E^*$ such that we have (3.9), (3.10) and (3.12). Thus, $3° \Rightarrow 1°$.

Assume again that G is not a semi-Čebyšev subspace, that is, for a suitable $x \in E$ there exist $g_1, g_2 \in \mathfrak{L}_G(y)$, $g_1 \neq g_2$. Put

$$y_1 = x - g_1, \; y_2 = x - g_2.$$

Then $y_1 \neq y_2$ and the argument of §2, the proof of the implication $1° \Rightarrow 4°$ of theorem 2.1, shows that both pairs (x, y_1) and (x, y_2) satisfy (3.7) and (3.8) for $\Gamma = G^{\perp}$, whence $G^{\perp}$ does not have the property $(\mathfrak{U}_*)$. Thus, $4° \Rightarrow 1°$.

Assume now that G is closed and that $G^{\perp}$ does not have the property $(\mathfrak{U}_*)$, i.e. for a suitable $x \in E$ there exist $y_1, y_2 \in E$, $y_1 \neq y_2$, both elements satisfying (3.7) and (3.8) for $\Gamma = G^{\perp}$. Then we have $x \in E \setminus \overline{G}$ (otherwise we would have $\|x\|_{\Gamma} = \|x\|_{G^{\perp}} = 0$, whence, by (3.8), $y_1 = 0 = y_2$, which contradicts $y_1 \neq y_2$). Put

$$g_1 = x - y_1, \; g_2 = x - y_2.$$

Then $g_1 \neq g_2$ and the argument of §2, the proof of the implication $4° \Rightarrow 1°$ of theorem 2.1, shows that $g_1, g_2 \in \mathfrak{L}_G(x)$, whence G is not a semi-Čebyšev subspace. Thus, if G is closed, then $1° \Rightarrow 4°$.

Assume now that G does not satisfy condition 5°, i.e. for a suitable $\Phi \in (G^{\perp})^*$ there exist $y_1, y_2 \in E$, $y_1 \neq y_2$, both elements satisfying (3.13) and (3.14). Then for $x = y_1$ the pairs (x, y_1) and (x, y_2) satisfy (3.7) and (3.8) with $\Gamma = G^{\perp}$ (since $\|y_i\| = \|\Phi\| = \sup_{\substack{f \in G^{\perp} \\ \|f\| \leqslant 1}} |\Phi(f)| = \sup_{\substack{f \in G^{\perp} \\ \|f\| \leqslant 1}} |f(y_i)| = \sup_{\substack{f \in G^{\perp} \\ \|f\| \leqslant 1}} |f(x)| = \|x\|_{G^{\perp}}$), whence $G^{\perp}$ does not have the property $(\mathfrak{U}_*)$. Thus, $4° \Rightarrow 5°$.

Finally, the implication $5^\circ \Rightarrow 4^\circ$ is obvious, by considering for every $x \in E$ the functional $\Phi \in (G^\perp)^*$ defined by

$$\Phi(f) = f(x) \qquad (f \in G^\perp),$$

and taking into account that $\|\Phi\| = \|x\|_{G^\perp}$, which completes the proof of theorem 3.2.

The equivalence $1^\circ \Leftrightarrow 2^\circ$ of theorem 3.2 has been given in the paper [212], p. 184. The equivalence $1^\circ \Leftrightarrow 3^\circ$ of the theorem has been given in [215], theorem 1, and the equivalence $1^\circ \Leftrightarrow 4^\circ$ in [228], theorem 3.

The characterizations $2^\circ - 5^\circ$, in theorem 3.2, of the semi-Čebyšev subspaces G, *are not intrinsic*, i.e. they involve also elements of $E \setminus G$. Nevertheless, they are convenient for applications since from these criteria one can deduce, as we shall see, intrinsic characterizations of semi-Čebyšev subspaces in the usual concrete normed linear spaces.

We shall now give various corollaries of theorem 3.2. Following R. R. Phelps [176], a linear subspace G of a normed linear space E is said to have *the property* (U), if every functional $\varphi \in G^*$ has a unique extension with the same norm to the whole space E.

COROLLARY 3.1. *Let E be a normed linear space and G a linear subspace of E.*

a) *If $G^\perp$ has the property (U) then G is a semi-Čebyšev subspace.*

b) *If G is proximinal and $G^\perp$ has the property (U) then G is a Čebyšev subspace.*

Proof. If $G^\perp$ has the property (U), then it has also the property $(\mathfrak{U}_*)$. Indeed, if for an $x \in E$ there existed $y_1, y_2 \in E$, $y_1 \neq y_2$, both satisfying (3.7) and (3.8) for $\Gamma = G^\perp$, then the functional $\Phi_x \in (G^\perp)^*$, defined by

$$\Phi_x(f) = f(x) \qquad (f \in G^\perp),$$

would have two distinct extensions with the same norm to the whole E^*, namely

$$\Psi_{y_1}(f) = f(y_1), \ \Psi_{y_2}(f) = f(y_2) \qquad (f \in E^*),$$

whence $G^\perp$ would not have the property (U). Thus, $G^\perp$ has the property $(\mathfrak{U}_*)$, whence, by virtue of implication $4^\circ \Rightarrow 1^\circ$ of theorem 3.2, it follows that G is a semi-Čebyšev subspace.

b) is an immediate consequence of a), which completes the proof.

Corollary 3.1 b) has been given by R. R. Phelps ([176], theorem 1.3) and the above proof has been given in [228], p. 360.

As has been remarked by R. R. Phelps ([176], p. 241 and 252) the condition that G be proximinal is essential in corollary 3.1 b), since from the fact that $G^\perp$ has the property (U)

it does not follow that G is proximinal. In fact, let $E = c_0$, let $f = \{\eta_n\} \in E^* \equiv l^1$ be such that $\eta_n \neq 0$ $(n = 1, 2 \ldots)$ and let $G = \{y \in E \mid f(y) = 0\}$. Then, obviously, f has no maximal element, whence by virtue of theorem 2.1 (implication $1° \Rightarrow 3°$), G is not proximinal. On the other hand, by Chap. III, corollary 3.1, the $\sigma(E^{**}, E^*)$-closed linear subspace $\Gamma = \{\Phi \in E^{**} \mid \Phi(f) = 0\}$ of E^{**} is a Čebyšev subspace, whence, by virtue of corollary 3.2 below, $G^{\perp} = \Gamma_{\perp} \subset E^*$ has the property (U).

R. R. Phelps has also observed ([176], p. 241 and 252) that the converse of the statement of corollary 3.1 b) is not valid, i.e. that in general there exist Čebyšev subspaces $G \subset E$ for which $G^{\perp}$ does not have the property (U). In fact, let $E = c_0$ and let G be a Čebyšev subspace of E of finite dimension $\geqslant 2$ (from the remark following the proof of theorem 2.2 of Chap. II it follows that such subspaces exist). Then the canonical image of G in $E^{**} \equiv l^{\infty}$ by the natural embedding $E \subset E^{**}$ is $\Gamma = G^{\perp\perp}$ and by virtue of corollary 2.2 of Chap. II, Γ is not a Čebyšev subspace, whence by corollary 3.2 below, $G^{\perp} = \Gamma_{\perp} \subset E^*$ does not have the property (U). Besides, by virtue of the criterion (2.46) of Chap. II and by the definition of c_0, this remark remains valid also for one dimensional Čebyšev subspaces of $E = c_0$.

LEMMA 3.1 ([228], p. 360 lemma). *Let E be a normed linear space and Γ a $\sigma(E^*, E)$-closed linear subspace of the conjugate space E^*. Then we have*

$$\|f\|_{\Gamma^{\perp}} = \|f|_{\Gamma_{\perp}}\| \qquad (f \in E^*). \tag{3.17}$$

Proof. For $f \in \Gamma$ the statement is obvious. Now let $f \in E \setminus \Gamma = E \setminus \bar{\Gamma}$. Then there exists, by virtue of corollary 2.5 of §2, a $\gamma_0 \in \mathfrak{L}_{\Gamma}(f)$. Consequently, by virtue of corollary 1.2 b) of §1, we have

$$\|f\|_{\Gamma^{\perp}} = \|f - \gamma_0\|_{\Gamma^{\perp}} \leqslant \|f - \gamma_0\| = \|(f - \gamma_0)|_{\Gamma_{\perp}}\| = \|f|_{\Gamma_{\perp}}\|.$$

Since the opposite inequality is obvious, it follows that we have (3.17), which completes the proof.

LEMMA 3.2 ([228], p.362). *Let E be a normed linear space and Γ a $\sigma(E^*, E)$-closed linear subspace of the conjugate space E^*. Then $\Gamma^{\perp}$ has the property $(\mathfrak{U}_*)$ if and only if $\Gamma_{\perp}$ has the property (U).*

Proof. Assume that $\Gamma_{\perp}$ does not have the property (U), i.e. for a suitable $\varphi \in (\Gamma_{\perp})^*$ there exist $f_1, f_2 \in E^*$, $f_1 \neq f_2$, such that

$$f_1(x) = f_2(x) = \varphi(x) \qquad (x \in \Gamma_{\perp}), \tag{3.18}$$

$$\|f_1\| = \|f_2\| = \|\varphi\|. \tag{3.19}$$

Since Γ is $\sigma(E^*, E)$-closed, it follows from (3.18), by applying the same argument as that of §2, the proof of corollary 2.5, that we have

$$\Phi(f_1) = \Phi(f_2) \qquad (\Phi \in \Gamma^{\perp}). \tag{3.20}$$

On the other hand, by (3.18) and (3.19) we have

$$\|f_j\| \geqslant \|f_j\|_{\Gamma^{\perp}} \geqslant \|f_j|_{\Gamma_{\perp}}\| = \|\varphi\| = \|f_j\| \qquad (j = 1, 2),$$

whence

$$\|f_1\| = \|f_1\|_{\Gamma^{\perp}} = \|f_2\|_{\Gamma^{\perp}} = \|f_2\|. \tag{3.21}$$

Consequently, for $f = f_1 \in E^*$ there exist two distinct elements, namely $f_1, f_2 \in E^*$, such that

$$\Phi(f) = \Phi(f_j) \qquad (\Phi \in \Gamma^{\perp}, j = 1, 2), \tag{3.22}$$

$$\|f\|_{\Gamma^{\perp}} = \|f_1\| = \|f_2\|, \tag{3.23}$$

i.e. $\Gamma^{\perp}$ does not have the property $(\mathfrak{U}_*)$.

Conversely, assume that $\Gamma^{\perp}$ does not have the property $(\mathfrak{U}_*)$, i.e. for a suitable $f \in E^*$ there exist $f_1, f_2 \in E^*$, $f_1 \neq f_2$, such that we have (3.22), (3.23). Define $\varphi \in (\Gamma_{\perp})^*$ by

$$\varphi = f|_{\Gamma_{\perp}}. \tag{3.24}$$

Then by $\Gamma_{\perp} \subset \Gamma^{\perp}$ and (3.22) we have

$$\varphi(x) = f_j(x) \qquad (x \in \Gamma_{\perp}, j = 1, 2),$$

and by lemma 3.1 and (3.23) we have

$$\|\varphi\| = \|f|_{\Gamma_{\perp}}\| = \|f\|_{\Gamma^{\perp}} = \|f_j\| \qquad (j = 1, 2),$$

whence φ has two distinct extensions with the same norm to the whole space E, i.e. $\Gamma_{\perp}$ does not have the property (U), which completes the proof.

Combining corollary 2.5, the equivalence 1°⇔4° of theorem 3.2 and lemma 3.2, we obtain

Corollary 3.2. *Let E be a normed linear space and Γ a $\sigma(E^*, E)$-closed linear subspace of the conjugate space E^*. In order that Γ be a Čebyšev subspace it is necessary and sufficient that the subspace $\Gamma_{\perp}$ of E have the property (U).*

Corollary 3.2 has been given by R. R. Phelps ([176], p. 240). The above proof of corollary 3.2 has been given in [228], p. 365.

It arises naturally the problem of characterization of the normed linear spaces E having the property that *all linear subspaces of E are semi-Čebyšev subspaces*. We recall for this purpose that, following J. A. Clarkson ([38], §4), a normed linear space E is said to be a *strictly convex space* *), if the relations

$$x, y \in E \setminus \{0\}, \quad \|x + y\| = \|x\| + \|y\| \tag{3.25}$$

imply the existence of a $c > 0$ such that

$$y = cx. \tag{3.26}$$

In the sequel we shall use the following well known **) characterization of strictly convex spaces (given by M. G. Krein [125]) : *A normed linear space E is strictly convex if and only if every functional $f \in E^*$ has at most one maximal element of norm 1.*

This being said, we now can prove the following corollary of theorem 3.2 :

COROLLARY 3.3. *Let E be a normed linear space. The following statements are equivalent* :

1° *All linear subspaces of E are semi-Čebyšev subspaces.*

2° *All linear subspaces of E of a certain fixed finite dimension n, where $1 \leqslant n \leqslant \dim E-1$, are semi-Čebyšev subspaces.*

3° *All closed linear subspaces of E of a certain fixed finite codimension m, where $1 \leqslant m \leqslant \dim E-1$, are semi-Čebyšev subspaces.*

4° *E is strictly convex.*

Proof. The implications 1°⇒2° and 1°⇒3° are obvious, taking into account corollary 2.2.

Assume now that E is not strictly convex, whence there exists an $f \in E^*$ with $\|f\| = 1$ which has two distinct maximal elements x_1, x_2 of norm $\|x_1\| = \|x_2\| = 1$. Put

$$G_1 = [x_1 - x_2] = \{\lambda(x_1 - x_2) \,|\, \lambda \text{ scalar}\}, \tag{3.27}$$

$$G_2 = \{y \in E \,|\, f(y) = 0\}, \tag{3.28}$$

and let G_1', G_2' be closed linear subspaces of E with $G_1 \subset G_1' \subset G_2$, $G_1 \subset G_2' \subset G_2$, $\dim G_1' = n$, $\operatorname{codim} G_2' = m$. Then we have

$$f(g) = 0 \qquad (g \in G_1',\ g \in G_2'), \tag{3.29}$$

$$f(x_2) = \|x_2\| = \|x_2 + (x_1 - x_2)\|, \tag{3.30}$$

whence, by virtue of the implication 1°⇒3° of theorem 3.2, it follows that G_1' and G_2' are not semi-Čebyšev subspaces, hence G_1' is not a Čebyšev subspace. Thus, 2°⇒4° and 3°⇒4°.

---

*) Some authors use for such spaces the term *strictly normed space* (M. G. Krein [125], N. I. Ahiezer [1] and others), or *rotund space* (M. M. Day [43] and others), but the term *strictly convex space* too, is used also in the present (G. Köthe [122], R. R. Phelps [176] and others).

**) See e.g. G. Köthe [122], p. 346, theorem (1).

Finally, assume that we have 4°, whence every $f \in E^*$ has at most one maximal element of norm 1. Then condition 3° of theorem 3.2 is satisfied for every linear subspace G of E, hence all linear subspaces of E are semi-Čebyšev subspaces. Thus, 4°⇒1°, which completes the proof.

The implication 4°⇒2° (hence in particular also 4°⇒1° as well as 4°⇒3°) of corollary 3.3 above is well known, being e.g. in N. I. Ahiezer ([1], Chap. I, § 11; see also § 9); it has been given, essentially, by M. G. Krein [125]. The implication 2°⇒4° with $n=1$ (hence, in particular, also 1°⇒4°), has been given by K. Tatarkiewicz ([243], §3) and found again by R. A. Hirschfeld ([88], theorem 3) and R. R. Phelps ([176], p. 240).

COROLLARY 3.4. *Let E be a Banach space. The following statements are equivalent:*

1° *All closed linear subspaces of E are Čebyšev subspaces.*

2° *All closed linear subspaces of E of a certain fixed finite codimension m, where $1 \leqslant m \leqslant \dim E-1$, are Čebyšev subspaces.*

3° *E is reflexive and strictly convex.*

Proof. The implication 1°⇒2° is obvious, and the implications 2°⇒3°⇒1° follow combining corollary 2.4 and corollary 3.3 above.

The implication 3°⇒1° has been given by M. M. Day ([42], p. 316, lemma). The equivalence 1°⇔3° of corollary 3.4 was noted by D. F. Cudia ([40], p. 93), and in the particular case when the conjugate space E^* is smooth, the equivalence 1°⇔2° with $m = 1$ has been given in the same paper ([40], p. 93); however, corollary 3.4 above shows that the equivalence 1°⇔2° too, is valid for arbitrary Banach spaces.

Since the spaces $E = L^p(T, \nu)$ $(1 < p < \infty)$, where (T, ν) is a positive measure space, as well as the complete inner product spaces $E = \mathcal{H}$, satisfy condition 3° of corollary 3.4, we have the following well known

COROLLARY 3.5. *Let $E = L^p(T, \nu)$ $(1 < p < \infty)$, where (T, ν) is a positive measure space, or let $E = \mathcal{H}$ = a (complete) Hilbert space. Then all closed linear subspaces of E are Čebyšev subspaces.*

We recall that a normed linear space E is said to be a *smooth space*, if all elements of norm 1 in E are normal *), i.e. if for every $x \in E$ with $\|x\| = 1$ there exists only one $f = f_x \in E^*$ such that

$$\|f\| = 1, \quad f(x) = 1. \tag{3.31}$$

COROLLARY 3.6. *Let E be a normed linear space. The following statements are equivalent:*

1° *All $\sigma(E^*, E)$-closed linear subspaces of the conjugate space E^* are Čebyšev subspaces.*

*) See § 1, section 1.1.

2° *All linear subspaces of* E *have the property* (U).

3° *All closed linear subspaces of* E *of a certain fixed finite codimension* m, *where* $1 \leqslant m \leqslant \dim E - 1$, *have the property* (U).

4° *All linear subspaces of* E^* *are semi-Čebyšev subspaces.*

5° *All linear subspaces of* E^*, *of a certain fixed finite dimension* n, *where* $1 \leqslant n \leqslant \dim E^* - 1$, *are Čebyšev subspaces.*

6° *All closed linear subspaces of* E^*, *of a certain fixed finite codimension* m, *where* $1 \leqslant m \leqslant \dim E^* - 1$, *are semi-Čebyšev subspaces.*

7° E^* *is strictly convex.*

These statements imply the following ones, which are equivalent to each other:

8° *All* $\sigma(E^*, E)$*-closed linear subspaces of* E^*, *of a certain fixed finite codimension* m, *where* $1 \leqslant m \leqslant \dim E^* - 1$, *are Čebyšev subspaces.*

9° *All linear subspaces of* E *of a certain fixed finite dimension* n, *where* $1 \leqslant n \leqslant \dim E - 1$, *have the property* (U).

10° E *is smooth.*

In the particular case when E *is a reflexive Banach space, all the statements* 1°—10° *are equivalent, even replacing in* 6° *"semi-Čebyšev" by "Čebyšev".*

Proof. Assume that we have 1° and let G be an arbitrary linear subspace of E. Then $G^\perp$ is $\sigma(E^*, E)$-closed, whence by 1°, it is a Čebyšev subspace. Consequently, by virtue of corollary 3.2, the subspace $\overline{G} = (G^\perp)_\perp$ of E has the property (U), whence G too has the property (U). Thus, 1°$\Rightarrow$2°.

The implication 2°$\Rightarrow$3° is obvious.

Assume now that we have 3° and let Γ be a linear subspace of E^* of finite dimension n. Then the closed linear subspace $\Gamma_\perp$ of E is of codimension n, whence by 3° it has the property (U). Consequently, by virtue of corollary 3.2, Γ is a Čebyšev subspace. Thus, 3°$\Rightarrow$5°.

The equivalences 5°$\Leftrightarrow$4°$\Leftrightarrow$6°$\Leftrightarrow$7° are nothing else than corollary 3.3 applied to E^*.

The implication 4°$\Rightarrow$1° is an immediate consequence of corollary 2.5. Thus, 1°$\Leftrightarrow$ ... $\Leftrightarrow$7°.

The implication 1°$\Rightarrow$8° is obvious. In the particular case when E is a reflexive Banach space, the implication 8°$\Rightarrow$6°, even with "Čebyšev" in 6°, is also obvious, since then every closed linear subspace of E^* is $\sigma(E^*, E)$-closed.

The equivalence 8°$\Leftrightarrow$9° is shown by the same argument as that applied above for proving the implications 1°$\Rightarrow$2° and 3°$\Rightarrow$5°, taking into account that if $\dim G = n$, then $G^\perp$ is a $\sigma(E^*, E)$-closed linear subspace of codimension n and conversely, if Γ is a $\sigma(E^*, E)$-closed linear subspace of codimension m, then $\dim \Gamma_\perp = m$.

Assume now that E is not smooth, i.e. for a suitable $x \in E$ with $\|x\| = 1$ there exist $f_1, f_2 \in E^*$, $f_1 \neq f_2$, such that

$$\|f_1\| = \|f_2\| = 1,\ f_1(x) = f_2(x) = 1.$$

Consider an n-dimensional linear subspace G of E containing the element x and such that $(f_1 - f_2)(g) = 0$ $(g \in G)$, and let $\varphi = f_1|_G \in G^*$.

Then we have

$$f_1(g) = f_2(g) = \varphi(g) \qquad (g \in G), \tag{3.32}$$

$$\|f_1\| = \|f_2\| = \|\varphi\|, \tag{3.33}$$

whence G does not have the property (U). Since $\dim G = n$, it follows that we do not have 9°. Thus, 9°⇒10°.

Finally, assume that we do not have 9°, i.e. there exists an n-dimensional linear subspace G of E which does not have the property (U). Then for a suitable $\varphi \in G^*$ with $\|\varphi\| = 1$ there exist $f_1, f_2 \in E^*$, $f_1 \neq f_2$, such that we have (3.32) and (3.33). Since $\dim G = n < \infty$, there exists an $x \in G$ such that

$$\|x\| = 1, \quad \varphi(x) = 1.$$

Then we have

$$\|f_1\| = \|f_2\| = \|\varphi\| = 1,\ f_1(x) = f_2(x) = \varphi(x) = 1 = \|x\|,$$

whence E is not smooth. Thus 10°⇒9°, which completes the proof of corollary 3.6.

The implication 7°⇒2° of corollary 3.6 has been given by A. E. Taylor ([244], theorem 6). In the particular case when E is a reflexive Banach space, the implication 2°⇒7°, too, has been given by A. E. Taylor ([244], theorem 6), and in the general case it has been given by S. R. Foguel [57]. The implications 3°⇔7°⇒1°⇒2°⇒3° with $m = 1$, therefore also the equivalence of these conditions, have been given by R. R. Phelps ([176], p. 240). The implication 7°⇒10° has been given by V. Klee ([110], theorem (A 1.1)), and the equivalences 8°⇔9°⇔10° with $m = 1$, $n = 1$ have been given by R. R. Phelps ([176], p. 242) and D. F. Cudia ([40], p. 93). In the particular case when E is a reflexive Banach space, the implication 10°⇒2° has been given by A. E. Taylor ([244], theorem 4), and D. F. Cudia ([40], p. 90) has remarked that in this case the statement 2° is equivalent to the following variant of 1° : All closed linear subspaces of E^* are Čebyšev subspaces.

Corollaries 3.3 and 3.4 show that in a strictly convex normed linear space E all linear subspaces are semi-Čebyšev subspaces, and in a strictly convex and reflexive Banach space E all closed linear subspaces are Čebyšev subspaces. It arises naturally the *problem whether the other extreme case is also possible.*

For linear semi-Čebyšev subspaces the answer is obviously negative, i.e. every normed linear space E has at least one semi-Čebyšev subspace; for instance, every linear subspace G, dense in E, with $G \neq E$, satisfies $\mathfrak{L}_G(x) = \emptyset (x \in E \setminus G)$, and hence is a semi-Čebyšev subspace.

However, for closed linear semi-Čebyšev subspaces of Banach spaces the problem is open, i.e. it is not known whether there exists a Banach space E which has no closed linear semi-Čebyšev subspace. Since every non-proximinal hyperplane G satisfies $\mathfrak{L}_G(x) = \emptyset (x \in E \setminus G)$, and hence is a semi-Čebyšev subspace, the space E must be sought, by virtue of corollary 2.4, only among reflexive Banach spaces (in other words, *every non-reflexive Banach space E has at least one semi-Čebyšev hyperplane*). Since every Banach space E in which the unit cell S_E has at least one exposed point *) obviously has a Čebyšev hyperplane, it follows that the space E must be sought only among the reflexive Banach spaces E in which the unit cell S_E has no exposed point. But up to the present**) it is not known whether such a space exists (V. Klee [111], p. 96, problem (4.6)).

As has remarked A. L. Garkavi ([69], theorem 1), in the case of Čebyšev subspaces of Banach spaces, the answer to the above problem is affirmative, e.g. *if I is a set of cardinality $\overline{\overline{I}} > \mathfrak{c}$, then the space $E = E(I)$ of all bounded families of scalars $x = \{\xi_\iota\}_{\iota \in I}$ which have at most a countable number of non-zero "co-ordinates" ξ_ι, endowed with the usual vector operations and with the norm $\|x\| = \sup_{\iota \in I} |\xi_\iota|$, has no Čebyšev subspace.* For, if G is a closed linear subspace of cardinality $\overline{\overline{G}} \leqslant \mathfrak{c}$, then the cardinality of the set of all indices $\iota \in I$ corresponding to the nonzero coordinates of the elements of G is $\leqslant \mathfrak{c}$, whence, by $\overline{\overline{I}} > \mathfrak{c}$, it follows that there exists an index $\iota_0 \in I$ such that

$$\gamma_{\iota_0} = 0 \qquad (g = \{\gamma_\iota\}_{\iota \in I} \in G).$$

*) We recall that any point $x \in E$ with $\|x\| = 1$ having the property that there exists a support hyperplane H of S_E such that $H \cap S_E = \{x\}$, is called an *exposed point* of S_E (E. Straszewicz [239]). Some authors use for such points different terms, e.g. D. P. Milman [150] calls them *accessible points*, and N. Bourbaki ([23], Chap. II, § 4, exercise 15) calls them *points of strict convexity*.

**) After this monograph had gone to print, the problem was solved by J. Lindenstrauss (Bull. Amer. Math. Soc., 72, 967—970 (1966)), who showed that in every reflexive Banach space E the unit cell has an exposed point. Consequently, *every Banach space E has at least one semi-Čebyšev hyperplane.*

Consequently, for the element $x = \{\xi_\iota\}_{\iota \in I} \in E$ defined by

$$\xi_\iota = \begin{cases} 1 \text{ for } \iota = \iota_0 \\ 0 \text{ for } \iota \in I \setminus \{\iota_0\}, \end{cases}$$

and for every $g = \{\gamma_\iota\}_{\iota \in I} \in G$, we have

$$\|x - g\| = \max\left(\sup_{\iota \in I \setminus \{\iota_0\}} |\gamma_\iota|, 1\right) \geqslant 1 = \|x\|,$$

with the equality sign whenever $\|g\| \leqslant 1$. Thus, $S_G \subset \mathfrak{P}_G(x)$, whence G is not a Čebyšev subspace.

On the other hand, if G is a closed linear subspace of cardinality $\overline{\overline{G}} > \mathfrak{c}$, there are two cases: *a*) If G is non-proximinal, then it is not a Čebyšev subspace. *b*) If G is proximinal, let $x = \{\xi_\iota\}_{\iota \in I} \in E \setminus G$ and $g_0 = \{\gamma_\iota^0\}_{\iota \in I} \in \mathfrak{P}_G(x)$ be arbitrary and let A be the at most countable set $\{\iota \in I \mid \xi_\iota - \gamma_\iota^0 \neq 0\}$. Since $\overline{\overline{l^\infty(A)}} \leqslant \mathfrak{c}$, and $\overline{\overline{G}} > \mathfrak{c}$, it follows that there exist two distinct elements of G which coincide on A, whence there exists also a $g_1 = \{\gamma_\iota^1\}_{\iota \in I} \in G \setminus \{0\}$ such that

$$\gamma_\iota^1 = 0 \qquad (\iota \in A).$$

Consequently, for every ε with $0 < \varepsilon \leqslant \frac{1}{\|g_1\|} \|x - g_0\|$ we have

$$\|x - g_0 + \varepsilon g_1\| = \sup_{\iota \in I} |\xi_\iota - \gamma_\iota^0 + \varepsilon \gamma_\iota^1| = \max\left(\sup_{\iota \in A} |\xi_\iota - \gamma_\iota^0|,\right.$$

$$\left.\sup_{\iota \in I \setminus A} |\gamma_\iota^1| \varepsilon\right) \leqslant \|x - g_0\| = \inf_{g \in G} \|x - g\|,$$

whence $g_0 - \varepsilon g_1 \in \mathfrak{P}_G(x)$, and thus G is not a Čebyšev subspace, which completes the proof of the statement above.

The Banach space E in the example above is non-separable. In connection with this fact A. L. Garkavi has observed ([69], theorem 2) *that the separable Banach space $E = c_0$ has no infinite dimensional Čebyšev subspace* (the proof is similar to the second part of the argument above, the set of indices A being replaced by the set of indices $\{1, \ldots, N\}$, where N is a positive integer such that $|\xi_n - \gamma_n^0| < \frac{1}{2} \|x - g_0\|$ for all $n > N$ and taking $\varepsilon \leqslant \frac{1}{2\|g_1\|} \|x - g_0\|$); however, as will be seen in Chap. II, it does have finite dimensional Čebyšev subspaces. On the other hand, in Chap. II and Chap. III it will be shown that the separable Banach space $E = L^1([0, 1])$ has no Čebyšev subspace of finite dimension or codimension, yet it contains Čebyšev subspaces. However, for arbitrary separable Banach spaces, the problem is open: *it is not known whether there exists a separable Banach space E which has no Čebyšev subspace.* Since in every separable conjugate space $E = B^*$ (and even in the dual of any separable Banach space) the unit cell has at least one exposed point (D. P. Milman [150]; V. Klee [111], p. 96—97), whence at least one Čebyšev hyperplane, the space E must be sought among the separable Banach spaces E which are not isometric to any conjugate space.

In connection with these problems, A. L. Garkavi has introduced [67], [69] the notion of an *almost Čebyšev subspace* of a normed linear space E, calling by this term any linear subspace with the property that the set of all elements $x \in E$ for which $\mathfrak{P}_G(x)$ does not consist of a single element, forms a set at most of the first category in E. As has been shown by A. L. Garkavi [69], the consideration of these subspaces has, among others, the following advantages with respect to the consideration of Čebyšev subspaces: a) *in every separable Banach space E there exist almost Čebyšev subspaces G of any finite dimension*; b) *for every closed linear subspace G of a reflexive separable Banach space E there exists an almost Čebyšev subspace G_1 of E and an isomorphism u of E into itself such that $u(G) = G_1$*; c) *in every separable conjugate space there exist almost Čebyšev subspaces of any finite codimension.* On the other hand, for such subspaces a part of the negative examples given above remain valid, e.g. as has been observed by A. L. Garkavi [69], *the Banach space $E(I)$ above has no almost Čebyšev subspace and the space c_0 has no almost Čebyšev subspace of infinite dimension.* Indeed, the arguments above show that all closed linear subspaces G of cardinality $\overline{\overline{G}} > \mathfrak{c}$ of $E(I)$ as well as all infinite dimensional closed linear subspaces of $E = c_0$, are "very" non-Čebyšev, in the sense that for no element $x \in E \setminus G$ does the set $\mathfrak{P}_G(x)$ consist of a single element. On the other hand, let G be a closed linear subspace of cardinality $\overline{\overline{G}} \leqslant \mathfrak{c}$ of $E(I)$, and let $\iota_0 \in I$ and $x = \{\xi_\iota\}_{\iota \in I} \in E(I)$ be as in the corresponding argument above. Then for every $x' = \{\xi'_\iota\}_{\iota \in I}$ in the cell $S\left(x, \frac{1}{4}\right)$ we have $|1 - \xi'_{\iota_0}| \leqslant \frac{1}{4}$, $|\xi'_\iota| \leqslant \frac{1}{4} (\iota \in I \setminus \{\iota_0\})$, whence $|\xi'_{\iota_0}| \geqslant \frac{3}{4} > \frac{1}{4} \geqslant |\xi'_\iota| (\iota \in I \setminus \{\iota_0\})$, whence for every $g = \{\gamma_\iota\}_{\iota \in I} \in G$ we have

$$\|x' - g\| = \max\,(\sup_{\iota \in I \setminus \{\iota_0\}} |\xi'_\iota - \gamma_\iota|, |\xi'_{\iota_0}|) \geqslant |\xi'_{\iota_0}| = \sup_{\iota \in I} |\xi'_\iota| = \|x'\|,$$

with the equality sign whenever $\|g\| \leqslant \frac{1}{2}$ (since then $|\xi'_\iota - \gamma_\iota| \leqslant$ $\leqslant |\xi'_\iota| + |\gamma_\iota| \leqslant \frac{3}{4} \leqslant |\xi_{\iota_0}|$ for every $\iota \in I \setminus \{\iota_0\}$). Thus,

$\frac{1}{2} S_G \subset \mathfrak{L}_G(x')$ for every $x' \in S\left(x, \frac{1}{4}\right)$, whence G is not an almost Čebyšev subspace, which completes the proof.

3.2. APPLICATIONS IN THE SPACES $C(Q)$ AND $C_R(Q)$

THEOREM 3.3. *Let $E = C(Q)$ (Q compact) and let G be a linear subspace of E. The following statements are equivalent:*

1° G is a semi-Čebyšev subspace.

2° There do not exist a Radon measure μ on Q and an element $g_0 \in G \setminus \{0\}$ such that

$$|\mu|(Q) = 1, \tag{3.34}$$

$$\int_Q g(q)\, d\mu(q) = 0 \qquad (g \in G), \tag{3.35}$$

$$\frac{d\mu}{d|\mu|} \in C(Q), \tag{3.36}$$

$$g_0(q) = 0 \qquad (q \in S(\mu)), \tag{3.37}$$

*where (3.36) is understood in the sense that $\frac{d\mu}{d|\mu|}$ can be made continuous *) on Q by changing its values on a set of $|\mu|$-measure zero and where $(S\mu)$ denotes the carrier of the measure μ.*

In the particular case when the scalars are real (i.e. when $E = C_R(Q)$), these statements are equivalent to the following one:

3° There do not exist two disjoint sets Y^+ and Y^- closed in Q, a real Radon measure μ on Q and an element $g_0 \in G \setminus \{0\}$, such that we have (3.34), (3.35), (3.37) and

$$\mu \text{ is non-decreasing on } Y^+, \text{ non-increasing on } Y^- \text{ and } Y^+ \cup Y^- \supset S(\mu). \tag{3.38}$$

Proof. Assume that we do not have 2°, i.e. there exist a Radon measure μ on Q and an element $g_0 \in G \setminus \{0\}$ such that we have (3.34)—(3.37); obviously, we may assume that $\|g_0\| \leqslant 1$. Define $f \in E^*$ by

$$f(x) = \int_Q x(q)\, d\mu(q) \qquad (x \in E). \tag{3.39}$$

Then, by (3.34) and (3.35) we have (3.9) and (3.10). On the other hand, by (3.36) we have

$$\left|\frac{d\mu}{d|\mu|}(q)\right| = 1 \qquad (q \in S(\mu)) \tag{3.40}$$

*) In the sequel by $\frac{d\mu}{d|\mu|}$ is denoted precisely this continuous function on Q.

(see § 1, formula (1.32)), whence

$$\operatorname{sign} \frac{d\mu}{d|\mu|}(q) = \frac{d\mu}{d|\mu|}(q) \qquad (q \in S(\mu)), \tag{3.41}$$

whence the function $\operatorname{sign} \left.\frac{d\mu}{d|\mu|}\right|_{S(\mu)}$ is continuous on $S(\mu)$. Since $S(\mu) \subset Q$ is closed, by virtue of the classical theorem of Tietze *) the functions $\operatorname{Re} \operatorname{sign} \left.\frac{d\mu}{d|\mu|}\right|_{S(\mu)}$ and $\operatorname{Im} \operatorname{sign} \left.\frac{d\mu}{d|\mu|}\right|_{S(\mu)}$, whence also the function $\operatorname{sign} \left.\frac{d\mu}{d|\mu|}\right|_{S(\mu)}$, can be then extended to the whole Q, so as to remain continuous. Consequently, there exists an $x_0 \in C(Q)$ such that

$$x_0(q) = \operatorname{sign} \frac{d\mu}{d|\mu|}(q) \qquad (q \in S(\mu)); \tag{3.42}$$

we may assume (dividing, if necessary, by $|x_0(q)|$ on the set $\{q \in Q \mid |x_0(q)| > 1\}$) that we have also

$$\max_{q \in Q} |x_0(q)| = 1. \tag{3.43}$$

Put

$$x(q) = x_0(q)(1 - |g_0(q)|) \qquad (q \in Q). \tag{3.44}$$

Then $x \in C(Q)$ and by (3.42), (3.37), (3.43), $\|g_0\| \leqslant 1$ and again (3.43), we have

$$x(q) = \operatorname{sign} \frac{d\mu}{d|\mu|}(q) \qquad (q \in S\mu)), \tag{3.45}$$

$$\max_{q \in Q} |x(q)| = 1, \tag{3.46}$$

$$|x(q)| + |g_0(q)| = |x_0(q)|(1 - |g_0(q)|) + |g_0(q)| \leqslant 1 \qquad (q \in Q). \tag{3.47}$$

From (3.34), (3.40), (3.45), (3.35) and again (3.34) and (3.47) it follows that we have

$$1 = |\mu|(Q) = \int_Q \left|\frac{d\mu}{d|\mu|}(q)\right| d|\mu|(q) = \int_Q x(q) \frac{d\mu}{d|\mu|}(q) d|\mu|(q) =$$

$$= \int_Q x(q)\, d\mu(q) = \int_Q [x(q) - g_0(q)]\, d\mu(q) \leqslant \|x - g_0\| \, |\mu|(Q) =$$

$$= \|x - g_0\| \leqslant \max_{q \in Q} (|x(q)| + |g_0(q)|) \leqslant 1,$$

*) See e.g. N. Dunford and J. Schwartz [49], p. 15, theorem 3.

whence, taking into account (3.46), we obtain

$$f(x) = \int_Q x(q)\, d\mu(q) = \|x - g_0\| = 1 = \|x\|, \tag{3.48}$$

i.e. (3.12). Consequently, by virtue of the implication 1°⇒3° of theorem 3.2, G is not a semi-Čebyšev subspace. Thus, 1°⇒2°.

Conversely, assume that G is not a semi-Čebyšev subspace, i.e. there exists an $x \in E \setminus G$ for which $\mathfrak{P}_G(x)$ contains two distinct elements g_1, g_2. Then there exists, by virtue of theorem 1.4 of § 1, a Radon measure μ on Q satisfying (1.25)—(1.27) and (1.33) for $M = \{g_1, g_2\}$. By (1.25)—(1.27) we have (3.34)—(3.36). On the other hand, by $g_1, g_2 \in \mathfrak{P}_G(x)$ we have $\max_{t \in Q} |x(t) - g_1(t)| = \max_{t \in Q} |x(t) - g_2(t)|$, whence, by (1.33),

$$g_1(q) - g_2(q) = 0 \qquad (q \in S(\mu)), \tag{3.49}$$

whence for $g_0 = g_1 - g_2 \in G \setminus \{0\}$ we have (3.37). Thus *), 2°⇒1°.

Finally, in the case when the scalars are real, the equivalence 1°⇔3° can be deduced from the previously proved equivalence 1°⇔2°, with an argument similar to that used in § 1, the first proof of theorem 1.5. It can be proved also directly ([229], p. 510—511) by means of a method similar to that used in the above proof of the equivalence 1°⇔2°, in which there occur the following differences: in the part 1°⇒3°, instead of the $x_0 \in C(Q)$ of the proof of the implication 1°⇒2° it is sufficient to take an $x_0 \in C_R(Q)$ satisfying

$$x_0(q) = \begin{cases} 1 & \text{for} \quad q \in S(\mu) \cap Y^+ \\ -1 & \text{for} \quad q \in S(\mu) \cap Y^- \end{cases} \tag{3.50}$$

and (3.43) (such an x_0 exists by virtue of the lemma of Uryson) while in the part 3°⇒1°, instead of theorem 1.4 of § 1 it is applied theorem 1.6 of § 1. This completes the proof of theorem 3.3

The equivalence 1°⇔3° of theorem 3.3 has been given in [226] (see [226], theorem), and the equivalence 1°⇔2°, with a different proof, has been given, essentially, by Z. S. Romanova ([198], theorem 2).

COROLLARY 3.7. *Let $E = C(Q)$ (Q compact) and let G be a linear subspace of E. If G is proximinal and if we have $S(\mu) = Q$*

---

*) We remark that in this part of the proof we have not applied theorem 3.2 but rather theorem 1.4 of § 1, since a proof based on theorem 3.2 would have required a repetition of the argument used in the proof of theorem 1.4 of § 1.

for every Radon measure μ *on* Q *satisfying (3.34)—(3.36) (or (3.34), (3.35), (3.37) and (3.38) in the case when the scalars are real), then* G *is a Čebyšev subspace.*

Proof. If the conditions of the corollary are satisfied, then obviously there exists no pair μ, g_0 (with $g_0 \in G \setminus \{0\}$) satisfying (3.34)—(3.37) (respectively (3.34), (3.35), (3.37) and (3.38) in the case when the scalars are real), whence G is a semi-Čebyšev subspace by virtue of theorem 3.3. Since G is by hypothesis proximinal, it follows that G is a Čebyšev subspace, which completes the proof.

In the particular case when the scalars are real, corollary 3.7 has been given, essentially, by R. R. Phelps ([177], corollary 4).

Obviously, in theorem 3.3 and corollary 3.7 the condition (3.34) may be replaced by $|\mu|(Q) \neq 0$ (considering the measure $\frac{1}{|\mu|(Q)}\mu$). A similar remark is valid also for the other results of the present and of the next paragraph, concerning concrete spaces.

3.3. APPLICATIONS IN THE SPACES $L^1(T, \nu)$ AND $L^1_R(T,\nu)$

We shall mention first that there are known some simple conditions in order that an element $g_0 \in G$ be *the only* element of best approximation of an $x \in L^1(T, \nu) \setminus \overline{G}$, where (T, ν) is a positive measure space. Namely, V. N. Nikolsky ([169], p. 106) has observed *) that *we have* $\mathfrak{P}_G(x) = \{g_0\}$ *if and only if we have* $g_0 \in \mathfrak{P}_G(x)$ *and*

$$\operatorname{Re} \int_{T \setminus Z(x-g)} [g(t) - g_0(t)] \operatorname{sign} [x(t) - g(t)] \, d\nu(t) < -\int_{Z(x-g)} |g(t) - g_0(t)| \, d\nu(t) \qquad (g \in G \setminus \{g_0\}). \tag{3.51}$$

This follows from the equivalence 1°⇔5° of theorem 3.1; besides, the sufficiency part follows immediately also from the implication 1°⇔5° of theorem 1.19, since by virtue of this latter, from condition (3.51) it follows that $g \notin \mathfrak{P}_G(x)$ $(g \in G \setminus \{g_0\})$.

*) Actually, V. N. Nikolsky has considered in [169] only the particular case when $T = [0, 1]$ and $\nu =$ the Lebesgue measure.

Furthermore, B. R. Kripke and T. J. Rivlin have remarked ([129], corollary 1.4) that if in condition (1.64) of theorem 1.7 we have the sign $<$ for every $g \in G \setminus \{0\}$, i.e. *if*

$$\int_{T \setminus Z(x-g_0)} g(t) \operatorname{sign}[x(t) - g_0(t)]\, d\nu(t) < \tag{3.52}$$
$$< \int_{Z(x-g_0)} |g(t)|\, d\nu(t) \qquad (g \in G \setminus \{0\}),$$

then $\mathfrak{P}_G(x) = \{g_0\}$. As has been observed in [232], p. 354, this implication can be proved directly, in the same way as the implication $3° \Rightarrow 1°$ of theorem 1.7.

We shall now consider the characterization of semi-Čebyšev subspaces of the spaces $L^1(T, \nu)$ and $L^1_R(T, \nu)$.

THEOREM 3.4. *Let $E = L^1(T, \nu)$, where (T, ν) is a positive measure space with the property that the dual $L^1(T, \nu)^*$ is canonically equivalent to $L^\infty(T, \nu)$, and let G be a linear subspace of E. The following statements are equivalent:*

1° G is a semi-Čebyšev subspace.

2° There do not exist a ν-measurable set $U \subset T$ with $\nu(U) > 0$, a $\beta \in L^\infty(T, \nu)$ and an element $g_0 \in G \setminus \{0\}$, such that

$$\operatorname*{ess\,sup}_{t \in T} |\beta(t)| = 1, \tag{3.53}$$

$$\int_T g(t)\, \beta(t)\, d\nu(t) = 0 \qquad (g \in G), \tag{3.54}$$

$$g_0(t) \neq 0 \qquad \nu\text{-a. e. on } U, \tag{3.55}$$

$$\beta(t) = \pm \operatorname{sign} g_0(t) \qquad \nu\text{-a. e. on } U, \tag{3.56}$$

$$g_0(t) = 0 \qquad \nu\text{-a.e. on } T \setminus U. \tag{3.57}$$

These statements are implied by — and in the particular case when the scalars are real (i.e. when $E = L^1_R(T, \nu)$), they are equivalent to — the following statement:

3° There do not exist a ν-measurable set $U \subset T$ with $\nu(U) > 0$, a $\beta \in L^\infty_R(T, \nu)$ and an element $g_0 \in G \setminus \{0\}$, such that we have (3.53), (3.54), (3.57) and

$$|\beta(t)| = 1 \qquad \nu\text{-a. e. on } U. \tag{3.58}$$

Proof. Assume that we do not have 2°, i.e. there exist U, β and g_0 such that we have (3.53)—(3.57). Define $f \in E^*$ by

$$f(y) = \int_T y(t)\, \beta(t)\, d\nu(t) \qquad (y \in E). \tag{3.59}$$

Then by (3.53) and (3.54) we have (3.9) and (3.10). Put

$$x(t)=\begin{cases}|g_0(t)|\operatorname{sign}\beta(t) & \text{for} \quad t\in U\\ 0 & \text{for} \quad t\in T\setminus U.\end{cases} \tag{3.60}$$

Then we have $x\in L^1(T, \nu)$, and since by (3.55) and (3.56) we have (3.58), it follows, taking into account (3.56) and (3.57), that we have

$$\beta(t)x(t)=\begin{cases}|g_0(t)|\,|\beta(t)| = |g_0(t)|>0 & \nu\text{-a.e.on } U\\ 0 & \nu\text{-a.e. on } T\setminus U,\end{cases} \tag{3.61}$$

$$\beta(t)[x(t)-g_0(t)]=$$

$$=\begin{cases}|g_0(t)|\,|\beta(t)|-g_0(t)\beta(t)=|g_0(t)|\mp|g_0(t)|\geqslant 0 \ \nu\text{-a. e.on } U\\ 0 \ \ \nu\text{-a. e. on } T\setminus U.\end{cases} \tag{3.62}$$

Consequently, taking again into account (3.58), we have

$$f(x)=\int_T x(t)\beta(t)\,d\nu(t)=\int_T|x(t)|\,d\nu(t) = \|x\|,$$

$$f(x-g_0)=\int_T [x(t)-g_0(t)]\beta(t)\,d\nu(t)=$$

$$=\int_T|x(t)-g_0(t)|\,d\nu(t)=\|x-g_0\|,$$

i.e. (3.11) for $x_1=x$, $x_2=x-g_0$, and hence, by virtue of the implication $1^\circ\Rightarrow 2^\circ$ of theorem 3.2, G is not a semi-Čebyšev subspace. Thus, $1^\circ\Rightarrow 2^\circ$.

Conversely, assume that we do not have 1°, i.e. there exists an $x\in E\setminus G$ for which $\mathfrak{P}_G(x)$ contains two distinct elements g_0', g_1'. Then there exist, by virtue of theorem 1.8 b) of § 1, a ν-measurable set $U\subset T$ with $\nu(U)>0$, and a $\beta\in L^\infty(T, \nu)$ such that we have (3.53)—(3.57) for $g_0=g_1'-g_0'\in G\setminus\{0\}$. Thus, $2^\circ\Rightarrow 1^\circ$.

The implication $3^\circ\Rightarrow 2^\circ$ is obvious.

Finally, assume that the scalars are real and that we do not have 3°, i.e. there exist U, β and g_0 such that we have (3.53), (3.54), (3.57) and (3.58). Then, as in the above proof of the implication $1^\circ\Rightarrow 2^\circ$, it follows that G is not a semi-Čebyšev subspace; in fact, the only difference is that in the case of real scalars, in order to obtain (3.62) it is no longer necessary to assume (3.55) and (3.56), since in this case we have obviously $|g_0(t)|\,|\beta(t)|-g_0(t)\beta(t)\geqslant 0$ ν-a. e. on U. Thus, $1^\circ\Rightarrow 3^\circ$ in the case of real scalars, which completes the proof of theorem 3.4.

3.4. APPLICATIONS IN THE SPACES $C^1(Q,\nu)$ AND $C^1_R(Q,\nu)$

THEOREM 3.5. *Let $E = C^1(Q, \nu)$, where Q is a compact space and ν a positive Radon measure on Q with the carrier $S(\nu) = Q$, and let G be a linear subspace of E. The following statements are equivalent:*

1° G is a semi-Čebyšev subspace.

2° There do not exist a non-void open set $U \subset Q$, a $\beta \in L^\infty(Q, \nu)$ with $\beta|_U$ continuous and an element $g_0 \in G \setminus \{0\}$, such that

$$\operatorname*{ess\,sup}_{q\in Q} |\beta(q)| = 1, \tag{3.63}$$

$$\int_Q g(q)\beta(q)\mathrm{d}\nu(q) = 0 \qquad (g\in G), \tag{3.64}$$

$$g_0(q) \neq 0 \qquad (q\in U), \tag{3.65}$$

$$\beta(q) = \pm \operatorname{sign} g_0(q) \qquad (q\in U), \tag{3.66}$$

$$g_0(q) = 0 \qquad (q\in Q\setminus U). \tag{3.67}$$

In the particular case when the scalars are real (i.e. when $E = C^1_R(Q, \nu)$), these statements are equivalent to the following:

3° There do not exist two disjoint sets U_1 and U_2 open in Q, with $U_1 \cup U_2 \neq \emptyset$, a ν-measurable real function α defined on $Q\setminus(U_1\cup U_2)$, and an element $g_0\in G\setminus\{0\}$, such that

$$|\alpha(q)| \leqslant 1 \qquad \nu\text{-a. e. on } Q\setminus(U_1\cup U_2), \tag{3.68}$$

$$\int_{Q\setminus(U_1\cup U_2)} g(q)\alpha(q)\mathrm{d}\nu(q) + \int_{U_1} g(q)\mathrm{d}\nu(q) - \int_{U_2} g(q)\mathrm{d}\nu(q) = 0 \quad (g\in G), \tag{3.69}$$

$$g_0(q) = 0 \qquad (q\in Q\setminus(U_1\cup U_2)). \tag{3.70}$$

Proof. Assume that we do not have 2°, i.e. there exist U, β and g_0 such that we have (3.63)—(3.67). Put

$$x(q) = \begin{cases} |g_0(q)| \operatorname{sign} \beta(q) & \text{for } q\in U \\ 0 & \text{for } q\in Q\setminus U. \end{cases} \tag{3.71}$$

Then by $g_0\in C^1(Q, \nu)$, (3.65), (3.66), the continuity of $\beta|_U$ and (3.67) we have $x\in C^1(Q, \nu)$. Consequently, as in the above proof of the implication 1°$\Rightarrow$2° of theorem 3.4, it follows that G is not a semi-Čebyšev subspace. Thus, 1°$\Rightarrow$2°.

The proof of the implication 2°$\Rightarrow$1° is similar to that of the implication 2°$\Rightarrow$1° of theorem 3.4, taking also into account the remark preceding theorem 1.10.

Assume now that the scalars are real and that we do not have 2°, i.e. there exist U, β and g_0 such that we have (3.63)—(3.67). Put

$$U_1 = U \cap \{q \in Q \mid \beta(q) = 1\}, \tag{3.72}$$

$$U_2 = U \cap \{q \in Q \mid \beta(q) = -1\}, \tag{3.73}$$

$$\alpha = \beta|_{Q \setminus (U_1 \cup U_2)} \tag{3.74}$$

Then $U_1 \cup U_2 = U \neq \emptyset$ and U_1, U_2 are disjoint. Since $\beta|_U$ is continuous, it follows that U_1 and U_2 are open in U, whence also in Q (since U is open in Q). At the same time, by (3.63), (3.64) and (3.67) we have (3.68)—(3.70). Thus, 3°⇒2° in the case of real scalars *).

Finally, assume that the scalars are real and that we do not have 3°, i.e. there exist U_1, U_2, α and g_0 such that we have (3.68)—(3.70). Put

$$\beta(q) = \begin{cases} 1 & \text{for} \quad q \in U_1 \\ -1 & \text{for} \quad q \in U_2 \\ \alpha(q) & \text{for} \quad q \in Q \setminus (U_1 \cup U_2) \end{cases} \tag{3.75}$$

and define $x(q)$ by (3.71). Then, as in the above, and taking into account the remark made in the proof of the implication 1°⇒3° of theorem 3.4, it follows that G is not a semi-Čebyšev subspace. Thus, 1°⇒3° in the case of real scalars, which completes the proof of theorem 3.5.

Under different conditions on Q and ν, namely when ν is a positive measure on a completely additive class of subsets of a separable metric space S, containing all the Borel sets, and $Q \subset S$ is a ν-measurable set "reduced" with respect to ν (that is, such that every nonvoid "portion" of Q is of ν-measure $\neq 0$), the equivalence 1°⇔2° of theorem 3.5 has been given, essentially, by S. Ya. Havinson ([87], theorem 3). In the particular case when $Q = [a, b]$ and ν is the Lebesgue measure, the equivalence 1°⇔3° has been given in [227], theorem 1, and in the particular case when $Q = [a, b]$, $\nu =$ the Lebesgue measure and $G =$ a finite dimensional subspace, it has been given by V. Pták [181]; actually, as remarked in [227], the proof given by V. Pták in [181], [182] does not make use of the hypothesis $\dim G < \infty$, and therefore it is also valid for arbitrary subspaces G.

*) One can also prove directly the implication 3°⇒1° in the case of real scalars, by applying theorem 1.10.

§ 4. k-DIMENSIONAL $\mathfrak{L}_G(x)$ SETS

4. 1. PRELIMINARIES

For an arbitrary non-void convex set A in a linear space E, we shall denote by $l(A)$ the linear manifold spanned by A, i.e.

$$l(A) = \{\alpha y + (1 - \alpha)z \,|\, y, z \in A, \ \alpha = \text{scalar}\}; \tag{4.1}$$

for every fixed $y \in A$ the set $l(A) - y = \{x - y \mid x \in l(A)\}$ is then a linear subspace of E, satisfying

$$l(A) - y = l(A - y). \tag{4.2}$$

The dimension (real or complex, as is the linear space E) of an arbitrary convex set $A \subset E$ is defined by

$$\dim A = \begin{cases} \dim l(A) & \text{if } A \neq \emptyset, \\ -1 & \text{if } A = \emptyset. \end{cases} \tag{4.3}$$

For every $y \in A$ we have then, taking into account (4.2),

$$\dim A = \dim l(A) = \dim [l(A) - y] = \dim l(A - y) = \\ = \dim (A - y). \tag{4.4}$$

We recall that $m + 1$ elements $x_0, x_1, \ldots, x_m$ of E are said to be *baricentrically independent* *), if the elements

$$x_1 - x_0, \ldots, x_m - x_0 \tag{4.5}$$

are linearly independent. This property is not dependent on the way in which the elements $x_0, x_1, \ldots, x_m$ are numbered; indeed, it is easy to check **) that the elements $x_0, x_1, \ldots, x_m$ are baricentrically independent if and only if from the relations

$$\sum_{i=0}^{m} \lambda_i x_i = 0, \sum_{i=0}^{m} \lambda_i = 0 \tag{4.6}$$

it follows that $\lambda_0 = \lambda_1 = \ldots = \lambda_m = 0$ (actually, this is the reason why such elements are called "baricentrically" independent).

LEMMA 4.1. *Let E be a normed linear space, A a non-void convex set in E and k an integer with $0 \leqslant k < \infty$. We have $\dim A = k$ if and only if A contains $k + 1$ baricentrically independent elements but does not contain $k + 2$ such elements.*

---

*) This term is used by E. Ya Remez [187]. Some authors use other terms, e.g. L. S. Pontryagin [179] calls such elements $x_0, x_1, \ldots, x_m$ "independent".
**) See e.g. L. S. Pontryagin [179], Chap. I, § 1, proposition A.

Proof. Assume that we have $\dim A = k$. Then for an arbitrary $x_0 \in A$ we have $A - x_0 \ni 0$ and $\dim (A - x_0) = \dim A = k$, whence $A - x_0$ contains k linearly independent elements $x_1 - x_0, \ldots, x_k - x_0$ but does not contain $k + 1$ such elements. Then the elements $x_0, x_1, \ldots, x_k \in A$ are baricentrically independent, but A does not contain $k + 2$ baricentrically independent elements.

Conversely, assume that A contains $k + 1$ baricentrically independent elements $x_0, x_1, \ldots, x_k$ but does not contain $k + 2$ such elements. Then the elements $x_1 - x_0, \ldots, x_k - x_0 \in A - x_0$ are linearly independent, but $A - x_0$ does not contain $k + 1$ such elements. Consequently, $\dim A = \dim (A - x_0) = k$, which completes the proof.

We shall say that a linear subspace G of a normed linear space E is a *k-semi-Čebyšev subspace*, respectively a *k-Čebyšev subspace* (where k is an integer with $0 \leqslant k < \infty$), if for every $x \in E$ we have*) $-1 \leqslant \dim \mathfrak{P}_G(x) \leqslant k$, respectively $0 \leqslant \dim \mathfrak{P}_G(x) \leqslant k$. The 0-semi-Čebyšev and 0-Čebyšev subspaces are nothing else than the semi-Čebyšev and, respectively, the Čebyšev subspaces. We shall now show that the results of § 3 can be extended to k-semi-Čebyšev and k-Čebyšev subspaces $(0 \leqslant k < \infty)$.

4.2. k-SEMI-ČEBYŠEV SUBSPACES IN GENERAL NORMED LINEAR SPACES

The following theorem gives a characterization of k-semi-Čebyšev subspaces:

THEOREM 4.1 ([225], theorem 3). *Let E be a normed linear space, G a linear subspace of E and k an integer with $0 \leqslant k < \infty$. In order that G be a k-semi-Čebyšev subspace it is necessary and sufficient that there do not exist $f \in E^*$, $x \in E$ and $k + 1$ linearly independent elements $g_0, g_1, \ldots, g_k \in G$ such that*

$$\|f\| = 1, \tag{4.7}$$

$$f(g) = 0 \qquad (g \in G), \tag{4.8}$$

$$f(x) = \|x\| = \|x - g_0\| = \|x - g_1\| = \ldots = \|x - g_k\|. \tag{4.9}$$

Proof. Assume that the condition of the theorem is not satisfied, that is, there exist f, x and $g_i (i = 0, 1, \ldots, k)$ such that we have (4.7)—(4.9). Then $x \in E \setminus G$ (otherwise from (4.8) and (4.9) one would obtain $0 = f(x) = \|x\|$, whence also $0 = f(x - g_0) = \|x - g_0\|$, whence $g_0 = x - (x - g_0) = 0$, which contradicts the hypothesis). Consequently, by theorem 1.1 we

---

*) The sets $\mathfrak{P}_G(x)$ are convex, since for $g_0, g_1 \in \mathfrak{P}_G(x)$ and $0 \leqslant \lambda \leqslant 1$ we have $\min_{g \in G} \|x - g\| \leqslant \|x - \lambda g_0 - (1 - \lambda) g_1\| \leqslant \lambda \|x - g_0\| + (1 - \lambda) \|x - g_1\| = \min_{g \in G} \|x - g\|$.

have $g_0, g_1, \ldots, g_k \in \mathfrak{P}_G(x)$. Since $g_0, g_1, \ldots, g_k$ are linearly independent, the elements $0, g_0, g_1, \ldots, g_k$ are baricentrically independent, hence by lemma 4.1 we have $\dim \mathfrak{P}_G(x) > k$, and thus G is not a k-semi-Čebyšev subspace.

Conversely, assume that G is not a k-semi-Čebyšev subspace, that is, there exists a $y \in E \setminus \overline{G}$ with $\dim \mathfrak{P}_G(y) > k$. Then, by lemma 4.1, $\mathfrak{P}_G(y)$ contains $k+2$ baricentrically independent elements $g'_0, g'_1, \ldots, g'_{k+1}$. Consequently, for

$$x = y - g'_0,\ g_0 = g'_1 - g'_0, \ldots, g_k = g'_{k+1} - g'_0 \tag{4.10}$$

we have

$$x \in E \setminus \overline{G},\ \mathfrak{P}_G(x) = \mathfrak{P}_G(y - g'_0) = \mathfrak{P}_G(y) - g'_0 \supset \{0, g_0, g_1, \ldots, g_k\}, \tag{4.11}$$

whence, by virtue of corollary 1.3, it follows that there exists an $f \in E^*$ such that we have (4.7)—(4.9). Since $g_0, g_1, \ldots, g_k \in G$ are linearly independent, this completes the proof of theorem 4.1.

In the particular case when $k = 0$, theorem 4.1 reduces to the equivalence $1^\circ \Leftrightarrow 3^\circ$ of theorem 3.2.

The number $r(G)$ defined by

$$r(G) = \sup_{x \in E} \dim \mathfrak{P}_G(x) \tag{4.12}$$

is called the *Čebyšev rank* of the subspace G. Theorem 4.1 above gives then a necessary and sufficient condition in order that $r(G) \leqslant k$.

It arises naturally the problem of characterizing the normed linear spaces E with the property that *all linear subspaces of E are k-semi-Čebyšev subspaces*. For this purpose it is convenient to introduce the notion of a k-strictly convex space. A normed linear space E is said to be *k-strictly convex*, where $1 \leqslant k < \infty$, if for any $k+1$ elements $x_0, x_1, \ldots, x_k \in E$ the equality

$$\left\| \sum_{i=0}^{k} x_i \right\| = \sum_{i=0}^{k} \|x_i\| \tag{4.13}$$

implies the linear dependence of these elements.

Obviously, the 1-strict convexity is nothing else but the usual strict convexity (since for $c \neq 0$ the relations (3.25) and (3.26) imply $|1 + c| = 1 + |c|$, whence $c > 0$). It is also clear that from the k-strict convexity of E it follows the m-strict convexity of E for all $m \geqslant k$; on the other hand, for every integer $k > 1$ there exist k-strictly convex normed linear spaces E which are not $(k-1)$-strictly convex.

In the following lemma there are given some characterizations of k-strictly convex spaces.

LEMMA 4.2 ([225], theorem 1). *Let E be a normed linear space and let k be an integer with $1 \leqslant k < \infty$. The following statements are equivalent:*

1° *E is k-strictly convex.*

2° *For any $k+1$ elements $x_0, x_1, \ldots, x_k \in E$ of norm $\|x_i\| = 1 (i = 0, 1, \ldots, k)$ the equality (4.13) implies the linear dependence of these elements.*

3° *For any $k+1$ linearly independent elements $x_0, x_1, \ldots, x_k \in E$ of norm $\|x_i\| = 1$ $(i = 0, 1, \ldots, k)$ we have*

$$\mathrm{Int}_l(\mathrm{co}\{x_0, x_1, \ldots, x_k\}) \subset \mathrm{Int}_l S_E, \tag{4.14}$$

where co A *denotes the convex hull of the set A, and* $\mathrm{Int}_l\, B$ *denotes the interior of the convex set $B \subset E$ with respect to the linear manifold $l(B)$ spanned*) by B.*

4° *The set*

$$\mathrm{Fr}\, S_E = \{x \in E \mid \|x\| = 1\} \tag{4.15}$$

contains no convex subset of dimension $> k-1$.

5° *For any $x_0 \in E$ and $r > 0$ the set*

$$\mathrm{Fr}\, S(x_0, r) = \{x \in E \mid \|x - x_0\| = r\} \tag{4.16}$$

contains no convex subset of dimension $> k-1$.

6° *For every functional $f \in E^*$ with $\|f\| = 1$ the convex set*

$$\mathfrak{M}_1 = \{x \in E \mid f(x) = 1, \|x\| = 1\} \subset \mathrm{Fr}\, S_E \tag{4.17}$$

is at most $(k-1)$-dimensional.

7° *For every $f \in E^*$ the convex cone**) with vertex in 0*

$$\mathcal{C}_f = \{x \in E \mid f(x) = \|f\|\, \|x\|\} \tag{4.18}$$

is at most k-dimensional.

Proof. The implication 1°⇒2° is obvious.

Assume now that we do not have 3°, i.e. there exist $k+1$ linearly independent elements $x_0, x_1, \ldots, x_k \in E$ of norm $\|x_i\| = 1$ $(i = 0, 1, \ldots, k)$ and $k+1$ numbers $\lambda_0, \lambda_1, \ldots, \lambda_k > 0$ with

---

*) Thus, since $x_0, x_1, \ldots, x_k$ are baricentrically independent,

$$\mathrm{Int}_l\, (\mathrm{co}\, \{x_0, x_1, \ldots, x_k\}) = \left\{ \sum_{i=0}^{k} \lambda_i x_i \,\middle|\, \lambda_0, \lambda_1, \ldots, \lambda_k > 0, \sum_{i=0}^{k} \lambda_i = 1 \right\}.$$

**) We recall that a set A in a linear space E is called *a cone with vertex in* $x_0 (\in E)$, if

$$x \in A \Rightarrow x_0 + \lambda(x - x_0) = \lambda x + (1 - \lambda) x_0 \in A \qquad (\lambda > 0).$$

$\sum_{i=0}^{k} \lambda_i = 1$ such that $\left\| \sum_{i=0}^{k} \lambda_i x_i \right\| = 1$; we may assume, without loss of generality, that $\frac{1}{\lambda_0} = \max_{0 \leqslant i \leqslant k} \frac{1}{\lambda_i}$. Then we have

$$\sum_{i=0}^{k} \|x_i\| + \sum_{i=0}^{k} \left(\frac{1}{\lambda_0} - \frac{1}{\lambda_i}\right) \|\lambda_i x_i\| = \frac{1}{\lambda_0} \sum_{i=0}^{k} \|\lambda_i x_i\| = \frac{1}{\lambda_0} =$$

$$= \frac{1}{\lambda_0} \left\| \sum_{i=0}^{k} \lambda_i x_i \right\| \leqslant \left\| \sum_{i=0}^{k} x_i \right\| + \left\| \sum_{i=0}^{k} \left(\frac{1}{\lambda_0} - \frac{1}{\lambda_i}\right) \lambda_i x_i \right\| \leqslant \sum_{i=0}^{k} \|x_i\| +$$

$$+ \sum_{i=0}^{k} \left(\frac{1}{\lambda_0} - \frac{1}{\lambda_i}\right) \|\lambda_i x_i\|,$$

whence

$$\left\| \sum_{i=0}^{k} x_i \right\| = \sum_{i=0}^{k} \|x_i\|,$$

which, together with the linear independence of $x_0, x_1, \ldots, x_k \in \operatorname{Fr} S_E$, contradicts 2°. Thus, 2°$\Rightarrow$3°.

Now assume that we do not have 4°, that is, $\operatorname{Fr} S_E$ contains a convex subset A with $\dim A > k - 1$. Then, by lemma 4.1, the set A contains at least $k + 1$ baricentrically independent elements $x_0, x_1, \ldots, x_k$. Since A is convex and $A \subset \operatorname{Fr} S_E$, we have then $\frac{1}{k+1} \sum_{i=0}^{k} x_i \in A \subset \operatorname{Fr} S_E$, whence

$$\frac{1}{k+1} \sum_{i=0}^{k} x_i \in \operatorname{Int}_l (\operatorname{co} \{x_0, x_1, \ldots, x_k\}), \qquad \frac{1}{k+1} \sum_{i=0}^{k} x_i \notin \operatorname{Int} S_E,$$

which, together with the linear independence*) of $x_0, x_1, \ldots, x_k$, contradicts 3°. Thus, 3°$\Rightarrow$4°.

Assume now that we do not have 5°, that is, there exist $x \in E$ and $r > 0$ such that $\operatorname{Fr} S(x, r)$ contains a convex subset of dimension $> k - 1$. Then the set $\operatorname{Fr} S_E = \operatorname{Fr} S(0, 1)$ contains the subset $\frac{1}{r}(A - x) = \left\{ \frac{1}{r}(y - x) \middle| y \in A \right\}$, and since tran-

*) If there existed $\lambda_0, \lambda_1, \ldots, \lambda_k$ with $\max_{0 \leqslant i \leqslant k} |\lambda_i| \neq 0$, $\sum_{i=0}^{k} \lambda_i x_i = 0$, then, by taking an $f \in E^*$ such that $\|f\| = 1, f(y) = 1$ $(y \in A)$, one would obtain $0 = f\left(\sum_{i=0}^{k} \lambda_i x_i\right) = \sum_{i=0}^{k} \lambda_i$, which contradicts the baricentric independence of $x_0, x_1, \ldots, x_k$.

slations and homotheties of positive ratio do not alter convexity and dimension, this latter set is convex and of dimension $> k-1$, which contradicts 4°. Thus $4° \Rightarrow 5°$.

The implication $5° \Rightarrow 6°$ is obvious via $5° \Rightarrow 4° \Rightarrow 6°$.

The implication $6° \Rightarrow 7°$ follows immediately from the fact that for every $f \in E^*$ we have, by $\mathcal{C}_f = \left\{ \lambda x \,|\, x \in \mathfrak{M}_{\frac{f}{\|f\|}}, \lambda \geqslant 0 \right\} = $

$= \mathrm{co}\left(\mathfrak{M}_{\frac{f}{\|f\|}} \cup \{0\}\right)$ and $l\left(\mathfrak{M}_{\frac{f}{\|f\|}}\right) \not\ni 0$,

$$\dim \mathcal{C}_f = \dim \mathfrak{M}_{\frac{f}{\|f\|}} + 1. \tag{4.19}$$

Finally, assume that we do not have 1°, that is, there exist $k+1$ linearly independent elements $x_0, x_1, \ldots, x_k \in E$ satisfying (4.13). Take a functional $f \in E^*$ such that $f\left(\sum_{i=0}^{k} x_i\right) = \|f\| \left\| \sum_{i=0}^{k} x_i \right\|$. Then the relations

$$\sum_{i=0}^{k} \|f\| \, \|x_i\| = \|f\| \left\| \sum_{i=0}^{k} x_i \right\| = f\left(\sum_{i=0}^{k} x_i\right) \leqslant \sum_{i=0}^{k} |f(x_i)| \leqslant \sum_{i=0}^{k} \|f\| \, \|x_i\|,$$

$$|f(x_i)| \leqslant \|f\| \, \|x_i\| \qquad (i = 0, 1, \ldots, k)$$

imply

$$f(x_i) = \|f\| \, \|x_i\| \qquad (i = 0, 1, \ldots, k),$$

whence $\frac{x_i}{\|x_i\|} \in \mathfrak{M}_{\frac{f}{\|f\|}}$ $(i = 0, 1, \ldots, k)$. These relations, together with the linear (whence also baricentric) independence of $\frac{x_0}{\|x_0\|}, \frac{x_1}{\|x_1\|}, \ldots, \frac{x_k}{\|x_k\|}$, imply, by lemma 4.1, $\dim \mathfrak{M}_{\frac{f}{\|f\|}} > k-1$, whence, by (4.19), $\dim \mathcal{C}_f > k$, which contradicts 7°. Thus, $7° \Rightarrow 1°$, which completes the proof.

This being said, we now can prove the following corollary of theorem 4.1:

COROLLARY 4.1. *Let E be a normed linear space and let k be an integer with $0 \leqslant k < \infty$. The following statements are equivalent:*

1° *All linear subspaces of E are k-semi-Čebyšev subspaces.*

2° *All linear subspaces of E of a given fixed finite dimension n, where $k+1 \leqslant n \leqslant \dim E - 1$, are k-Čebyšev subspaces.*

3° *All closed linear subspaces of E of a given fixed finite co-dimension m, where $1 \leqslant m \leqslant \dim E - k - 1$, are k-semi-Čebyšev subspaces.*

4° *E is $(k+1)$-strictly convex.*

Proof. The implications $1^\circ \Rightarrow 2^\circ$ and $1^\circ \Rightarrow 3^\circ$ are obvious, taking into account corollary 2.2.

Assume now that E is not $(k+1)$-strictly convex. Then, by the implication $6^\circ \Rightarrow 1^\circ$ of lemma 4.2, there exists an $f \in E^*$ with $\|f\| = 1$ for which the set $\mathfrak{M}_f$ is of a dimension $> k$. Take an arbitrary convex subset $A \subset \mathfrak{M}_f$ with $\dim A = k+1$. Then, by lemma 4.1, A contains $k+2$ baricentrically independent element $x_0, x_1, \ldots, x_{k+1}$. Put

$$G_1 = l(A) - x_0 = \{x - x_0 \mid x \in l(A)\}, \tag{4.20}$$

$$G_2 = \{y \in E \mid f(y) = 0\}, \tag{4.21}$$

and let G_1', G_2' be closed linear subspaces of E with $G_1 \subset G_1' \subset G_2$, $G_1 \subset G_2' \subset G_2$, $\dim G_1' = n$, $\operatorname{codim} G_1' = m$. Then we have

$$f(g) = 0 \qquad (g \in G_1',\ g \in G_2'), \tag{4.22}$$

$$f(x_0) = \|x_0\| = \|x_0 + (x_i - x_0)\| \quad (i = 1, \ldots, k+1), \tag{4.23}$$

where $g_0 = x_1 - x_0, \ldots, g_k = x_{k+1} - x_0 \in G_1$ and they are linearly independent. Consequently, by virtue of the necessity part of theorem 4.1, G_1' and G_2' are not k-semi-Čebyšev subspaces, whence G_1' is not a k-Čebyšev subspace. Thus, $2^\circ \Rightarrow 4^\circ$ and $3^\circ \Rightarrow 4^\circ$.

Finally, assume that we have 4°. Then by the implication $1^\circ \Rightarrow 6^\circ$ of lemma 4.2, for every functional $f \in E^*$ with $\|f\| = 1$ the set $\mathfrak{M}_f$ is of dimension $\leqslant k$. Consequently, all linear subspaces G of E satisfy the condition of theorem 4.1, whence, by virtue of the sufficiency part of that theorem, they are k-semi-Čebyšev subspaces. Thus, $4^\circ \Rightarrow 1^\circ$, which concludes the proof.

The equivalence $1^\circ \Leftrightarrow 4^\circ$ of coro)lary 4.1 has been given in the paper [225] ([225], theorem 2), with a different proof.

In the particular case when $k = 0$, from corollary 4.1 above we find again corollary 3.3. We leave to the reader the similar extension of corollary 3.6 to the case when k is an arbitrary integer with $0 \leqslant k < \infty$.

4.3. APPLICATIONS IN THE SPACES $C(Q)$ AND $C_R(Q)$

THEOREM 4.2. *Let $E = C(Q)$ (Q compact), let G be a linear subspace of E and k an integer with $0 \leqslant k < \infty$. The following statements are equivalent:*

1° G is a k-semi-Čebyšev subspace.

2° *There do not exist a Radon measure* μ *on* Q *and* $k+1$ *linearly independent elements* $g_0, g_1, \ldots, g_k \in G$ *such that*

$$|\mu|(Q) = 1, \tag{4.24}$$

$$\int_Q g(q)\mathrm{d}\mu(q) = 0 \qquad (g \in G), \tag{4.25}$$

$$\frac{\mathrm{d}\mu}{\mathrm{d}|\mu|} \in C(Q), \tag{4.26}$$

$$g_i(q) = 0 \quad (q \in S(\mu);\ i = 0, 1, \ldots, k), \tag{4.27}$$

where (4.26), $\frac{\mathrm{d}\mu}{\mathrm{d}|\mu|}$ *and* $S(\mu)$ *are understood in the same sense as in theorem 3.3.*

In the particular case when the scalars are real (i.e. when $E = C_R(Q)$*), these statements are equivalent to the following:*

3° *There do not exist two disjoint sets* Y^+ *and* Y^- *closed in* Q*, a real Radon measure* μ *on* Q *and* $k+1$ *linearly independent elements* $g_0, g_1, \ldots, g_k \in G$ *such that we have (4.24), (4.25), (4.27) and*

$$\mu \text{ is non-decreasing on } Y^+, \text{ non-increasing on } Y^-, \text{ and } Y^+ \cup Y^- \supset S(\mu). \tag{4.28}$$

In the particular case when $k = 0$, theorem 4.2 reduces to theorem 3.3. In the general case the proof of theorem 4.2 is similar to that of theorem 3.3, with the following differences: in the part $1° \Rightarrow 2°$, instead of the function (3.44) one considers the function defined by

$$x(q) = x_0(q)\,(1 - \sum_{i=0}^{k} |g_i(q)|) \qquad (q \in Q), \tag{4.29}$$

and instead of (3.47) one makes use of the relations

$$|x(q)| + |g_j(q)| \leqslant |x(q)| + \sum_{i=0}^{k} |g_i(q)| = |x(q)|\,(1 - \sum_{i=0}^{k} |g_i(q)|) + \sum_{i=0}^{k} |g_i(q)| \leqslant 1 \qquad (q \in Q;\ j = 0, 1, \ldots, k), \tag{4.30}$$

and the necessity part of theorem 4.1 is applied instead of the implication $1° \Rightarrow 3°$ of theorem 3.2; in the part $2° \Rightarrow 1°$, $\mathfrak{P}_G(x)$ contains, by the hypothesis $\dim \mathfrak{P}_G(x) > k$ and by lemma 4.1, $k+2$ baricentrically independent elements $g_0', g_1', \ldots, g_{k+1}'$, and the relations (1.33) for $M = \{g_0', g_1', \ldots, g_{k+1}'\}$ and

$\max_{t\in Q}|x(t)-g'_0(t)| = \ldots = \max_{t\in Q}|x(t)-g'_{k+1}(t)|$ imply (4.27) with $g_0 = g'_1 - g'_0, \ldots, g_k = g'_{k+1} - g'_0$, which belong to G and are linearly independent; finally, in the part 1°⇔3° (in the case of real scalars) there are similar changes.

The equivalence 1°⇔3° of theorem 4.2 has been given in [225], theorem 6, and the equivalence 1°⇔2°, with a proof different from the above, has been given, essentially, by Z. S. Romanova ([198], theorem 2).

4.4. APPLICATIONS IN THE SPACES $L^1(T,\nu)$, $L^1_R(T,\nu)$, $C^1(Q,\nu)$ AND $C^1_R(Q,\nu)$

THEOREM 4.3. *Let $E = L^1(T,\nu)$, where (T,ν) is a positive measure space with the property that the dual $L^1(T,\nu)^*$ is canonically equivalent to $L^\infty(T,\nu)$, let G be a linear subspace of E and let k be an integer with $0 \leqslant k < \infty$. The following statements are equivalent:*

1° G is a k-semi-Čebyšev subspace.

2° There do not exist a ν-measurable set $U \subset T$ with $\nu(U) > 0$, a $\beta \in L^\infty(T,\nu)$ and $k+1$ linearly independent elements $g_0, g_1, \ldots, g_k \in G$ such that

$$\operatorname*{ess\,sup}_{t\in T} |\beta(t)| = 1, \tag{4.31}$$

$$\int_T g(t)\beta(t)\,d\nu(t) = 0 \qquad (g\in G), \tag{4.32}$$

$$\sum_{i=0}^{k} |g_i(t)| \neq 0 \qquad \nu\text{-a.e. on } U, \tag{4.33}$$

$$\beta(t) = \pm \operatorname{sign} g_i(t) \quad \nu\text{-a.e. on } T\setminus Z(g_i) \quad (i = 0,1,\ldots,k), \tag{4.34}$$

$$g_i(t) = 0 \quad (i = 0,1,\ldots,k) \quad \nu\text{-a.e. on } T\setminus U. \tag{4.35}$$

These statements are implied by — and in the particular case when the scalars are real (i.e. when $E = L^1_R(T,\nu)$), equivalent to — the following statement:

3° There do not exist a ν-measurable set $U \subset T$ with $\nu(U) > 0$, a $\beta \in L^\infty_R(T,\nu)$ and $k+1$ linearly independent elements $g_0, g_1, \ldots, g_k \in G$ such that we have (4.31), (4.32), (4.35) and

$$|\beta(t)| = 1 \qquad \nu\text{-a.e. on } U. \tag{4.36}$$

In the particular case when $k = 0$, theorem 4.3 reduces to theorem 3.4. In the general case, the proof of theorem 4.3 is similar to that of theorem 3.4, with the following differences: in the part 1°⇒2°, instead of the function (3.60) one considers the function defined by

$$x(t) = \begin{cases} \left(\sum_{j=0}^{k} |g_j(t)|\right) \operatorname{sign}\beta(t) & \text{for } t\in U, \\ 0 & \text{for } t\in T\setminus U, \end{cases} \tag{4.37}$$

whence instead of (3.61) and (3.62) we obtain then

$$\beta(t)x(t) = \begin{cases} \left(\sum_{j=0}^{k} |g_j(t)|\right)|\beta(t)| = \sum_{j=0}^{k} |g_j(t)| > 0 & \nu\text{-a.e. on } U \\ 0 & \nu\text{-a.e. on } T\setminus U, \end{cases}$$

$$\beta(t)[x(t) - g_i(t)] = \begin{cases} \left(\sum_{j=0}^{k} |g_j(t)|\right)|\beta(t)| - g_i(t)\beta(t) = \\ \quad = \sum_{j=0}^{k} |g_j(t)| \mp |g_i(t)| > 0 & \nu\text{-a.e. on } T\setminus Z(g_i) \\ \beta(t)x(t) > 0 & \nu\text{-a.e. on } Z(g_i), \end{cases}$$

and instead of the implication 1°⇒2° of theorem 3.2 one make use of the necessity part of theorem 4.1; in the part 2°⇒1°, $\mathfrak{L}_G(x)$ contains, by the hypothesis dim $\mathfrak{L}_G(x) > k$ and by lemma 4.1, $k+2$ baricentrically independent elements $g'_0, g'_1, \ldots, g'_{k+1}$ and by theorem 1.8 b) there exist then a ν-measurable set $U \subset T$ with $\nu(U) > 0$ and a $\beta \in L^\infty(T, \nu)$ such that we have (4.31) — (4.35) for $g_0 = g'_1 - g'_0, \ldots, g_k = g'_{k+1} - g'_0$ which belong to G and are linearly independent; in the part 1°⇒3° (in the case of real scalars), there are similar changes; finally, the implication 3°⇒2° is immediate, since the relations (4.33) and (4.35) imply

$$U = T \setminus \bigcap_{i=0}^{k} Z(g_i) = \bigcup_{i=0}^{k} T\setminus Z(g_i),$$

whence, taking into account (4.34), it follows (4.36).

Similarly, theorem 3.5 admits the following extension to the case when k is an arbitrary integer with $0 \leqslant k < \infty$:

THEOREM 4.4. *Let $E = C^1(Q, \nu)$, where Q is a compact space and ν a positive Radon measure on Q with the carrier $S(\nu) = Q$, let G be a linear subspace of E and let k be an integer with $0 \leqslant k < \infty$. The following statements are equivalent:*

1° *G is a semi-Čebyšev subspace.*

2° *There do not exist a non-void open set $U \subset Q$, a $\beta \in L^\infty(Q, \nu)$ with $\beta|_U$ continuous and $k+1$ linearly independent elements $g_0, g_1, \ldots, g_k \in G$ such that*

$$\operatorname*{ess\,sup}_{q \in Q} |\beta(q)| = 1, \tag{4.38}$$

$$\int_Q g(q)\,\beta(q)\,d\nu(q) = 0 \qquad (g \in G), \tag{4.39}$$

$$\sum_{i=0}^{k} g_i(q) \neq 0 \qquad (q \in U), \tag{4.40}$$

$$\beta(q) = \pm \operatorname{sign} g_i(q) \quad (q \in Q \setminus Z(g_i);\ i = 0,1,\ldots,k), \tag{4.41}$$

$$g_i(q) = 0 \qquad (q \in Q \setminus U; \quad i = 0, 1,\ldots,k). \tag{4.42}$$

In the particular case when the scalars are real (i.e. when $E = C_R^1(Q,\nu)$), these statements are equivalent to the following:

3° *There do not exist two disjoint sets U_1 and U_2, open in Q, with $U_1 \cup U_2 \neq \emptyset$, a ν-measurable function α defined on $Q \setminus (U_1 \cup U_2)$, and $k+1$ linearly independent elements $g_0, g_1, \ldots, g_k \in G$ such that*

$$|\alpha(q)| \leqslant 1 \qquad \nu\text{-a.e. on } Q \setminus (U_1 \cup U_2), \tag{4.43}$$

$$\int_{Q \setminus (U_1 \cup U_2)} g(q)\,\alpha(q)\,\mathrm{d}\nu(q) + \int_{U_1} g(q)\,\mathrm{d}\nu(q) - \int_{U_2} g(q)\,\mathrm{d}\nu(q) = 0 \qquad (g \in G), \tag{4.44}$$

$$g_0(q) = g_1(q) = \ldots = g_k(q) = 0 \quad (q \in Q \setminus (U_1 \cup U_2)). \tag{4.45}$$

Under different conditions on Q and ν, namely those mentioned in the final part of section 3.4 (in connection with the equivalence $1° \Leftrightarrow 2°$ of theorem 3.5), the equivalence $1° \Leftrightarrow 2°$ of theorem 4.4 has been given, essentially, by S. Ya. Havinson ([86], theorem 2).

§ 5. INTERPOLATIVE BEST APPROXIMATION, BEST APPROXIMATION BY ELEMENTS OF LINEAR MANIFOLDS AND THEIR EQUIVALENCE TO BEST APPROXIMATION BY ELEMENTS OF LINEAR SUBSPACES

Let E be a normed linear space, G a linear subspace o E, $\varphi_1, \ldots, \varphi_m \in G^*$, $c_1 \ldots, c_m$ m scalars*) and $x \in E$. We call [222] *interpolatory element of best approximation* of the element x any element $g_0 \in G$ with the properties

$$\varphi_i(g_0) = c_i \qquad (i = 1,\ldots,m), \tag{5.1}$$

$$\|x - g_0\| = \inf_{\substack{g \in G \\ \varphi_i(g) = c_i (i=1,\ldots,m)}} \|x - g\|. \tag{5.2}$$

*) We shall assume, without any special mention, that m is an integer with $1 \leqslant m < \infty$.

In the present paragraph we shall be concerned with *interpolative best approximation*, i.e. with problems related to interpolatory elements of best approximation. We shall show that the interpolative best approximation is equivalent to best approximation by elements of linear manifolds and this, although apparently more general, reduces in turn, by a simple translation, to best approximation by elements of linear subspaces. Using these observations, we shall deduce from the foregoing results on best approximation by elements of linear subspaces, the solutions of the corresponding problems for the interpolative best approximation (respectively for the best approximation by elements of linear manifolds).

Throughout the sequel we shall make the following hypothesis:

(I) *The functionals* $\varphi_1, \ldots, \varphi_m \in G^*$ *are linearly independent.*

This hypothesis does not restrict the generality. For, if $\sum_{i=1}^{m} a_i \varphi_i = 0$, $\sum_{i=1}^{m} |a_i| \neq 0$ and $\sum_{i=1}^{m} a_i c_i \neq 0$, then the equations (5.1) are incompatible, while if $\sum_{i=1}^{m} a_i c_i = 0$, then these equations are linearly dependent, whence by omitting a suitable finite number of them we obtain an equivalent linearly independent subsystem; but the latter is a system of the same type, with a smaller m and with linearly independent $\varphi_1, \ldots, \varphi_m$.

LEMMA 5.1 ([222], lemma 1.1). *Let E be a normed linear space, G a linear subspace of E, $\varphi_1, \ldots, \varphi_m \in G^*$ linearly independent and $c_1, \ldots, c_m$ m scalars. Then the set V defined by*

$$V = \{g \in G \mid \varphi_i(g) = c_i \ (i = 1, \ldots, m)\} \tag{5.3}$$

is a non-void linear manifold in E, closed in G.

Proof. The set V is a linear manifold in E, closed in G, since it is the intersection of m hyperplanes of G, closed in G:

$$V = \bigcap_{i=1}^{m} \{g \in G \mid \varphi_i(g) = c_i\}. \tag{5.4}$$

Let us show, finally, that $V \neq \emptyset$. We claim that there exist m elements $g_1, \ldots, g_m \in G$ such that $\det(\varphi_i(g_l))_{i,l=1,\ldots,m} \neq 0$. In fact, the existence of a $g_1 \in G$ with the property $\varphi_1(g_1) \neq 0$ is obvious, since otherwise one would have $\varphi_1 = 0$, which contradicts the hypothesis of the linear independence of $\varphi_1, \ldots, \varphi_m$; assume now that for $j - 1 < m$ there exist $g_1, \ldots, g_{j-1} \in G$ such that $\det(\varphi_i(g_l))_{i,l=1,\ldots,j-1} \neq 0$. Then, by virtue of the hypothesis of the linear independence of $\varphi_1, \ldots, \varphi_m$, we cannot have

$$\begin{vmatrix} \varphi_1(g_1) & \ldots & \varphi_j(g_1) \\ \cdot & \cdot & \cdot \\ \varphi_1(g_{j-1}) & \ldots & \varphi_j(g_{j-1}) \\ \varphi_1(g) & \ldots & \varphi_j(g) \end{vmatrix} = 0 \qquad (g \in G),$$

whence there exists an element $g_j \in G$ such that $\det(\varphi_i(g_l))_{i,l=1,\ldots,j} \neq 0$, which proves the statement above.

Consequently, the system of linear equations

$$\sum_{l=1}^{m} a_l \varphi_i(g_l) = c_i \qquad (i = 1, \ldots, m) \tag{5.5}$$

has a solution $a_1, \ldots, a_m$. But then the element $g = \sum_{l=1}^{m} a_l g_l \in G$ belongs to the linear manifold V, whence $V \neq \emptyset$, which completes the proof of lemma 5.1.

Now let W be an arbitrary linear manifold in E and let $x \in E$. Any element $w_0 \in W$ with the property

$$\|x - w_0\| = \inf_{w \in W} \|x - w\| \tag{5.6}$$

is called *an element of best approximation* of the element x (by means of the elements of the linear manifold W).

Using the foregoing, it is easy to see that *interpolative best approximation is equivalent to best approximation by elements of linear manifolds*. In fact, if we are given a normed linear space E, a linear subspace G of E, linearly independent $\varphi_1, \ldots, \varphi_m \in G^*$, m scalars $c_1, \ldots, c_m$, and $x \in E$, then in order that a $g_0 \in G$ be an interpolatory element of best approximation of x, obviously it is necessary and sufficient that it be an element of best approximation of x by means of the elements of the linear manifold V of lemma 5.1. Conversely, if we are given a normed linear E space, a linear manifold $W \subset E$ with $\overline{W} \neq E$ and an $x \in E$, then there exists a linear subspace G of E such that W is a closed hyperplane of the normed linear space G (if W is not a linear subspace of E, then G = the linear subspace of E generated by W, while if W is a linear subspace of E, one may take for G the linear subspace $W \oplus [y]$ of E, where y is an arbitrary element of $E \setminus \overline{W}$), whence there exist a functional $\varphi_1 \in G^* \setminus \{0\}$ and a scalar c_1 such that

$$W = \{g \in G \mid \varphi_1(g) = c_1\}; \tag{5.7}$$

but then in order that a $g_0 \in W$ be an element of best approximation of x by means of the elements of the linear manifold

W it is necessary and sufficient that it be an interpolatory element of best approximation of x with respect to G, φ_1 and c_1. Thus, interpolative best approximation is equivalent to best approximation by elements of linear manifolds.

Since every linear subspace is a linear manifold, the problem of best approximation by elements of linear manifolds is more general than the problem of best approximation by elements of linear subspaces. Nevertheless, *the best approximation by elements of linear manifolds reduces, by means of a simple translation, to best approximation by elements of linear subspaces.* In fact, if we are given a normed linear space E, a linear manifold $W \subset E$ and an $x \in E$, then for every $w \in W$ *the set* $W - w$ *is a linear subspace of* E, *and in order that an element* $w_0 \in W$ *be an element of best approximation of* x *by means of the elements of the linear manifold* W, *obviously it is necessary and sufficient that* $0 \in \mathfrak{L}_{W-w_0}(x - w_0)$.

The above remarks allow us to obtain immediately, from the results on best approximation by elements of linear subspaces, the solutions of the corresponding problems for best approximation by elements of linear manifolds (respectively for interpolative best approximation). In fact, from these remarks and from theorem 1.1 of § 1 it follows immediately.

THEOREM 5.1 ([222], theorem 3.1). *Let* E *be a normed linear space,* V *a linear manifold in* E *(or, what is equivalent, a linear manifold of the form (5.3), where* G, φ_i, c_i *are as in lemma 5.1),* $x \in E \setminus \overline{V}$ *and* $v_0 \in V$. *We have*)* $v_0 \in \mathfrak{L}_V(x)$ *if and only if there exists an* $f \in E^*$ *with the following properties:*

$$\|f\| = 1, \tag{5.8}$$

$$f(v - v_0) = 0 \qquad (v \in V), \tag{5.9}$$

$$f(x - v_0) = \|x - v_0\|. \tag{5.10}$$

Concerning the problem of existence of interpolatory elements of best approximation, it follows from the remarks above and from the results of § 2 that *in a normed linear space* E *every linear manifold* V *of finite dimension (respectively, every* n*-dimensional linear manifold* V *of the form (5.3), where* G, φ_i, c_i *are as in lemma 5.1 and* $m < n$*) is proximinal* ([222], theorem 2.2). Also, *in order that all closed linear manifolds in a Banach space* E *be proximinal, it is necessary and sufficient that* E *be reflexive.*

Concerning the problems of uniqueness, we mention e.g. the following result, which is obtained immediately from the remarks above and from theorem 3.2:

---

*) We denote by $\mathfrak{L}_V(x)$ the set of all elements of best approximation of x by means of the elements of the linear manifold V.

THEOREM 5.2 ([222], theorem 4.1). *Let E be a normed linear space and let V be a linear manifold in E (or, what is equivalent, a linear manifold of the form (5.3), where G, φ_i, c_i are as in lemma 5.1). In order that V be semi-Čebyšev it is necessary and sufficient that there do not exist $f \in E^*$, $x \in E$ and $v_0, v_1 \in V$, $v_0 \neq v_1$, such that we have (5.8), (5.9) and*

$$f(x - v_0) = \|x - v_0\| = \|x - v_1\|. \tag{5.11}$$

We leave to the reader the applications of the above results to various concrete spaces. For example, for applications of theorem 5.1 in the spaces $L^1(T,\nu)$ and in inner product spaces, see [222], theorem 3.2 and theorem 3.3 respectively.

Finally, we mention that a generalization of the interpolative best approximation, in which the scalars $c_1,\ldots,c_m$ depend on the elements $\bar{x} = x + \{\xi_i\}_{i\in I}$ of the direct sum $E \oplus E_{m_1}$ (where I is a fixed subset of the set of indices $\{1,\ldots,m\}$, containing m_1 elements, $0 \leqslant m_1 \leqslant m$), has been also studied in the paper [222], both in general normed linear spaces, and in the spaces $C(Q)$ (Q compact); in the particular case when $Q = [a,b]$, the scalars are real, G is finite dimensional and the functionals φ_i are evaluations at certain points, in [222] there have been found again the results of S. Paszkowski [173].

§ 6. THE OPERATORS π_G AND THE FUNCTIONALS e_G. DEVIATIONS. ELEMENTS OF ε-APPROXIMATION

Let E be a normed linear space and G a linear subspace of E. Then one defines, in a natural way, a mapping $\pi_G : D(\pi_G) \to G$ by the condition

$$\pi_G(x) \in \mathfrak{P}_G(x) \qquad (x \in D(\pi_G)), \tag{6.1}$$

and a functional e_G on E by

$$e_G(x) = \inf_{g\in G} \|x-g\| = \rho(x, G) \qquad (x\in E). \tag{6.2}$$

In general $D(\pi_G) \neq E$ and the mapping π_G is multivalued on $D(\pi_G)\setminus G$, but we have always $G \subset D(\pi_G)$ and the restriction of the mapping π_G to G is one-valued. We have $D(\pi_G) = E$ if and only if G is proximinal; on the other hand, π_G is one-valued on $D(\pi_G)$ if and only if G is a semi-Čebyšev subspace. Even in the case when these conditions are simultaneously satisfied (i.e. G is a Čebyšev subspace), the mapping π_G is, in general, non-linear on $E\setminus G$, but the restriction of π_G to G

is always linear. Also, the functional e_G is non-linear on $E \setminus G$, but $e_G|_G = 0$. In the present paragraph we shall give various elementary properties of the operators π_G and the functionals e_G.

6.1. THE OPERATORS π_G

THEOREM 6.1. *Let E be a normed linear space and G a linear subspace of E.*

a) *If $x \in D(\pi_G)$, then $\pi_G(x) \in D(\pi_G)$ and we have*

$$\pi_G^2(x) = \pi_G(x) \qquad (x \in D(\pi_G)), \tag{6.3}$$

i.e. the mapping π_G is idempotent.

b) *We have*

$$| \|x - \pi_G(x)\| - \|y - \pi_G(y)\| | \leqslant \|x - y\| \quad (x, y \in D(\pi_G)), \tag{6.4}$$

$$\|x - \pi_G(x)\| \leqslant \|x\| \qquad (x \in D(\pi_G)), \tag{6.5}$$

$$\|\pi_G(x)\| \leqslant 2\|x\| \qquad (x \in D(\pi_G)). \tag{6.6}$$

c) *π_G is continuous at the origin.*

d) *If G_1 is a linear subspace of G, we have*

$$\|x - \pi_G(x)\| \leqslant \|x - \pi_{G_1}(x)\| \qquad (x \in D(\pi_G) \cap D(\pi_{G_1})). \tag{6.7}$$

If, in addition, the mapping π_G is one-valued on $D(\pi_G)$ (i.e. G is a semi-Čebyšev subspace), then

e) *If $x \in D(\pi_G)$ and $g \in G$, then $x + g \in D(\pi_G)$ and we have*

$$\pi_G(x + g) = \pi_G(x) + \pi_G(g) = \pi_G(x) + g \quad (x \in D(\pi_G), \quad g \in G), \tag{6.8}$$

i.e. π_G is quasi-additive.

f) *If $x \in D(\pi_G)$ and α is an arbitrary scalar, then $\alpha x \in D(\pi_G)$ and we have*

$$\pi_G(\alpha x) = \alpha \pi_G(x) \quad (x \in D(\pi_G), \quad \alpha = \text{scalar}), \tag{6.9}$$

i.e. π_G is homogeneous.

Proof. a) For every $g \in G$ we have, uniquely, $\pi_G(g) = g$ whence, for an arbitrary $x \in D(\pi_G)$ and $g = \pi_G(x) \in G$ we obtain

$$\pi_G^2(x) = \pi_G[\pi_G(x)] = \pi_G(x).$$

b) We have

$$\|x - \pi_G(x)\| \leqslant \|x - \pi_G(y)\| \leqslant \|x - y\| + \|y - \pi_G(y)\| \quad (x, y \in D(\pi_G)),$$

whence

$$\|x - \pi_G(x)\| - \|y - \pi_G(y)\| \leqslant \|x - y\| \qquad (x, y \in D(\pi_G)).$$

Changing in these relations x by y and y by x, we obtain the inequalities

$$\|y - \pi_G(y)\| - \|x - \pi_G(x)\| \leqslant \|x - y\| \qquad (x, y \in D(\pi_G)),$$

which, together with the preceding inequalities, give (6.4).

The relation (6.5) is obvious from the definition (6.1) of π_G, taking into account that $0 \in G$; it also follows immediately from (6.4), by taking $y = 0$ and taking into account that $\pi_G(0) = 0$.

Finally, from (6.5) we obtain

$$\|\pi_G(x)\| \leqslant \|\pi_G(x) - x\| + \|x\| \leqslant 2\,\|x\| \quad (x \in D\,(\pi_G)),$$

i.e. (6.6).

c) is an immediate consequence of (6.6).

d) If G_1 is a linear subspace of G, we have

$$\|x - \pi_G(x)\| = \inf_{g\in G} \|x - y\| \leqslant \inf_{g\in G_1} \|x - g\| = \|x - \pi_{G_1}(x)\|$$

$$(x \in D(\pi_G) \cap D(\pi_{G_1})),$$

i.e. (6.7). In this connection we remark that if G_1 is a linear subspace of G, then, in general, there is no inclusion relation between $D(\pi_G)$ and $D(\pi_{G_1})$. In fact, e.g. for $G \neq E$ dense in E, $G_1 \subset G$ of finite dimension and $x \in E \setminus G$, we have $x \in E \setminus D(\pi_G) = D(\pi_{G_1}) \setminus D(\pi_G)$ and for $G_1 \subset G$, $G_1 \neq G$ dense in G and $g \in G \setminus G_1$ we have $g \in D(\pi_G) \setminus D(\pi_{G_1})$.

e) Assume now that π_G is one-valued on $D(\pi_G)$, i.e. G is a semi-Čebyšev subspace and let $x \in D(\pi_G)$, $g \in G$. Then for every $g' \in G$ we have

$$\|x + g - g'\| \geqslant \|x - \pi_G(x)\| = \|x + g - [\pi_G(x) + g]\|\,,$$

whence $\pi_G(x) + g \in \mathfrak{L}_G(x + g)$. Consequently $x + g \in D(\pi_G)$, and since π_G is one-valued on $D(\pi_G)$, we have

$$\pi_G(x + g) = \pi_G(x) + g,$$

i.e. (6.8).

f) Assuming again that π_G is one-valued on $D(\pi_G)$, let $x \in D(\pi_G)$ and α a scalar $\neq 0$. Then for every $g' \in G$ we have

$$\|\alpha x - g'\| = |\alpha| \left\| x - \frac{1}{\alpha} g' \right\| \geqslant |\alpha|\, \|x - \pi_G(x)\| = \|\alpha x - \alpha \pi_G(x)\|,$$

whence $\alpha\pi_G(x) \in \mathfrak{L}_G(\alpha x)$. Consequently $\alpha x \in D(\pi_G)$, and since π_G is one-valued on $D(\pi_G)$, we have

$$\pi_G(\alpha x) = \alpha \pi_G(x),$$

which proves (6.9) for $\alpha \neq 0$. On the other hand, for $\alpha = 0$ the relation (6.9) is obvious since by $0 \in G$ we have $\pi_G(0) = 0$. This completes the proof of theorem 6.1.

The properties of π_G given in theorem 6.1 are well known, they may be found e.g. in the papers of N. Aronszajn and K. T. Smith [6] and R. A. Hirschfeld [89]*). These papers also contain, implicitly, the following result on the additivity of the mappings π_G:

THEOREM 6.2. *Let E be a normed linear space and G a proximinal hyperplane in E passing through 0 (hence $D(\pi_G) = E$). Then π_G admits an additive selection on E, i.e. for every $x \in E$ one can choose a $\pi_G^0(x) \in \mathfrak{L}_G(x)$ in such a way that we have*

$$\pi_G^0(x + y) = \pi_G^0(x) + \pi_G^0(y) \qquad (x, y \in E). \tag{6.10}$$

Consequently, if the mapping π_G is one-valued on $D(\pi_G) = E$ (i.e. if the hyperplane G is a Čebyšev subspace), then it is additive.

Proof. Let us fix an arbitrary $x' \in E \setminus G$ and an arbitrary $\pi_G(x') \in \mathfrak{L}_G(x')$. Since G is a hyperplane in E, we have then

$$E = G \oplus [x' - \pi_G(x')]. \tag{6.11}$$

Let p_G be the linear projection of E onto G which vanishes on $[x' - \pi_G(x')]$. Then every $x \in E$ can be written, uniquely, in the form

$$x = p_G(x) + \alpha(x' - \pi_G(x')). \tag{6.12}$$

Hence, if $\alpha \neq 0$, then for every $g \in G$ we have

$$\|x - p_G(x)\| = \|\alpha(x' - \pi_G(x'))\| = |\alpha| \, \|x' - \pi_G(x')\| \leqslant$$

$$\leqslant |\alpha| \, \|x' - \pi_G(x') + \frac{1}{\alpha} p_G(x) - \frac{1}{\alpha} g\| =$$

$$= \|p_G(x) + \alpha(x' - \pi_G(x')) - g\| = \|x - g\|,$$

and thus $p_G(x) \in \mathfrak{L}_G(x)$. On the other hand, if $\alpha = 0$, then $x = p_G(x) \in G$, whence $p_G(x) \in \mathfrak{L}_G(x)$. Consequently, we have

$$p_G(x) \in \mathfrak{L}_G(x) \qquad (x \in E), \tag{6.13}$$

which, taking into account the linearity of p_G, completes the proof.

We remark that in the particular case when the mapping π_G is one-valued on $D(\pi_G) = E$, the additivity of π_G follows immediately from theorem 6.1. Indeed, since G is a hyperplane in E, every $x \in E$ and $y \in E$ can be written, uniquely, in the form

$$x = g_1 + \alpha_1 x', \qquad y = g_2 + \alpha_2 x'$$

*) The property e) has been given by M. Nicolescu ([161], proposition 3).

where $x' \in E$ is arbitrarily fixed, g_1, $g_2 \in G$, and α_1, α_2 are scalars. Consequently, by theorem 6.1 e) and f) we have

$$\pi_G(x+y) = \pi_G[(\alpha_1+\alpha_2)\,x'+g_1+g_2] = \pi_G[(\alpha_1+\alpha_2)\,x']+g_1+g_2 =$$
$$= (\alpha_1+\alpha_2)\pi_G(x')+g_1+g_2 = \pi_G(\alpha_1 x')+g_1+\pi_G(\alpha_2 x')+g_2 =$$
$$= \pi_G(\alpha_1 x'+g_1)+\pi_G(\alpha_2 x'+g_2) = \pi_G(x)+\pi_G(y),$$

which completes the proof.

We also remark that if $D(\pi_G) = E$, then every additive selection π_G^0 of π_G is *continuous on the whole space* E, by virtue of (6.6).

Consider now the sets $\pi_G^{-1}(g_0) \subset D(\pi_G)$, defined for an arbitrary linear subspace G of a normed linear space E and for an arbitrary $g_0 \in G$ by

$$\pi_G^{-1}(g_0) = \{x \in D(\pi_G) \mid g_0 = \pi_G(x)\} = \{x \in D(\pi_G) \mid g_0 \in \mathfrak{L}_G(x)\}. \tag{6.14}$$

Obviously we have the relation

$$D(\pi_G) = \bigcup_{g_0 \in G} \pi_G^{-1}(g_0). \tag{6.15}$$

THEOREM 6.3. *Let E be a normed linear space, G a linear subspace of E and $g_0 \in G$. Then the set $\pi_G^{-1}(g_0)$ is closed and*

$$x \in \pi_G^{-1}(g_0) \Rightarrow \alpha x + (1-\alpha) g_0 \in \pi_G^{-1}(g_0) \qquad (\alpha = \text{scalar}). \tag{6.16}$$

Proof. Let $x_n \in \pi_G^{-1}(g_0)$ and $x \in E$ such that $\lim_{n\to\infty} x_n = x$. By virtue of theorem 6.5 e) of the next section we have then

$$|\,\|x_n - g_0\| - e_G(x)\,| = |\,e_G(x_n) - e_G(x)\,| \leqslant \|x_n - x\| < \varepsilon \,(n > N(\varepsilon))$$

whence

$$\|x - g_0\| \leqslant \|x - x_n\| + \|x_n - g_0\| \leqslant e_G(x) + 2\varepsilon \quad (n > N(\varepsilon)).$$

Since these inequalities hold for every $\varepsilon > 0$, it follows that we have

$$\|x - g_0\| \leqslant e_G(x),$$

whence $x \in D(\pi_G)$ and $g_0 \in \mathfrak{L}_G(x)$, i.e. $x \in \pi_G^{-1}(g_0)$. Thus, the set $\pi_G^{-1}(g_0)$ is closed.*)

Now let $x \in \pi_G^{-1}(g_0)$ be arbitrary, let α be an arbitrary scalar and let

$$y = \alpha x + (1-\alpha)\, g_0. \tag{6.17}$$

*) We remark that this statement follows also directly from the relation

$$\pi_G^{-1}(g_0) = \bigcap_{g \in G} \{x \in E \mid \|x - g_0\| \leqslant \|x - g\|\}.$$

If $\alpha = 0$, then $y = g_0 \in \pi_G^{-1}(g_0)$ by virtue of the idempotence of π_G. On the other hand, if $\alpha \neq 0$, then for every $g \in G$ we have, taking into account that $x \in \pi_G^{-1}(g_0)$,

$$\|y - g\| = \|\alpha x + (1 - \alpha) g_0 - g\| = |\alpha| \, \|x -$$

$$-\left(1 - \frac{1}{\alpha}\right) g_0 - \frac{1}{\alpha} g\| \geqslant |\alpha| \, \|x - g_0\| = \|y - g_0\|,$$

and thus $g_0 \in \mathfrak{L}_G(y)$, i.e. $y \in \pi_G^{-1}(g_0)$. Thus, we have (6.16), which completes the proof.

The first statement of theorem 6.3 has been given by R. R. Phelps ([175], p. 867). The second statement of theorem 6.3 implies, in particular, that $\pi_G^{-1}(g_0)$ *is a cone*) with vertex in* g_0. This result has been given by V. Klee [109] (see also Appendix I, lemma 2.1).

We remark that in the particular case when the mapping π_G is one-valued on $D(\pi_G)$, the second statement of theorem 6.3 follows also from theorem 6.1, since for every $x \in \pi_G^{-1}(g_0)$ and every scalar α we have in this case

$$\pi_G(\alpha x + (1 - \alpha) g_0) = \pi_G(\alpha x) + \pi_G((1 - \alpha) g_0) =$$

$$= \alpha \pi_G(x) + (1 - \alpha) \pi_G(g_0) = \alpha g_0 + (1 - \alpha) g_0 = g_0,$$

whence $\alpha x + (1 - \alpha) g_0 \in \pi_G^{-1}(g_0)$.

THEOREM 6.4. *Let E be a normed linear space and G a linear subspace of E. The following statements are equivalent:*

1° *For every $g_0 \in G$ the set $\pi_G^{-1}(g_0)$ is a closed linear manifold passing through g_0.*

2° *$\pi_G^{-1}(0)$ is a closed linear subspace of E.*

3° *π_G is one-valued and additive on $D(\pi_G)$.*

Proof. The equivalence 1° $\Leftrightarrow$ 2° follows from the relations (which are obvious by the quasi-additivity of π_G)

$$\pi_G^{-1}(g_0) = g_0 + \pi_G^{-1}(0) \qquad (g_0 \in G), \tag{6.18}$$

and from the fact that $\pi_G^{-1}(g_0)$ passes through g_0, for every $g_0 \in G$.

Now assume that we have 2°. Then π_G is one-valued on $D(\pi_G)$, since the relations $x \in E$, $g_0, g_1 \in \mathfrak{L}_G(x)$ imply $x - g_i \in \pi_G^{-1}(0)$, whence, by 2°, $g_1 - g_0 \in \pi_G^{-1}(0) \cap G$, whence $g_1 = g_0$.

Let $x, y \in D(\pi_G)$ be arbitrary and let $\pi_G(x) = g_1$, $\pi_G(y) = g_2$. Then we have $x - g_1 \in \pi_G^{-1}(0)$, $y - g_2 \in \pi_G^{-1}(0)$, whence, by

---

*) See § 4, section 4.2.

2°, $x + y - g_1 - g_2 \in \pi_G^{-1}(0)$, whence, by taking again into account the quasi-additivity of π_G, we obtain

$$\pi_G(x + y) - g_1 - g_2 = \pi_G(x + y - g_1 - g_2) = 0 = \pi_G(x) - g_1 + \pi_G(x) - g_2,$$

and consequently $\pi_G(x + y) = \pi_G(x) + \pi_G(y)$. Thus, 2° $\Rightarrow$ 3°.

Finally, assume that we have 3°. Then for every x, $y \in \pi_G^{-1}(0)$ and any scalars α, β we have

$$\pi_G(\alpha x + \beta y) = \pi_G(\alpha x) + \pi_G(\beta y) = \alpha\pi_G(x) + \beta\pi_G(y) = 0,$$

whence $\alpha x + \beta y \in \pi_G^{-1}(0)$, and thus $\pi_G^{-1}(0)$ is a linear subspace of E, closed by virtue of theorem 6.3. Thus 3° $\Rightarrow$ 2°, which completes the proof of theorem 6.4.

By virtue of theorem 6.2, the condition 3° of theorem 6.4 is satisfied *if G is a Čebyšev hyperplane in E, passing through* 0.

The foregoing show the connection between various properties of the mappings π_G and the corresponding properties of the sets $\pi_G^{-1}(0)$.

In the particular case when E is*) an inner product space $\mathcal{H}$, whence strictly convex, the mapping π_G is, by virtue of corollary 3.3, one-valued on $D(\pi_G)$, for every linear subspace G of E. Moreover, in this case the mapping π_G is additive (whence, taking into account 6.1, linear and continuous) on $D(\pi_G)$, which is now a linear subspace of E. For, if $x, y \in D(\pi_G)$ and $\pi_G(x) = g_1$, $\pi_G(y) = g_2$, then by theorem 1.11′ we have

$$(x + y - g_1 - g_2, g) = (x - g_1, g) + (y - g_2, g) = 0 \quad (g \in G),$$

whence, again by theorem 1.11′, $x + y \in D(\pi_G)$ and

$$\pi_G(x + y) = g_1 + g_2 = \pi_G(x) + \pi_G(y). \tag{6.19}$$

By virtue of theorem 6.4, the set $\pi_G^{-1}(0)$ is in this case a closed linear subspace of E.

Furthermore, in this case we have

$$\|\pi_G(x)\| \leqslant \|x\| \qquad (x \in D(\pi_G)). \tag{6.20}$$

In fact, if $x \in D(\pi_G)$ and $\pi_G(x) = g_0 \neq 0$, then by theorem 1.11′ we have

$$\|x\| \, \|g_0\| \geqslant |(x, g_0)| = |(x - g_0, g_0) + (g_0, g_0)| = (g_0, g_0) = \|g_0\|^2,$$

whence (6.20) follows.

In Chap. II, § 5, it will be shown that each of the above properties characterize the inner product spaces.

---

*) We recall (see § 1, section 1.7) that the completeness of the space $\mathcal{H}$ is not assumed.

If, in particular, E is a *complete* inner product spage $\mathcal{H}$, then it is reflexive and strictly convex, whence by virtue of corollary 3.4 we have $D(\pi_G) = E$ and π_G is one-valued on $D(\pi_G) = E$, for every *closed* linear subspace G of E. Moreover, in this case $\pi_G(x) =$ *the orthogonal projection of x onto G*, that is,

$$\pi_G(x) = \begin{cases} x \text{ if } x \in G, \\ 0 \text{ if } (x,g) = 0 \ (g \in G); \end{cases} \tag{6.21}$$

in fact, by theorem 1.11′ we have

$$(x - \pi_G(x),g) = 0 \qquad (g \in G), \tag{6.22}$$

and thus, if $(x, g) = 0$ $(g \in G)$, then for $g = \pi_G(x)$ we obtain $(\pi_G(x), \pi_G(x)) = 0$, i.e. $\pi_G(x) = 0$, which proves (6.21). In particular, it follows that in this case $\pi_G^{-1}(0)$ is the orthogonal complement of G.

We mention that in the case when G is a Čebyšev subspace of a normed linear space E, the mapping π_G is sometimes called the *metric projection* or the *normal projection* of the space E onto the subspace G; as has been remarked above, this is, in general, a non-linear projection. The study of the non-linear projections presents interest also for other applications (see e.g. J. Lindenstrauss [136]).

Now let E be a strictly convex reflexive Banach space (hence, by corollary 3.4, $D(\pi_G) = E$ and π_G is one-valued on E) and let G_1 and G_2 be two closed linear subspaces of E. Consider the following alternating procedure of approximating an arbitrary element $x \in E$, with the aid of the metric projections π_{G_1} and π_{G_2}: project x onto G_1, obtaining the element $\pi_{G_1}(x)$ and a remainder $r_1(x) = x - \pi_{G_1}(x)$; project the remainder $r_1(x)$ onto G_2, obtaining the element $\pi_{G_2}[r_1(x)]$ and a remainder $r_2(x) = r_1(x) - \pi_{G_2}[r_1(x)]$; project $r_2(x)$ onto $G_1 \dots$; project $r_3(x)$ onto $G_2 \dots$; continue so indefinitely. Following R. A. Hirschfeld [89], *the space E is said to admit the alternating method*, if for every G_1, G_2 and x the above process converges to $\pi_G(x)$, where $G =$ the closed linear subspace of E spanned by G_1 and G_2, that is, if we have

$$\pi_G(x) = \pi_{G_1}(x) + \pi_{G_2}[r_1(x)] + \pi_{G_1}[r_2(x)] + \dots$$
$$\dots + \pi_{G_2}[r_{(2j-1)}(x)] + \pi_G[r_{(2j)}(x)] + \dots \quad (x \in E). \tag{6.23}$$

As has been shown by J. von Neumann ([160], p. 56, corollary; see also N. Wiener and P. Masani [262], § 3), *the complete inner product spaces admit the alternating method.* R. A. Hirschfeld [89] has raised the problem, whether the converse is also true, i.e. whether every strictly convex reflexive Banach space which admits the alternating method is equivalent to a complete inner product space. V. Klee [114] has shown that for 2-dimensional spaces E the answer is negative and has given a simple characterization of 2-dimensional Banach spaces which admit the alternating method. For spaces E of finite dimension $n \geqslant 3$ the answer is negative too; namely, as has shown W. J. Stiles [238 a], *every strictly convex and smooth finite dimensional Banach space admits the alternating method.* The infinite dimensional problem is still open.

The condition that a strictly convex reflexive Banach space E admit the alternating method can be expressed also as follows: for any closed linear subspaces G_1 and G_2 of E we have

$$\lim (I - \pi_{G_1})(I - \pi_{G_2}) \dots (I - \pi_{G_1})(I - \pi_{G_2})(x) = (I - \pi_G)(x) \quad (x \in E), \tag{6.24}$$

where G = the closed linear subspace of E spanned by G_1 and G_2 and I = the identical mapping of E onto itself. It arises naturally the problem, what happens with $\lim (\pi_{G_1} \pi_{G_2} \dots \pi_{G_1} \pi_{G_2})(x)$. As has been shown by J. von Neumann ([159], p. 475, lemma 22 and [160], p. 55, theorem 13.7) and, independently, by N. Wiener ([261], p. 101, lemma; for the extension of this result to a finite number of subspaces see I. Halperin [83]), *if E is a complete inner product space, then for any closed linear subspces G_1, G_2 of E we have*

$$\lim (\pi_{G_1} \pi_{G_2} \dots \pi_{G_1} \pi_{G_2})(x) = \pi_{G_1 \cap G_2}(x) \quad (x \in E). \tag{6.25}$$

W. J. Stiles has shown ([238], theorem 2.2) that conversely, *if E is a strictly convex reflexive Banach space of dimension $\geqslant 3$ and if for every pair of 2-dimensional linear subspaces G_1, G_2 of E we have (6.25), then E is a complete inner product space*; however an analogous result is no longer valid for $\dim E = 2$, since every strictly convex and smooth 2-dimensional space E satisfies the condition of the above statement (W. J. Stiles [238], theorem 3.2). For other results related to the above, see W. J. Stiles [238], [238 b].

6.2. THE FUNCTIONALS e_G

THEOREM 6.5 (M. Nicolescu [161], propositions 3, 2, 4 and 1). *Let E be a normed linear space and G a linear subspace of E. Then*

a) *We have*

$$0 \leqslant e_G(x) = e_{\bar{G}}(x) < \infty \qquad (x \in E), \tag{6.26}$$

$$e_G(g) = 0 \qquad (g \in G), \tag{6.27}$$

$$e_G(x) = \|x - \pi_G(x)\| \qquad (x \in D(\pi_G)). \tag{6.28}$$

b) *If G_1 is a linear subspace of G, we have*

$$e_G(x) \leqslant e_{G_1}(x) \qquad (x \in E). \tag{6.29}$$

c) *We have*

$$e_G(x+g) = e_G(x) + e_G(g) = e_G(x) \quad (x \in E,\ g \in G), \tag{6.30}$$

$$e_G(x+y) \leqslant e_G(x) + e_G(y) \qquad (x, y \in E), \tag{6.31}$$

that is, e_G is quasi-additive and sub-additive.

d) *We have*

$$e_G(\alpha x) = |\alpha|\, e_G(x) \qquad (x \in E,\ \alpha = \text{scalar}), \tag{6.32}$$

that is, e_G is positively homogeneous.

e) *We have*

$$|e_G(x) - e_G(y)| \leqslant \|x - y\| \qquad (x, y \in E), \tag{6.33}$$

$$e_G(x) \leqslant \|x\| \qquad (x \in E). \tag{6.34}$$

f) *e_G is continuous on E.*

Proof. a) and b) are obvious by the definitions (6.2) and (6.3) of e_G and π_G respectively.

c) Let $x \in E$, $g \in G$ and $\varepsilon > 0$ be arbitrary. By the definition of e_G there exists a $g_0 \in G$ such that

$$\|x - g_0\| \leqslant e_G(x) + \varepsilon. \tag{6.35}$$

Consequently, we have

$$e_G(x+g) \leqslant \|x + g - (g_0 + g)\| = \|x - g_0\| \leqslant e_G(x) + \varepsilon,$$

whence, since $x \in E$, $g \in G$ and $\varepsilon > 0$ were arbitrary, we obtain

$$e_G(x+g) \leqslant e_G(x) \qquad (x \in E, g \in G). \tag{6.36}$$

Applying these relations for $x + g \in E$ instead of x and $-g \in G$ instead of g, we obtain

$$e_G(x) \leqslant e_G(x+g) \qquad (x \in E, g \in G), \tag{6.37}$$

and from the relations (6.36) and (6.37) it follows (6.30).

Now let $x, y \in E$ and $\varepsilon > 0$ be arbitrary. By the definition of e_G there exist elements $g_1, g_2 \in G$ such that

$$\|x - g_1\| \leqslant e_G(x) + \frac{\varepsilon}{2}, \quad \|y - g_2\| \leqslant e_G(y) + \frac{\varepsilon}{2}.$$

Consequently, we have

$$e_G(x + y) \leqslant \|x + y - (g_1 + g_2)\| \leqslant \|x - g_1\| + \|y - g_2\| \leqslant$$
$$\leqslant e_G(x) + e_G(y) + \varepsilon,$$

whence, since $x, y \in E$ and $\varepsilon > 0$ were arbitrary, (6.31) follows.

d) Let $x \in E$, $\alpha =$ a scalar $\neq 0$ and $\varepsilon > 0$ be arbitrary and take a $g_0 \in G$ satisfying

$$\|x - g_0\| \leqslant e_G(x) + \frac{\varepsilon}{|\alpha|}.$$

Then we have

$$e_G(\alpha x) \leqslant \|\alpha x - \alpha g_0\| = |\alpha|\, \|x - g_0\| \leqslant |\alpha|\, e_G(x) + \varepsilon,$$

whence, since x, $\alpha \neq 0$ and ε were arbitrary, and since $e_G(0) = 0$, it follows (6.32).

e) Let $x, y \in E$ and $\varepsilon > 0$ be arbitrary and take a $g_0 \in G$ satisfying

$$\|y - g_0\| \leqslant e_G(y) + \varepsilon.$$

Then we have

$$e_G(x) \leqslant \|x - g_0\| \leqslant \|x - y\| + \|y - g_0\| \leqslant \|x - y\| + e_G(y) + \varepsilon,$$

whence, since x, y and ε were arbitrary, there follows

$$e_G(x) - e_G(y) \leqslant \|x - y\| \qquad (x, y \in E). \qquad (6.38)$$

Changing in these relations x by y and y by x and comparing the relations thus obtained with (6.38), there follows (6.33).

Relation (6.34) is obvious by the definition (6.2) of e_G, taking into account that $0 \in G$; besides, it follows immediately also from (6.33), putting $y = 0$ and taking into account that $e_G(0) = 0$.

Finally, f) is an immediate consequence of (6.33), which completes the proof.

COROLLARY 6.1. *Let E be normed linear space, G a linear subspace of E and $y, z \in E$. Then the real function*

$$\varphi(\lambda) = e_G(y + \lambda z) \qquad (-\infty < \lambda < \infty) \qquad (6.39)$$

is convex. Consequently, φ is continuous and, if $z \notin \overline{G}$, we have

$$\lim_{\lambda \to \pm\infty} \varphi(\lambda) = \infty.$$

Proof. For any real numbers λ and α we have, by theorem 6.5 d) and c),

$$2\varphi(\lambda) = 2e_G(y + \lambda z) = e_G[2(y + \lambda z)] = e_G[y + (\lambda + \alpha) z + y + $$
$$+ (\lambda - \alpha)z] \leqslant e_G[y + (\lambda+\alpha)z] + e_G[y + (\lambda-\alpha) z] = \varphi(\lambda + \alpha) +$$
$$+ \varphi(\lambda - \alpha),$$

whence φ is convex. If φ is constant, from $|\lambda| e_G(z) = e_G(\lambda z) \leqslant \leqslant e_G(y) + \varphi(\lambda)$ $(-\infty < \lambda < \infty)$ it follows that $e_G(z) = 0$, whence $z \in \overline{G}$. Consequently, if $z \notin \overline{G}$, then φ is non-constant, whence $\lim\limits_{\lambda \to \pm\infty} \varphi(\lambda) = \infty$, which completes the proof.

Corollary 6.1 is well known (see e.g. S. Bernstein [16]).

Concerning the properties of the functionals e_G for the weak topology $\sigma(E, E^*)$, we have the following well known result*):

THEOREM 6.6. *Let E be a normed linear space and G a linear subspace of E. Then e_G is lower semi-continuous for the weak topology $\sigma(E, E^*)$.*

Proof. By (6.27) and (6.26) we have $e_G(x) = 0$ $(x \in \overline{G})$, whence e_G is lower semi-continuous on $\overline{G}$. Now let $x_0 \in E \setminus \overline{G}$ be arbitrary, whence $d = e_G(x_0) = \rho(x_0, G) > 0$. Then, by virtue of a corollary of the Hahn-Banach theorem, there exists an $f_0 \in E^*$ such that $\|f_0\| = \frac{1}{d}$, $f_0(g) = 0$ $(g \in G)$ and $f_0(x_0) = 1$, hence the functional $f = df_0 \in E^*$ satisfies

$$\|f\| = 1, \tag{6.40}$$

$$f(x_0 - g) = d \qquad (g \in G). \tag{6.41}$$

Now let $\varepsilon > 0$ be arbitrary and let x be an arbitrary element in the weak neighbourhood $V_{f;\varepsilon}(x_0) = \{y \in E \mid |f(y) - - f(x_0)| < \varepsilon\}$ of x_0.

Then we have

$$|f(x_0 - g)| - |f(x - g)| \leqslant |f(x_0 - x)| < \varepsilon \qquad (g \in G),$$

whence, taking into account (6.41),

$$|f(x - g)| > d - \varepsilon = e_G(x_0) - \varepsilon \qquad (g \in G),$$

and consequently, by (6.40),

$$e_G(x) = \inf_{g \in G} \|x - g\| \geqslant \inf_{g \in G} |f(x - g)| \geqslant e_G(x_0) - \varepsilon,$$

which completes the proof.

---

*) See e.g. N. Bourbaki [23], Chap. IV, p. 117, exercise 1 a), or G. Kötbe [122], p. 348, theorem 7.

From theorem 6.6 one finds again corollaries 2.1 and 2.1′, by applying the following property of the functionals e_G, which is also useful in other circumstances:

LEMMA 6.1. *Let E be a normed linear space and G a linear subspace of E. Then we have*

$$e_G(x) = \inf_{\substack{g \in G \\ \|g\| \leqslant 2\|x\|}} \|x - g\| \qquad (x \in E). \tag{6.42}$$

Proof. If $x \in E$, $g_0 \in G$ and $\|g_0\| > 2\|x\|$, then, taking into account (6.34), we have

$$\|x - g_0\| \geqslant \|g_0\| - \|x\| > 2\|x\| - \|x\| = \|x\| \geqslant e_G(x),$$

whence, considering the definition (6.2) of e_G, the statement follows.

Lemma 6.1 is well known, e.g. it is contained, essentially, in the monograph of N. I. Ahiezer [1], § 8.

6.3. THE FUNCTIONALS e_{G_n} FOR INCREASING OR DECREASING SEQUENCES $\{G_n\}$ OF CLOSED LINEAR SUBSPACES

Let E be a Banach space and let $\{G_n\}$ be an increasing sequence of closed linear subspaces of E. It arises naturally the problem of the existence of elements $x \in E$ with prescribed values $e_{G_n}(x)$ $(n = 1,2,\ldots)$, called by S. N. Bernstein*) "the inverse problem of the theory of best approximation".

Following V. N. Nikolsky [171], a Banach space E is said to have the (absolute) *property* (B), if for every increasing sequence

$$0 = G_0 \subset G_1 \subset G_2 \subset \ldots \tag{6.43}$$

of distinct closed linear subspaces of E and every sequence of numbers $\{e_n\}$ with the properties

$$e_0 \geqslant e_1 \geqslant e_2 \geqslant \ldots \geqslant 0, \qquad \lim_{n\to\infty} e_n = 0 \tag{6.44}$$

there exists an element $x \in E$ such that

$$e_{G_k}(x) = e_k \qquad (k = 0,\ 1,\ 2,\ldots). \tag{6.45}$$

THEOREM 6.7 (I. S. Tyuremskih [251]). *Every complete inner product space $\mathcal{H}$ has the absolute property (B).*

Proof. Let (6.43) be an arbitrary increasing sequence of distinct closed linear subspaces of $\mathcal{H}$ and let $\{e_n\}$ be a sequence

*) For the particular case considered by S. N. Bernstein see Chap. II, § 5 section 5.3.

of numbers with the properties (6.44). Then every G_{n+1} is a complete inner product space, whence

$$G_{n+1} = G_n \oplus Z_n \qquad (n = 1,2,\ldots), \tag{6.46}$$

where Z_n is the orthogonal complement of G_n in G_{n+1}. Take arbitrary elements $x_1 \in G_1, x_2 \in Z_1, x_3 \in Z_2, \ldots$ with

$$\|x_n\| = \sqrt{e_{n-1}^2 - e_n^2} \qquad (n = 1,2,\ldots). \tag{6.47}$$

Then we have

$$x_j \perp G_k \qquad (j > k; j, k = 1,2,\ldots), \tag{6.48}$$

since $G_k \subset G_{j-1} \perp Z_{j-1} \ni x_j$. Consequently, in particular

$$x_j \perp x_k \qquad (j > k; j, k = 1,2,\ldots) \tag{6.49}$$

(because $x_k \in Z_{k-1} \subset G_k$), whence, taking into account (6.47),

$$\left\|\sum_{i=n}^{m} x_i\right\|^2 = \sum_{i=n}^{m} \|x_i\|^2 = \sum_{i=n}^{m} (e_{i-1}^2 - e_i^2) = e_{n-1}^2 - e_m^2 \to 0 \ (m > n \to \infty),$$

and thus, since $\mathcal{H}$ is complete, the series $\sum_{i=1}^{\infty} x_i$ converges. We claim that the element $x = \sum_{i=1}^{\infty} x_i \in \mathcal{H}$ satisfies (6.45). In fact, by (6.49) and (6.47) we have

$$e_0(x) = \|x\| = \sqrt{\left\|\sum_{i=1}^{\infty} x_i\right\|^2} = \sqrt{\sum_{i=1}^{\infty} \|x_i\|^2} = \sqrt{\sum_{i=1}^{\infty} (e_{i-1}^2 - e_i^2)} = e_0.$$

On the other hand, by (6.48) we have

$$x - \sum_{i=1}^{k} x_i = \sum_{i=k+1}^{\infty} x_i \perp G_k \qquad (k = 1,2,\ldots),$$

whence, by theorem 1.11′, $\sum_{i=1}^{k} x_i \in \mathfrak{P}_{G_k}(x)$ $(k = 1,2,\ldots)$, and thus

$$e_{G_k}(x) = \left\|x - \sum_{i=1}^{k} x_i\right\| = \left\|\sum_{i=k+1}^{\infty} x_i\right\| = \sqrt{\sum_{i=k+1}^{\infty} \|x_i\|^2} =$$

$$= \sqrt{\sum_{i=k+1}^{\infty} (e_{i-1}^2 - e_i^2)} = e_k \qquad (k = 1,2,\ldots),$$

which completes the proof.

There are not known, up to the present, other Banach spaces with the absolute property (B); in any case, such a space must be reflexive, as shown by theorem 6.8 below.

Following V. N. Nikolsky [171], a Banach space E is said to have *the property (B) with respect to a family* ($\mathfrak{S}$) *of closed linear subspaces of* E, if for every increasing sequence (6.43) of distinct closed linear subspaces belonging to the family ($\mathfrak{S}$) and every sequence of numbers $\{e_n\}$ with the properties (6.44), there exists an element $x \in E$ satisfying (6.45). V. N. Nikolsky [171] and I. S. Tyuremskih [252] have given various examples of Banach spaces E having the property *(B)* with respect to certain families ($\mathfrak{S}$) of closed linear subspaces.

Another variant of the property *(B)*, considered by V. N. Nikolsky [170], is the following: a Banach space E is said to have *the property* (B_f) if for every increasing sequence (6.43) of distinct closed linear subspaces of E and every sequence of numbers $\{e_n\}$ with the properties

$$e_0 \geqslant e_1 \geqslant e_2 \geqslant \dots \geqslant e_n \geqslant e_{n+1} = e_{n+2} = \dots = 0 \tag{6.50}$$

there exists an element $x \in E$ satisfying (6.45).

THEOREM 6.8. (V. N. Nikolsky [170], p. 123). *A Banach space E has the property (B_f) if and only if it is reflexive.*

Proof. Assume that E is not reflexive. Then E has, by virtue of a theorem of V. Klee*), a non-reflexive closed linear subspace E_0 of infinite codimension, and by corollary 2.4, E_0 has a non-proximinal hyperplane H passing through 0. Put $G_0 = 0$, $G_1 = H$, $G_2 = E_0$ and take an arbitrary increasing sequence $G_3 \subset G_4 \subset \dots$ of distinct closed linear subspaces of E, containing E_0 (such a sequence exists since E_0 is of infinite codimension); take also a sequence of numbers $\{e_n\}$ with $e_0 = e_1 > 0$, $e_2 = e_3 = \dots = 0$. Then there exists no element $x \in E$ satisfying (6.45), hence E does not have the property (B_f), since otherwise one would have $\|x\| = e_0(x) = e_0$ and $\rho(x, G_1) = e_{G_1}(x) = e_1 = e_0$, whence $0 \in \mathfrak{L}_{G_1}(x)$, and also $x \in G_2$ (by $e_2 = 0$), which, by virtue of lemma 2.1, contradicts the hypothesis that $G_1 = H$ is non-proximinal in $G_2 = E_0$. Thus, the condition is necessary.

Now assume that E is reflexive and let (6.43) be an arbitrary increasing sequence of distinct closed linear subspaces of E and $\{e_n\}$ a sequence of numbers with the properties (6.50). Take an arbitrary $x_{n+1} \in G_{n+1} \setminus G_n$. Then for $\lambda_{n+1} = \dfrac{e_n}{e_{G_n}(x_{n+1})}$ we have

$$e_{G_n}(\lambda_{n+1} x_{n+1}) = e_n. \tag{6.51}$$

*) See e.g. G. Köthe [122], p. 322, theorem 2.

We shall show that there exist an $x_n \in G_n$ and a real number λ_n scuh that

$$e_{G_k}(\lambda_{n+1} x_{n+1} + \lambda_n x_n) = e_k \quad (k = n-1, n). \tag{6.52}$$

Indeed, since E is reflexive, by corollary 2.4 there exists a $g_0 \in \mathfrak{P}_{G_n}(\lambda_{n+1}x_{n+1})$, say $g_0 \in G_m \setminus G_{m-1}$. The following cases are possible :

a) $m = n$, $e_n \leqslant e_{n-1}$. In this case take $x_n = -g_0$. Then, since $x_n \notin G_{n-1}$ and since we have

$$e_{G_{n-1}}(\lambda_{n+1} x_{n+1} + x_n) \leqslant \| \lambda_{n+1} x_{n+1} - g_0 \| = e_{G_n}(\lambda_{n+1} x_{n+1}) =$$
$$= e_n \leqslant e_{n-1},$$

by virtue of corollary 6.1 there exists a $\lambda = \lambda_n$ such that

$$e_{G_{n-1}}(\lambda_{n+1} x_{n+1} + \lambda_n x_n) = e_{n-1}. \tag{6.53}$$

b) $m < n$, $e_n = e_{n-1}$. In this case take again $x_n = -g_0$. Then by $x_n \in G_m \subset G_{n-1}$ and $-x_n \in \mathfrak{P}_{G_n}(\lambda_{n+1}x_{n+1})$ we have

$$e_{G_{n-1}}(\lambda_{n+1} x_{n+1} + x_n) = e_{G_{n-1}}(\lambda_{n+1} x_{n+1}) = e_{G_n}(\lambda_{n+1} x_{n+1}) = e_n = e_{n-1},$$

whence (6.53) with $\lambda_n = 1$.

c) $m < n$, $e_n < e_{n-1}$. In this case, since $g_0 \in G_m \subset G_n$ and $\|\lambda_{n+1} x_{n+1} - g_0\| = e_{G_n}(\lambda_{n+1} x_{n+1}) = e_n < e_{n-1}$, we can take an $x_n \in G_n \setminus G_{n-1}$ so near to $-g_0$ that we have

$$\| \lambda_{n+1} x_{n+1} + x_n \| < e_{n-1}.$$

Then, since $x_n \notin G_{n-1}$ and since

$$e_{G_{n-1}}(\lambda_{n+1} x_{n+1} + x_n) \leqslant \| \lambda_{n+1} x_{n+1} + x_n \| < e_{n-1},$$

by virtue of corollary 6.1 there exists a $\lambda = \lambda_n$ such that we have (6.53).

Thus, in any case there exist an $x_n \in G_n$ and a λ_n such that we have (6.53). At the same time, by theorem 6.5 c) and by (6.51) we have

$$e_{G_n}(\lambda_{n+1} x_{n+1} + \lambda_n x_n) = e_{G_n}(\lambda_{n+1} x_{n+1}) = e_n,$$

which, together with (6.53), means that we have (6.52).

Repeating this argument for $\lambda_{n+1} x_{n+1} + \lambda_n x_n$ instead of $\lambda_{n+1} x_{n+1}$ and for $n-1$ instead of n, it follows that there exist an $x_{n-1} \in G_{n-1}$ and a λ_{n-1} such that

$$e_{G_k}(\lambda_{n+1} x_{n+1} + \lambda_n x_n + \lambda_{n-1} x_{n-1}) = e_k \quad (k = n-2, n-1, n).$$

Continuing in this way, we obtain an element $x = \sum_{i=1}^{n+1} \lambda_i x_i \in G_{n+1}$ such that

$$e_{G_k}(x) = e_k \qquad (k = 0,1,\ldots,n). \tag{6.54}$$

Since $x \in G_{n+1} \subset G_{n+2} \subset \ldots$, we have $e_{G_{n+1}}(x) = e_{G_{n+2}}(x) = \ldots = 0$, and thus, by (6.50) and (6.54), the element $x \in E$ satisfies (6.45), which completes the proof of theorem 6.8.

For decreasing sequences of subspaces we have

THEOREM 6.9 (I. S. Tyuremskih [252], theorem 4). *A Banach space E has the property that for every decreasing sequence of distinct closed linear subspaces*

$$E \supset G_1 \supset G_2 \supset \ldots \tag{6.55}$$

and every sequence of numbers $\{e_n\}$ such that

$$0 \leqslant e_1 \leqslant e_2 \leqslant \ldots \leqslant c < \infty \tag{6.56}$$

there exists an element $x \in E$ satisfying

$$\|x\| = \lim_{n\to\infty} e_n, \; e_{G_k}(x) = e_k \qquad (k = 1,2,\ldots), \tag{6.57}$$

if and only if E is reflexive.

Proof. Assume that E is not reflexive. Then by corollary 2.4 E has a non-proximinal hyperplane H passing through 0. Put $G_1 = H$ and take an arbitrary decreasing sequence $G_2 \supset G_3 \supset \ldots$ of distinct closed linear subspaces of G_1, and a sequence of numbers $\{e_n\}$ with $0 < e_1 = e_2 = e_3 = \ldots$ Then there exists no element $x \in E$ satisfying (6.57), since otherwise one would have $\|x\| = \lim_{n\to\infty} e_n = e_1$ and $\rho(x, G_1) = e_{G_1}(x) = e_1$, whence $0 \in \mathfrak{L}_{G_1}(x)$, which, by virtue of lemma 2.1, contradicts the hypothesis that $G_1 = H$ is non-proximinal in E. Thus, the condition is necessary.

Now assume that E is reflexive and let (6.55) be an arbitrary decreasing sequence of distinct closed linear subspaces of E and $\{e_n\}$ a sequence of numbers with the properties (6.56). Then for every positive integer n we have $G_n \subset G_{n-1} \subset \ldots \subset G_2 \subset G_1 \subset E$ and $e_n \geqslant e_{n-1} \geqslant \ldots \geqslant e_2 \geqslant e_1$, whence for every n there exists, by theorem 6.8, an element $y_n \in E$ such that

$$\|y_n\| = e_n, \; e_{G_k}(y_n) = e_k \qquad (k = 1,\ldots,n). \tag{6.58}$$

As shows the above proof of theorem 6.8, these elements can be taken so that

$$y_{n+1} \in y_n + G_n \qquad (n = 1,2,\ldots); \tag{6.59}$$

indeed, if $x_1 \in E \setminus G_1$, $x_i \in G_{i-1} (i = 2,\dots,n+1)$ and $\lambda_i (i = 1,\dots$ $\dots, n+1)$ are taken so that the element $y_n = \sum_{i=1}^{n+1} \lambda_i x_i$ satisfies (6.58), then one can find $x'_{n+1} \in G_n$, $x_{n+2} \in G_{n+1}$ and λ'_{n+1}, λ_{n+2} such that the element $y_{n+1} = \sum_{i=1}^{n} \lambda_i x_i + \lambda'_{n+1} x'_{n+1} + \lambda_{n+2} x_{n+2}$ satisfies (6.58) with $n+1$ instead of n, and we have then $y_{n+1} - y_n = \lambda'_{n+1} x'_{n+1} + \lambda_{n+2} x_{n+2} - \lambda_{n+1} x_{n+1} \in G_n$, whence (6.59).

Consequently,

$$y_{n+1} + G_{n+1} \subset y_n + G_n \qquad (n = 1,2,\dots), \tag{6.60}$$

and by (6.58), (6.55) and theorem 6.5 c) every element $x \in y_n + G_n$ satisfies $e_{G_k}(x) = e_k (k = 1,\dots,n)$. Now let $e = \lim_{n\to\infty} e_n$. Then $e \geqslant e_n = \|y_n\|$, whence the intersection $(y_n + G_n) \cap S(0, e)$ contains y_n and thus it is non-void $(n = 1,2\dots)$; the space E being reflexive, these sets are weakly compact. Since we have, by (6.60),

$$[(y_1 + G_1) \cap S(0, e)] \supset [(y_2 + G_2) \cap S(0, e)] \supset \dots$$

it follows that the intersection $\bigcap_{n=1}^{\infty} [(y_n + G_n) \cap S(0, e)]$ is non-void. Obviously, every element x of this intersection satisfies $e_{G_k}(x) = e_k (k = 1,2,\dots)$; since then $\|x\| \geqslant e_{G_k}(x) = e_k (k = 1,2,\dots)$, it follows that we have $\|x\| \geqslant \lim_{k\to\infty} e_k = e$, which, together with $x \in S(0, e)$, gives $\|x\| = \lim_{k\to\infty} e_k$, whence we have (6.57), and thus theorem 6.9 is proved.

We remark that *in the sufficiency parts of theorems 6.8 and 6.9 the condition that E be reflexive may be replaced by the condition that E be the conjugate space B^* of a Banach space B, if we assume that the linear subspaces $G_n (n = 1,2,\dots)$ are closed for the weak topology* $\sigma(B^*, B)$. Indeed, by corollary 2.5 the subspaces G_n are then proximinal, and thus the above proofs remain valid (using the compactness of $S_E = S_{B^*}$ for the weak topology $\sigma(B^*, B)$).

6.4. THE DEVIATION OF A SET FROM A LINEAR SUBSPACE

We recall that the *deviation* of a set $A \subset E$ from a set $B \subset E$ is defined by

$$\delta(A, B) = \sup_{x\in A} \rho(x, B) = \sup_{x\in A} \inf_{y\in B} \|x - y\|. \tag{6.61}$$

In particular, if B is a linear subspace G of E, we have

$$\delta(A, G) = \sup_{x \in A} e_G(x) = \sup_{x \in A} \inf_{g \in G} \|x - g\|. \tag{6.62}$$

THEOREM 6.10. *Let E be a normed linear space, G a linear subspace of E and A a set in E. Then*

a) *We have*

$$0 \leqslant \delta(A, G) = \delta(\overline{A}, G) = \delta(A, G) \leqslant \infty. \tag{6.63}$$

b) *If A_1 is a set in E, we have*

$$\delta(A_1, G) - \delta(A_1, A) \leqslant \delta(A, G); \tag{6.64}$$

consequently, if $A_1 \subset A$, we have

$$\delta(A_1, G) \leqslant \delta(A, G). \tag{6.65}$$

c) *If $(A_\iota)_{\iota \in I}$ is a family of sets in E, we have*

$$\delta(\bigcup_{\iota \in I} A_\iota, G) = \sup_{\iota \in I} \delta(A_\iota, G). \tag{6.66}$$

d) *For every scalar α we have*

$$\delta(\alpha A, G) = |\alpha| \, \delta(A, G); \tag{6.67}$$

consequently, for the circled hull $\tau(A) = \{\alpha x \,|\, x \in A, |\alpha| = 1\}$ of A we have

$$\delta(\tau(A), G) = \delta(A, G). \tag{6.68}$$

e) *We have**)

$$\delta(\operatorname{co} A, G) = \delta(A, G). \tag{6.69}$$

f) *If G_1 is a linear subspace of G, we have*

$$\delta(A, G) \leqslant \delta(A, G_1). \tag{6.70}$$

Proof. a) follows immediately from (6.62) and (6.26).

b) Let $\varepsilon > 0$ be arbitrary. Take $x_0 \in A_1$ such that

$$\sup_{x \in A_1} e_G(x) \leqslant e_G(x_0) + \frac{\varepsilon}{2}$$

and $y_0 \in A$ such that

$$\|x_0 - y_0\| \leqslant \rho(x_0, A) + \frac{\varepsilon}{2}.$$

---

*) We recall (see § 4, section 4.2) that by co A we denote the convex hull of the set A.

Then we have, taking into account theorem 6.5 (formulas (6.27) and (6.33)),

$$\sup_{x\in A_1} e_G(x) - \sup_{y\in A} e_G(y) \leqslant e_G(x_0) + \frac{\varepsilon}{2} - e_G(y_0) \leqslant \|x_0 - y_0\| + \frac{\varepsilon}{2} \leqslant$$

$$\leqslant \rho(x_0, A) + \varepsilon \leqslant \sup_{x\in A_1} \rho(x, A) + \varepsilon,$$

whence, since $\varepsilon > 0$ was arbitrary, it follows (6.64). In particular, if $A_1 \subset A$, then by (6.61) we have $\delta(A_1, A) = 0$ and thus we obtain (6.65).

c) If $(A_\iota)_{\iota\in I}$ is a family of sets in E, we have

$$\delta(\bigcup_{i\in I} A_\iota, G) = \sup_{x\in \bigcup_{\iota\in I} A_\iota} e_G(x) = \sup_{\iota\in I} \sup_{x\in A_\iota} e_G(x) = \sup_{\iota\in I} \delta(A_\iota, G).$$

d) For every scalar α we have, by theorem 6.5 d),

$$\delta(\alpha A, G) = \sup_{x\in\alpha A} e_G(x) = \sup_{y\in A} e_G(\alpha y) = |\alpha| \sup_{y\in A} e_G(y) = |\alpha|\ \delta(A, G)$$

i.e. (6.67). In particular, taking into account

$$\tau(A) = \bigcup_{|\alpha|=1} \alpha A,$$

and c), we obtain

$$\delta(\tau(A), G) = \delta(\bigcup_{|\alpha|=1} \alpha A, G) = \sup_{|\alpha|=1} \delta(\alpha A, G) = \sup_{|\alpha|=1} |\alpha| \delta(A, G) = \delta(A, G),$$

i.e. (6.68).

e) Let $x \in \operatorname{co} A$ be arbitrary. Then there exist a finite number of elements $x_1, \ldots, x_m \in A$ and scalars $\lambda_1, \ldots, \lambda_m > 0$ with $\sum_{i=1}^{m} \lambda_i = 1$, such that $x = \sum_{i=1}^{m} \lambda_i x_i$, whence we have, taking into account theorem 6.5 (formulas (6.31) and (6.32)),

$$e_G(x) = e_G\left(\sum_{i=1}^{m} \lambda_i x_i\right) \leqslant \sum_{i=1}^{m} e_G(\lambda_i x_i) = \sum_{i=1}^{m} \lambda_i e_G(x_i) \leqslant$$

$$\leqslant \sum_{i=1}^{m} \lambda_i\ \delta(A, G) = \delta(A, G),$$

whence

$$\delta(\operatorname{co} A, G) = \sup_{x\in\operatorname{co} A} e_G(x) \leqslant \delta(A, G).$$

On the other hand, by $A \subset \operatorname{co} A$ and (6.65) we have

$$\delta(A, G) \leqslant \delta(\operatorname{co} A, G),$$

which, together with the preceding inequality, gives (6.69).

Finally, f) follows immediately from theorem 6.5 b), which completes the proof.

Theorem 6.10 has been given, esentially, by A. L. Brown ([27], p. 585).

A well known useful geometrical characterization of the deviation is given by

THEOREM 6.11. *Let E be a normed linear space, G a liuear subspace of E and A a set in E. Then we have*

$$\delta(A, G) = \inf_{\substack{\varepsilon>0 \\ A\subset G+\varepsilon S_E}} \varepsilon. \tag{6.71}$$

Proof. Let $\varepsilon > 0$ be such that $A \subset G + \varepsilon S_E$ and let $x \in A$ be arbitrary. Then $x = g + \varepsilon y$, with suitable $g \in G$ and $y \in S_E$, whence

$$e_G(x) \leqslant \|x - g\| = \|\varepsilon y\| \leqslant \varepsilon,$$

and consequently

$$\delta(A, G) = \sup_{x\in A} e_G(x) \leqslant \inf_{\substack{\varepsilon>0 \\ A\subset G+\varepsilon S_E}} \varepsilon. \tag{6.72}$$

In order to prove the opposite inequality, let $x_0 \in A$ and $\eta > 0$ be arbitrary and let $g_0 \in G$ be such that

$$\|x_0 - g_0\| \leqslant e_G(x_0) + \eta.$$

Then we have $x_0 = g_0 + (x_0 - g_0)$, whence

$$x_0 \in G + \|x_0 - g_0\| S_E \subset G + [\sup_{x\in A} e_G(x) + \eta] S_E,$$

and thus

$$A \subset G + [\sup_{x\in A} e_G(x) + \eta] S_E,$$

whence

$$\delta(A, G) + \eta = \sup_{x\in A} e_G(x) + \eta \geqslant \inf_{\substack{\varepsilon>0 \\ A\subset G+\varepsilon S_E}} \varepsilon.$$

Since $\eta > 0$ has been arbitrary, it follows that we have

$$\delta(A, G) \geqslant \inf_{\substack{\varepsilon>0 \\ A\subset G+\varepsilon S_E}} \varepsilon,$$

which, together with (6.72), completes the proof.

An element $x_0 \in A$ is said to be *extremal* (with respect to G) if we have

$$\delta(A, G) = e_G(x_0), \tag{6.73}$$

that is, if x_0 is "farthest" from G among the elements of A. The problem of determining the deviation $\delta(A, G)$ and the extremal elements $x_0 \in A$ has been considered by V. M. Tihomirov [249].

In the particular case when $G = \{0\}$, the extremal elements $x_0 \in A$ in the sense (6.73) are nothing else but the elements of A which are *farthest* from 0 among the elements of the set A. Such elements have been studied by T. S. Motzkin, E. G. Straus and F. A. Valentine [156], who have analyzed the sets A in a real two-dimensional space E with the property that every $y \in E$ has a unique farthest element $x_0 \in A$ and have characterized the sets A in a two dimensional real inner product space E with the property that every $y \in E$ has at least two farthest elements $x_0, x_0' \in A$.

Sometimes, instead of the deviation $\delta(A, G)$ of a set $A \subset E$ from a linear subspace (or, more generally, from a linear manifold) $G \subset E$, defined by (6.62), one considers the *linear deviation* $\delta''(A, G)$, defined by

$$\delta''(A, G) = \inf_{v(E) \subset G} \sup_{x \in A} \|x - v(x)\|, \tag{6.74}$$

where the inf is taken over all "non homogeneous linear mappings" of E into G, i.e. over all mappings $v: E \to G$ of the form

$$v(x) = u(x) + y, \tag{6.75}$$

with $u: E \to G$ linear and $y \in E$. Since

$$\|x - v(x)\| \geqslant e_G(x), \qquad (x \in A,\ v: E \to G),$$

we have

$$\sup_{x \in A} \|x - v(x)\| \geqslant \sup_{x \in A} e_G(x) = \delta(A, G) \qquad (v: E \to G),$$

whence

$$\delta''(A, G) \geqslant \delta(A, G). \tag{6.76}$$

As has been shown by V. M. Tihomirov ([247], p. 118—119), in general one can also have the sign $>$ in (6.76).

In the case when G is a Čebyšev subspace of E, the usual deviations $\delta(A, G)$ may be written in the form

$$\delta(A, G) = \sup_{x \in A} \|x - \pi_G(x)\| \quad (A \subset E). \tag{6.77}$$

In connection with this, S. M. Nikolsky ([163], p. 3—4) has raised the problem whether in (6.77) one can replace π_G

by a continuous linear mapping $u: E \to G$, that is, whether there exists a continuous linear mapping $u: E \to G$ such that

$$\sup_{x \in A} e_G(x) = \sup_{x \in A} \|x - u(x)\| \qquad (A \subset E). \tag{6.78}$$

In the particular case when A runs over all sets consisting of a single point $\{x\}$ the problem obviously reduces to the problem of the linearity of the mapping π_G, mentioned in section 6.1.

The deviation $\delta(A, G)$ defined by (6.61) is a "non-symmetric distance" in the set 2^E of all non-void bounded closed subsets of E (the equivalence $\delta(A, B) = 0 \Leftrightarrow A \subset B$ is immediate and the triangle inequality is shown in the same way as formula (6.64) of theorem 6.10), whence

$$\Delta(A, B) = \max(\delta(A, B), \delta(B, A)) \tag{6.79}$$

is a metric*) in the set 2^E; we mention that there are also other metrics which are introduced in this set in a natural way.

If A is a linear subspace $G_1 \not\subset \overline{G}$ of E, then obviously the deviation $\delta(G_1, G)$ is infinite. If G_1, G are linear subspaces of E, then instead of $\delta(G_1, G)$ one considers**)

$$\delta(\Sigma_{G_1}, G) = \sup_{\substack{x \in G_1 \\ \|x\| = 1}} \rho(x, G), \tag{6.80}$$

where $\Sigma_{G_1} = \{x \in G_1 \mid \|x\| = 1\}$. The number

$$\theta(G_1, G) = \begin{cases} \max(\delta(\Sigma_{G_1}, G), \delta(\Sigma_G, G_1)) & \text{if } G_1 \neq \{0\},\ G_2 \neq \{0\}, \\ 1 & \text{otherwise} \end{cases} \tag{6.81}$$

is called the *opening* between the subspaces G_1 and G. There are various properties of G_1, for instance the dimension, which are conserved by the linear subspaces G for which the opening $\theta(G_1, G)$ is sufficiently small; for some results of this type see Chap. II, § 6, section 6.1.

The opening $\theta(G_1, G)$ defined by (6.81) is not a metric in the set $\mathfrak{S}(E)$ of all closed linear subspaces of the space E, since in general it does not satisfy the triangle inequality.

---

*) This metric has been introduced by D. Pompeiu [178]; see also F. Hausdorff [84], Chap. VIII, § 6.

**) Let us also mention the number $\widehat{G_1, G} = \inf_{\substack{x \in G_1 \\ \|x\| = 1}} \rho(x, G) = \rho(\Sigma_{G_1}, G)$, called the *inclination* of G_1 to G. There are other terms also in the literature, such as "index of disjunction" (E. R. Lorch [138], D. del Pasqua [172]).

For this reason I. Ts. Gohberg and A. S. Markus [76] have modified the definition of the opening, putting

$$\tilde{\theta}(G_1, G_2) = \begin{cases} \Delta(\Sigma_{G_1}, \Sigma_{G_2}) \text{ if } G_1 \neq \{0\},\ G_2 \neq \{0\}, \\ 1 \text{ otherwise} \end{cases} \tag{6.82}$$

where $\Sigma_{G_i} = \{x \in G_i |\ \|x\| = 1\}$ $(i = 1,2)$, and Δ is the distance defined by (6.79). The opening modified in this way constitutes a metric in the set $\mathfrak{S}(E)$. Some properties of the set $\mathfrak{S}(E)$ endowed with the metric $\tilde{\theta}(G_1, G)$ have been given by I. Ts. Gohberg and A. S. Markus [76]; they have also remarked [76] that for every pair of linear subspaces $G_1, G \in \mathfrak{S}(E)$ we have

$$\theta(G_1, G) \leqslant \tilde{\theta}(G_1, G) \leqslant 2\theta(G_1, G), \tag{6.83}$$

whence θ determines on $\mathfrak{S}(E)$ the same uniformity as the metric $\tilde{\theta}$ (the uniformity determined by θ has as a basis the sets

$$\{\{G_1, G\} \in \mathfrak{S}(E) \times \mathfrak{S}(E) \mid \theta(G_1, G) < r\}, \tag{6.84}$$

where $r > 0$). We mention that there also exist other metrics which are introduced in a natural way in the set $\mathfrak{S}$(E); for relations between these metrics and the metric $\tilde{\theta}$ see e.g. E. Berkson [14].

6.5. ELEMENTS OF ε-APPROXIMATION

In the proofs given in the preceding sections for theorems 6.5 and 6.10 as well as in other arguments (e.g. in the proof of the remark at the end of section 1.13 of §1) we have seen the usefulness of considering elements $g_0 \in G$ with the property (6.35). If E is a normed linear space, G a linear subspace of E, $x \in E$ and $\varepsilon > 0$, an element $g_0 \in G$ is said to be *an element of* ε-*approximation* of x (by means of the elements of G) if we have (6.35), and we shall denote by $\mathfrak{L}_G(x, \varepsilon)$ the set of all elements of ε-approximation of x. In particular, for $\varepsilon = 0$ we find again the elements of best approximation of x and respectively the set $\mathfrak{L}_G(x)$. One of the advantages of considering the sets $\mathfrak{L}_G(x, \varepsilon)$ with $\varepsilon > 0$, instead of the sets $\mathfrak{L}_G(x)$, is that the sets $\mathfrak{L}_G(x, \varepsilon)$ are always non-void for $\varepsilon > 0$.

Explicitly the elements of ε-approximation and the sets $\mathfrak{L}_G(x, \varepsilon)$ have been introduced by R. C. Buck [29]*), who

---

*) Instead of the term "element of ε-approximation" R. C. Buck [29] uses the term "element of good approximation".

has also given a characterization of the elements of ε-approximation. We shall give here another characterization of these elements, which is more convenient for applications in concrete spaces.

THEOREM 6.12. *Let E be a normed linear space, G a linear subspace of E, $x \in E \setminus \bar{G}$, $g_0 \in G$ and $\varepsilon > 0$. We have $g_0 \in \mathfrak{L}_G(x, \varepsilon)$ if and only if there exists an $f \in E^*$* with the following properties:

$$\|f\| = 1, \tag{6.85}$$

$$f(g) = 0 \qquad (g \in G), \tag{6.86}$$

$$f(x - g_0) \geqslant \|x - g_0\| - \varepsilon. \tag{6.87}$$

Proof. Assume that $g_0 \in \mathfrak{L}_G(x, \varepsilon)$. Since by our hypothesis $d = \rho(x, G) > 0$, by virtue of a corollary of the Hahn-Banach theorem there exists an $f_0 \in E^*$ such that $\|f_0\| = \frac{1}{d}$, $f_0(g) = 0$ $(g \in G)$ and $f_0(x) = 1$, hence the functional $f = df_0 \in E^*$ satisfies (6.85), (6.86) and $f(x - g_0) = d$. Since $g_0 \in \mathfrak{L}_G(x, \varepsilon)$, we have then

$$\|x - g_0\| \leqslant d + \varepsilon = f(x - g_0) + \varepsilon,$$

whence also (6.87).

Conversely, assume that there exists an $f \in E^*$ satisfying (6.85), (6.86) and (6.87). Then for every $g \in G$ we have

$$\|x - g_0\| \leqslant |f(x - g_0)| + \varepsilon = |f(x - g)| + \varepsilon \leqslant$$
$$\leqslant \|f\| \|x - g\| + \varepsilon = \|x - g\| + \varepsilon,$$

whence

$$\|x - g_0\| \leqslant \inf_{g \in G} \|x - g\| + \varepsilon,$$

and thus $g_0 \in \mathfrak{L}_G(x, \varepsilon)$, which completes the proof of theorem 6.12.

In the particular case when $\varepsilon = 0$, theorem 6.12 reduces to theorem 1.1 on the characterization of elements of best approximation. In the general case the above proof of theorem 6.12 is similar to that of theorem 1.1. As in the case of theorem 1.1, one can give equivalent variants of the conditions of theorem 6.12, corresponding to those of corollary 1.1. We leave to the reader the geometrical interpretation of theorem 6.12 as well as the application of theorem 6.12 in various concrete spaces.

CHAPTER II

BEST APPROXIMATION IN NORMED LINEAR SPACES BY ELEMENTS OF LINEAR SUBSPACES OF FINITE DIMENSION

An important particular case of best approximation in normed linear spaces E by elements of linear subspaces G is that when the dimension (real or complex, according to the space E) of G is finite : $\dim G = n < \infty$. In this case*) $G = [x_1, \ldots, x_n]$, where $x_1, \ldots, x_n$ are linearly independent elements of G, and every element $g \in G$ can be written, uniquely, in the form

$$g = \sum_{k=1}^{n} \alpha_k x_k, \tag{1.1}$$

where $\alpha_1, \ldots, \alpha_n$ are scalars (real or complex, according to the space E); the linear combinations (1.1) are also called *polynomials* (in $x_1, \ldots, x_n$).

Naturally, the results of the preceding chapter are applicable, in particular, also in the case of polynomial best approximation. However, by making use of the fact that**) $\dim G = n$, one can obtain additional results, which we shall present in this chapter.

Since the problem of existence of polynomials of best approximation has an affirmative answer (Chap. I, corollary 2.2), we shall have to consider only the other problems of polynomial best approximation.

---

*) We denote by $[x_1, \ldots, x_n]$ the linear subspace of E spanned by the elements $x_1, \ldots, x_n$. Naturally, in all results concerning best approximation by elements of n-dimensional linear subspaces, we shall understand, without any special mention, that $n + 1 \leqslant \dim E \leqslant \infty$ (in order that the hypotheses $G \subset E$, $\dim G = n$, $x \in E \setminus G$ have sense).

**) Whenever we write $\dim X = N$, we shall understand that $1 \leqslant N < \infty$.

§1. CHARACTERIZATIONS OF POLYNOMIALS OF BEST APPROXIMATION

1.1. PRELIMINARY LEMMAS

LEMMA 1.1 ([215], proposition 1*)). *Let E be a Banach space of finite dimension k and let $\varphi \in E^*$, $\|\varphi\| = 1$. Then there exist h extremal points $\varphi_1, \ldots, \varphi_h$ of the unit cell $S_{E^*} \subset E^*$, where $1 \leqslant h \leqslant k$ if the scalars are real and $1 \leqslant h \leqslant 2k - 1$ if the scalars are complex, and h numbers $\lambda_1, \ldots, \lambda_h > 0$ with $\sum_{j=1}^{h} \lambda_j = 1$, such that*

$$\varphi = \sum_{j=1}^{h} \lambda_j \varphi_j. \tag{1.2}$$

Proof. Assume first that the scalars are real. Let Γ be the smallest extremal subset of S_{E^*} containing φ and let**) $h = \dim \Gamma + 1$. Since $\dim E^* = k$, we have $\dim \Gamma \leqslant k - 1$, whence $h \leqslant k$. We shall prove now the statement of the lemma by induction on $\dim \Gamma$.

If $\dim \Gamma = 0$, then by the definition of Γ we have $\varphi \in \mathcal{E}(S_{E^*})$, hence the statement is true (taking $\lambda_1 = 1$). Assume now that it is true for all Γ_1 with $\dim \Gamma_1 < m$ (where $m \leqslant k - 1$), and let $\dim \Gamma = m$. By the definition of Γ we have $\varphi \in \operatorname{Int}_l \Gamma$, hence we may consider the Minkowsky***) functional $p = p_{\Gamma, \varphi}$ of Γ with respect to φ. Let $\varphi_0 \in \mathcal{E}(\Gamma) = \mathcal{E}(S_{E^*}) \cap \Gamma$ be arbitrary and let

$$\chi(\lambda) = p[\lambda \varphi_0 + (1 - \lambda) \varphi] \qquad (-\infty < \lambda < \infty). \tag{1.3}$$

---

*) Actually, in [215] only the case of real scalars has been considered; for the method of passing to complex scalars, used in the present proof of lemma 1.1, see [232], p. 347–348. In [215] it has also been remarked that $\varphi_1, \ldots, \varphi_h$ are necessarily non-opposite two by two, and in the next paragraph we shall see (lemma 2.2) that they may be taken even "real-linearly independent"; however, we shall not use these properties in the present paragraph. For $k = 3$ the real case of lemma 1.1 has been given, essentially, by H. Minkowsky [151] and for an arbitrary k it has been given in [215], proposition 1 and, independently, by E. Ya. Remez ([187], p. 431, theorem 12), with a similar proof.

**) For the definitions of $\dim \Gamma$, $\operatorname{Int}_l \Gamma$, $\operatorname{Fr}_l \Gamma$ and $l(\Gamma)$, see Chap. I, § 4.

***) We recall that $p = p_{\Gamma, \varphi}$ is defined on the linear manifold $l(\Gamma)$ spanned by Γ, by

$$p(\psi) = \inf_{\substack{\mu > 0 \\ \frac{1}{\mu}(\psi - \varphi) \in \Gamma - \varphi}} \mu \qquad (\psi \in l(\Gamma)),$$

whence we have

$$\operatorname{Int}_l \Gamma = \{\psi \in l(\Gamma) \mid p(\psi) < 1\},$$

$$\operatorname{Fr}_l \Gamma = \{\psi \in l(\Gamma) \mid p(\psi) = 1\}.$$

Then we have, taking into account the definition of p,

$$\chi(0) = p(\varphi) = 0, \quad \lim_{\lambda \to -\infty} \chi(\lambda) = +\infty.$$

Since χ is continuous, it follows that there exists a $\lambda' < 0$, and hence also a $\psi = \lambda'\varphi_0 + (1 - \lambda')\,\varphi \in E^*$ with $\lambda' < 0$, such that

$$1 = \chi(\lambda') = p[\lambda'\varphi_0 + (1 - \lambda')\varphi] = p(\psi), \tag{1.4}$$

i.e. such that $\psi \in \mathrm{Fr}_l \Gamma$. Let Γ_1 be the smallest extremal subset of Γ, whence also of S_{E^*} (since Γ is an extremal subset of S_{E^*}), containing ψ. Then*) $\dim \Gamma_1 < \dim \Gamma = m$ (since $\Gamma_1 \subset H \cap \Gamma$, where H is a hyperplane in $l(\Gamma)$ supporting Γ at the point ψ). Consequently, by virtue of the induction hypothesis there exist $h_1 = \dim \Gamma_1 + 1 \leq m$ extremal points $\varphi_1, \ldots, \varphi_{h_1} \in \mathcal{E}(S_{E^*})$ and h_1 numbers $\mu_1, \ldots, \mu_{h_1} > 0$ with $\sum_{j=1}^{h_1} \mu_j = 1$, such that

$$\psi = \sum_{j=1}^{h_1} \mu_j \varphi_j,$$

whence

$$\varphi = \frac{1}{1-\lambda'}\psi + \frac{-\lambda'}{1-\lambda'}\varphi_0 = \sum_{j=1}^{h_1} \frac{\mu_j}{1-\lambda'}\varphi_j + \frac{-\lambda'}{1-\lambda'}\varphi_0.$$

Putting $h = h_1 + 1$, $\varphi_h = \varphi_0$ and

$$\lambda_j = \frac{\mu_j}{1-\lambda'} \quad (j = 1, \ldots, h_1), \quad \lambda_h = \frac{-\lambda'}{1-\lambda'},$$

and taking into account $h = h_1 + 1 \leq m + 1 = \dim \Gamma + 1$, $\lambda' < 0$, it follows that the statement is true also for $\dim \Gamma = m$, which completes the proof of lemma 1.1 in the case when the scalars are real.

Now assume that the scalars are complex and let Γ again be the smallest extremal subset of S_{E^*} containing φ. Since $\dim E < \infty$, there exists an $x_0 \in E$ with $\|x_0\| = 1$ such that $\varphi(x_0) = 1$. Then, by Chap. I, corollary 1.8, the set $\mathfrak{M}_{x_0} = \{\psi \in E^* \mid \|\psi\| = 1,\ \psi(x_0) = 1\}$ is an extremal subset of S_{E^*}, containing φ, hence we have

$$\Gamma \subset \mathfrak{M}_{x_0} \subset \{\psi \in E^* \mid \psi(x_0) = 1\},$$

whence $\dim \Gamma \leq k - 1$. Consequently, considering E as a $2k$-dimensional real Banach space $E_{(r)}$ in the usual way, we have $\dim \Gamma \leq 2k - 2$, whence, taking into account $\|\mathrm{Re}\, \varphi\| =$

*) This is the point where the hypothesis that the scalars are real is used essentially.

$= \|\varphi\| = 1$ *) and lemma 1.1 proved above for real scalars, it follows that there exist h extremal points $\Phi_1, \ldots, \Phi_h$ of Γ, whence also of $S_{(E_{(r)})^*}$, where $1 \leqslant h = \dim \Gamma + 1 \leqslant 2k - 1$, and h numbers $\lambda_1, \ldots, \lambda_h > 0$ with $\sum_{j=1}^{h} \lambda_j = 1$, such that we have

$$\operatorname{Re} \varphi = \lambda_1 \Phi_1 + \ldots + \lambda_h \Phi_h. \tag{1.5}$$

Put

$$\varphi_j(x) = \Phi_j(x) - \mathrm{i}\, \Phi_j(\mathrm{i}x) \quad (x \in E,\ j = 1, \ldots, h), \tag{1.6}$$

where $\mathrm{i} = \sqrt{-1}$. Since $\Phi \in \mathcal{E}(S_{(E\ \ )^*})$ $(j = 1, \ldots, h)$ and since the mapping $\Phi \to \psi$, where $\psi(x) = \Phi(x) - \mathrm{i}\, \Phi(\mathrm{i}x)$ $(x \in E)$, is a linear isometry of $(E_{(r)})^*$ onto $(E^*)_{(r)}$, we have $\varphi_j \in \mathcal{E}(S_{(E^*)_{(r)}})$, whence

$$\varphi_j \in \mathcal{E}(S_{E^*}) \qquad (j = 1, \ldots, h).$$

Since by (1.5) and (1.6) we have

$$\varphi(x) = \operatorname{Re}\varphi(x) - \mathrm{i} \operatorname{Re}\varphi(\mathrm{i}x) = \sum_{j=1}^{h} \lambda_j \Phi_j(x) - \mathrm{i} \sum_{j=1}^{h} \lambda_j \Phi_j(\mathrm{i}x) =$$

$$= \sum_{j=1}^{h} \lambda_j [\Phi_j(x) - \mathrm{i}\, \Phi_j(\mathrm{i}x)] = \sum_{j=1}^{h} \lambda_j \varphi_j(x) \quad (x \in E),$$

it follows that we have (1.2) also in the case of complex scalars, which completes the proof of lemma 1.1.

LEMMA 1.2 ([209], theorem 5 and [232], theorem 2). *Let E be a normed linear space, G a linear subspace of E and φ an extremal point of the unit cell $S_{G^*} \subset G^*$. Then there exists an extension of φ to an extremal point of the unit cell $S_{E^*} \subset E^*$.*

Proof. We shall show first that the set $\mathcal{A}_\varphi$ defined by

$$\mathcal{A}_\varphi = \{f \in E^* \,|\, f|_G = \varphi, \|f\| = 1\} \tag{1.7}$$

is a non-void extremal subset of the cell S_{E^*} endowed with $\sigma(E^*, E)$.

Indeed, the set $\mathcal{A}_\varphi$ is non-void by virtue of the Hahn-Banach theorem. Also, it is convex and $\sigma(E^*, E)$-closed, since

$$\mathcal{A}_\varphi = \bigcap_{x \in G} \{f \in E^* \,|\, f(x) = \varphi(x)\} \cap S_{E^*}. \tag{1.8}$$

Finally, let $f_1, f_2 \in S_{E^*}$, $f_1 \neq f_2$ and $0 \leqslant \lambda \leqslant 1$ be such that

$$\lambda f_1 + (1 - \lambda) f_2 \in \mathcal{A}_\varphi. \tag{1.9}$$

---

*) See e.g. N. Dunford and J. Schwartz [49], p. 64. We recall that $\operatorname{Re} \varphi \in (E_{(r)})^*$ is defined by $(\operatorname{Re} \varphi)(x) = \operatorname{Re} \varphi(x)$ $(x \in E_{(r)})$.

Put

$$\varphi_1 = f_1|_G, \quad \varphi_2 = f_2|_G. \tag{1.10}$$

Then $\varphi_1, \varphi_2 \in S_{G^*}$ and by (1.9) and (1.7) we have

$$\varphi = \lambda\varphi_1 + (1 - \lambda)\varphi_2.$$

Taking into account the hypothesis $\varphi \in \mathcal{E}(S_{G^*})$, it follows that we have $\varphi_1 = \varphi_2 = \varphi$, whence, by (1.10), $\|f_1\| = \|f_2\| = 1$*) and (1.7) we obtain $f_1, f_2 \in \mathcal{A}_\varphi$, which proves that $\mathcal{A}_\varphi$ is a non-void extremal subset of the cell S_{E^*} endowed with $\sigma(E^*, E)$.

But, by virtue of the Krein-Milman theorem, the set $\mathcal{A}_\varphi$ has then an extremal point f, which is necessarily an extremal point of S_{E^*} too, completing the proof of lemma 1.2.

As has been remarked in [209], p. 106, the converse of teh statement of lemma 1.2 is not true: an extension $f \in E^*$ of a $\varphi \in G^*$ of norm $\|\varphi\| = 1$ may be in $\mathcal{E}(S_{E^*})$ even if $\varphi \notin \mathcal{E}(S_{G^*})$. Indeed, this happens for instance when $E = C(Q)$, $G = \left\{y \in E \mid y(q_0) = \frac{y(q_1) + y(q_2)}{2}\right\}$, where q_0, q_1, q_2 are distinct points of Q, $\varphi(y) = \frac{y(q_1) + y(q_2)}{2}$ $(y \in G)$ and $f(x) = x(q_0)$ $(x \in E)$.

Since the functional $\varphi \in S_{G^*} \setminus \mathcal{E}(S_{G^*})$ in this example admits also non-extremal extensions of norm 1, e.g. $f'(x) = \frac{x(q_1) + x(q_2)}{2}$ $(x \in E)$, we remark that for the linear subspace $G = \{y \in E \mid y(q_0) = 0\}$ of $E = C(Q)$ the zero functional $\varphi_0 = 0 \in S_{G^*} \setminus \mathcal{E}(S_{G^*})$ does not admit a non-extremal extension f of norm $\|f\| = 1$. Indeed, for any extension $f \in E^*$ of φ_0 the hyperplane $\{y \in E \mid f(y) = 0\}$ coincides with the hyperplane G, whence every extension $f \in E^*$ of φ_0 is of the form

$$f(y) = \alpha y(q_0) \qquad (y \in E),$$

where α is a scalar of modulus $|\alpha| = \|f\|$, and for $|\alpha| = 1$ these functionals are, as we have seen in Chap. I, extremal points of S_{E^*}.

This being said, we can now prove the following generalization of lemma 1.1 which will play in the sequel a fundamental role:

LEMMA 1.3 ([213], proposition 4). *Let E be a normed linear space, E_k a k-dimensional linear subspace of E and $f \in E^*$, $\|f|_{E_k}\| = 1$. Then there exist h extremal points $f_1, \ldots, f_h$ of the unit cell $S_{E^*} \subset E^*$, where $1 \leqslant h \leqslant k$ if the scalars are real and*

*) Indeed $1 = \|\lambda f_1 + (1 - \lambda)f_2\| \leqslant \lambda\|f_1\| + (1 - \lambda)\|f_2\| \leqslant 1$.

$1 \leqslant h \leqslant 2k - 1$ *if the scalars are complex, and* h *numbers* $\lambda_1, \ldots, \lambda_h > 0$ *with* $\sum_{j=1}^{h} \lambda_j = 1$, *such that*

$$f(y) = \sum_{j=1}^{h} \lambda_j f_j(y) \qquad (y \in E_k). \tag{1.11}$$

Proof. Put

$$\varphi = f|_{E_k}. \tag{1.12}$$

Then $\varphi \in E_k^*$, $\|\varphi\| = 1$, whence by virtue of lemma 1.1 there exist h extremal points $\varphi_1, \ldots, \varphi_h \in \mathcal{E}(S_{E_k^*})$, where $1 \leqslant h \leqslant k$ if the scalars are real and $1 \leqslant h \leqslant 2k - 1$ if the scalars are complex, and h numbers $\lambda_1, \ldots, \lambda_h > 0$ with $\sum_{j=1}^{h} \lambda_j = 1$, such that we have (1.2). But, by virtue of lemma 1.2, the extremal functionals $\varphi_1, \ldots, \varphi_h \in \mathcal{E}(S_{E_k^*})$ can be extended to extremal functionals $f_1, \ldots, f_h \in \mathcal{E}(S_{E^*})$, which, by (1.2) and (1.12), will satisfy (1.11), which completes the proof.

1.2. CHARACTERIZATIONS OF POLYNOMIALS OF BEST APPROXIMATION IN GENERAL NORMED LINEAR SPACES

THEOREM 1.1 ([212], theorem 3.1 and [232] theorem 4). *Let E be a normed linear space, $G = [x_1, \ldots, x_n]$ an n-dimensional linear subspace of E, $x \in E \setminus G$ and $g_0 \in G$. The following statements are equivalent:*

1° $g_0 \in \mathfrak{L}_G(x)$.

2° *There exist h extremal points $f_1, \ldots, f_h$ of S_{E^*}, where $1 \leqslant h \leqslant n + 1$ if the scalars are real and $1 \leqslant h \leqslant 2n + 1$ if the scalars are complex and h numbers $\lambda_1, \ldots, \lambda_h > 0$ with $\sum_{j=1}^{h} \lambda_j = 1$, such that*

$$\sum_{j=1}^{h} \lambda_j f_j(g) = 0 \qquad (g \in G), \tag{1.13}$$

$$\sum_{j=1}^{h} \lambda_j f_j(x - g_0) = \|x - g_0\|. \tag{1.14}$$

3° *There exist h extremal points $f_1, \ldots, f_h$ of S_{E^*}, where $1 \leqslant h \leqslant n + 1$ if the scalars are real and $1 \leqslant h \leqslant 2n + 1$ if the scalars are complex, and h numbers $\lambda_1, \ldots, \lambda_h > 0$ with $\sum_{j=1}^{h} \lambda_j = 1$, such that we have (1.13) and*

$$f_j(x - g_0) = \|x - g_0\| \qquad (j = 1, \ldots, h). \tag{1.15}$$

4° *There exist* h *extremal points* $f_1, \ldots, f_h$ *of* S_{E^*}, *where* $1 \leqslant h \leqslant n+1$ *if the scalars are real and* $1 \leqslant h \leqslant 2n+1$ *if the scalars are complex, and* h *numbers* $\mu_1, \ldots \mu_h \neq 0$ *with* $\sum_{j=1}^{h} |\mu_j| = 1$, *such that*

$$\sum_{j=1}^{h} \mu_j f_j(x_k) = 0 \quad (k = 1, \ldots, n), \tag{1.16}$$

$$\sum_{j=1}^{h} \mu_j f_j(x - g_0) = \|x - g_0\|. \tag{1.17}$$

5° *There exist* h *extremal points* $f_1, \ldots, f_h$ *of* S_{E^*}, *where* $1 \leqslant h \leqslant n+1$ *if the scalars are real and* $1 \leqslant h \leqslant 2n+1$ *if the scalars are complex, and* h *numbers* $\mu_1, \ldots, \mu_h \neq 0$ *with* $\sum_{j=1}^{h} |\mu_j| = 1$, *such that we have (1.16) and*

$$f_j(x - g_0) = (\operatorname{sign} \mu_j) \|x - g_0\| \qquad (j = 1, \ldots, h). \tag{1.18}$$

Proof. Assume that we have 1°. Then there exists, by virtue of theorem 1.1 of Chap. I, a functional $f \in E^*$ satisfying the conditions (1.2), (1.3) and (1.4) of Chap. I. Let E_{n+1} be the $(n+1)$-dimensional linear subspace

$$E_{n+1} = [x_1, \ldots, x_n, x] = G \oplus [x] \tag{1.19}$$

of E. Then the conditions (1.2) and (1.4) of Chap. I imply $\|f|_{E_{n+1}}\| = 1$. Consequently, by virtue of lemma 1.3 there exist h extremal points $f_1, \ldots, f_h$ of S_{E^*}, where $1 \leqslant h \leqslant n+1$ if the scalars are real and $1 \leqslant h \leqslant 2n+1$ if the scalars are complex, and h numbers $\lambda_1, \ldots, \lambda_h > 0$ with $\sum_{j=1}^{h} \lambda_j = 1$, such that

$$f(y) = \sum_{j=1}^{h} \lambda_j f_j(y) \quad (y \in E_{n+1}). \tag{1.20}$$

Taking into account the conditions (1.3) and (1.4) of Chap. I, it follows that we have (1.13) and (1.14). Thus, 1°⇒2°.

Assume now that we have 2° and that there exists a j_0 with $1 \leqslant j_0 \leqslant h$, such that

$$\operatorname{Re} f_{j_0}(x - g_0) < \|x - g_0\|. \tag{1.21}$$

We have then, taking into account (1.14),

$$\|x - g_0\| = \operatorname{Re} \sum_{j=1}^{h} \lambda_j f_j(x - g_0) = \sum_{j=1}^{h} \lambda_j \operatorname{Re} f_j(x - g_0) <$$

$$< \sum_{j=1}^{h} \lambda_j \|x - g_0\| = \|x - g_0\|, \tag{1.22}$$

which is absurd. Consequently, (1.21) is not possible, i.e. we have

$$\|x - g_0\| = \operatorname{Re} f_j(x - g_0) \leqslant |f_j(x - g_0)| \leqslant \|x - g_0\|$$

$$(j = 1, \ldots, h), \tag{1.23}$$

whence we have (1.15). Thus, $2° \Rightarrow 3°$.

The implication $3° \Rightarrow 5°$ is obvious, with $\mu_j = \lambda_j$ $(j = 1, \ldots, h)$. The implication $5° \Rightarrow 4°$ is also obvious.

Finally, assume that we have 4°. Then the functional $f \in E^*$ defined by

$$f = \sum_{j=1}^{h} \mu_j f_j \tag{1.24}$$

satisfies the conditions of theorem 1.1 of Chap. I, whence we have $g_0 \in \mathfrak{P}_G(x)$. Thus, $4° \Rightarrow 1°$, which completes the proof.

The additional information to theorem 1.1 of Chap. I, which we obtain in theorem 1.1 above for linear subspaces G of finite dimension n, is that for such subspaces the functional f of theorem 1.1 of Chap. I can be taken of the special form (1.24), where h, μ_j and f_j are as in 4° above, or even of the form of a "convex combination"

$$f = \sum_{j=1}^{h} \lambda_j f_j, \tag{1.25}$$

where h, λ_j and f_j are as in 2° above. From this remark it follows also the geometrical interpretation of theorem 1.1 above, namely that in the case of linear subspaces G of finite dimension n, the hyperplane H of theorem 1.2 of Chap. I may be taken as a convex combination of h extremal hyperplanes $H_1, \ldots, H_h$ (where $1 \leqslant h \leqslant n + 1$ if the scalars are real and $1 \leqslant h \leqslant 2n + 1$ if the scalars are complex), each of them supporting the cell $S(x, \|x - g_0\|)$ at the point g_0 (and H supports $S(x, \|x - g_0\|)$ in the set $H_1 \cap \ldots \cap H_h \cap S(x, \|x - g_0\|)$).

Using theorem 1.1 we shall now give, as we announced in Chap. I, the:

Third proof of theorem 1.12 of Chap. I ([232], p. 348). Let $g_0 \in \mathfrak{P}_G(x)$. Then, as has been remarked in the first proof of theorem 1.12, there exists an $f_0 \in E^*$ such that $f_0 \in \mathcal{E}(S_{E^*})$, $f_0(x - g_0) = \|x - g_0\|$. Assume that for all $f_0 \in E^*$ satisfying these conditions we have

$$\operatorname{Re} f_0(g_0) < 0.$$

Let G_0 be an arbitrary finite dimensional linear subspace of G, containing g_0 (e.g. one can take as G_0 the one-dimensional linear subspace of G spanned by g_0), say $\dim G_0 = n$. Then

$$\inf_{g \in G_0} \|x - g\| \leqslant \|x - g_0\| = \inf_{g \in G} \|x - g\| \leqslant \inf_{g \in G_0} \|x - g\|,$$

whence $g_0 \in \mathfrak{P}_{G_0}(x)$. Consequently, by virtue of theorem 1.1, there exist $f_1, \ldots, f_h \in \mathcal{E}(S_{E^*})$, where $1 \leqslant h \leqslant 2n + 1$, and $\lambda_1, \ldots, \lambda_h > 0$ with $\sum_{j=1}^{h} \lambda_j = 1$, such that we have (1.13) for G_0 and (1.15). Then by $f_1, \ldots, f_h \in \mathcal{E}(S_{E^*})$ and (1.15) we have, by virtue of our hypothesis,

$$\operatorname{Re} f_j(g_0) < 0 \qquad (j = 1, \ldots, h),$$

whence, by $\lambda_1, \ldots, \lambda_h > 0$, we obtain

$$\operatorname{Re} \sum_{j=1}^{h} \lambda_j f_j(g_0) < 0,$$

which contradicts (1.13). This completes the proof.

From theorem 1.1 it follows that in the case when $\dim G = n$, corollary 1.10 of Chap. I can be sharpened. Namely, we shall prove

COROLLARY 1.1. *Let E be a normed linear space, $G = [x_1, \ldots, x_n]$ an n-dimensional linear subspace of E, $x \in E \setminus G$ and $g_0 \in G$. Then there exist h extremal points $f_1, \ldots, f_h$ of S_{E^*}, where $1 \leqslant h \leqslant n + 1$ if the scalars are real and $1 \leqslant h \leqslant 2n + 1$ if the scalars are complex, such that*

$$\min_{g \in G} \|x - g\| = \min_{g \in G} \max_{1 \leqslant j \leqslant h} |f_j(x) - f_j(g)|, \tag{1.26}$$

and for every $g_0 \in \mathfrak{P}_G(x)$ the min *in the right side is attained (hence among the elements $g_0 \in G$ for which the* min *in the right side is attained, there exists at least one $g_0 \in \mathfrak{P}_G(x)$), and $f_1, \ldots, f_h$ may be chosen to be in the set $\mathcal{E}(\mathfrak{M}_{x, g_0})$, where*

$$\mathfrak{M}_{x, g_0} = \{f \in E^* \mid \|f\| = 1,\ f(x - g_0) = \|x - g_0\|\}. \tag{1.27}$$

Proof. Let $g_0 \in \mathfrak{P}_G(x)$ be arbitrary. Then, by virtue of theorem 1.1, there exist h extremal points $f_1, \ldots, f_h$ of S_{E^*}, where $1 \leqslant h \leqslant n + 1$ if the scalars are real and $1 \leqslant h \leqslant 2n + 1$ if the scalars are complex, and h numbers $\lambda_1, \ldots, \lambda_h > 0$ with $\sum_{j=1}^{h} \lambda_j = 1$, such that we have (1.13) and (1.15), whence also $f_j \in \mathcal{S}(\mathfrak{M}_{x, g_0})$ $(j = 1, \ldots, h)$. Fixing these $f_1, \ldots, f_h$, let E_1 be the space *) of all the h-tuples $\{f_1(y), \ldots, f_h(y)\} (y \in E)$, endowed with the usual vector operations and with the norm

$$\|\{f_1(y), \ldots, f_h(y)\}\|_{E_1} = \max_{1 \leqslant j \leqslant h} |f_j(y)|. \tag{1.28}$$

Define $\varphi \in E_1^*$ by

$$\varphi(\{f_1(y), \ldots, f_h(y)\}) = \sum_{j=1}^{h} \lambda_j f_j(y) \quad (\{f_1(y), \ldots, f_h(y)\} \in E_1). \tag{1.29}$$

We have then, by $\lambda_1, \ldots, \lambda_h > 0$, $\sum_{j=1}^{h} \lambda_j = 1$ and (1.28),

$$|\varphi(\{f_1(y), \ldots, f_h(y)\})| \leqslant \max_{1 \leqslant j \leqslant h} |f_j(y)| =$$

$$= \|\{f_1(y), \ldots, f_h(y)\}\|_{E_1} \quad (\{f_1(y), \ldots, f_h(y)\} \in E_1),$$

whence

$$\|\varphi\| \leqslant 1. \tag{1.30}$$

From (1.13), (1.30) and (1.15) we obtain, taking into account (1.29) and (1.28),

$$\varphi(\{f_1(g), \ldots, f_h(g)\}) = \sum_{j=1}^{h} \lambda_j f_j(g) = 0 \quad (g \in G), \tag{1.31}$$

$$\|\{f_1(x), \ldots, f_h(x)\} - \{f_1(g_0), \ldots, f_h(g_0)\}\|_{E_1} \geqslant |\varphi(\{f_1(x - g_0), \ldots, f_h(x - g_0)\})| = \left|\sum_{j=1}^{h} \lambda_j f_j(x - g_0)\right| = \sum_{j=1}^{h} \lambda_j f_j(x - g_0) = \|x - g_0\| \geqslant$$

$$\geqslant \|\{f_1(x), \ldots, f_h(x)\} - \{f_1(g_0), \ldots, f_h(g_0)\}\|_{E_1},$$

whence

$$\varphi(\{f_1(x), \ldots, f_h(x)\} - \{f_1(g_0), \ldots, f_h(g_0)\}) = \sum_{j=1}^{h} \lambda_j f_j(x - g_0) =$$

$$= \|\{f_1(x), \ldots, f_h(x)\} - \{f_1(g_0), \ldots, f_h(g_0)\}\|_{E_1}, \tag{1.32}$$

*) In other words, $E_1 = u(E) \subset l_h^\infty$, where the mapping $u: E \to l_h^\infty$ is defined by $u(y) = \{f_1(y), \ldots, f_h(y)\}$ $(y \in E)$.

and thus, taking into account (1.30),

$$\|\varphi\| = 1. \tag{1.33}$$

Consequently, applying theorem 1.1 of Chap. I to the linear subspace $G_1 = \{\{f_1(g),\ldots, f_h(g)\} \mid g \in G\}$ of E_1, it follows that we have $\{f_1(g_0),\ldots,f_h(g_0)\} \in \mathfrak{P}_{G_1}(\{f_1(x),\ldots,f_h(x)\})$. But this means, taking into account (1.15) and (1.28), that we have

$$\min_{g\in G} \|x-g\| = \|x-g_0\| = \max_{1\leqslant j\leqslant h} |f_j(x-g_0)| = \|\{f_1(x),\ldots$$

$$\ldots, f_h(x)\} - \{f_1(g_0),\ldots,f_h(g_0)\}\|_{E_1} = \min_{g\in G} \|\{f_1(x),\ldots,f_h(x)\} -$$

$$- \{f_1(g),\ldots,f_h(g)\}\|_{E_1} = \min_{g\in G} \max_{1\leqslant j\leqslant h} |f_j(x) - f_j(g)|,$$

which completes the proof of corollary 1.1.

In the particular case when E is a real Banach space, corollary 1.1 has been given, essentially, by A. L. Garkavi ([62], theorem 1); in the general case, it has been given, with the above proof, in the paper [233], theorem 3.

Using corollary 1.1, we shall now give, as we have announced in Chap. I, another proof of corollary 1.10 of Chap. I:

Second proof of corollary 1.10 of Chap. I. For dim $G = 1$ the statement follows immediately from corollary 1.1. For dim $G \geqslant 2$, let G_0 be an arbitrary 2-dimensional linear subspace of G, containing g_0. Then

$$\min_{g\in G_0}\|x-g\| \leqslant \|x-g_0\| = \min_{g\in G} \|x-g\| \leqslant \min_{g\in G_0} \|x-g\|,$$

whence $g_0 \in \mathfrak{P}_{G_0}(x)$. Consequently, by virtue of corollary 1.1, there exist $f_1,\ldots,f_h \in \mathcal{E}(\mathfrak{M}_{x,g_0})$, where $1 \leqslant h \leqslant 5$, such that we have (1.26) with G replaced by G_0, whence

$$\|x-g_0\| = \min_{g\in G_0} \|x-g\| = \min_{g\in G_0} \max_{1\leqslant j\leqslant h} |f_j(x) - f_j(g)| \leqslant$$

$$\leqslant \min_{g\in G_0} \max_{f\in\mathcal{E}(\mathfrak{M}_{x,g_0})} |f(x) - f(g)| \leqslant \min_{g\in G_0} \|x-g\|.$$

Since G_0 has been an arbitrary 2-dimensional linear subspace of G containing g_0, it follows that we have the equalities (1.126) of Chap. I, which completes the proof. This proof of corollary 1.10 of Chap. I was given in [233].

Now consider the problem of simultaneous characterization of a set of polynomials of best approximation. From theorem 1.1 and the fact that $\|x-g_0\| = \|x-g_0'\|$ for every pair of elements g_0, $g_0' \in \mathfrak{P}_G(x)$, there follows

COROLLARY 1.2. *Let E be a normed linear space, $G = [x_1, \ldots, x_n]$ an n-dimensional linear subspace of E, $x \in E \setminus G$ and $M \subset G$. The following statements are equivalent:*

1° $M \subset \mathfrak{L}_G(x)$.

2° *There exist h extremal points $f_1, \ldots, f_h$ of S_{E^*}, where $1 \leqslant h \leqslant n+1$ if the scalars are real and $1 \leqslant h \leqslant 2n+1$ if the scalars are complex, and h numbers $\lambda_1, \ldots, \lambda_h > 0$ with $\sum_{j=1}^{h} \lambda_j = 1$, such that we have (1.13) and*

$$f_j(x - g_0) = \|x - g_0\| \qquad (g_0 \in M). \tag{1.34}$$

Consequently, in this case we have

$$f_j(g_0 - g_0') = 0 \qquad (g_0, g_0' \in M,\ j = 1, \ldots, h). \tag{1.35}$$

The relation (1.35) has been given, essentially, in [211], p. 126, remark 2.

Finally, we mention the following theorem which gives a lower bound for $\min_{g \in G} \|x - g\|$:

THEOREM 1.2 ([223], theorem 1 and [233], theorem 2). *Let E be a normed linear space, $G = [x_1, \ldots, x_n]$ an n-dimensional linear subspace of E, $x \in E \setminus G$ and $g_0 \in G$ with the property that there exist h extremal points $f_1, \ldots, f_h$ of S_{E^*}, where $1 \leqslant h \leqslant n+1$ if the scalars are real and $1 \leqslant h \leqslant 2n+1$ if the scalars are complex, and h numbers $\mu_1, \ldots, \mu_h \neq 0$ with $\sum_{j=1}^{h} |\mu_j| = 1$, such that*

$$\sum_{j=1}^{h} \mu_j f_j(x_k) = 0 \qquad (k = 1, \ldots, n), \tag{1.36}$$

$$\operatorname{sign} \mu_1 f_1(x - g_0) = \ldots = \operatorname{sign} \mu_h f_h(x - g_0). \tag{1.37}$$

We have then

$$\min_{1 \leqslant j \leqslant h} |f_j(x - g_0)| \leqslant \min_{g \in G} \|x - g\|, \tag{1.38}$$

and the equality holds only if

$$|f_1(x - g_0)| = \ldots = |f_h(x - g_0)|. \tag{1.39}$$

Proof. Assume the contrary, i.e. that we have (1.36), (1.37) and

$$|f_j(x - g_0)| > \min_{g \in G} \|x - g\| \qquad (j = 1, \ldots, h). \tag{1.40}$$

Then, by (1.37), (1.40) and $\sum_{j=1}^{h} |\mu_j| = 1$,

$$\left| \sum_{j=1}^{h} \mu_j f_j (x - g_0) \right| = \sum_{j=1}^{h} |\mu_j| \, |f_j(x - g_0)| > \min_{g \in G} \|x - g\|. \quad (1.41)$$

On the other hand, by (1.36), $\|f_j\| = 1 \ (j = 1, \ldots, h)$ and $\sum_{j=1}^{h} |\mu_j| = 1$ we have

$$\left| \sum_{j=1}^{h} \mu_j f_j(x - g_0) \right| = \left| \sum_{j=1}^{h} \mu_j f_j (x - g) \right| \leqslant \left\| \sum_{j=1}^{h} \mu_j f_j \right\| \|x - g\| \leqslant$$

$$\leqslant \| x - g \| \quad (g \in G),$$

whence

$$\left| \sum_{j=1}^{h} \mu_j f_j (x - g_0) \right| \leqslant \min_{g \in G} \| x - g \|, \quad (1.42)$$

which contradicts (1.41). This proves that we have (1.38),

Now assume that in (1.38) the equality sign holds. Then. by (1.37) and $\sum_{j=1}^{h} |\mu_j| = 1$ we have

$$\left| \sum_{j=1}^{h} \mu_j f_j (x - g_0) \right| = \sum_{j=1}^{h} |\mu_j| \, | f_j(x - g_0) | \geqslant \min_{g \in G} \| x - g \|, \quad (1.43)$$

whence, taking into account (1.42) and (1.38) with the sign of equality, it follows that we have (1.39), which completes the proof of theorem 1.2.

Let us remark that even in the case when in (1.38) the equality sign holds, it may happen that $g_0 \notin \mathfrak{L}_G (x)$. In fact, let for instance $E = l_2^\infty =$ the space of all pairs of scalars $y = \{\eta_1, \eta_2\}$ endowed with the usual vector operations and with the norm $\|y\| = \max (|\eta_1|, |\eta_2|)$, and let $G = \{y \in E \mid \eta_2 = 0\}$, $x = \{0, 1\} \in E \setminus G$, $g_0 = \{2, 0\} \in G$. Then for $f_1 \in E^*$ defined by $f_1(y) = \eta_2 (y \in E)$ and for $\mu_1 = 1$ we have $f_1 \in \mathcal{E}(S_{E^*})$, (1.36), (1.37) and (1.38) with the equality sign, namely $f_1 (x - g_0) = = 1 = \min_{g \in G} \|x - g\|$, but at the same time

$$\|x - g_0\| = 2 > 1 = \min_{g \in G} \| x - g \|,$$

whence $g_0 \notin \mathfrak{L}_G(x)$.

However, if in (1.38) the equality sign holds and if, in addition, there exists an index j_0 with $1 \leqslant j_0 \leqslant h$, such that $|f_{j_0}(x - g_0)| = \|x - g_0\|$, then $g_0 \in \mathfrak{P}_G(x)$, since in this case we have, taking into account (1.42), (1.37), (1.39) and $\sum_{j=1}^{h} |\mu_j| = 1$,

$$\min_{g \in G} \|x - g\| \geqslant \left| \sum_{j=1}^{h} \mu_j f_j(x - g_0) \right| = \sum_{j=1}^{h} |\mu_j| |f_j(x - g_0)| = \|x - g_0\|.$$

Finally, we remark that in the particular case when the scalars are real, the condition (1.37) of theorem 1.2 may be replaced by the condition

$$\operatorname{sign} f_j(x - g_0) = \operatorname{sign} \mu_j \qquad (j = 1, \ldots, h). \quad (1.44)$$

Indeed, in this case the implication (1.44) $\Rightarrow$ (1.37) is obvious. Conversely, if we have (1.37) and sign $\mu_1 f_1(x - g_0) \geqslant 0$, then we have (1.44); on the other hand, if we have (1.37) and sign $\mu_1 f_1(x - g_0) < 0$, then, changing each μ_j by $-\mu_j$ $(j = 1, \ldots, h)$ (which does not alter (1.36) and (1.37)), we shall have again (1.44).

1.3. APPLICATIONS IN THE SPACES $C(Q)$, $C_R(Q)$ AND $C_0(T)$

THEOREM 1.3. *Let $E = C(Q)$ (Q compact*)), $G = [x_1, \ldots, x_n]$ an n-dimensional linear subspace of E, $x \in E \setminus G$ and $g_0 \in G$. The following statements are equivalent:*

1° $g_0 \in \mathfrak{P}_G(x)$.

2° *There exist h points $q_1, \ldots, q_h \in Q$, where $1 \leqslant h \leqslant n + 1$ if the scalars are real and $1 \leqslant h \leqslant 2n + 1$ if the scalars are complex, and h numbers $\mu_1, \ldots, \mu_h \neq 0$ with $\sum_{j=1}^{h} |\mu_j| = 1$, such that*

$$\sum_{j=1}^{h} \mu_j x_k(q_j) = 0 \qquad (k = 1, \ldots, n), \quad (1.45)$$

$$\sum_{j=1}^{h} \mu_j [x(q_j) - g_0(q_j)] = \max_{t \in Q} |x(t) - g_0(t)|. \quad (1.46)$$

3° *There exist h points $q_1, \ldots, q_h \in Q$, where $1 \leqslant h \leqslant n + 1$ if the scalars are real and $1 \leqslant h \leqslant 2n + 1$ if the scalars are*

*) Since by our hypothesis $n + 1 \leqslant \dim E \leqslant \infty$, Q must contain at least $n + 1$ distinct points, but we shall not mention explicitly this condition in the statements of the results which follow.

complex, and h numbers $\mu_1, \ldots, \mu_h \neq 0$ *with* $\sum_{j=1}^{h} |\mu_j| = 1$, *such that we have (1.45) and*

$$x(q_j) - g_0(q_j) = (\operatorname{sign} \mu_j) \max_{t \in Q} |x(t) - g_0(t)| \quad (j = 1, \ldots, h). \tag{1.47}$$

These statements are implied by — and in the case when the scalars are real, equivalent to — the following statement:

4° *There exist h points* $q_1, \ldots, q_h \in Q$, *where* $1 \leqslant h \leqslant n + 1$, *with the following properties:*

a) *the matrix*

$$\begin{pmatrix} x_1(q_1) & \ldots & x_1(q_h) \\ \ldots & \ldots & \ldots \\ x_n(q_1) & \ldots & x_n(q_h) \\ x(q_1) & \ldots & x(q_h) \end{pmatrix}$$

is of rank h;

b) *all minors of order h of this matrix which do not contain the last row of the matrix, are* $= 0$;

c) *among the minors of order h of this matrix, there exists at least a minor* $\Delta \neq 0$ *in which all cofactors* Δ_j *of the elements* $x(q_j)$ $(j = 1, \ldots, h)$ *are* $\neq 0$;

d) *We have*

$$x(q_j) - g_0(q_j) = \left(\operatorname{sign} \frac{\Delta_j}{\Delta} \right) \max_{t \in Q} |x(t) - g_0(t)| \; (j = 1, \ldots, h). \tag{1.48}$$

Proof. The equivalences $1° \Leftrightarrow 2° \Leftrightarrow 3°$ follow immediately from theorem 1.1, taking into account the general form of the extremal points of the unit cell $S_{C(Q)^*}$ (see Chap. I, section 1.10, formula (1.135)).

Now assume that we have 4° and that the minor Δ of c) is

$$\Delta = \begin{vmatrix} x_1(q_1) & \ldots & x_1(q_h) \\ \ldots & \ldots & \ldots \\ x_{h-1}(q_1) & \ldots & x_{h-1}(q_h) \\ x(q_1) & \ldots & x(q_h) \end{vmatrix};$$

this does not restrict the generality, since we can renumber, if necessary, the functions $x_1, \ldots, x_n$. Put

$$\mu_j = \frac{\Delta_j}{\Delta} \max_{t \in Q} |x(t) - g_0(t)| \qquad (j = 1, \ldots, h). \tag{1.49}$$

Then by c) we have $\mu_1, \ldots, \mu_h \neq 0$. Furthermore, we have

$$\sum_{j=1}^{h} \mu_j \, x_k(q_j) = \frac{1}{\Delta} \max_{t \in Q} |x(t) - g_0(t)| \sum_{j=1}^{h} x_k(q_j) \, \Delta_j = 0$$

$$(k = 1, \ldots, n),$$

i.e. (1.45); indeed, for $k = 1, \ldots, h-1$ these equalities follow from the definition of Δ and Δ_j, and for $k = h, \ldots, n$ they follow from b).

On the other hand, obviously we also have the equalities

$$\sum_{j=1}^{h} \mu_j [x(q_j) - g_0(q_j)] = \frac{1}{\Delta} \max_{t \in Q} |x(t) - g_0(t)| \sum_{j=1}^{h} x(q_j) \, \Delta_j =$$

$$= \max_{t \in Q} |x(t) - g_0(t)|,$$

i.e. (1.46). Finally, from (1.46) and d) it follows that we have

$$\max_{t \in Q} |x(t) - g_0(t)| = \sum_{j=1}^{h} \mu_j \, [x(q_j) - g_0(q_j)] =$$

$$= \left(\sum_{j=1}^{h} \mu_j \operatorname{sign} \frac{\Delta_j}{\Delta} \right) \max_{t \in Q} |x(t) - g_0(t)|,$$

whence, taking into account (1.49), we obtain

$$\sum_{j=1}^{h} |\mu_j| = \sum_{j=1}^{h} \mu_j \operatorname{sign} \mu_j = \sum_{j=1}^{h} \mu_j \operatorname{sign} \frac{\Delta_j}{\Delta} = 1.$$

Thus, $4^\circ \Rightarrow 2^\circ$.

Conversely, assume now that we have 2° and that the scalars are real. Then by (1.45) we have b). Now consider *) the system of h linear equations formed of the first $h-1$ relations (1.45) and of (1.46). Since this system admits by our hypothesis the solution $\mu_1, \ldots, \mu_h \neq 0$, it follows that the determinant Δ of this system is $\neq 0$ (hence we have a)) and, by Cramer's rule, that we have (1.49), whence also c). Finally, from (1.47) and (1.49) it follows that we have d). Thus, $2^\circ \Rightarrow 4^\circ$ for real scalars, which completes the proof of theorem 1.3.

The equivalence $1^\circ \Leftrightarrow 4^\circ$ of theorem 1.3 has been given, essentially, by E. Ya. Remez [184], and in the above form, by S. I. Zuhovitzky ([267], p. 156—157). The equivalence $1^\circ \Leftrightarrow 3^\circ$

---

*) Considering all systems of h linear equations formed of $h-1$ relations (1.45) and of (1.46), we see that c) may be replaced also by the following condition:

c′) *all minors* $\neq 0$ *of order* h *of the matrix in* a) *have the property that all cofactors of the elements* $x(q_j)$ $(j = 1, \ldots, h)$ *are* $\neq 0$.

has been given, essentially, by E. Ya. Remez [185], [186] (see also V. K. Ivanov [94], [95], [96], V. S. Vidensky [256], [257] and V. I. Smirnov—N. A. Lebedev [235], p. 401, theorem 2′). The above proofs of the equivalences 2°⇔4° and 1°⇔2°⇔3° of theorem 1.3 have been given in the papers [212], p. 187—188 and [217], p. 258 respectively.

The additional information to theorem 1.3 of Chap. I, which is obtained in theorem 1.3 above for subspaces G of finite dimension n, consists in that for such subspaces the Radon measure μ of theorem 1.3 of Chap. I may be taken with finite carrier $S(\mu) = \{q_1, \ldots, q_h\} \subset Q$, that is, of the form

$$\int_Q x(q)\, \mathrm{d}\mu(q) = \sum_{j=1}^{h} \mu_j x(q_j) \qquad (x \in E),$$

where h and μ_j are as above. For such a measure the condition (1.27) of Chap. I is satisfied, since as $\frac{\mathrm{d}\mu}{\mathrm{d}|\mu|}$ one may take any continuous function on Q which takes the value sign $\bar{\mu}_j$ at the point $q_j (j = 1, \ldots, h)$, and condition (1.28) of Chap. I is then equivalent to the condition (1.47) above, since we have

$$\operatorname{sign} \frac{\mathrm{d}\mu}{\mathrm{d}|\mu|}(q_j) = \operatorname{sign} \operatorname{sign} \bar{\mu}_j = \operatorname{sign} \frac{\mu_j}{|\mu_j|} = \operatorname{sign} \mu_j \ (j = 1, \ldots, h).$$

As has been remarked in [232], § 5, from theorem 1.3 above it follows immediately the following "variational lemma" of T. J. Rivlin and H. S. Shapiro ([194], theorem 1) which is similar to the necessity part of the theorem of E. W. Cheney and A. A. Goldstein given in Chap. I, the final part of section 1.4: *Let $E = C(Q)$ (Q compact), $G = [x_1, \ldots, x_n]$ an n-dimensional linear subspace of E and x an element of $E \setminus G$ such that $0 \in \mathfrak{P}_G(x)$. Then the element 0 of the usual n-dimensional linear space is in the convex hull of the points*

$$\{\overline{x(q)} x_1(q), \ldots, \overline{x(q)}\, x_n(q)\},$$

where q runs over the set

$$Y_x = \{q \in Q \mid |x(q)| = \|x\|\}.$$

Indeed, by (1.47) for $g_0 = 0$, we have

$$\mu_j = |\mu_j| \operatorname{sign} \bar{\mu}_j = |\mu_j| \frac{\overline{x(q_j)}}{\|x\|} \qquad (j = 1, \ldots, h),$$

whence, by (1.45),

$$\sum_{j=1}^{h} |\mu_j| \overline{x(q_j)} x_k(q_j) = 0,$$

which, taking into account that $\sum_{j=1}^{h} |\mu_j| = 1$ and that $q_j \in Y_x(j = 1,\ldots,h)$, completes the proof.

For a closed set $A \subset Q$ we shall denote by I_A the linear subspace

$$I_A = \{x \in C(Q) \,|\, x(q) = 0 \;\; (q \in A)\} \tag{1.50}$$

of $C(Q)$; in particular, $I_\varnothing = C(Q)$. *Theorem 1.3 is valid also for the spaces* $E = I_A$, *replacing the condition* $q_1,\ldots,q_h \in Q$ *by* $q_1,\ldots,q_h \in Q \setminus A$. Indeed, in this case every $f \in \mathcal{E}(S_{E^*})$ is*) of the form

$$f(y) = \alpha y(q) \qquad (y \in E = I_A),$$

where $|\alpha| = 1$ and $q \in Q \setminus A$, and in the rest the proof is the same as for theorem 1.3.

Now let T be a locally compact space and let $C_0(T)$ be the space of all continuous functions on T which vanish at the infinity (i.e. of all continuous functions x with the property that for every $\varepsilon > 0$ the set $\{t \in T \,|\, |x(t)| \geqslant \varepsilon\}$ is compact), endowed with the norm $\|x\| = \sup_{t \in T} |x(t)|$. Compactifying T by the addition of a point ∞, we obtain a compact space $Q = T \cup \{\infty\}$, and by an obvious embedding $C_0(T)$ may be considered as the linear subspace $I_{\{\infty\}}$ of $C(Q)$. Hence, by virtue of the above remark, *theorem 1.3 holds also for the spaces* $C_0(T)$**) (*T locally compact*); a direct proof of this latter result has been given by J. Bram [26].

Returning to the space $E = C(Q)$, we shall now consider the important particular case when the functions $x_1,\ldots,x_n$ which generate the linear subspace G form a Čebyšev system. We recall that a system of n functions $x_1,\ldots,x_n \in C(Q)$ (Q compact) is called ***) a *Čebyšev system* (on Q) if every non-zero polynomial (1.1) has at most $n-1$ zeros on Q; obviously, this happens if and only if for any n distinct points $q_1,\ldots,q_n \in Q$ we have

$$\begin{vmatrix} x_1(q_1) & \ldots & x_1(q_n) \\ \ldots & \ldots & \ldots \\ x_n(q_1) & \ldots & x_n(q_n) \end{vmatrix} \neq 0.$$

THEOREM 1.4. *Let* $E = C(Q)$ (Q *compact*), *let* $G = [x_1,\ldots,x_n]$ *be an* n*-dimensional linear subspace of* E *such that* $x_1,\ldots,x_n$

*) See Chap. I, section 1.11, formula (1.145).

**) Consequently, in particular, also for the space c_0, which is obtained by taking as T a countable set endowed with the discrete topology.

***) Following S. N. Bernstein [15].

form a Čebyšev system, and let $x \in E \setminus G$ *and* $g_0 \in G$. *The following statements are equivalent:*

1° $g_0 \in \mathfrak{P}_G(x)$.

2° *There exist* h *distinct points* $q_1, \ldots, q_h \in Q$, *where* $h = n + 1$ *if the scalars are real and* $n + 1 \leqslant h \leqslant 2n + 1$ *if the scalars are complex, and* h *numbers* $\mu_1, \ldots, \mu_h \neq 0$ *with* $\sum_{j=1}^{h} |\mu_j| = 1$, *such that we have (1.45) and (1.46).*

These statements are implied by — and in the case when the scalars are real, equivalent to — the following statement:

3° *There exist* $n + 1$ *distinct points* $q_1, \ldots, q_{n+1} \in Q$ *such that*

$$x(q_j) - g_0(q_j) = \left(\operatorname{sign} \frac{\Delta_j}{\Delta}\right) \max_{t \in Q} |x(t) - g_0(t)| \quad (j = 1, \ldots, n + 1), \tag{1.51}$$

where

$$\Delta = \begin{vmatrix} x_1(q_1) & \ldots & x_1(q_{n+1}) \\ \ldots & \ldots & \ldots \\ x_n(q_1) & \ldots & x_n(q_{n+1}) \\ x(q_1) & \ldots & x(q_{n+1}) \end{vmatrix} \neq 0 \tag{1.52}$$

and where Δ_j *is the cofactor of the element* $x(q_j)$ *in* Δ.

Proof: Assume that we have 1°. Then, by virtue of theorem 1.3, there exist h points $q_1, \ldots, q_h \in Q$, where $1 \leqslant h \leqslant n + 1$ if the scalars are real and $1 \leqslant h \leqslant 2n + 1$ if the scalars are complex, and h numbers $\mu_1, \ldots, \mu_h \neq 0$ with $\sum_{j=1}^{h} |\mu_j| = 1$, such that we have (1.45) and (1.46). We claim that the points $q_1, \ldots, q_h$ may be assumed to be distinct *). Indeed, let h be the smallest number with the property 2° of theorem 1.3; if we had $q_{j_1} = q_{j_2}$, then by (1.47) it would follow that $\operatorname{sign} \mu_{j_1} = \operatorname{sign} \mu_{j_2}$, whence $|\mu_{j_1} + \mu_{j_2}| = |\mu_{j_1}| + |\mu_{j_2}|$, whence putting $\mu' = \mu_{j_1} + \mu_{j_2}$, we would obtain relations of the form (1.45) and (1.46) for $h - 1$ points of Q and $h - 1$ non-zero numbers with the sum of their modules $= 1$, which contradicts the minimality of h. Thus, taking distinct $q_1, \ldots, q_h$, assume now that we have $h < n + 1$. Since by (1.45) all minors of order h of the matrix $(x_k(q_j))_{\substack{j=1,\ldots,h \\ k=1,\ldots,n}}$ are then zero, choosing arbitrarily distinct points **) $q_{h+1}, \ldots$

*) A similar remark is valid (by virtue of the same argument) for the other results of the present paragraph, concerning the existence of h extremal points $f_h \in \mathcal{E}(S_{E^*})$, or of h points $q_1, \ldots, q_h \in Q$, etc.

**) We recall that by hypothesis Q contains at least $n + 1$ distinct points.

$\ldots, q_n \in Q$, different from $q_1, \ldots, q_h$, it follows that $\det (x_k(q_j))_{\substack{j=1,\ldots,n \\ k=1,\ldots,n}} = 0$, which contradicts the hypothesis that $x_1, \ldots, x_n$ form a Čebyšev system. Consequently, we have $h \geqslant n+1$, whence $h = n+1$ if the scalars are real and $n+1 \leqslant h \leqslant 2n+1$ if the scalars are complex. Thus, $1^\circ \Rightarrow 2^\circ$.

The implications $3^\circ \Rightarrow 2^\circ \Rightarrow 1^\circ$ are obvious from theorem 1.3.

Finally, assume that we have 2° and that the scalars are real. Then $h = n+1$ in (1.45) and (1.46), whence, by Cramer's rule,

$$\mu_j = \frac{\Delta_j}{\Delta} \max_{t \in Q} |x(t) - g_0(t)| \qquad (j = 1, \ldots, n+1),$$

which, together with (1.47), implies (1.51). Thus, $2^\circ \Rightarrow 3^\circ$ for real scalars, which completes the proof of theorem 1.4.

Theorem 1.4 has been given, essentially, by E. Ya. Remez [184], [186] (see also V. K. Ivanov [94], [95], [96], V. S. Vidensky [256], [257] and V. I. Smirnov—N. A. Lebedev [235], p. 409, lemma 1). In the above form the equivalence $1^\circ \Leftrightarrow 3^\circ$ has been given by S. I. Zuhovitzky ([267], p. 152). The above proof of theorem 1.4 has been given in the paper [212], p. 187—188.

In the particular case when $Q = [a, b]$ and the scalars are real, the following classical "alternation theorem" of P. L. Čebyšev—S. N. Bernstein ([15], p. 3) results from theorem 1.4:

COROLLARY 1.3. *Let $E = C_R([a, b])$, let $G = [x_1, \ldots, x_n]$ be an n-dimensional linear subspace of E such that $x_1, \ldots, x_n$ form a Čebyšev system, and let $x \in E \setminus G$ and $g_0 \in G$. We have $g_0 \in \mathfrak{L}_G(x)$ if and only if there exist $n+1$ points $q_1 < q_2 < \cdots < q_{n+1}$ of $[a, b]$, at which the difference $x(q) - g_0(q)$ takes the value $\max_{t \in [a, b]} |x(t) - g_0(t)|$ with alternating signs (i.e. with opposite signs at consecutive points $q_j, q_{j+1} (j = 1, \ldots, n)$).*

This corollary is deduced from theorem 1.4 with the aid of the following lemma due to E. Ya. Remez [184]:

LEMMA 1.4. *Let $x_1, \ldots, x_n \in C_R([a, b])$ be a Čebyšev system on $[a, b]$. Then for any $n+1$ points $q_1 < q_2 < \cdots < q_{n+1}$ of $[a, b]$ we have*

$$0 \neq \operatorname{sign} \Delta_j = - \operatorname{sign} \Delta_{j-1} \quad (j = 2, \ldots, n+1), \tag{1.53}$$

where Δ_j is the cofactor of the element $x(q_j)$ in the determinant Δ defined by (1.52).

Proof (S. I. Zuhovitzky [267], p. 154—155). Since $x_1, \ldots, x_n$ is a Čebyšev system on $[a, b]$, we have

$$\Delta_j \neq 0 \qquad (j = 1, \ldots, n+1). \tag{1.54}$$

Let $n = 1$. Then we have

$$\Delta_1 = -x_1(q_2), \quad \Delta_2 = x_1(q_1),$$

whence, taking into account the hypothesis that x_1 forms a Čebyšev system on $[a, b]$, we infer (1.53) for $n = 1$.

Now let $n \geqslant 2$. Consider for every integer j with $2 \leqslant j \leqslant n + 1$ the polynomial

$$D(q_1, \ldots, q_{j-2}, q_{j+1}, \ldots, q_{n+1}) = \\ = \begin{vmatrix} x_1(q_1) & \ldots & x_1(q_{j-2}) & x_1(q) & x_1(q_{j+1}) & \ldots & x_1(q_{n+1}) \\ \cdot & \cdot & \cdot & \cdot & \cdot & \cdot & \cdot \\ x_n(q_1) & \ldots & x_n(q_{j-2}) & x_n(q) & x_n(q_{j+1}) & \ldots & x_n(q_{n+1}) \end{vmatrix} \quad (q \in [a, b]). \tag{1.55}$$

Obviously, the zeros of $D(q_1, \ldots, q_{j-2}, q, q_{j+1}, \ldots, q_{n+1})$ on $[a, b]$ are $q_1, \ldots, q_{j-2}, q_{j+1}, \ldots, q_{n+1}$ and only these points (since by our hypothesis $x_1, \ldots, x_n$ form a Čebyšev system on $[a, b]$). Consequently, this polynomial does not vanish on the interval (q_{j-2}, q_{j+1}), whence, taking into account that $q_{j-2} < q_{j-1} < q_j < q_{j+1}$, we infer

$$\operatorname{sign} D(q_1, \ldots, q_{j-2}, q_{j-1}, q_{j+1}, \ldots, q_{n+1}) = \\ = \operatorname{sign} D(q_1, \ldots, q_{j-2}, q_j, q_{j+1}, \ldots, q_{n+1}) \quad (j = 2, \ldots, n+1). \tag{1.56}$$

Since by definition

$$\Delta_j = (-1)^{n+1} D(q_1, \ldots, q_{j-2}, q_{j-1}, q_{j+1}, \ldots, q_{n+1}) \quad (j = 1, \ldots, n+1),$$

it follows that we have (1.53), which completes the proof of lemma 1.4.

This being said, we can now give the

Proof of corollary 1.3. Assume that we have $g_0 \in \mathfrak{L}_G(x)$. Then, by virtue of theorem 1.4 there exist $n + 1$ distinct points $q_1, \ldots, q_{n+1} \in [a, b]$ such that we have (1.51). Renumbering, if necessary, the points q_j, we may assume that we have $q_1 < q_2 < \ldots < q_{n+1}$. Then, by (1.52) and lemma 1.4, we have

$$\operatorname{sign} \frac{\Delta_j}{\Delta} = -\operatorname{sign} \frac{\Delta_{j-1}}{\Delta} \qquad (j = 2, \ldots, n+1),$$

and thus, taking into account (1.51), the condition of corollary 1.3 is satisfied.

Conversely, assume that the condition of corollary 1.3 is satisfied, i.e. that there exist $q_1 < q_2 < \ldots < q_{n+1}$ in $[a, b]$ and $\varepsilon = \pm 1$ such that

$$x(q_j) - g_0(q_j) = (-1)^j \varepsilon \max_{t \in [a,b]} |x(t) - g_0(t)| \quad (j = 1, \ldots, n+1). \tag{1.57}$$

By virtue of lemma 1.4 we have then

$$\Delta = \sum_{j=1}^{n+1} \Delta_j [x(q_j) - g_0(q_j)] = \sum_{j=1}^{n+1} \Delta_j (-1)^j \varepsilon \max_{t \in [a,b]} |x(t) - g_0(t)| =$$

$$= \varepsilon \max_{t \in [a,b]} |x(t) - g_0(t)| \sum_{j=1}^{n+1} |\Delta_j| \neq 0,$$

and we have either

$$(-1)^j \varepsilon = \operatorname{sign} \frac{\Delta_j}{\Delta} \qquad (j = 1, \ldots, n+1) \tag{1.58}$$

or

$$(-1)^j \varepsilon = - \operatorname{sign} \frac{\Delta_j}{\Delta} \qquad (j = 1, \ldots, n+1). \tag{1.59}$$

But, in the first case, from (1.58) and (1.57) it follows that we have (1.51) for $Q = [a, b]$, whence, by theorem 1.4, $g_0 \in \mathfrak{L}_G(x)$. In the second case, from (1.59) and (1.57) it follows that we have

$$x(q_j) - g_0(g_j) = -\left(\operatorname{sign} \frac{\Delta_j}{\Delta}\right) \max_{t \in (a,b]} |x(t) - g_0(t)|$$

$$(j = 1, \ldots, n+1), \tag{1.60}$$

whence, as in the proof of the implication $3° \Rightarrow 2°$ of theorem 1.4, it follows that the numbers $\mu_1, \ldots, \mu_{n+1} \neq 0$ defined by

$$\mu_j = -\frac{\Delta_j}{\Delta} \max_{t \in [a,b]} |x(t) - g_0(t)| \qquad (j = 1, \ldots, n+1) \tag{1.61}$$

satisfy (1.45) and (1.46) with $h = n + 1$ and $\sum_{j=1}^{n+1} |\mu_j| = 1$, and thus, by theorem 1.4, $g_0 \in \mathfrak{L}_G(x)$. This completes the proof of corollary 1.3.

Taking into account the general form of the extremal points of the unit cell $S_{C(Q)^*}$, from corollary 1.1 it follows

COROLLARY 1.4. *Let $E = C(Q)$ (Q compact), let $G = [x_1, \ldots, x_n]$ be an n-dimensional linear subspace of E, $x \in E \setminus G$ and $g_0 \in G$. Then there exist h points $q_1, \ldots, q_h \in Q$, where $1 \leqslant h \leqslant n + 1$ if the scalars are real and $1 \leqslant h \leqslant 2n + 1$ if the scalars are complex, such that*

$$\min_{g \in G} \|x - g\| = \min_{g \in G} \max_{1 \leqslant j \leqslant h} |x(q_j) - g(q_j)|, \tag{1.62}$$

and for every $g_0 \in \mathfrak{L}_G(x)$ *the* min *in the right side is attained* (*hence, among the elements* $g_0 \in G$ *for which the* min *in the right side is attained, there exists at least one* $g_0 \in \mathfrak{L}_G(x)$), *and* $q_1, \ldots, q_h$ *may be choosen such that we have*

$$|x(q_j) - g_0(q_j)| = \max_{t \in Q} |x(t) - g_0(t)| \qquad (j = 1, \ldots, h). \tag{1.63}$$

In the case when the scalars are real this result has been given, essentially, by E. Ya. Remez [184] (see also L. G. Schnirelman [241] and S. I. Zuhovitzky [267]); in the particular case when $Q = [a, b]$, $x_k(t) = t^{k-1}$ ($t \in [a, b]$, $k = 1, \ldots, n$), and the scalars are real, it has been given by Ch. J. de la Vallée-Poussin [255]. In the case of complex scalars it has been given by E. Ya. Remez [185], [186] (see also V. K. Ivanov [94], [95], [96], V. S. Videnský [256], [257] and V. I. Smirnov — N. A. Lebedev [235], p. 406, corollary 3).

Taking into account the general form of the extremal points of $S_{C(Q)^*}$, from theorem 1.2 there follows

THEOREM 1.5. *Let* $E = C(Q)$ (Q *compact*), *let* $G = [x_1, \ldots, x_n]$ *be an* n*-dimensional linear subspace of* E, $x \in E \setminus G$ *and* $g_0 \in G$ *with the property that there exist* h *points* $q_1, \ldots, q_h \in Q$, *where* $1 \leqslant h \leqslant n + 1$ *if the scalars are real and* $1 \leqslant h \leqslant 2n + 1$ *if the scalars are complex, and* h *numbers* $\mu_1, \ldots, \mu_h \neq 0$ *with* $\sum_{j=1}^{h} |\mu_j| = 1$, *such that*

$$\sum_{j=1}^{h} \mu_j x_k(q_j) = 0 \qquad (k = 1, \ldots, n), \tag{1.64}$$

$$\operatorname{sign} \mu_1 [x(q_1) - g_0(q_1)] = \ldots = \operatorname{sign} \mu_h [x(q_h) - g_0(q_h)]. \tag{1.65}$$

Then we have

$$\min_{1 \leqslant j \leqslant h} |x(q_j) - g_0(q_j)| \leqslant \min_{g \in G} \|x - g\|, \tag{1.66}$$

and the equality holds only if

$$|x(q_1) - g_0(q_1)| = \ldots = |x(q_h) - g_0(q_h)|. \tag{1.67}$$

The first part of theorem 1.5 has been given in [223], p. 320 (see also [233]). For real scalars, theorem 1.5 has been given, essentially, by E. Ya. Remez [184] and a condition of the type of condition 4° of theorem 1.3, equivalent to that of theorem 1.5 (hence sufficient for (1.66)), has been given by S. I. Zuhovitzky ([267], p. 158). In the case of complex scalars, another, more restrictive condition than that of theorem 1.5, sufficient for (1.66), has been given by E. Ya. Remez [186] (see also V. K. Ivanov [94], [95], [96], V. S. Videnský [256], [257]).

As has been remarked in section 1.1, even in the case when we have the sign of equality in (1.66), whence also (1.67), it may happen that $g_0 \notin \mathfrak{L}_G(x)$; indeed, the example of section 1.1 was actually in the space $E = C(Q)$, where Q consists of two points. However, it may be shown that *if in addition to the hypotheses of theorem 1.5, $x_1, \ldots, x_n$ form a Čebyšev system, $h = n + 1$ and the scalars are real, then* *) *we can have equality in (1.66) only if $g_0 \in \mathfrak{L}_G(x)$.* This statement results immediately from the following property of Čebyšev systems:

LEMMA 1.5. *Let $x_1, \ldots, x_n \in C_R(Q)$ (Q compact) be a Čebyšev system on Q. Then no polynomial $g = \sum_{i=1}^{n} \alpha_i x_i \neq 0$ may have at $n + 1$ points $q_1, \ldots, q_{n+1} \in Q$ the signs*

$$\operatorname{sign} g(q_j) = \varepsilon \operatorname{sign} \Delta_j \text{ or } 0 \qquad (j = 1, \ldots, n + 1), \quad (1.68)$$

where $\varepsilon = \pm 1$ and Δ_j is the cofactor of the element $x(q_j)$ in the determinant Δ defined by (1.52).

Lemma 1.5, without the hypothesis that sign $g(q_j)$ may be also zero, has been given by S. I. Zuhovitzky ([267], p. 153). However, it is just this additional hypothesis which allows us to deduce the statement above from lemma 1.5**). Indeed, if we have the sign of equality in (1.66), whence also (1.67), and if $g_1 \in \mathfrak{L}_G(x)$, where $G = [x_1, \ldots, x_n]$, then

$$|x(q_j) - g_0(q_j)| = \min_{g \in G} \|x - g\| = \|x - g_1\| \geqslant |x(q_j) - g_1(q_j)|$$
$$(j = 1, \ldots, n + 1),$$

and thus the polynomial

$$g = g_1 - g_0 = (x - g_0) - (x - g_1) \in G \quad (1.69)$$

has at the points $q_1, \ldots, q_{n+1}$ the signs

$$\operatorname{sign} g(q_j) = \operatorname{sign} [x(q_j) - g_0(q_j)] \quad \text{or} \quad 0 \quad (j = 1, \ldots, n + 1).$$

Consequently, taking into account (1.65) and that by (1.64) (with $h = n + 1$) we have $\mu_j = c\Delta_j$ $(j = 1, \ldots, n + 1)$, it follows that we have

$$\operatorname{sign} g(q_j) = \operatorname{sign} \mu_j \text{ or } 0 = (\operatorname{sign} c) \operatorname{sign} \Delta_j \text{ or } 0$$
$$(j = 1, \ldots, n + 1),$$

*) We observe that in the example of section 1.1, mentioned above, the subspace G was not spanned by a Čebyšev system and we had $h = 1 < 2 = n + 1$ (even in the particular case of real scalars).

**) By S. I. Zuhovitzky [267], p. 155 and 156, this assertion is stated without proof. The present proof is, essentially, an adaptation of the proof of theorem 1.5, given in S. I. Zuhovitzky's paper [267].

i.e. (1.68) with $\varepsilon = \text{sign } c$, whence, by lemma 1.5, we obtain $g \equiv 0$, hence $g_0 = g_1 \in \mathfrak{L}_G(x)$, which completes the proof of our statement above. For the sake of completeness, let us give now the

Proof of lemma 1.5. We have

$$\sum_{j=1}^{n+1} \Delta_j g(q_j) = \begin{vmatrix} x_1(q_1) & \dots & x_1(q_{n+1}) \\ \cdot & \cdot\ \cdot\ \cdot & \cdot \\ x_n(q_1) & \dots & x_n(q_{n+1}) \\ g(q_1) & \dots & g(q_{n+1}) \end{vmatrix} = 0. \tag{1.70}$$

Now assume that we have (1.68). Then

$$\text{sign } \Delta_j g(q_j) = \varepsilon \quad \text{or} \quad 0 \quad (j = 1, \dots, n+1),$$

whence, by (1.70),

$$\Delta_j g(q_j) = 0 \qquad (j = 1, \dots, n+1).$$

Since $x_1, \dots, x_n$ form a Čebyšev system on Q, we have $\Delta_j \neq 0$ $(j = 1, \dots, n+1)$, whence

$$g(q_j) = 0 \qquad (j = 1, \dots, n+1),$$

whence, taking again into account that $x_1, \dots, x_n$ form a Čebyšev system on Q, we obtain $g \equiv 0$, which contradicts the hypothesis. This completes the proof of lemma 1.5.

In the particular case when $Q = [a, b]$, $x_1, \dots, x_n$ form a Čebyšev system, $h = n + 1$ and the scalars are real, from theorem 1.5 and the above statement it results the following classical theorem of Ch. J. de la Vallée-Poussin — S. N. Bernstein ([15], p. 4):

COROLLARY 1.5. *Let $E = C_R([a, b])$, let $G = [x_1, \dots, x_n]$ be an n-dimensional linear subspace of E such that $x_1, \dots, x_n$ form a Čebyšev system and let $x \in E$ and $g_0 \in G$ with the property that there exist $n + 1$ points $q_1 < q_2 < \dots < q_{n+1}$ in $[a, b]$ on which the difference $x(q) - g_0(q)$ takes values $\neq 0$ with alternating signs (i.e. with opposite signs at consecutive points q_j, q_{j+1} $(j = 1, \dots, n)$). Then we have (1.66), and the equality holds only if we have (1.67); in this latter case we have $g_0 \in \mathfrak{L}_G(x)$.*

Proof. By virtue of the hypothesis and of lemma 1.4 we have

$$0 \neq \text{sign } [x(q_j) - g_0(q_j)] = \varepsilon \text{ sign } \Delta_j \ (j = 1, \dots, n+1), \tag{1.71}$$

where $\varepsilon = \pm 1$ and Δ_j is defined as in lemma 1.4.

Put

$$\mu_j = \frac{\varepsilon \Delta_j}{\sum_{i=1}^{n+1} |\Delta_i|} \qquad (j = 1, \ldots, n+1). \tag{1.72}$$

Then we have $\mu_1, \ldots, \mu_{n+1} \neq 0$, $\sum_{j=1}^{n+1} |\mu_j| = 1$, and

$$\sum_{j=1}^{n+1} \mu_j x_k(q_j) = \varepsilon \sum_{j=1}^{n+1} \Delta_j x_k(q_j) = 0 \qquad (k = 1, \ldots, n),$$

$$\operatorname{sign} \mu_j [x(q_j) - g_0(q_j)] = \operatorname{sign} (\varepsilon^2 \Delta_j^2) = 1 \qquad (j = 1, \ldots, n+1),$$

i.e. (1.64) and (1.65) with $h = n + 1$, and one applies theorem 1.5, respectively the statement made after the proof of theorem 1.5. This completes the proof of corollary 1.5.

Finally, we mention that the problem of best approximation in the spaces $C(Q)$ by elements of finite dimensional linear subspaces is equivalent to the problem of best approximate solution of (finite or infinite) systems of incompatible linear equations with a finite number of unknowns. Indeed, the problem of finding the n-tuples of scalars $\{\zeta_1^{(0)}, \ldots, \zeta_n^{(0)}\}$ for which

$$\max_{q \in Q} |x(q) - \sum_{i=1}^{n} \zeta_i x_i(q)| = \min, \tag{1.73}$$

where $x \in C(Q)$ and $x_1, \ldots, x_n \in C(Q)$ are given, may be also stated (see e.g. E. Ya. Remez [186]) in the following way (putting $x(q) = l^{(q)}$, $x_i(q) = \alpha_i^{(q)} (q \in Q\,;\, i = 1, \ldots, n)$): Given the system of incompatible linear equations

$$\sum_{i=1}^{n} \alpha_i^{(q)} \zeta_i = l^{(q)} \qquad (q \in Q), \tag{1.74}$$

where Q is a compact space and $l^{(q)}$, $\alpha_i^{(q)}$ $(i = 1, \ldots, n)$ are continuous functions of q, find the best approximate solutions of this system, i.e. the n-tuples of scalars $\{\zeta_1^{(0)}, \ldots, \zeta_n^{(0)}\}$ for which

$$\max_{q \in Q} |l^{(q)} - \sum_{i=1}^{n} \alpha_i^{(q)} \zeta_i| = \min. \tag{1.75}$$

Since $l^{(q)}$, $\alpha_i^{(q)}$ $(i = 1, \ldots, n)$ are continuous functions of q, the set

$$\mathfrak{M} = \{\{\alpha_1^{(q)}, \ldots, \alpha_n^{(q)}, l^{(q)}\} \mid q \in Q\} \tag{1.76}$$

in the (complex or real) $(n+1)$-dimensional euclidean space $\mathcal{H}_{n+1}$ is bounded in this case.

Conversely, the problem of finding the best approximate solution of a given system of incompatible linear equations

$$\sum_{i=1}^{n} \alpha_i^{(j)} \zeta_i = l^{(j)} \qquad (j \in J), \qquad (1.77)$$

where J is an arbitrary set of indices and where the points $s^{(j)} = \{\alpha_1^{(j)}, \ldots, \alpha_n^{(j)}, l^{(j)}\} \in \mathcal{H}_{n+1}$ run over a bounded set $\mathfrak{M} \subset \mathcal{H}_{n+1}$ may be stated (see e.g. N. I. Ahiezer [1], § 46) also as follows (since the coordinates $\alpha_1^{(j)}, \ldots, \alpha_n^{(j)}, l^{(j)}$ are continuous function of the point $s^{(j)}$): Given $n + 1$ continuous functions $x_1(s), \ldots$ $\ldots, x_n(s), x(s)$ on a bounded set $\mathfrak{M} \subset \mathcal{H}_{n+1}$, find the n-tuples of scalars $\{\zeta_1^{(0)}, \ldots, \zeta_n^{(0)}\}$ for which

$$\sup_{s \in \mathfrak{M}} | x(s) - \sum_{i=1}^{n} \zeta_i x_i(s) | = \max_{s \in \overline{\mathfrak{M}}} | x(s) - \sum_{i=1}^{n} \zeta_i x_i(s) | = \min, \quad (1.78)$$

where $x_i (i = 1, \ldots, n)$ and x are extended by continuity to the closure $\overline{\mathfrak{M}}$ of the set $\mathfrak{M}$. Since $S = \overline{\mathfrak{M}}$ is compact, we find thus again the problem of best approximation in the space $C(S)$ by elements of the linear subspace spanned by $x_1, \ldots, x_n$.

Some authors (e.g. E. Ya. Remez [187]) prefer the language of the systems of incompatible linear equations with a finite number of unknowns for the reason that this language eliminates the argument q. Actually, in this case *each $q \in Q$ is replaced by an equation of the system* (1.74), and with this interchanging of roles the results concerning the spaces $C(Q)$ are translated into corresponding results for systems of equations, and conversely. In the sequel we shall continue to use the language of the spaces $C(Q)$.

From the above it also follows that the problem of best approximation in the spaces $C(Q)$ by elements of n-dimensional linear subspaces is equivalent to the same problem in the spaces $C(S)$, where $S \subset \mathcal{H}_{n+1}$. However, one can also see this fact directly, by considering $x_1, \ldots, x_n, x$ as continuous functions on the unit cell S of the conjugate space E_{n+1}^*, where $E_{n+1} =$ the linear subspace of $C(Q)$ spanned by $x_1 \ldots, x_n, x$.

1.4. THE CONJUGATE SPACE OF THE SPACE $C_E(Q)$ AND THE EXTREMAL POINTS OF ITS UNIT CELL

Let Q be a compact space and E a (real or complex) Banach space. We shall denote by $C_E(Q)$ the space of all functions defined on Q and with values in E, strongly continuous*) on Q, endowed

*) I.e. $\lim_{q \to q_0} \|x(q) - x(q_0)\|_E = 0 \ (x(.) \in C_E(Q))$.

with the usual vector operations and with the norm *) $\|x(\cdot)\|_{C_E(Q)} = \max_{q \in Q} \|x(q)\|_E$. In the particular case when $\dim E = 1$, we find again the space $C(Q)$, and for $\dim E = 1$ and real scalars we find again the space $C_R(Q)$. On the other hand, in the particular case when Q consists of one single point, we find again the space E itself.

In order to apply the preceding general results to the case of the spaces $C_E(Q)$, it will be necessary to give first the general form of the elements of the conjugate space $C_E(Q)^*$ and the general form of the extremal points of the unit cell $S_{C_E(Q)^*}$.

We recall that a set function f_e defined on the Borel sets $e \subset Q$ and with values in the conjugate space E^* is called *completely additive*, if for every sequence of pairwise disjoint Borel sets $\{e_n\} \subset Q$ we have

$$f_{\bigcup_{n=1}^{\infty} e_n} = \sum_{n=1}^{\infty} f_{e_n}, \tag{1.79}$$

in the sense of the strong convergence in E^*. The function f_e is said to be *weakly* completely additive* if for every $x \in E$ the set function $f_e(x)$ with numerical values is completely additive in the usual sense. The function f_e is said to be (weakly*) *regular*, if for every $x \in E$ the numerical-valued set function $f_e(x)$ is regular in the usual sense, i.e. if for every $x \in E$, every Borel set $e \subset Q$ and every $\varepsilon > 0$ there exist an open set $U_x \supset e$ and a compact set $K_x \subset e$ such that $|f_e(x) - f_{e'}(x)| < \varepsilon$ for all Borel sets $e' \subset Q$ with $K_x \subset e' \subset U_x$. The *variation* of f_e on a Borel set $A \subset Q$ is defined by

$$\operatorname*{Var}_{e \subset A} f_e = \sup \sum_{i=1}^{n} \|f_{e_i}\|, \tag{1.80}$$

where the sup is taken over all finite partitions of A into pairwise disjoint Borel subsets e_i. The function f_e is said to be *of bounded variation*, if $\operatorname*{Var}_{e \subset Q} f_e < \infty$.

The Gowurin integral [77] of a function $x(\cdot) \in C_E(Q)$ with respect to a set function f_e defined on the Borel sets $e \subset Q$, with values in E^*, completely additive and of bounded variation, is defined as follows: if $z(\cdot)$ is a *simple function*, i.e. a function defined on Q with values in E, of the form **)

$$z(\cdot) = \sum_{i=1}^{n} x_i \varphi_{e_i}(\cdot), \tag{1.81}$$

---

*) We shall use the notation $x(.)$ for the elements of $C_E(Q)$, in order to distinguish them from the elements $x \in E$.

**) In general, $z(.) \notin C_E(Q)$.

where $e_i \subset Q$ are pairwise disjoint Borel sets such that $\bigcup_{i=1}^{n} e_i = Q$ and where $\varphi_{e_i}(\cdot) =$ the characteristic function of the set e_i, and $x_i \in E$ $(i = 1, \ldots, n)$, then one puts

$$\int_Q \langle z(q), \mathrm{d}f_q \rangle = \sum_{i=1}^{n} f_{e_i}(x_i). \tag{1.82}$$

Since every $x(\cdot) \in C_E(Q)$ is the limit of a uniformly convergent sequence of simple functions $z_n(.)$ and since f_e is of bounded variation, the limit $\lim_{n\to\infty} \int_Q \langle z_n(q), \mathrm{d}f_q \rangle$ exists and does not depend on the choice of the sequence $\{z_n(\cdot)\}$; one puts, by definition

$$\int_Q \langle x(q), \mathrm{d}f_q \rangle = \lim_{n\to\infty} \int_Q \langle z_n(q), \mathrm{d}f_q \rangle. \tag{1.83}$$

From this definition it follows immediately the inequality

$$\left| \int_Q \langle x(q), \mathrm{d}f_q \rangle \right| \leqslant \max_{q \in Q} \|x(q)\|_E \operatorname*{Var}_{e \subset Q} f_e \qquad (x(\cdot) \in C_E(Q)). \tag{1.84}$$

LEMMA 1.6 ([216], theorem 1 and [224], p. 398, theorem). *Let Q be a compact space and E a Banach space. Then the conjugate space $C_E(Q)^*$ is equivalent to the space of all set functions f_e, defined on the Borel sets $e \subset Q$, with values in E^*, completely additive, of bounded variation and regular, endowed with the usual vector operations and with the norm*

$$|||f_e||| = \operatorname*{Var}_{e \subset Q} f_e. \tag{1.85}$$

The equivalence $\Phi \to f_e$ between these spaces is given by the relation

$$\Phi[x(.)] = \int_Q \langle x(q), \mathrm{d}f_q \rangle \qquad (x(.) \in C_E(Q)). \tag{1.86}$$

Proof. Let $\Phi \in C_E(Q)^*$ be arbitrary. Since for every fixed $x \in E$ the subspace

$$C_x = \{\alpha(.)x \mid \alpha(.) \in C(Q)\} \tag{1.87}$$

of $C_E(Q)$ is equivalent to $C(Q)$ (by the mapping $\alpha(.)x \to \alpha(.)\,\|x\|$), there exists a Radon measure μ_x on Q (real or complex, according to the space E), such that

$$\Phi[\alpha(.)x] = \int_Q \alpha(q)\, \mathrm{d}\mu_x(q) \qquad (\alpha(.) \in C(Q),\ x \in E). \tag{1.88}$$

Put, for every Borel set $e \subset Q$ and every $x \in E$,

$$f_e(x) = \mu_x(e). \tag{1.89}$$

Then for every $\alpha(.) \in C(Q)$ and $x_1, x_2 \in E$ we have

$$\int_Q \alpha(q)\, d\mu_{x_1+x_2}(q) = \Phi[\alpha(.)\,(x_1 + x_2)] = \Phi[\alpha(.)x_1] + \Phi[\alpha(.)x_2] =$$

$$= \int_Q \alpha(q)\, d\mu_{x_1}(q) + \int_Q \alpha(q)\, d\mu_{x_2}(q) = \int_Q \alpha(q)\, d(\mu_{x_1} + \mu_{x_2})\,(q),$$

whence, for every Borel set $e \subset Q$,

$$f_e(x_1 + x_2) = \mu_{x_1+x_2}(e) = \mu_{x_1}(e) + \mu_{x_2}(e) = f_e(x_1) + f_e(x_2),$$

and, on the other hand, for every $x \in E$ and Borel set $e \subset Q$ we have

$$|f_e(x)| = |\mu_x(e)| \leqslant \operatorname*{Var}_{e \subset Q} \mu_x(e) = \|\Phi|_{C_x}\|\,\|x\| \leqslant \|\Phi\|\,\|x\|.$$

Consequently, for every fixed Borel set $e \subset Q$ we have

$$f_e \in E^*. \tag{1.90}$$

By its definition (1.89), the set function f_e is weakly* completely additive and regular. We shall show that it is also of bounded variation.

Let $e_1, \ldots, e_n$ be pairwise disjoint Borel subsets of Q, such that $\bigcup_{i=1}^{n} e_i = Q$, and let ε be arbitrary with $0 < \varepsilon < \sum_{i=1}^{n} \|f_{e_i}\|$. Then there exist*) linearly independent elements $x_1, \ldots, x_n \in E$ such that

$$\|x_i\| = 1 \qquad (i = 1, \ldots, n), \tag{1.91}$$

$$0 < \sum_{i=1}^{n} \|f_{e_i}\| - \varepsilon \leqslant \sum_{i=1}^{n} f_{e_i}(x_i). \tag{1.92}$$

On the other hand, since $C(Q)$ is dense in every $L^1(Q, \mu)$, there exists a sequence $\alpha_i^{(k)}(.) \in C(Q)$ $(i = 1, \ldots, n;\ k = 1, 2, \ldots)$ such that

$$\lim_{k\to\infty} \int_Q |\alpha_i^{(k)}(q) - \varphi_{e_i}(q)|\, d\mu_{x_i}(q) = 0 \qquad (i = 1, \ldots, n) \tag{1.93}$$

*) Indeed, there exist $x_i' \in E$ with $\|x_i'\| = 1$, $|f_{e_i}(x_i')| \geqslant \|f_{e_i}\| - \frac{\varepsilon}{n}$, and we have only to put $x_i = [\operatorname{sign} f_{e_i}(x_i')]\, x_i'$ $(i = 1, \ldots, n)$. Obviously, one may assume that $x_1, \ldots, x_n$ are linearly independent (replacing them, if necessary, by sufficiently near elements).

But, by (1.93) we can extract a subset of $\{\alpha_i^{(k)}(.)\}$, which we shall denote again by $\{\alpha_i^{(k)}(.)\}$, such that for every $i = 1, \ldots, n$ we have

$$\lim_{k\to\infty} \alpha_i^{(k)}(q) = \varphi_{e_i}(q) \text{ a.e. with respect to } \mu_{x_1}, \ldots, \mu_{x_n} \text{ (simultaneously).} \tag{1.94}$$

Put

$$\beta_i^{(k)}(q) = \frac{\alpha_i^{(k)}(q)}{\frac{1}{k} + \|\sum_{i=1}^{n} \alpha_i^{(k)}(q) x_i\|_E} \quad (q \in Q,\ i = 1, \ldots, n;\ k = 1, 2, \ldots). \tag{1.95}$$

Then $\beta_i^{(k)}(.) \in C(Q)$ $(i = 1, \ldots, n;\ k = 1, 2, \ldots)$ and we have

$$\left\| \sum_{i=1}^{n} \beta_i^{(k)}(q)\, x_i \right\|_E = \frac{\left\| \sum_{i=1}^{n} \alpha_i^{(k)}(q) x_i \right\|_E}{\frac{1}{k} + \left\| \sum_{i=1}^{n} \alpha_i^{(k)}(q) x_i \right\|_E} \leqslant 1 \qquad (q \in Q,\ k = 1, 2, \ldots). \tag{1.96}$$

On the other hand, by virtue of (1.95), (1.94) and (1.91), for each $i = 1, \ldots, n$ we have

$$\lim_{k\to\infty} \beta_i^{(k)}(q) = \frac{\varphi_{e_i}(q)}{\left\| \sum_{i=1}^{n} \varphi_{e_i}(q) x_i \right\|_E} = \varphi_{e_i}(q) \text{ a.e. with respect to } \mu_{x_1}, \ldots, \mu_{x_n} \text{ (simultaneously).} \tag{1.97}$$

From (1.96) and the linear independence of $x_1, \ldots, x_n$ it follows that there exists a constant $M > 0$ such that $|\beta_i^{(k)}(q)| \leqslant \leqslant M$ $(i = 1, \ldots, n;\ k = 1, 2, \ldots)$, and consequently, from (1.97) it follows (by the theorem of Lebesgue) that we have

$$\lim_{k\to\infty} \int_Q \beta_i^{(k)}(q)\, \mathrm{d}\mu_{x_i}(q) = \int_Q \varphi_{e_i}(q) \mathrm{d}\mu_{x_i}(q),$$

whence, taking into account (1.92), (1.89), (1.88) and (1.96),

$$0 < \sum_{i=1}^{n} \|f_{e_i}\| - \varepsilon \leqslant \left| \sum_{i=1}^{n} f_{e_i}(x_i) \right| = \left| \sum_{i=1}^{n} \mu_{x_i}(e_i) \right| =$$

$$= \left| \sum_{i=1}^{n} \int_Q \varphi_{e_i}(q)\, \mathrm{d}\mu_{x_i}(q) \right| = \lim_{k\to\infty} \left| \sum_{i=1}^{n} \int_Q \beta_i^{(k)}(q)\, \mathrm{d}\mu_{x_i}(q) \right| =$$

$$= \lim_{k\to\infty} \left| \Phi\left[\sum_{i=1}^{n} \beta_i^{(k)}(.)\, x_i \right] \right| \leqslant \|\Phi\|.$$

Since $\{e_1, \ldots, e_n\}$ was an arbitrary partition of Q into pairwise disjoint Borel sets and since ε was an arbitrary number satisfying $0 < \varepsilon < \sum_{i=1}^{n} \|f_{e_i}\|$, it follows that we have

$$\operatorname*{Var}_{e \subset Q} f_e \leqslant \|\Phi\|; \tag{1.98}$$

being, by the above, weakly* completely additive and of bounded variation, the function f_e is completely additive.

Since by (1.88) and by the definition of the Gowurin integral we have

$$\Phi\left[\sum_{i=1}^{n} \alpha_i(.)x_i\right] = \sum_{i=1}^{n} \Phi[\alpha_i(.)x_i] = \sum_{i=1}^{n} \int_Q \alpha_i(q) d\mu_{x_i}(q) =$$

$$= \sum_{i=1}^{n} \int_Q \langle \alpha_i(q)x_i, df_q\rangle = \int_Q \left\langle \sum_{i=1}^{n} \alpha_i(q)x_i, df_q \right\rangle$$

$$(\alpha_i(.) \in C(Q),\ x_i \in E,\ i = 1, \ldots, n),$$

and since the set of all functions of the form $\sum_{i=1}^{n} \alpha_i(.)x_i$ $(\alpha_i(.) \in C(Q),$ $x_i \in E,\ i = 1, \ldots, n)$ is dense in $C_E(Q)$, taking into account (1.98) it follows that we have (1.86). Thus, for every $\Phi \in C_E(Q)^*$ the set function f_e defined by (1.89) is regular and completely additive and satisfies (1.86) and (1.98); being regular, it is the only set function with these properties.

Conversely, if f_e is a set function defined on the Borel sets $e \subset Q$, with values in E^*, completely additive, regular and of bounded variation, then the functional Φ defined by (1.86) is linear on $C_E(Q)$ and by virtue of (1.84) it satisfies

$$\|\Phi\| \leqslant \operatorname*{Var}_{e \subset Q} f_e, \tag{1.99}$$

whence $\Phi \in C_E(Q)^*$. By virtue of (1.98) and (1.99) the above mapping $\Phi \to f_e$ is an isometry, which completes the proof of lemma 1.6.

An important particular class of functionals (1.86) is that in which the set function f_e is of the form

$$f_e = \begin{cases} f & \text{for} \quad e \ni q_0 \\ 0 & \text{for} \quad e \not\ni q_0, \end{cases} \tag{1.100}$$

where f is a fixed element in E^* and q_0 a fixed point in Q. In this case we have

$$\Phi[x(.)] = \int_Q \langle x(q), df_q\rangle = f[x(q_0)] \qquad (x(.) \in C_E(Q)), \tag{1.101}$$

and since the set function (1.100) is regular,

$$\|\Phi\| = \operatorname*{Var}_{e \subset Q} f_e = \|f\|. \tag{1.102}$$

LEMMA 1.7 ([217], theorem 1.1). *In order that a functional (1.86) be an extremal point of the unit cell $S_{C_E(Q)^*}$, it is necessary and sufficient that it be of the form (1.101), where f is a fixed extremal point of S_{E^*} and q_0 a fixed point in Q.*

Proof. Assume that $\Phi \in \mathscr{E}(S_{C_{E}(Q)^*})$ and Φ is not of the form (1.101), where $f \in \mathscr{E}(S_{E^*})$ and $q_0 \in Q$. Let f_e be the set function defined on the Borel sets $e \subset Q$, with values in E^*, completely additive, of bounded variation and regular, which corresponds by lemma 1.6 to Φ. Then there exists at least one point $q_0 \in Q$ with the property that on every neighborhood *) U of q_0 we have $\operatorname*{Var}_{e \subset U} f_e > 0$, since otherwise every point $q \in Q$ would have a neighborhood U_q on which $\operatorname*{Var}_{e \subset U_q} f_e = 0$, and selecting from the covering $\{U_q | q \in Q\}$ of Q a finite subcovering $\{U_{q_i}\}_{i=1}^n$ (since Q is compact), we would obtain $\operatorname*{Var}_{e \subset Q} f_e \leqslant \sum_{i=1}^{n} \operatorname*{Var}_{e \subset U_{q_i}} f_e = 0$, which contradicts the hypothesis $\operatorname*{Var}_{e \subset Q} f_e = \|\Phi\| = 1$.

We claim that there exists at least one point $q_0' \in Q \setminus \{q_0\}$ with the same property. Indeed, assume the contrary, i.e. that every point $q \in Q \setminus \{q_0\}$ has a neighborhood U_q on which $\operatorname*{Var}_{e \subset U_q} f_e = 0$. We shall show first that in this case the relations

$$x(.) \in C_E(Q), \quad x(q_0) = 0 \tag{1.103}$$

imply $\Phi[x(.)] = 0$. Indeed, if we have (1.103), then for every positive integer n there exists a neighborhood $U_{q_0}^{(n)}$ of q_0 such that $\|x(q)\| < \frac{1}{n}$ $(q \in U_{q_0}^{(n)})$. Since the set $Q \setminus U_{q_0}^{(n)}$ is compact, by selecting from the covering $\{U_q | q \in Q\}$ of $Q \setminus U_{q_0}^{(n)}$ a finite subcovering $\{U_{q_i}\}_{i=1}^n$, we obtain $\operatorname*{Var}_{e \subset Q \setminus U_{q_0}^{(n)}} f_e \leqslant \sum_{i=1}^{n} \operatorname*{Var}_{e \subset U_{q_i}} f_e = 0$. Consequently, taking a simple function (1.81) such that $x_1 = x(q_0) = 0$, $e_1 = U_{q_0}^{(n)}$ and $\|x(.) - z(.)\| < \frac{1}{n}$, we shall have $\int_Q \langle z(q), \, df_q \rangle = 0$, whence, for $n \to \infty$, we obtain $\Phi[x(.)] =$

*) By "neighborhood" we shall mean throughout this proof: open neighborhood.

$\int_Q \langle x(q), \mathrm{d}f_q\rangle = 0$. Thus, the relations (1.103) imply $\Phi[x(.)] = 0$. Consequently, the relations

$$x(.) \in C_E(Q), \qquad f[x(q_0)] = 0 \qquad (f \in S_{E^*}) \tag{1.104}$$

also imply $\Phi[x(.)] = 0$, whence the functional $\Phi \in C_E(Q)^*$ belongs to the $\sigma(C_E(Q)^*, C_E(Q))$-closed convex subset $\Omega(A)$ of $C_E(Q)^*$, spanned by the functionals of the form

$$\Psi_f[x(.)] = f[x(q_0)] \qquad (x(.) \in C_E(Q)), \tag{1.105}$$

where $f \in S_{E^*}$. But this set $\Omega(A)$ coincides with the set

$$A = \{\Psi_f \in C_E(Q)^* \,|\, f \in S_{E^*}\}, \tag{1.106}$$

i.e. the set A itself is convex and $\sigma(C_E(Q)^*, C_E(Q))$-closed; indeed, A is the image of S_{E^*} by the mapping

$$f \to \Psi_f \tag{1.107}$$

and since S_{E^*} is convex and compact for $\sigma(E^*, E)$, and the mapping (1.107) is linear and continuous *) for $\sigma(E^*, E)$, $\sigma(C_E(Q)^*, C_E(Q))$, it follows that A is convex and compact (whence closed) for $\sigma(C_E(Q)^*, C_E(Q))$.

Consequently, we have then $\Phi \in A$, i.e. there exists an $f \in S_{E^*}$ such that

$$\Phi[x(.)] = f[x(q_0)] \qquad (x(.) \in C_E(Q)).$$

Since by our hypothesis $\Phi \in \mathcal{E}(S_{C_E(Q)^*})$, it follows that we have $f \in \mathcal{E}(S_{E^*})$, and thus Φ is of the form (1.101) with $q_0 \in Q$ and $f \in \mathcal{E}(S_{E^*})$, which contradicts the hypothesis made at the beginning of the proof. This proves that there exist at least two distinct points $q_0, q_0' \in Q$ with the property that on every neighborhood U of each of them we have $\operatorname{Var}_{e \subset U} f_e > 0$. But, taking then two disjoint neighborhoods U_1, U_2 of q_0 and q_0', and taking into account that $\operatorname{Var}_{e \subset Q} f_e = \|\Phi\| = 1$, we obtain

$$0 < V_i < 1 \qquad (i = 1, 2), \tag{1.108}$$

where $V_i = \operatorname{Var}_{e \subset U_i} f_e$.

*) Indeed, $|(f - f_0)[x_i(q_0)]| < \varepsilon$ implies $|\Psi_f[x_i(.)] - \Psi_{f_0}[x_i(.)]| < \varepsilon$ $(i = 1, \ldots, n)$.

Consider now the functional $\Phi' \in C_E(Q)^*$ defined by *)

$$\Phi'[x(.)] = V_1 \int_{U_2} \langle x(q), df_q \rangle - V_2 \int_{U_1} \langle x(q), df_q \rangle \quad (x(.) \in C_E(Q)). \tag{1.109}$$

We have obviously $\|\Phi'\| = 2 V_1 V_2 > 0$. Put

$$\Phi_1 = \Phi + \Phi', \qquad \Phi_2 = \Phi - \Phi'.$$

Then we have

$$|\Phi_1[x(.)]| = \left| \int_Q + V_1 \int_{U_2} - V_2 \int_{U_1} \right| =$$

$$= \left| \int_{Q \setminus (U_1 \cup U_2)} + (1 + V_1) \int_{U_2} + (1 - V_2) \int_{U_1} \right| \leqslant 1 - V_1 - V_2 +$$

$$+ (1 + V_1) V_2 + (1 - V_2) V_1 = 1 \qquad (x(.) \in S_{C_E(Q)}),$$

whence $\Phi_1 \in S_{C_E(Q)^*}$ and, similarly, $\Phi_2 \in S_{C_E(Q)^*}$. Since $\Phi = \frac{\Phi_1 + \Phi_2}{2}$, and $\Phi_1 \neq \Phi_2$, it follows that $\Phi \notin \mathcal{E}(S_{C_E(Q)^*})$, which contradicts the hypothesis. This proves that every $\Phi \in \mathcal{E}(S_{C_E(Q)^*})$ is of the form (1.101), where $f \in \mathcal{E}(S_{E^*})$ and $q_0 \in Q$.

Conversely, assume that there exist an $f \in \mathcal{E}(S_{E^*})$ and a $q_0 \in Q$ such that the functional $\Phi \in C_E(Q)^*$, defined by (1.101), is not an extremal point of $S_{C_E(Q)^*}$. Then there exist $\Phi_1, \Phi_2 \in S_{C_E(Q)^*}$, $\Phi_1 \neq \Phi \neq \Phi_2$, such that

$$\Phi = \frac{\Phi_1 + \Phi_2}{2}. \tag{1.110}$$

However, by virtue of lemma 1.6 there exist two set functions $f_e^{(1)}, f_e^{(2)}, f_e^{(1)} \not\equiv f_e \not\equiv f_e^{(2)}$, defined on the Borel sets $e \subset Q$, with values in E^*, completely additive, of bounded variation and regular, such that

$$\Phi_i[x(.)] = \int_Q \langle x(q), df_q^{(i)} \rangle \quad (x(.) \in C_E(Q), i = 1, 2), \tag{1.111}$$

$$\operatorname*{Var}_{e \subset Q} f_e^{(i)} = \|\Phi_i\| = 1 \qquad (i = 1, 2). \tag{1.112}$$

*) For an open set $U \subset Q$ the set $Q \setminus U$ is compact and the integral $\int_U \langle x(q), df_q \rangle$ can be defined e.g. by $\int_Q \langle x(q), df_q \rangle - \int_{Q \setminus U} \langle x(q), df_q \rangle$.

By (1.110), (1.101), (1.111) and the regularity of the functions f_e, $f_e^{(1)}$, $f_e^{(2)}$ we have then

$$f_e = \frac{f_e^{(1)} + f_e^{(2)}}{2}. \tag{1.113}$$

Consequently, taking into account (1.110) and (1.112), for every neighborhood U of q_0 we have then

$$1 = \operatorname*{Var}_{e \subset U} f_e = \operatorname*{Var}_{e \subset U} \frac{f_e^{(1)} + f_e^{(2)}}{2} \leqslant \frac{1}{2} (\operatorname*{Var}_{e \subset U} f_e^{(1)} + \operatorname*{Var}_{e \subset U} f_e^{(2)}) \leqslant 1,$$

whence

$$\operatorname*{Var}_{e \subset U} f_e^{(1)} = \operatorname*{Var}_{e \subset U} f_e^{(2)} = 1. \tag{1.114}$$

From this, by the same argument as that used in the above proof of necessity part, it follows that there exist $f^{(1)} \in S_{E^*}$ and $f^{(2)} \in S_{E^*}$, $f^{(1)} \neq f^{(2)}$, such that

$$f_e^{(i)} = \begin{cases} f^{(i)} & \text{for} \quad e \ni q_0 \\ 0 & \text{for} \quad e \not\ni q_0 \end{cases} \qquad (i = 1, 2). \tag{1.115}$$

But from (1.115) and (1.113) it follows that we have

$$f = \frac{f^{(1)} + f^{(2)}}{2},$$

whence $f \notin \mathcal{E}(S_{E^*})$, which contradicts the hypothesis. Thus, $\Phi \in \mathcal{E}(S_{C_E(Q)^*})$, which completes the proof of lemma 1.7.

In the particular case when $\dim E = 1$, from lemma 1.7 one finds again the well known result on the general form of the extremal points of $S_{C(Q)^*}$ (see Chap. I, section 1.10, formula (1.135)); the above proof of the necessity part of lemma 1.7 is, essentially, an extension of the proof of this particular result, given by R. F. Arens and J. L. Kelley [5]. On the other hand, using this particular result and lemma 1.2 on the extremal extension of extremal functionals, one can give the following short proof of the necessity part of lemma 1.7 (this proof has been given, essentially in the paper [219], p. 75):

Let $\Phi \in \mathcal{E}(S_{C_E(Q)^*})$. Since the space $C_E(Q)$ can be identified (linearly isometrically) with a linear subspace of the space $C(Q \times S_{E^*})$ (where S_{E^*} is endowed with the weak topology $\sigma(E^*, E)$, and $Q \times S_{E^*}$ with the product topology), by the mapping $x(.) \to \beta_{x(.)}(.,.)$, where

$$\beta_{x(.)}(q, f) = f[x(q)] \qquad (q \in Q, f \in S_{E^*}), \tag{1.116}$$

lemma 1.2 implies the possibility of extending Φ to a $\widetilde{\Phi} \in \mathcal{E}(S_{C(Q \times S_{E^*})^*})$. Consequently, by virtue of the result on the general form of the elements of $\mathscr{E}(S_{C(Q \times S_{E^*})^*})$, there exist a $q_0 \in Q$, an $f_0 \in S_{E^*}$ and a scalar α_0 of modulus 1 such that

$$\Phi[x(.)] = \Phi[\beta_{x(.)}(.,.)] = \alpha_0 \beta_{x(.)}(q_0, f_0) = \alpha_0 f_0[x(q_0)] \quad (x(.) \in C_E(Q))$$

Writing $f = \alpha_0 f_0$, we have then

$$\Phi[x(.)] = f[x(q_0)] \qquad (x(.) \in C_E(Q)).$$

Since $\Phi \in \mathscr{E}(S_{C_E(Q)^*})$, it follows that we have $f \in \mathscr{E}(S_{E^*})$, which completes the proof.

1.5. APPLICATIONS IN THE SPACES $C_E(Q)$

Combining theorem 1.1 (equivalence 1°⇔5°) and lemma 1.7, one obtains the following characterization of polynomials of best approximation in the spaces $C_E(Q)$:

THEOREM 1.6 ([217], theorem 3.1). *Let Q be a compact space, E a Banach space, $G = [x_1(.), \ldots, x_n(.)]$ an n-dimensional linear subspace of $C_E(Q)$, $x(.) \in C_E(Q) \setminus G$ and $g_0(.) \in G$. We have $g_0(.) \in \mathfrak{L}_G(x(.))$ if and only if there exist h extremal points $f_1, \ldots, f_h$ of S_{E^*}, where $1 \leq h \leq n+1$ if the scalars are real and $1 \leq h \leq 2n+1$ if the scalars are complex, h points $q_1, \ldots, q_h \in Q$, and h numbers $\mu_1, \ldots, \mu_h \neq 0$ with $\sum_{j=1}^{h} |\mu_j| = 1$, such that*

$$\sum_{j=1}^{h} \mu_j f_j [x_k(q_j)] = 0 \qquad (k = 1, \ldots, n), \tag{1.117}$$

$$f_j[x(q_j) - g_0(q_j)] = (\operatorname{sign} \mu_j) \max_{t \in Q} \|x(t) - g_0(t)\|_E \qquad (j = 1, \ldots, h). \tag{1.118}$$

In the particular case when dim $E = 1$, from theorem 1.6 one finds again theorem 1.3 (equivalence 1°⇔3°).

We shall consider now the important particular case when the functions $x_1(.), \ldots, x_n(.)$ which generate the linear subspace G have the interpolation property (P_m) for an $m \leq n$. A system of n functions $x_1(.), \ldots, x_n(.) \in C_E(Q)$ (Q a compact space, E a Banach space) is said to have the *property* (P_m), for an $m \leq n$, if for any set of m distinct points $q_j \in Q$ and m elements $y_j \in E$ $(j = 1, \ldots, m)$ there exists at least one polynomial $g(.) = \sum_{i=1}^{n} \alpha_i x_i(.)$ such that

$$g(q_j) = y_j \qquad (j = 1, \ldots, m). \tag{1.119}$$

THEOREM 1.7 ([217], theorem 3.2). *Let Q be a compact space, E a Banach space, $G = [x_1(\cdot), \ldots, x_n(\cdot)]$ an n-dimensional linear subspace of $C_E(Q)$ such that the system $x_1(\cdot), \ldots, x_n(\cdot)$ has the property (P_m) for an $m \leqslant n$, and let $x(\cdot) \in C_E(Q) \setminus G$ and $g_0(\cdot) \in G$. We have $g_0(\cdot) \in \mathfrak{P}_G(x(\cdot))$ if and only if there exist h extremal points $f_1, \ldots, f_h$ of S_{E^*}, where $m + 1 \leqslant h \leqslant n + 1$ if the scalars are real and $m + 1 \leqslant h \leqslant 2n + 1$ if the scalars are complex, h distinct points $q_1, \ldots, q_h \in Q$, and h numbers $\mu_1, \ldots, \mu_h \neq 0$ with $\sum_{j=1}^{h} |\mu_j| = 1$, such that we have (1.117) and (1.118).*

Proof. Assume that we have $g_0(\cdot) \in \mathfrak{P}_G(x(\cdot))$. Then, by virtue of theorem 1.6 there exist h extremal points $f_1, \ldots, f_h$ of S_{E^*}, where $1 \leqslant h \leqslant n + 1$ if the scalars are real and $1 \leqslant h \leqslant 2n + 1$ if the scalars are complex, h points $q_1, \ldots, q_h \in Q$, and h numbers $\mu_1, \ldots, \mu_h \neq 0$ with $\sum_{j=1}^{h} |\mu_j| = 1$, such that we have (1.117) and (1.118). But the relations (1.117) imply

$$\sum_{j=1}^{h} \mu_j f_j[g(q_j)] = 0 \qquad (g(\cdot) \in G), \tag{1.120}$$

which, together with the hypothesis (P_m), ensures that among the points $q_1, \ldots, q_h$ there exist at least $m + 1$ distinct points; indeed, if among the points $q_1, \ldots, q_h$ there existed at most m distinct points, say $q_1, \ldots, q_m$, then taking $y_1, \ldots, y_h$ such that $y_i = y_j$ whenever $q_i = q_j$ and that $\sum_{j=1}^{h} \mu_j f_j(y_j) \neq 0$, there would exist no $g(\cdot) \in G$ satisfying (1.119), which would contradict the hypothesis that the system $x_1(\cdot), \ldots, x_n(\cdot)$ has the property (P_m). This proves the necessity of the conditions of theorem 1.7.

On the other hand, the sufficiency of the conditions of theorem 1.7 is an immediate consequence of the sufficiency part of theorem 1.6, which completes the proof of theorem 1.7.

Let us denote by l the number of distinct points of the set $\{q_1, \ldots, q_h\}$ occurring in theorem 1.7. Then by theorem 1.7 we have $m + 1 \leqslant l \leqslant h$ ($\leqslant n + 1$ or $\leqslant 2n + 1$, according to E being real or complex). As shown by the example $Q = \{q_1\}$ = the set consisting of one single point q_1 (whence $m = 0$, $l = 1$ and $C_E(Q) \equiv E$), both the case $l = h$ (e.g. if the conjugate space E^* is strictly convex) and the case $l < h$ (e.g. if $E = C([0, 1])$; in this case we even have $h \geqslant n + 1$, by virtue of theorem 1.4), may occur.

On the other hand, we remark that in the particular case when $\dim E = 1$ and $m = n$, from theorem 1.7 one finds again theorem 1.4 (equivalence $1° \Leftrightarrow 2°$), since in this case the condition

(P_n) is obviously equivalent to the condition that the functions $x_1(.), \ldots, x_n(.)$ form a Čebyšev system on Q.

Finally, let us mention that from theorem 1.7 there follows immediately

COROLLARY 1.6. *Let Q be a compact space, E a Banach space, $G = [x_1(.), \ldots, x_n(.)]$ an n-dimensional linear subspace of $C_E(Q)$ such that the system $x_1(.), \ldots, x_n(.)$ has the property (P_m) for an $m \leqslant n$, and let $x(.) \in C_E(Q) \setminus G$ and $g_0(.) \in \mathfrak{P}_G(x(.))$. Then there exist h distinct points $q_1, \ldots, q_h \in Q$, where $m + 1 \leqslant h \leqslant n + 1$ if the scalars are real and $m + 1 \leqslant h \leqslant 2n + 1$ if the scalars are complex, such that*

$$\|x(q_j) - g_0(q_j)\|_E = \max_{t \in Q} \|x(t) - g_0(t)\|_E \quad (j = 1, \ldots, h). \tag{1.121}$$

In the particular case when the scalars are real, corollary 1.6 has been given by S. I. Zuhovitzky and S. B. Stečkin ([273], theorem 2).

1.6. APPLICATIONS IN*) THE SPACES $L^1(T,\nu), L^1_R(T,\nu), C^1(Q,\nu)$ AND $C^1_R(Q,\nu)$

Taking into account the general form **) of the extremal points of the unit cell $S_{L^1(T,\nu)^*}$, from theorem 1 it results the following theorem (the case $E = l^1(I)$ with complex scalars, of this theorem, follows immediately from Chap. I, theorem 1.7, equivalence 1°⇔5°, by observing that every complex number β with $|\beta| \leqslant 1$ may be written as $\beta = \frac{\beta_1 + \beta_2}{2}$, where β_1, β_2 are complex numbers with $|\beta_1| = |\beta_2| = 1$) :

THEOREM 1.8. *Let $E = L^1(T, \nu)$, where (T, ν) is a positive measure space with the property that the conjugate space $L^1(T, \nu)^*$ is canonically equivalent to $L^\infty(T, \nu)$ and let $G = [x_1, \ldots, x_n]$ be an n-dimensional linear subspace of E, $x \in E \setminus G$ and $g_0 \in G$. We have $g_0 \in \mathfrak{P}_G(x)$ if and only if there exist h ν-measurable and ν-essentially bounded functions $\beta_1, \ldots, \beta_h$ with*

$$|\beta_j(t)| = 1 \qquad (j = 1, \ldots h) \qquad \nu\text{-a. e. on } T, \tag{1.122}$$

where $1 \leqslant h \leqslant n + 1$ if the scalars are real and $1 \leqslant h \leqslant 2n + 1$

*) In the case of the spaces $E = L^p(T, \nu)$ $(1 < p < \infty)$ the space E^* is strictly convex, whence every $f \in E^*$ with $\|f\| = 1$ is an extremal point of S_{E^*}, hence in theorem 1.1 one may take, in this case, $h = 1$ and thus we find again theorem 1.11 of Chap. I. Therefore, among the spaces $L^p(T, \nu)$ $(1 \leqslant p < \infty)$ only the case $p = 1$ requires a separate treatment.

**) See Chap. I, lemma 1.13.

if the scalars are complex, and h numers $\mu_1, \ldots, \mu_h \neq 0$ *with* $\sum_{j=1}^{h} |\mu_j| = 1$, *such that*

$$\sum_{j=1}^{h} \mu_j \int_T x_k(t)\, \beta_j(t)\, d\nu(t) = 0 \qquad (k = 1, \ldots, n), \tag{1.123}$$

$$\int_T [x(t) - g_0(t)]\beta_j(t) d\nu(t) = (\operatorname{sign} \mu_j) \int_T | x(t) - \\ - g_0(t) | d\nu(t) \qquad (j = 1, \ldots, h); \tag{1.124}$$

in the case when $E = l^1(I)$ *and the scalars are complex, one may take* $h = 2$ *and* $\mu_1 = \mu_2 = \frac{1}{2}$.

As shown by this theorem, for linear subspaces G of finite dimension n of $L^1_R(T, \nu)$ the function $\beta \in L^\infty_R(T, \nu)$ of Chap. I, theorem 1.7, condition 5°, may be taken of the form

$$\beta = \sum_{j=1}^{h} \mu_j \beta_j, \tag{1.125}$$

with $\beta_j, \mu_j (j = 1, \ldots, h)$ as above. Since for real scalars every β_j is obviously of the form $\beta_j = \beta_{M_j}$, where M_j is a ν-measurable subset of T and where

$$\beta_{M_j}(t) = \begin{cases} 1 & \nu\text{-a.e.on } M_j \\ -1 & \nu\text{-a.e.on } T \setminus M_j \end{cases} \qquad (j = 1, \ldots, h), \tag{1.126}$$

it follows that in this case the above function β may be taken to be a simple function (i.e. a linear combination of characteristic functions corresponding to a finite partition of T into ν-measurable subsets).

Combining lemma 1.13 and theorem 1.8 of Chap. I with lemma 1.3, we obtain that *for linear subspaces G of finite dimension of* $L^1(T, \nu)$ *the function* $\beta \in L^\infty(T, \nu)$ *of Chap. I, theorem 1.8, may be taken of the form (1.125), where* $\beta_j, \mu_j (j = 1, \ldots, h$ *respectively* $j = 1,2)$ *are as above; consequently in the particular case when the scalars are real, it may be taken to be a simple function.*

Similarly, combining lemma 1.13 and the remark made before theorem 1.10 of Chap. I with lemma 1.3, it follows that *if* $T = Q$ *(compact) and* ν *is a positive Radon measure on* Q *with the carrier* $S(\nu) = Q$, *then for* $x \in C^1(Q, \nu)$ *and* $G \subset C^1(Q, \nu)$ *the sets* U, U_0 *of theorem 1.8 of Chap. I may be taken to be open, and the function* β *defined by (1.125) may be taken to be continuous on* U. In the particular case of real scalars we obtain

THEOREM 1.9. *Let $E = C^1_R(Q, \nu)$, where Q is a compact space and ν a positive Radon measure on Q with the carrier $S(\nu) = Q$, and let $G = [x_1, \ldots, x_n]$ be an n-dimensional linear subspace of E, $x \in E \setminus G$ and $M = \{g'_0, g'_1, \ldots, g'_{k+1}\} \subset \mathfrak{P}_G(x)$, where k is an integer with $0 \leqslant k < \infty$. Then there exist two disjoint sets U_1 and U_2 open in Q, with $U_1 \cup U_2 \neq \emptyset$, h ν-measurable subsets $M_1, \ldots, M_h$ of Q, where $1 \leqslant h \leqslant n+1$, and h numbers $\lambda_1, \ldots, \lambda_h > 0$ with $\sum_{j=1}^{h} \lambda_j = 1$, such that*

$$\sum_{j=1}^{h} \lambda_j \int_{Q \setminus (U_1 \cup U_2)} x_k(q) \alpha_{M_j}(q) \, d\nu(q) + \int_{U_1} x_k(q) d\nu(q) - \int_{U_2} x_k(q) d\nu(q) = 0 \qquad (k = 1, \ldots, n), \tag{1.127}$$

$$M_j \supset U_1, \qquad Q \setminus M_j \supset U_2 \qquad (j = 1, \ldots, h), \tag{1.128}$$

$$g'_0(q) = g'_1(q) = \ldots = g'_{k+1}(q) = x(q) \qquad (q \in Q \setminus (U_1 \cup U_2)), \tag{1.129}$$

where

$$\alpha_{M_j}(q) = \beta_{M_j}|_{Q \setminus (U_1 \cup U_2)}(q) = \begin{cases} 1 & \nu\text{-a.e. on } (Q \setminus U_1) \cap M_j \\ -1 & \nu\text{-a.e. on } Q \setminus (U_2 \cup M_j) \end{cases} \qquad (j = 1, \ldots, h). \tag{1.130}$$

Proof. By virtue of theorem 1.10 of Chap. I, there exist U_1, U_2 and α such that we have the relations (1.89)—(1.91) of Chap. I. But by lemma 1.13 of Chap. I and lemma 1.3, the function β of the proof of theorem 1.10 may be taken of the form

$$\beta = \sum_{j=1}^{h} \lambda_j \beta_{M_j}, \tag{1.131}$$

where $1 \leqslant h \leqslant n+1$, $M_1, \ldots, M_h$ are ν-measurable subsets of Q and $\lambda_j > 0$ $(j = 1, \ldots, h)$, $\sum_{j=1}^{h} \lambda_j = 1$. Consequently, by virtue of the formulas (1.90) and (1.91) of Chap. I we have (1.127) and (1.129), and by virtue of formula (1.95) of Chap. I we have

$$\sum_{j=1}^{h} \lambda_j \beta_{M_j}(q) = \begin{cases} 1 & \nu\text{-a.e. on } U_1 \\ -1 & \nu\text{-a.e. on } U_2, \end{cases}$$

whence, taking into account $\lambda_j > 0$ $(j = 1, \ldots, h)$, $\sum_{j=1}^{h} \lambda_j = 1$ and the definition (1.126) of β_{M_j}, it follows that we also have (1.128), which completes the proof.

§ 2. UNIQUENESS OF POLYNOMIALS OF BEST APPROXIMATION

From corollary 1.1 one immediately obtains a characterization of n-dimensional Čebyšev subspaces of a normed linear space E (in the same way as in Chap. I, from corollary 1.3 we have obtained theorem 3.2). In the present paragraph we shall show that this result can be improved, namely the conditions $1 \leqslant h \leqslant n+1$ for real scalars and $1 \leqslant h \leqslant 2n+1$ for complex scalars may be replaced by $1 \leqslant h \leqslant n$ and $1 \leqslant h \leqslant 2n-1$ respectively, and we shall then give applications of the characterization thus obtained, in various concrete spaces; we shall also give some direct applications of the results of Chap. I concerning best approximation in these concrete spaces.

2.1. PRELIMINARY LEMMAS

LEMMA 2.1 ([215], proposition 2). *If p is an integer $\geqslant 2$ and $\lambda_1, \ldots, \lambda_p, \mu_1, \ldots, \mu_p$ are $2p$ numbers > 0, then there exists an index j_1, $1 \leqslant j_1 \leqslant p$, such that*

$$\frac{\lambda_j}{\lambda_{j_1}} \geqslant \frac{\mu_j}{\mu_{j_1}} \qquad (j = 1, \ldots, p). \tag{2.1}$$

Proof. For $p = 2$ lemma 2.1 is obvious, since if $\dfrac{\lambda_1}{\lambda_2} \geqslant \dfrac{\mu_1}{\mu_2}$, we have (2.1) with $j_1 = 2$, and if $\dfrac{\lambda_1}{\lambda_2} < \dfrac{\mu_1}{\mu_2}$, we have (2.1) with $j_1 = 1$.

Assume now that lemma 2.1 is valid for $2(p-1)$ arbitrary numbers $\lambda_1, \ldots, \lambda_{p-1}, \mu_1, \ldots, \mu_{p-1} > 0$ and let $\lambda_1, \ldots \lambda_p, \mu_1, \ldots, \mu_p$ be $2p$ arbitrary numbers > 0. By the induction hypothesis there exists an index k_1, $1 \leqslant k_1 \leqslant p-1$, such that

$$\frac{\lambda_j}{\lambda_{k_1}} \geqslant \frac{\mu_j}{\mu_{k_1}} \qquad (j = 1, \ldots, p-1). \tag{2.2}$$

If $\frac{\lambda_p}{\lambda_{k_1}} \geqslant \frac{\mu_p}{\mu_{k_1}}$, then we have (2.1) with $j_1 = k_1$. But if $\frac{\lambda_p}{\lambda_{k_1}} < \frac{\mu_p}{\mu_{k_1}}$, then $\frac{\lambda_{k_1}}{\lambda_p} > \frac{\mu_{k_1}}{\mu_p}$, whence, taking into account (2.2), we obtain

$$\frac{\lambda_j}{\lambda_p} \geqslant \frac{\mu_j}{\mu_p} \qquad (j = 1, \ldots, p-1),$$

i.e. (2.1) with $j_1 = p$, which completes the proof of lemma 2.1.

Now we can prove the following lemma, which will play a fundamental role in the sequel:

LEMMA 2.2 ([215], p. 239). *In lemma 1.1 the extremal points $\varphi_1, \ldots, \varphi_h$ may be assumed to be real-linearly independent (i.e. linearly independent with respect to real coefficients). Consequently, in theorem 1.1 and corollary 1.2, the extremal points $f_1, \ldots, f_h$ may be assumed to have the restrictions $f_1|_{E_0}, \ldots, f_h|_{E_0}$, real-linearly independent, where E_0 is the $(n+1)$-dimensional linear subspace $G \oplus [x] = [x_1, \ldots, x_n, x]$ of E.*

Proof. Assume first that the scalars are real and let h be the smallest positive integer with the property of lemma 1.1. We shall show that in this case $\varphi_1, \ldots, \varphi_h$ are linearly independent.

Indeed, assume the contrary, i.e. that there exist real numbers $\mu_1, \ldots, \mu_h$ such that

$$\sum_{j=1}^{h} |\mu_j| \neq 0, \tag{2.3}$$

$$\sum_{j=1}^{h} \mu_j \varphi_j = 0. \tag{2.4}$$

Since $\|\varphi\| = 1$ and $\dim E = k < \infty$, there exists an $x \in E$ such that

$$\|x\| = 1, \qquad \varphi(x) = 1.$$

Then from the relations

$$\sum_{j=1}^{h} \lambda_j = 1, \ \sum_{j=1}^{h} \lambda_j \varphi_j(x) = \varphi(x) = 1, \ \lambda_j > 0,$$

$$|\varphi_j(x)| \leqslant \|x\| = 1 \qquad (j = 1, \ldots, h)$$

it follows that we have

$$\varphi_j(x) = 1 \qquad (j = 1, \ldots, h),$$

whence, taking into account (2.4), we obtain

$$\sum_{j=1}^{h} \mu_j = \sum_{j=1}^{h} \mu_j \varphi_j(x) = 0. \tag{2.5}$$

From (2.3) and (2.5) it follows that the sets of indices

$$P = \{j \mid \mu_j > 0\}, \quad N = \{j \mid \mu_j < 0\}, \tag{2.6}$$

are non-void, hence the relations (2.4) and (2.5) may be written in the form

$$\sum_{j \in P} \mu_j \varphi_j = \sum_{j \in N} (-\mu_j) \varphi_j, \tag{2.7}$$

$$\sum_{j \in P} \mu_j = \sum_{j \in N} (-\mu_j), \tag{2.8}$$

in which there occur only numbers > 0.

Let us denote by p the number of elements of the set P and by p' the number of elements of the set N. In the case when $h = 2$ (whence $p = 1$ and $p' = 1$), by (2.7) and (2.8) we have $\varphi_2 = -\frac{\mu_1}{\mu_2} \varphi_1 = \varphi_1$, whence $\varphi = (\lambda_1 + \lambda_2)\varphi_1 = \varphi_1$, which contradicts the hypothesis of minimality of h. On the other hand, if $h \geqslant 3$, then either $p \geqslant 2$ or $p' \geqslant 2$; we may assume that $p \geqslant 2$, since in the case $p' \geqslant 2$ the argument is similar. Since $\lambda_j, \mu_j > 0$ $(j \in P)$, there exists then by lemma 2.1 an index $j_1 \in P$ such that we have (2.1). But, taking into account (2.7), we have

$$\varphi = \sum_{j=1}^{h} \lambda_j \varphi_j = \sum_{j \in P \cup N} \lambda_j \varphi_j + \sum_{j \notin P \cup N} \lambda_j \varphi_j =$$

$$= \sum_{j \in P \cup N} \left(\lambda_j - \frac{\lambda_{j_1} \mu_j}{\mu_{j_1}}\right) \varphi_j + \sum_{j \notin P \cup N} \lambda_j \varphi_j = \sum_{j=1}^{h} \nu_j \varphi_j,$$

where

$$\nu_j = \begin{cases} \lambda_j - \dfrac{\lambda_{j_1} \mu_j}{\mu_{j_1}} & (j \in P \cup N) \\ \lambda_j & (j \notin P \cup N). \end{cases}$$

From the inequalities $\lambda_j > 0$ $(j = 1, \ldots, h)$, $\mu_{j_1} > 0$, $\mu_j < 0$ $(j \in N)$ and (2.1) it follows that we have

$$\nu_j \geqslant 0 \qquad (j = 1, \ldots, h).$$

On the other hand, since $j_1 \in P$, we have

$$\nu_{j_1} = \lambda_{j_1} - \frac{\lambda_{j_1} \mu_{j_1}}{\mu_{j_1}} = 0.$$

Finally, taking into account (2.8), we have

$$\sum_{j=1}^{h} \nu_j = \sum_{j \in P \cup N} \left(\lambda_j - \frac{\lambda_{j_1} \mu_j}{\mu_{j_1}} \right) + \sum_{j \notin P \cup N} \lambda_j =$$

$$= \sum_{j=1}^{h} \lambda_j - \frac{\lambda_{j_1}}{\mu_{j_1}} \sum_{j \in P \cup N} \mu_j = \sum_{j=1}^{h} \lambda_j = 1.$$

Thus, we have $\varphi = \sum_{j=1}^{h} \nu_j \varphi_j$, where $\nu_j \geqslant 0$ $(j = 1, \ldots, h)$, $\nu_{j_1} = 0$ and $\sum_{j=1}^{h} \nu_j = 1$, which contradicts the hypothesis of minimality of h. This proves lemma 2.2 in the case when the scalars are real.

Now assume that the scalars are complex. Then, considering E as a $2k$-dimensional real Banach space $E_{(r)}$ in the usual way, the extremal points $\Phi_1, \ldots, \Phi_h$ of $S_{(E_{(r)})^*}$ in the proof of lemma 1.1 may be taken to be linearly independent, by virtue of the real case of lemma 2.2 (proved above). In this case the extremal points φ_j of S_{E^*} defined by (1.6) will obviously be real-linearly independent, which completes the proof of lemma 2.2.

The above proof of lemma 2.2 has been given in [215], p. 239—240. Let us also mention the following proof of lemma 2.2 (the real case), which does not make use of lemma 2.1: Let h be the smallest positive integer with the property of lemma 1.1. Assuming that $\varphi_1, \ldots, \varphi_h$ are linearly dependent, hence (2.3) and (2.4), it follows, as in the above proof of lemma 2.2, that we have (2.5), i.e. $\varphi_1, \ldots, \varphi_h$ are also baricentrically dependent. But by virtue of (2.5) we can write relation (2.4) in the form

$$\left(- \sum_{j=2}^{h} \mu_j \right) \varphi_1 + \sum_{j=2}^{h} \mu_j \varphi_j = 0,$$

whence

$$\sum_{j=2}^{h} \mu_j (\varphi_j - \varphi_1) = 0. \tag{2.9}$$

Consequently, since by (2.3) and (2.5) we have $\sum_{j=2}^{h} |\mu_j| \neq 0$, it follows that the system $\varphi_2 - \varphi_1, \ldots, \varphi_h - \varphi_1$ is linearly dependent. Thus we have $\dim [\varphi_2 - \varphi_1, \ldots, \varphi_h - \varphi_1] \leqslant h - 2$, whence for the convex hull $\operatorname{co} \{\varphi_1, \ldots, \varphi_h\}$ of the points $\varphi_1, \ldots, \varphi_h$ we obtain

$$\dim \operatorname{co} \{\varphi_1, \ldots, \varphi_h\} \leqslant h - 2. \tag{2.10}$$

Since the extremal points of $\operatorname{co}\{\varphi_1, \ldots, \varphi_h\}$ are $\varphi_1, \ldots, \varphi_h$ (because they are extremal points of $S_{E^*} \supset \operatorname{co}\{\varphi_1, \ldots, \varphi_h\}$) and since $\varphi = \sum_{j=1}^{h} \lambda_j \varphi_j \in \operatorname{co}\{\varphi_1, \ldots, \varphi_h\}$, it follows *) from (2.10) and the proof of lemma 1.1 (the real case) that there exist a subsystem $\varphi_{i_1}, \ldots, \varphi_{i_{h_0}}$ of the system $\varphi_1, \ldots, \varphi_h$, where $1 \leqslant h_0 \leqslant h-1$, and h_0 numbers $\nu_1, \ldots, \nu_{h_0} > 0$ with $\sum_{j=1}^{h_0} \nu_j = 1$, such that $\varphi = \sum_{j=1}^{h_0} \nu_j \varphi_{i_j}$, contradicting the minimality of h, which completes the proof.

22. FINITE DIMENSIONAL ČEBYŠEV SUBSPACES IN GENERAL NORMED LINEAR SPACES

THEOREM 2.1 ([211], p. 127, theorem 2.2 and [215], p. 237, theorem 2). *Let E be a normed linear space and let $G = [x_1, \ldots, x_n]$ be an n-dimensional linear subspace of E. The following statements are equivalent:*

1° *G is a Čebyšev subspace.*

2° *There do not exist h extremal points $f_1, \ldots, f_h$ of the unit cell S_{E^*}, where $1 \leqslant h \leqslant n$ if the scalars are real and $1 \leqslant h \leqslant 2n-1$ if the scalars are complex, h numbers $\lambda_1, \ldots, \lambda_h > 0$ with $\sum_{j=1}^{h} \lambda_j = 1$, and $x \in E$, $g_0 \in G \setminus \{0\}$, such that*

$$\sum_{j=1}^{h} \lambda_j f_j(x_k) = 0 \qquad (k = 1, \ldots, n), \tag{2.11}$$

$$\sum_{j=1}^{h} \lambda_j f_j(x) = \|x\| = \|x - g_0\|. \tag{2.12}$$

3° *There do not exist h extremal points $f_1, \ldots, f_h$ of S_{E^*}, where $1 \leqslant h \leqslant n$ if the scalars are real and $1 \leqslant h \leqslant 2n-1$*

*) Indeed, by the argument of the proof of lemma 1.1 (the real case) we have actually proved the following more general proposition: If Γ is a bounded closed convex set in a finite-dimensional real normed linear space, then for every $\varphi \in \Gamma$ there exist h_0 extremal points $\psi_1, \ldots, \psi_{h_0} \in \mathcal{E}(\Gamma)$, where $1 \leqslant h_0 \leqslant \dim \Gamma + 1$ (in particular, if $\varphi \in \operatorname{Int}_l \Gamma$, then $h_0 = \dim \Gamma + 1$) and h_0 numbers $\nu_1, \ldots, \nu_{h_0} > 0$ with $\sum_{j=1}^{h_0} \nu_j = 1$, such that $\varphi = \sum_{j=1}^{h_0} \nu_j \psi_j$.

if the scalars are complex, h numbers $\lambda_1, \ldots, \lambda_h > 0$ with $\sum_{j=1}^{h} \lambda_j = 1$, and $x \in E$, $g_0 \in G \setminus \{0\}$, such that we have (2.11) and

$$f_j(x) = \|x\| = \|x - g_0\| \qquad (j = 1, \ldots, h). \tag{2.13}$$

Proof. If condition 2° is not satisfied, then the functional $f \in E^*$ defined by

$$f = \sum_{j=1}^{h} \lambda_j f_j, \tag{2.14}$$

and the elements $x \in E$, $g_0 \in G \setminus \{0\}$ satisfy the conditions (3.9), (3.10) and (3.12) of Chap. I, whence, by virtue of theorem 3.2 of Chap. I, it follows that G is not a Čebyšev subspace. Thus, $1° \Rightarrow 2°$.

The implication $2° \Rightarrow 3°$ is obvious.

Finally, assume that condition 1° is not satisfied. Then for a suitable $y \in E \setminus G$ there exist $g_1, g_2 \in \mathfrak{P}_G(y)$, $g_1 \neq g_2$. Hence, putting

$$x = y - g_1, \quad g_0 = g_2 - g_1,$$

we shall have $x \in E \setminus G$, $g_0 \in G \setminus \{0\}$ and $0, g_0 \in \mathfrak{P}_G(x)$. By virtue of corollary 1.2 and of lemma 2.2 there exist then h extremal points $f_1, \ldots, f_h$ of S_{E^*}, where $1 \leqslant h \leqslant n+1$ if the scalars are real and $1 \leqslant h \leqslant 2n+1$ if the scalars are complex, with the restrictions $f_1|_{E_0}, \ldots, f_h|_{E_0}$ real-linearly independent, where $E_0 = G \oplus [x]$, and h numbers $\lambda_1, \ldots, \lambda_h > 0$ with $\sum_{j=1}^{h} \lambda_j = 1$, such that we have (2.11) and (2.13), whence also

$$f_j(g_0) = 0 \qquad (j = 1, \ldots, h). \tag{2.15}$$

Now assume that the scalars are real and $h = n+1$. Then, since $\dim E_0 = n+1$, the functionals $f_1|_{E_0}, \ldots, f_h|_{E_0}$ form a basis of E_0, whence, by (2.15), $g_0 = 0$, which contradicts our hypothesis. Consequently, we have (2.11) and (2.13) with $1 \leqslant h \leqslant n$, which proves that $3° \Rightarrow 1°$ if the scalars are real.

Finally, assume that the scalars are complex and $h \geqslant 2n$. Then, considering E as a real space $E_{(r)}$ in the usual way, we have $\dim E_0 = 2n+2$. Let E_1 be the hyperplane $[x_1, ix_1, \ldots \ldots, x_n, ix_n, x]$ in E_0 (where $i = \sqrt{-1}$), hence $\dim E_1 = 2n+1$. Then $f_1|_{E_1}, \ldots, f_h|_{E_1}$ are real-linearly independent, since if we had $\sum_{j=1}^{h} \alpha_j f_j(y) = 0$ $(y \in E_1)$, $\sum_{j=1}^{h} |\alpha_j| \neq 0$, it would follow that $\sum_{j=1}^{h} \alpha_j f_j(ix) = i \sum_{j=1}^{h} \alpha_j f_j(x) = 0$, whence $\sum_{j=1}^{h} \alpha_j f_j|_{E_0} = 0$, which contradicts the real-linear independence of $f_1|_{E_0}, \ldots, f_h|_{E_0}$. Consequently, for $G_1 = \{y \in E_1 | f_1(y) = \ldots = f_h(y) = 0\}$ we have

$$\dim G_1 = \begin{cases} 1 \text{ if } h = 2n \\ 0 \text{ if } h = 2n+1. \end{cases} \tag{2.16}$$

On the other hand, by (2.15) we have

$$f_j(ig_0) = if_j(g_0) = 0 \qquad (j = 1, \ldots, h),$$

whence $g_0, ig_0 \in G_1$, which, taking into account the linear independence of g_0 and ig_0 in $E_{(r)}$, contradicts (2.16). Thus, $3° \Rightarrow 1°$ if the scalars are complex, which completes the proof of theorem 2.1.

As has been observed in the paper [211], p. 133, theorem 2.5, in the particular case when $n = 1$ theorem 2.1 can be sharpened, namely it is sufficient to consider only the extremal points f_1 of S_{E^*} for which the set $\{x \in E | f_1(x) = 1, \|x\| = 1\}$ is a face of the unit cell S_E of the space E (i.e. a maximal convex subset of Fr S_E). In [211] it has also been observed that in the case when $n \geqslant 2$ this statement is no longer valid, namely the condition thus modified is no longer sufficient for G to be a Čebyšev subspace.

Let us mention that another characterization of finite dimensional Čebyšev subspaces in general normed linear spaces has been given by K. Tatarkiewicz ([243], theorem 4.5). However, this characterization is nothing else than a geometric description of Čebyšev subspaces, which does not make use of the elements of the conjugate space E^*, but only of elements of the space E (essentially, it amounts to the condition that for every $x \in E$ the set $G \cap S(x, \inf_{g \in G} \|x - g\|)$ should contain one single element), and for this reason it has no applications in concrete spaces.

From theorem 2.1 we shall now deduce sufficient conditions in order that an n-dimensional linear subspace $G = [x_1, \ldots, x_n]$ of a normed linear space E be a Čebyšev subspace.

For this purpose we shall prove first

LEMMA 2.3 ([210], p. 34, theorem 2). *If E is a normed linear space of (finite or infinite) dimension $\geqslant n$, where $1 \leqslant n < \infty$, then the unit cell S_{E^*} has at least n linearly independent extremal points.*

Proof. By virtue of lemma 1.2 on the extremal extension of extremal functionals, it is sufficient to consider the case when $\dim E = n$. Assume then that the maximal number of linearly independent extremal points of S_{E^*} is $l < n$ and let $f_1, \ldots, f_l$ be an arbitrary system of l linearly independent extremal points of S_{E^*}. Since by lemma 1.1 every $f \in E^*$ with $\|f\| = 1$, whence also every $f \in E^*$, can be written as a linear combination of

$h \leqslant 2n - 1 < \infty$ extremal points of S_{E^*} and since by our hypothesis every extremal point of S_{E^*} is a linear combination of $f_1, \ldots, f_l$, it follows that every $f \in E^*$ can be written in the form

$$f = \sum_{j=1}^{l} \nu_j f_j.$$

But then $\dim E^* = l < n$, which contradicts the hypothesis $\dim E = n$. This completes the proof of lemma 2.3.

This being said, we can now prove

COROLLARY 2.1 ([211], p. 130, the consequence of theorem 2.2 and remark 3, and p. 132, lemma 2.3). *Let E be a real normed linear space and let $G = [x_1, \ldots, x_n]$ be an n-dimensional linear subspace of E. Each of the following equivalent conditions is sufficient in order that G be a Čebyšev subspace:*

1° *There do not exist n linearly independent extremal points $f_1, \ldots, f_n$ of the unit cell S_{E^*} and n numbers $\lambda_1, \ldots, \lambda_n \geqslant 0$ with $\sum_{=1}^{n} \lambda_j \neq 0$, such that*

$$\sum_{j=1}^{n} \lambda_j f_j(x_k) = 0 \qquad (k = 1, \ldots, n). \tag{2.17}$$

2° *There do not exist n linearly independent extremal points $f_1, \ldots, f_n$ of the unit cell S_{E^*} and n numbers $\mu_1, \ldots, \mu_n$ with $\sum_{j=1}^{n} |\mu_j| \neq 0$, such that*

$$\sum_{j=1}^{n} \mu_j f_j(x_k) = 0 \qquad (k = 1, \ldots, n). \tag{2.18}$$

3° *For any n linearly independent extremal points $f_1, \ldots, f_n$ of the unit cell S_{E^*} we have*

$$\begin{vmatrix} f_1(x_1) \ldots f_n(x_1) \\ \ldots\ldots\ldots\ldots\ldots \\ f_1(x_n) \ldots f_n(x_n) \end{vmatrix} \neq 0. \tag{2.19}$$

4° *For any n linearly independent extremal points $f_1, \ldots, f_n$ of S_{E^*} and any n numbers $\gamma_1, \ldots, \gamma_n$ there exists exactly one polynomial $g \in G$ such that*

$$f_j(g) = \gamma_j \qquad (j = 1, \ldots, n). \tag{2.20}$$

5° *There do not exist n linearly independent extremal points $f_1, \ldots, f_n$ of S_{E^*} and $g_0 \in G \setminus \{0\}$ such that*

$$f_j(g_0) = 0 \qquad (j = 1, \ldots, n). \tag{2.21}$$

The above statement remains also valid for complex normed linear spaces, replacing everywhere "linearly independent" by "pairwise linearly independent".

Proof. The equivalence of conditions $1^\circ - 5^\circ$ being obvious, we shall show that they are sufficient for G to be a Čebyšev subspace.

Assume that G is not a Čebyšev subspace. Then by virtue of theorem 2.1 there exist h extremal points $f_1, \ldots, f_h$ of S_{E^*}, where $1 \leqslant h \leqslant n$ if the scalars are real and $1 \leqslant h \leqslant 2n-1$ if the scalars are complex, h numbers $\lambda_1, \ldots, \lambda_h > 0$ with $\sum_{j=1}^{h} \lambda_j = 1$, and $x \in E, g_0 \in G \setminus \{0\}$, such that we have (2.11) and (2.13); as we have seen in the above proof of theorem 2.1, $f_1, \ldots, f_h$ may be assumed to be real-linearly independent. If now $h < n$, then by lemma 2.3 we can choose extremal points $f_{h+1}, \ldots, f_n$ of S_{E^*} such that $f_1, \ldots, f_n$ be real-linearly independent, whence, putting $\lambda_{h+1} = \ldots = \lambda_n = 0$, by virtue of (2.11) we shall have (2.17). On the other hand, if $h \geqslant n$, then by virtue of (2.11) and (2.13) we shall have (2.21), which completes the proof in the case when the scalars are real.

Now assume that the scalars are complex and that among the extremal points $f_1, \ldots, f_h$ there exist two, say f_1, f_2, which are linearly dependent, whence there exist complex numbers μ_1, μ_2 with $|\mu_1| + |\mu_2| \neq 0$, such that $\mu_1 f_1 + \mu_2 f_2 = 0$. Then, as in the above proof of lemma 2.2, we obtain $\mu_1 + \mu_2 = 0$, whence $\mu_1(f_1 - f_2) = 0$. Since $\mu_1 \neq 0$, it follows that we have $f_1 = f_2$, which contradicts the real-linear independence of $f_1, \ldots, f_h$. Consequently, $f_1, \ldots, f_h$ are pairwise linearly independent. Continuing now as in the first part of the proof, we obtain $f_{h+1}, \ldots, f_n \in \mathcal{E}(S_{E^*})$ such that $f_1, \ldots, f_n$ are pairwise linearly independent and satisfy (2.17) when $h < n$, and (2.21) when $h \geqslant n$, which completes the proof of corollary 2.2.

Obviously, when $n \geqslant 3$, the complex case of corollary 2.2 presents interest only in spaces E with the property that the pairwise linear independence of the extremal points $f_1, \ldots, f_n \in \mathcal{E}(S_{E^*})$ implies their linear independence; such a space is e.g. $E = C(Q)$ (Q compact).

The conditions of corollary 2.1 above are not necessary for an n-dimensional linear subspace $G = [x_1, \ldots, x_n]$ of a normed linear space E to be a Čebyšev subspace, as shown by the following example ([211], p. 133): Let $E = l_3^1 =$ the space of all triplets of scalars $x = \{\xi_1, \xi_2, \xi_3\}$ endowed with the usual vector operations and with the norm $\|x\| = \sum_{j=1}^{3} |\xi_j|$, and let $G = \{x \in E \mid \xi_3 = 0\}$. Then G is a Čebyšev subspace, namely, for every $x = \{\xi_1, \xi_2, \xi_3\} \in E$ we have $\mathfrak{P}_G(x) = \{g_x\}$, where $g_x = \{\xi_1, \xi_2, 0\}$ (since for $g = \{\eta_1, \eta_2, 0\} \in G$ we have $\|x - g\| = |\xi_1 - \eta_1| + |\xi_2 - \eta_2| + |\xi_3| \geqslant |\xi_3| = \|x - g_x\|$). On the other hand, G does not satisfy the conditions of corollary 2.1, since for $f_1 = \{1, 1, 1\}$, $f_2 = \{1, 1, -1\} \in l_3^\infty \equiv E^*$ and $g_0 = \{1, -1, 0\} \in G \setminus \{0\}$ we have $f_1, f_2 \in \mathcal{E}(S_{E^*})$, $\sum_{j=1}^{2} \alpha_j f_j \neq 0$ $\left(\sum_{j=1}^{2} |\alpha_j| = 0\right)$ and $f_j(g_0) = 0$ $(j = 1,2)$.

In connection with the existence of Čebyšev subspaces, let us mention the following problem raised by V. Klee [110 a], which has remained unsolved until the present: *Does every m-dimensional Banach space E, where $3 \leqslant m < \infty$, have a Čebyšev subspace G of dimension 1?* For $m = 3$ the answer is affirmative (T. J. McMinn [145 a]; see also A. S. Besicovitch [16 a]); for $m \geqslant 3$ an incorrect proof has been given by G. Ewald [54 b].

2.3. APPLICATIONS IN THE SPACES $C(Q)$, $C_R(Q)$, $C_0(T)$ AND $L^\infty(T,\nu)$

We shall now deduce from the preceding general results the following classical theorem of characterization of finite-dimensional Čebyšev subspaces in the spaces $C(Q)$:

THEOREM 2.2. *Let $E = C(Q)$ (Q compact) and let $G = [x_1, \ldots, x_n]$ be an n-dimensional linear subspace of E. In order that G be a Čebyšev subspace it is necessary and sufficient that $x_1, \ldots, x_n$ form a Čebyšev system* *).

Theorem 2.2 has been given, in the case when the scalars are real, by A. Haar [80], and in the case when the scalars are complex by A. N. Kolmogorov ([118], theorem 3).

First proof ([211], p. 130–131). Assume that the condition of theorem 2.2 is not satisfied, i.e. that there exist a $g_0 \in G \setminus \{0\}$ and n distinct points $q_1, \ldots, q_n \in Q$ such that

$$g_0(q_1) = \ldots = g_0(q_n) = 0; \tag{2.22}$$

obviously, we may assume that $\|g_0\| \leqslant 1$. Then the transposed system of equations

$$\sum_{j=1}^{n} \mu_j x_k(q_j) = 0 \qquad (k = 1, \ldots, n), \tag{2.23}$$

*) For the definition of the notion of a Čebyšev system see §1, section 1.2.

too, has a non-trivial solution $\mu_1, \ldots, \mu_n$, whence also a solution with $\sum_{j=1}^{n} |\mu_j| = 1$. We may assume (renumbering, if necessary, the points $q_1, \ldots, q_n$) that we have

$$\mu_1, \ldots, \mu_h \neq 0, \quad \mu_{h+1} = \ldots = \mu_n = 0 \tag{2.24}$$

for a suitable h with $1 \leqslant h \leqslant n$.

Then by the lemma of Uryson there exist $x', x'' \in C_R(Q)$ such that $x'(q_j) = \operatorname{Re} \operatorname{sign} \mu_j$, $x''(q_j) = \operatorname{Im} \operatorname{sign} \mu_j$ $(j = 1, \ldots, h)$, whence also an $x_0 \in E$ such that

$$x_0(q_j) = \operatorname{sign} \mu_j \qquad (j = 1, \ldots, h); \tag{2.25}$$

we may assume (dividing, if necessary, by $|x_0(q)|$ on the set $\{q \in Q \mid |x_0(q)| > 1\}$) that we also have

$$\max_{q \in Q} |x_0(q)| = 1. \tag{2.26}$$

Put

$$x(q) = x_0(q)\,(1 - |g_0(q)|) \quad (q \in Q). \tag{2.27}$$

Then $x \in E$ and by (2.25), (2.22), (2.26) and $\|g_0\| \leqslant 1$ we have

$$x(q_j) = \operatorname{sign} \mu_j \qquad (j = 1, \ldots, h), \tag{2.28}$$

$$\max_{q \in Q} |x(q)| = 1, \tag{2.29}$$

$$1 = \max_{q \in Q} |x(q) - g_0(q)| \leqslant \max_{q \in Q} (|x(q)| + |g_0(q)|) =$$

$$= \max_{q \in Q} [|x_0(q)|\,(1 - |g_0(q)|) + |g_0(q)|] = 1. \tag{2.30}$$

Define $f_j \in \mathcal{E}(S_{E^*})$ by

$$f_j(y) = (\operatorname{sign} \bar{\mu}_j) y(q_j) \quad (y \in E, j = 1, \ldots, h) \tag{2.31}$$

and the numbers $\lambda_j > 0$ $(j = 1, \ldots, h)$ with $\sum_{j=1}^{h} \lambda_j = 1$ by

$$\lambda_j = |\mu_j| \qquad (j = 1, \ldots, h). \tag{2.32}$$

Then by (2.23), (2.28), (2.29) and (2.30) we have (2.11) and (2.13), whence, by virtue of theorem 2.1, it follows that G is not a Čebyšev subspace.

Conversely, assume that G is not a Čebyšev subspace. Then there exist, by virtue of condition 5° of corollary 2.1, n distinct points $q_1, \ldots, q_n \in Q$, n scalars $\beta_1, \ldots, \beta_n$ of modulus 1, and a $g_0 \in G \setminus \{0\}$, such that

$$\beta_j g_0(q_j) = 0 \quad (j = 1, \ldots, n);$$

since $|\beta_j| = 1$ $(j = 1, \ldots, n)$, it follows that we have (2.22), whence $x_1, \ldots, x_n$ do not form a Čebyšev system, which completes the proof.

Second proof. Assume that the condition of theorem 2.2 is not satisfied, i.e. that there exist a $g_0 \in G \setminus \{0\}$ and n distinct points $q_1, \ldots, q_n \in Q$ such that we have (2.22). As has been observed in the first proof above, there exist then scalars $\mu_1, \ldots, \mu_n$ with $\sum_{j=1}^{n} |\mu_j| = 1$, satisfying (2.23), and we may assume that we have (2.24).

Define a Radon measure μ on Q by

$$\int_Q y(q) \mathrm{d}\mu(q) = \sum_{j=1}^{h} \mu_j y(q_j) \qquad (y \in E). \tag{2.33}$$

Then $S(\mu) = \{q_1, \ldots, q_h\}$, and for $\frac{\mathrm{d}\mu}{\mathrm{d}|\mu|}$ we may take any continuous function on Q which has the value sign $\overline{\mu}_j$ at the point q_j $(j = 1, \ldots, h)$. Consequently, taking into account $\sum_{j=1}^{h} |\mu_j| = 1$, (2.23) and (2.22), the Radon measure μ and the polynomial $g_0 \in G \setminus \{0\}$ satisfy the conditions (3.34)—(3.37) of Chap. I, whence, by virtue of theorem 3.3 of Chap. I, it follows that G is not a Čebyšev subspace.

Conversely, assume that G is not a Čebyšev subspace. Then there exist, by virtue of theorem 3.3 of Chap. I, a Radon measure μ on Q and an element $g_0 \in G \setminus \{0\}$ such that we have (3.34)—(3.37) of Chap. I. But if $S(\mu)$ contains at least n distinct points $q_1, \ldots, q_n$, then by (3.37) we have (2.22), whence $x_1, \ldots, x_n$ do not form a Čebyšev system. On the other hand, if $S(\mu)$ consists of h distinct points $q_1, \ldots, q_h$, where $1 \leqslant h \leqslant n - 1$, then μ is *) of the form

$$\int_Q y(q)\, \mathrm{d}\mu(q) = \sum_{j=1}^{h} \mu_j y(q_j) \qquad (y \in E); \tag{2.34}$$

in this case, by (3.34) we have $\sum_{j=1}^{h} |\mu_j| = |\mu|(Q) = 1$, and by (3.35) we have

$$\sum_{j=0}^{h} \mu_j x_k(q_j) = 0 \qquad (k = 1, \ldots, n), \tag{2.35}$$

---

*) See e.g. N. Bourbaki [22], Chap. III, p. 74, proposition 12.

which imply

$$\begin{vmatrix} x_1(q_1) & \dots & x_1(q_n) \\ \cdot & \cdot & \cdot \\ x_n(q_1) & \dots & x_n(q_n) \end{vmatrix} = 0,$$

where $q_{h+1}, \dots, q_n$ are arbitrary distinct *) points of Q, different from $q_1, \dots, q_h$. Consequently $x_1, \dots, x_n$ do not form a Čebyšev system, which completes the proof.

In the particular case when the scalars are real, one can give a third proof of theorem 2.2 (see [226], p. 167—168), similar to the second proof above, applying theorem 3.4 of Chap. I instead of theorem 3.3 of Chap. I. For other proofs of theorem 2.2 see N. I. Ahiezer [1], V. Pták [180], I. J. Schoenberg [204], T. J. Rivlin and H. S. Shapiro [193], [194].

As has remarked R. R. Phelps ([176], lemma 3.5), *theorem 2.2 remains valid also for the spaces* **) $E = I_A$, *replacing its condition by the condition that* $x_1, \dots, x_n$ *form a Čebyšev system on the set* $Q \setminus A$ (i.e. that every polynomial $\sum_{k=1}^{n} \alpha_k x_k \neq 0$ have at most $n-1$ zeros on $Q \setminus A$); indeed, this follows from the form of the extremal points of the unit cell $S_{(I_A)^*}$ and from the fact that by the lemma of Uryson the function x_0 with the properties (2.25) can be choosen to be in I_A and then the function (2.27) will be also in I_A.

In particular, from this remark it follows (R. R. Phelps [176], theorem 3.6) that *theorem 2.2 remains valid also in the spaces* $C_0(T)$ (T *locally compact*).

Theorem 1.4 and especially theorem 2.2 show the importance of the notion of Čebyšev system. Already A. Haar [80] has observed that *if Q is the closure of a bounded connected open set in the euclidean plane, then on Q there exists no Čebyšev system $x_1, \dots, x_n$ with $n \geqslant 2$.* Thus, the following problem arises naturally: what are the compact spaces Q which admit Čebyšev systems $x_1, \dots, x_n$ (or, what is equivalent, for which $C(Q)$ has n-dimensional Čebyšev subspaces) with ***) $n \geqslant 2$? In the case of real scalars, the answer is given by

THEOREM 2.3. *If a compact space Q admits a real Čebyšev system $x_1, \dots, x_n$ (or, what is equivalent: if $C_R(Q)$ has an n-di-*

*) We recall that by hypothesis Q contains at least $n+1$ distinct points.

**) For the definitions of the sp aces I_A and $C_0(T)$ and for the form of the extremal points of the unit cell $S_{(I_A)^*}$, see § 1, section 1.3.

***) The condition $n \geqslant 2$ is essential, since every compact space admits, obviously, Čebyšev systems consisting of one single element x_1, e.g. the function $x_1 \equiv 1$ (hence, for every compact space Q, the space $C(Q)$ has Čebyšev subspaces of dimension 1).

mensional Čebyšev subspace) with $n \geqslant 2$, then Q is homeomorphic to a subset of a circumference Γ.

This result has been conjectured by S. Mazur and proved for the first time by J. C. Mairhuber [142] under the additional hypothesis that Q is a subset of a k-dimensional euclidean space. For general compact spaces this theorem has been proved by K. Sieklucki [208] and, independently, by P. C. Curtis [41]. We shall present here a proof of theorem 2.3 given by I. J. Schoenberg and C. T. Yang [205]. The principal topological tool of this proof is

LEMMA 2.4 (I. J. Schoenberg — C. T. Yang [205], lemma 1). *Let Q be a compact space with the property that for every non-void open set $U \subset Q$ the set $Q \setminus U$ can be homeomorphically embedded into a circumference* Γ. *Then either Q can be embedded into Γ, or Q is homeomorphic to a union $\Gamma \cup \{\alpha\}$, where α is a point outside* Γ.

Proof. Case 1° : *Q is not connected.* In this case we have

$$Q = A \cup B, \qquad A \cap B = \varnothing, \tag{2.36}$$

where A, B are non-void closed subsets of Q. Since B is also open, it follows from the hypothesis of the lemma that A can be embedded into Γ; similarly, B can be embedded into Γ. If neither A nor B is homeomorphic to Γ, then each can be embedded into an arc of Γ; consequently, if C and D are two disjoint arcs of Γ, then A can be embedded into C and B can be embedded into D, whence Q can be embedded into Γ.

On the other hand, if one of the sets A, B, say A, is homeomorphic to Γ, let $\alpha \in B$, $\alpha \in U \subset B$, where U is open in Q. Then, by the hypothesis of the lemma, the set $A \cup (B \setminus U) = Q \setminus U$ can be embedded into Γ, whence, since A is homeomorphic to Γ and $A \cap B = \varnothing$, it follows that we have $B \setminus U = \varnothing$. Since U has been an arbitrary open set with the property $\alpha \in U \subset B$, it follows that we have $B = \{\alpha\}$, whence $Q = A \cup \{\alpha\}$ is homeomorphic to $\Gamma \cup \{\alpha\}$.

Case 2° : *Q is connected.* If Q consists of one single point, then Q can be embedded into Γ.

Assume now that Q contains two distinct points q_1 and q_2. Let U be an open set such that

$$q_1 \in U, \qquad q_2 \in Q \setminus \overline{U}. \tag{2.37}$$

Since Q is connected, the component C of $Q \setminus U$ which contains q_2 must intersect $\overline{U} \setminus U$, and thus it does not reduce to a single point. Since by the hypothesis of the lemma $Q \setminus U$ can be embedded into Γ, it follows that C is a closed arc. The component of $C \setminus \overline{U}$ containing q_2 does not reduce to a single point, whence it contains an open arc W. Since the set W

is open in Q (because $Q \setminus \overline{U}$ can be embedded into Γ), $Q \setminus W$ can be embedded into Γ, whence every component of $Q \setminus W$ is either an arc or a point. On the other hand, since Q is connected, every component of $Q \setminus W$ contains at least one of the endpoints a, b of the open arc W. Let us denote by C_a, C_b the components of $Q \setminus W$ containing a and b respectively; by the preceding we have thus $Q \setminus W = C_a \cup C_b$. There are two possibilities: 1) C_a and C_b are distinct components, hence disjoint. In this case C_a is either $= \{a\}$, or an arc having a as an endpoint and in the same way, C_b is either $= \{b\}$, or an arc having b as an endpoint. Consequently, $Q = C_a \cup W \cup C_b$ is an arc, hence it can be embedded into Γ. 2) We have $C_a = C_b$. In this case C_a is an arc having a and b as endpoints. Consequently, $Q = W \cup C_a$ is a closed simple curve and can be embedded into Γ, which completes the proof of lemma 2.4.

LEMMA 2.5 (I. J. Schoenberg — C. T. Yang [205], lemma 2). *Let $Q = \Gamma \cup \{\alpha\}$, where Γ is a circumference and α a point outside Γ. Then Q admits no real Čebyšev system $x_1, \ldots, x_n$ with $n \geqslant 2$.*

Proof. Assume the contrary, i.e. that Q admits a real Čebyšev system $x_1, \ldots, x_n$ with $n \geqslant 2$. Choose n distinct points $q_1, \ldots, q_n \in \Gamma$ in cyclic order, so that they be the vertices of a regular n-gon $\langle q_1, \ldots, q_n \rangle$ and consider the determinant

$$D(q_1, \ldots, q_n) = \begin{vmatrix} x_1(q_1) & \ldots & x_1(q_n) \\ \cdot & \cdot\ \cdot\ \cdot & \cdot \\ x_n(q_1) & \ldots & x_n(q_n) \end{vmatrix} \neq 0. \tag{2.38}$$

Now rotate the system of points by the angle $\frac{2\pi}{n}$. The determinant becomes then

$$D(q_2, \ldots, q_n, q_1) = (-1)^{n-1} D(q_1, \ldots, q_n). \tag{2.39}$$

Since for distinct arguments the determinant D is always $\neq 0$ (because $x_1, \ldots, x_n$ form by hypothesis a Čebyšev system on Q), it follows that n must be *odd*, since otherwise (by (2.39) and continuity) D would vanish somewhere during the rotation.

Now choose $q_1 = \alpha$ and place $q_2, \ldots, q_n$ on Γ at the vertices of a regular $(n-1)$-gon. If we rotate the system of points $q_2, \ldots, q_n$ by the angle $\frac{2\pi}{n-1}$, the determinant (2.38) becomes

$$D(q_1, q_3, q_4, \ldots, q_2) = (-1)^{n-2} D\,(q_1, \ldots, q_n). \tag{2.40}$$

Since $n-2$ is odd, it follows (by (2.40) and continuity) that D must vanish somewhere during the rotation, which contradicts the hypothesis that $x_1, \ldots, x_n$ form a real Čebyšev system on Q. This completes the proof of lemma 2.5.

This being said, we can now give the

Proof of theorem 2.3. We shall proceed by induction on n. First, let $n = 2$ and let x_1, x_2 be a real Čebyšev system on Q. Then

$$q \to \{x_1(q), x_2(q)\} \tag{2.41}$$

is a one-to-one continuous mapping of Q into the real projective line, since for any distinct $q_1, q_2 \in Q$ we have by hypothesis

$$\begin{vmatrix} x_1(q_1) & x_1(q_2) \\ x_2(q_1) & x_2(q_2) \end{vmatrix} \neq 0; \tag{2.42}$$

consequently, since Q is compact, the mapping (2.41) is actually a homeomorphism. But the real projective line being homeomorphic to a circumference Γ, it follows that Q can be embedded into Γ, which proves theorem 2.3 in the case $n = 2$.

Now let $n > 2$. Assume that theorem 2.3 is true for Čebyšev systems of $n - 1$ real functions and let $x_1, \ldots, x_n$ be a real Čebyšev system on Q. We claim that then the hypothesis of lemma 2.4 is satisfied, i.e. for every non-void open set $U \subset Q$ the set $Q \setminus U$ can be embedded homeomorphically into a circumference Γ.

Indeed, take a fixed $q_1 \in U$ and let $q_2, \ldots, q_n$ be arbitrary distinct points of $Q \setminus U$. We have by hypothesis

$$D(q_1, \ldots, q_n) = \begin{vmatrix} x_1(q_1) & \ldots & x_1(q_n) \\ \cdot & \cdot\ \cdot\ \cdot & \cdot \\ x_n(q_1) & \ldots & x_n(q_n) \end{vmatrix} \neq 0, \tag{2.43}$$

whence $\max\limits_{1 \leqslant k \leqslant n} |x_k(q_1)| \neq 0$; therefore we may assume that $x_1(q_1) \neq 0$. Put

$$\gamma_j = \frac{x_j(q_1)}{x_1(q_1)} \qquad (j = 2, \ldots, n). \tag{2.44}$$

Multiplying the first row of the determinant (2.43) by γ_j and subtracting it from the j-th row for $j = 2, \ldots, n$, we obtain

$$\begin{vmatrix} x_1(q_1) & x_1(q_2) & \ldots & x_1(q_n) \\ 0 & x_2(q_2) - \gamma_2 x_1(q_2) & \ldots & x_2(q_n) - \gamma_2 x_1(q_n) \\ \cdot & \cdot & \cdot\ \cdot\ \cdot & \cdot \\ 0 & x_n(q_2) - \gamma_n x_1(q_2) & \ldots & x_n(q_n) - \gamma_n x_1(q_n) \end{vmatrix} =$$

$$= D(q_1, \ldots, q_n) \neq 0,$$

whence

$$\begin{vmatrix} x_2(q_2) - \gamma_2 x_1(q_2) \dots x_2(q_n) - \gamma_2 x_1(q_n) \\ \dots\dots\dots\dots\dots\dots \\ x_n(q_2) - \gamma_n x_1(q_2) \dots x_n(q_n) - \gamma_n x_1(q_n) \end{vmatrix} \neq 0 \qquad (2.45)$$

for any distinct $q_2, \dots, q_n \in Q \setminus U$, whence the functions $y_2, \dots, y_n \in C_R(Q \setminus U)$ defined by

$$y_j(q) = x_j(q) - \gamma_j x_1(q) \qquad (q \in Q \setminus U, j = 2, \dots, n)$$

form a Čebyšev system of $n - 1$ real functions on $Q \setminus U$. Consequently, by virtue of the induction hypothesis, $Q \setminus U$ can be embedded homeomorphically into Γ.

Now, by lemma 2.4, it follows that either Q can be embedded into Γ, or Q is homeomorphic to $\Gamma \cup \{\alpha\}$, where α is a point outside Γ. But the second alternative is excluded by virtue of lemma 2.5, hence Q can be embedded into Γ, which completes the proof of theorem 2.3.

In the case of complex scalars there is known only the following partial answer to the above problem, obtained by I. J. Schoenberg and C. T. Yang ([205], theorem 4) : *If Q admits a complex Čebyšev system $x_1, \dots, x_n$ with $n \geqslant 2$ and if Q is homeomorphic to the space of a finite polyhedron, then Q is homeomorphic to a subset of the euclidean plane.*

Theorem 2.3 presents interest also from the following point of view : Although the classical theorem of Banach-Stone *) shows, theoretically, that the metric-linear properties of the spaces $C_R(Q)$ (and of the spaces $C(Q)$) are completely determined by the topological properties of the compact spaces Q and conversely, yet the effective, explicit study of this interdependence still presents many open problems, and theorem 2.3 may be regarded as a contribution to this study. Other results of the same nature, concerning the connections between best approximation in the spaces $C_R(Q)$ by elements of closed linear subspaces of finite codimension and the topological properties of compact spaces Q, will be given in Chap. III, § 2, section 2.2.

From theorems 2.2 and 2.3 there follows

COROLLARY 2.2 (R. R. Phelps [176], p. 251). *Let $E = L_R^\infty(T, \nu)$, where (T, ν) is a σ-finite positive measure space such that* $\dim E = \infty$. *Then E has no Čebyšev subspace of finite dimension $\geqslant 2$.*

Proof. There exists **) an extremally disconnected compact space (i.e. with the property that the closure of every open

*) See e.g. M. M. Day [43], Chap. V, § 3, theorem 2 for $C_R(Q)$ and N. Dunford and J. Schwartz [49], p. 442, theorem 8 for $C(Q)$.

**) See e.g. N. Dunford and J. Schwartz [49], p. 445, theorem 11 and p. 398.

set in Q is open), such that $E = L_R^\infty(T, \nu)$ is linearly isometric to $C_R(Q)$; since dim $E = \infty$, it follows that the set Q is infinite. Now assume that E has a Čebyšev subspace of finite dimension $\geqslant 2$. Then by the isometry above the space $C_R(Q)$ also has such a subspace, whence, by virtue of theorem 2.2, Q admits a real Čebyšev system $y_1, \ldots, y_n$ with $n \geqslant 2$. Consequently, by virtue of theorem 2.3, Q can be embedded homeomorphically into a circumference Γ. But this contradicts the fact that Q is an extremally disconnected infinite compact space (obviously, no infinite compact subset of Γ can be extremally disconnected), which completes the proof of corollary 2.2.

On the other hand, the spaces L^∞ admit Čebyšev subspaces of dimension 1, even in the case of complex scalars. Namely, as has been observed by R. R. Phelps and M. Jerison ([176], p. 251), a 1-dimensional linear subspace $G = [x_1]$ of $L^\infty(T, \nu)$ (where (T, ν) is a positive measure space) is a Čebyšev subspace if and only if

$$\operatorname*{ess\,inf}_{t \in T} |x_1(t)| > 0. \tag{2.46}$$

Indeed, G is a Čebyšev subspace if and only if its image by the above isometry $L^\infty(T, \nu) \equiv C(Q)$ is a 1-dimensional Čebyšev subspace of $C(Q)$; by virtue of theorem 2.2 this happens if and only if the image $y_1 \in C(Q)$ of x_1 is $\neq 0$ on the whole Q, hence if and only if no extremal point of the unit cell $S_{L^\infty(T,\nu)^*}$ vanishes at x_1. But the extremal points of $S_{L^\infty(T,\nu)^*}$ are *) precisely the functionals of the form αf, where $|\alpha| = 1$ and $f \neq 0$ is multiplicative on $L^\infty(T, \nu)$ (i.e. $f(xz) = f(x)f(z)$ for all $x, z \in L^\infty(T, \nu)$); obviously, such a functional f takes the value 1 at the element $1 \in L^\infty(T, \nu)$ (the function $\equiv 1$). Consequently, if we have (2.46), then $x_1^{-1} = \dfrac{1}{x_1} \in L^\infty(T, \nu)$, whence

$$f(x_1)f(x_1^{-1}) = f(x_1 x_1^{-1}) = f(1) = 1,$$

whence $\alpha f(x_1) \neq 0$ $(|\alpha| = 1)$, i.e. no extremal point of $S_{L^\infty(T,\nu)^*}$ vanishes at x_1. On the other hand, if we do not have (2.46), then $x_1^{-1} \notin L^\infty(T, \nu)$ and therefore the ideal $\{zx_1 \mid z \in L^\infty(T, \nu)\}$ does not contain 1, whence it is contained in a maximal ideal I. Since every such maximal ideal is**) of the form $I = f^{-1}(0)$ for some multiplicative functional $f \in L^\infty(T, \nu)$, it follows that we have

$$f(z)f(x_1) = f(zx_1) = 0 \qquad (z \in L^\infty(T, \nu)),$$

whence $f(x_1) = 0$, which completes the proof.

*) See e.g. N. Dunford and J. Schwartz [49], p. 443, lemma 9.
**) See e.g. M. A. Naimark [157], p. 176.

In particular, corollary 2.2 and the above remark are valid for the space $E = l^\infty$; condition (2.46) for $x_1 = \{\xi_n^{(1)}\} \in l^\infty$ becomes

$$\inf_{1 \leqslant n < \infty} |\xi_n^{(1)}| > 0. \tag{2.47}$$

Theorem 2.3 (and in fact even the example of A. Haar mentioned after the proof of theorem 2.2) shows that the classical theory of best approximation of real functions of one real variable cannot be extended to the case of functions of more than one variable. However, the remark that the continuous function (2.27) in the proof of the necessity part of theorem 2.2 is not differentiable leads to the following problem, raised by L. Collatz [39]: if $x_1, \ldots, x_n (n \geqslant 2)$ is not a Čebyšev system, could it yet happen that we have the uniqueness of polynomials of best approximation of *differentiable* functions? L. Collatz has shown [39] that *if a real function $x(s, t)$ of two real variables has continuous partial derivatives of the first order in the interior of a strictly convex closed region $\mathfrak{B}$ in the plane, then $x(s, t)$ has one single linear polynomial $g_0(s, t) = \alpha_1 + \alpha_2 s + \alpha_3 t$ of best approximation in $\mathfrak{B}$.* This result has been extended by T.J. Rivlin and H. S. Shapiro ([193], theorem 2) to best approximation by linear polynomials $\alpha_1 + \sum_{i=1}^{n} \alpha_{i+1} s_i$ of real functions $x(s_1, \ldots, s_n)$ of n real variables with continuous partial derivatives of the first order in the interior of a compact region $\mathfrak{B}$ of the n-dimensional real euclidean space, with a "smooth" boundary and such that no hyperplane touches its boundary at $n + 1$ distinct points; on the other hand they have shown ([193], theorem 3) that in the case when there exists a hyperplane tangent to the boundary of $\mathfrak{B}$ at $n + 1$ distinct points, there exists a function $x(s_1, \ldots, s_n)$ even in the class $C^\infty(\mathfrak{B})$, having infinitely many linear polynomials of best approximation. T. J. Rivlin and H. S. Shapiro ([193], corollary of theorem 6) have also shown that independently of the region $\mathfrak{B}$, the linearity of the approximating polynomials is essential in the foregoing, since *for every plane region $\mathfrak{B}$ there exists a function even in $C^\infty(\mathfrak{B})$, having infinitely many quadratic polynomials of best approximation.* For other results on best approximation of functions of more than one variable, see also J. R. Rice [189].

Finally, let us mention some results of A. L. Garkavi [67], [74] on finite dimensional almost Čebyšev *) subspaces in the spaces $C_R(Q)$. For an arbitrary open subset $U \subset Q$ we shall denote by $N_n(U)$ the number of elements of U if this number is $\leqslant n$; otherwise we shall put $N_n(U) = n$. *In order that*

*) For the definition of the notion of almost Čebyšev subspace, see Chap. I, § 3, the final part of section 3.1.

an n-dimensional linear subspace $G = [x_1, \ldots, x_n]$ *of* $E = C_R(Q)$ *(Q compact) be an almost Čebyšev subspace, it is necessary and sufficient that on every open subset* $U \subset Q$ *vanish identically at most* $N_n(U)$ *linearly independent polynomials of G* (A. L. Garkavi [74], theorem 1). *For any compact metric space Q, there exists an* $x \in C_R(Q)$ *such that for every positive integer n the n-dimensional linear subspace* $[1, x, x^2, \ldots, x^{n-1}]$ *of* $C_R(Q)$ *is an almost Čebyšev subspace* (A. L. Garkavi [67], theorem 6 and [74], theorem 2). However, this result is no longer valid for arbitrary compact spaces Q, e.g. *if the set of the isolated points of Q is of cardinality* $> \mathfrak{c}$, *then the space* $C_R(Q)$ *has no almost Čebyšev subspace of finite dimension* $n \geqslant 2$ (A. L. Garkavi [74], p. 44). *If the n-dimensional linear subspace* $G = [x_1, \ldots, x_n]$ *of* $C_R(Q)$ *is an almost Čebyšev subspace, then the set of all n-tuples* $\{q_1, \ldots, q_n\} \in Q^n$ *with the property that for any n real numbers* $\gamma_1, \ldots, \gamma_n$ *there exists exactly one polynomial* $g \in G$ *satisfying* $g(q_j) = \gamma_j$ $(j = 1, \ldots, n)$, *is of the second category in* Q^n (A. L. Garkavi [67], theorem 7 and [74], theorem 3); this result shows that in interpolation problems on compact spaces, the n-dimensional almost Čebyšev subspaces may compensate, to a certain extent, the eventual lack of n-dimensional Čebyšev subspaces (the usefulness of these latters in interpolation problems follows from the remark that $G = [x_1, \ldots, x_n]$ is a Čebyšev subspace if and only if the above set of n-tuples $\{q_1, \ldots, q_n\} \in Q^n$ coincides with Q^n).

2.4. APPLICATIONS IN THE SPACES $C_E(Q)$

Combining theorem 2.1 and lemma 1.7, we obtain the following characterization of finite dimensional Čebyšev subspaces in the spaces $C_E(Q)$:

THEOREM 2.4 ([217], theorem 2.2). *Let Q be a compact space, E a Banach space and* $G = [x_1(.), \ldots, x_n(.)]$ *an n-dimensional linear subspace of* $C_E(Q)$. *In order that G be a Čebyšev subspace, it is necessary and sufficient that there do not exist h extremal points* $f_1, \ldots, f_h$ *of* S_{E^*}, *where* $1 \leqslant h \leqslant n$ *if the scalars are real and* $1 \leqslant h \leqslant 2n - 1$ *if the scalars are complex, h points* $q_1, \ldots, q_h \in Q$, *h numbers* $\lambda_1, \ldots, \lambda_h > 0$ *with* $\sum_{j=1}^{h} \lambda_j = 1$, *and* $x(.) \in C_E(Q)$, $g_0(.) \in G \setminus \{0\}$, *such that*

$$\sum_{j=1}^{h} \lambda_j f_j[x_k(q_j)] = 0 \qquad (k = 1, \ldots, n), \tag{2.48}$$

$$f_j[x(q_j)] = \max_{t \in Q} \|x(t)\|_E = \max_{t \in Q} \|x(t) - g_0(t)\|_E \quad (j = 1, \ldots, h). \tag{2.49}$$

In the paper [217], p. 253—257, it has been shown that from theorem 2.4 and from Chap. I, theorem 3.2, there follow the theorems of S. I. Zuhovitzky and S. B. Stečkin [272], [273], hence also the theorems of S. I. Zuhovitzky and M. G. Krein [271], on the finite dimensional Čebyšev subspaces of the spaces $C_E(Q)$, where E is finite dimensional or strictly convex or Hilbertian.

Finally, let us also mention the following generalization of the problem of best polynomial approximation in the spaces $C_E(Q)$, studied by S. I. Zuhovitzky [266], S. B. Stečkin [236], S. I. Zuhovitzky and G. I. Eskin [269], [270]: Let Q be a compact space, E_1, E two Banach spaces, and $A(.)$ a function defined on Q whose values are closed linear operators from E_1 into E, having the same domain of definition $D_{A(q)} = D$ $(q \in Q)$, dense in E_1, and having the properties: (a) for every $z \in D$ the function $q \to A(q)z$ belongs to $C_E(Q)$; (b) the relations $z \in D$, $A(q)z = 0$ $(q \in Q)$ imply $z = 0$. The problem of *best approximation of a function* $x(.) \in C_E(Q)$ *by the operator function* $A(.)$ consists in finding an element $z_0 \in D$ with the property

$$\max_{q \in Q} \| x(q) - A(q)z_0 \|_E = \inf_{z \in D} \max_{q \in Q} \| x(q) - A(q)z \|_E . \quad (2.50)$$

In the paper [217], p. 261, it has been observed that for every Q, E_1, E and $A(.)$ as above, the set of all functions in $C_E(Q)$ of the form $q \to A(q)z$ $(z \in D)$ is a *linear subspace* of $C_E(Q)$ (because of the linearity of the operators $A(q)$ and of the fact that D is a linear space), and hence the results of the papers mentioned above on best approximation by operator functions can be also deduced from the foregoing general results.

In the particular case when $\dim E_1 = n$, taking into account the general form of the linear operators $E_1 \to E$ and the conditions (a), (b), we have, for every $z = \{\alpha_1, \ldots, \alpha_n\} \in E_1$,

$$A(q)z = A(q)\{\alpha_1, \ldots, \alpha_n\} = \sum_{k=1}^{n} \alpha_k x_k(q) \qquad (q \in Q),$$

with suitable linearly independent $x_1(.), \ldots, x_n(.) \in C_E(Q)$, and thus we find again the problem of best polynomial approximation in the spaces $C_E(Q)$, considered in the preceding.

2.5. APPLICATIONS IN*) THE SPACES $L^1(T, \nu)$, $L^1_R(T, \nu)$, $C^1(Q, \nu)$ AND $C^1_R(Q, \nu)$

THEOREM 2.5. *Let* $E = L^1(T, \nu)$, *where* (T, ν) *is a positive measure space with the property that the conjugate space* $L^1(T, \nu)^*$

*) In the case of the spaces $E = L^p(T, \nu)$ $(1 < p < \infty)$, by Chap. I, corollary 3.5, all closed linear subspaces of E, hence in particular also all finite dimensional linear subspaces of E, are Čebyšev subspaces. Consequently, among the spaces $L^p(T, \nu)$ $(1 \leqslant p < \infty)$ only the case $p = 1$ requires a separate treatment.

is canonically equivalent to $L^\infty(T, \nu)$, *and let* $G = [x_1, \ldots, x_n]$ *be an* n*-dimensional linear subspace of* E. *The following statements are equivalent:*

1° G *is a Čebyšev subspace.*

2° *There do not exist a* ν*-measurable set* $U \subset T$ *with* $\nu(U) > 0$, h ν*-measurable and* ν*-essentially bounded functions* $\beta_1, \ldots, \beta_h$ *with*

$$|\beta_j(t)| = 1 \qquad (j = 1, \ldots, h) \ \nu\text{-a.e. on } T, \qquad (2.51)$$

where $1 \leq h \leq n$ *if the scalars are real and* $1 \leq h \leq 2n - 1$ *if the scalars are complex,* h *numbers* $\lambda_1, \ldots, \lambda_h > 0$ *with* $\sum_{j=1}^{h} \lambda_j = 1$ $\Big($*in the case when* $E = l^1(I)$ *and the scalars are complex it is sufficient to take* $h = 2$ *and* $\lambda_1 = \lambda_2 = \frac{1}{2}\Big)$ *and* $g_0 \in G \setminus \{0\}$, *such that*

$$\sum_{j=1}^{h} \lambda_j \int_T x_k(t) \beta_j(t) \mathrm{d}\nu(t) = 0 \qquad (k = 1, \ldots, n), \qquad (2.52)$$

$$g_0(t) \neq 0 \qquad \nu\text{-a.e. on } U, \qquad (2.53)$$

$$\sum_{j=1}^{h} \lambda_j \beta_j(t) = \pm \operatorname{sign} g_0(t) \qquad \nu\text{-a.e. on } U, \qquad (2.54)$$

$$g_0(t) = 0 \qquad \nu\text{-a.e. on } T \setminus U. \qquad (2.55)$$

In the particular case when the scalars are real (i.e. when $E = L^1_R(T, \nu)$*), these statements are equivalent to the following:*

3° *There do not exist a* ν*-measurable set* $U \subset T$ *with* $\nu(U) > 0$, h ν*-measurable subsets* $M_1, \ldots, M_h$ *of* T, *where* $1 \leq h \leq n$, h *numbers* $\lambda_1, \ldots, \lambda_h > 0$ *with* $\sum_{j=1}^{h} \lambda_j = 1$ *and* $g_0 \in G \setminus \{0\}$, *such that we have (2.55) and*

$$\sum_{j=1}^{h} \lambda_j \int_T x_k(t) \beta_{M_j}(t) \mathrm{d}\nu(t) = 0 \qquad (k = 1, \ldots, n), \qquad (2.56)$$

$$\sum_{j=1}^{h} \lambda_j \beta_{M_j}(t) = 1 \qquad \nu\text{-a.e. on } U, \qquad (2.57)$$

where β_{M_j} $(j = 1, \ldots, h)$ *are defined by*

$$\beta_{M_j}(t) = \begin{cases} 1 & \nu\text{-a.e. on } M_j \\ -1 & \nu\text{-a.e. on } T \setminus M_j \end{cases} \qquad (j = 1, \ldots, h)^{*}). \quad (2.58)$$

Proof. Assume that we do not have 2°, i.e. that there exist U, β_j, $\lambda_j (j = 1, \ldots, h)$ and g_0 such that we have (2.51)—(2.55). Then, putting $\beta = \sum_{j=1}^{h} \lambda_j \beta_j$, we have (3.54)—(3.57) of Chap. I. Since $\nu(U) > 0$, it follows from (2.53), (2.54), $\lambda_j > 0 \, (j=1, \ldots, h)$, $\sum_{j=1}^{h} \lambda_j = 1$ and (2.51) that we have also (3.53) of Chap. I. Consequently, by virtue of the implication 1° ⇒ 2° of Chap. I, theorem 3.4, G is not a Čebyšev subspace. Thus, 1° ⇒ 2°.

Now assume that we do not have 1°. Then by theorem 2.1 and Chap. I, lemma 1.13, there exist h ν-measurable and ν-essentially bounded real functions $\beta_1, \ldots, \beta_h$ satisfying (2.51), where $1 \leqslant h \leqslant n$ if the scalars are real and $1 \leqslant h \leqslant 2n - 1$ if the scalars are complex, h numbers $\lambda_1, \ldots, \lambda_h > 0$ with $\sum_{j=1}^{h} \lambda_j = 1$, and $x \in E$, $g_0 \in G \setminus \{0\}$, such that we have (2.52) and

$$\sum_{j=1}^{h} \lambda_j \int_T x(t) \beta_j(t) d\nu(t) = \int_T |x(t)| d\nu(t) = \int_T |x(t) - g_0(t)| d\nu(t). \quad (2.59)$$

Then by (2.52), (2.59) and $g_0 \in G \setminus \{0\}$ we have $x \in E \setminus G$. Putting $\beta = \sum_{j=1}^{h} \lambda_j \beta_j$, $g_0' = 0$, $g_1' = g_0$, it follows, in the same way as in Chap. I, the proof of theorem 1.8 b), that there exists a ν-measurable set $U \subset T$ with $\nu(T) > 0$, such that we have (2.53)—(2.55). In the case when $E = l^1(I)$ and the scalars are complex, the implication 2° ⇒ 1° with $h = 2$ and $\lambda_1 = \lambda_2 = \frac{1}{2}$ follows immediately from Chap. I, theorem 3.4, observing that every complex number β with $|\beta| \leqslant 1$ can be written as $\beta = \frac{\beta_1 + \beta_2}{2}$, where β_1, β_2 are complex numbers with $|\beta_1| = |\beta_2| = 1$.

---

*) After this monograph has been submitted to the publisher, it has appeared the paper [177 a] of R. R. Phelps, in which there is given, for the case of real scalars, the following equivalent condition:

4° *There does not exist a* $\beta_1 \in L_R^\infty(T, \nu)$ *satisfying* (2.61), *such that the set* $\{t \in T \mid |\beta_1(t)| < \|\beta_1\|\}$ *(defined modulo a set of measure zero) be purely atomic and contain at most* $n - 1$ *atoms.*

Finally, the implication $3° \Rightarrow 2°$ for real scalars follows from the fact that in this case each β_j is obviously of the form (2.58), and the implication $1° \Rightarrow 3°$ for real scalars follows in the same way as the implication $1° \Rightarrow 2°$, using the implication $1° \Rightarrow 3°$ of Chap. I, theorem 3.4. This completes the proof of theorem 2.5.

COROLLARY 2.3. *Let $E = L^1_R(T, \nu)$, where (T, ν) is a positive measure space with the property that the conjugate space $L^1_R(T, \nu)^*$ is canonically equivalent to $L^\infty_R(T, \nu)$, and let $G = [x_1, \ldots, x_n]$ be an n-dimensional linear subspace of E. In order that G be a Čebyšev subspace it is necessary that there exist no ν-measurable function β_1 with*

$$|\beta_1(t)| = 1 \qquad \nu\text{-a.e. on } T, \tag{2.60}$$

such that

$$\int_T x_k(t)\beta_1(t)\mathrm{d}\nu(t) = 0 \qquad (k = 1, \ldots, n). \tag{2.61}$$

In the particular case when $n = 1$, this condition is necessary and sufficient in order that G be a Čebyšev subspace.

Proof. The first statement follows immediately taking $h = 1$ in theorem 2.5 and the second statement follows from corollary 2.1 and the general form of the extremal points of the unit cell $S_{L^1_R(T,\nu)^*}$ (Chap. I, lemma 1.13).

In the particular case when T is a Lebesgue measurable bounded set on the real axis and ν is the Lebesgue measure, the first statement of corollary 2.3 has been given by M. G. Krein [125], and in the case when (T, ν) is σ-finite, it occurs in the paper of R. R. Phelps [176], lemma 2.4, with the mention that it is due to the referee of the paper.

In the case when $n > 1$, the condition of corollary 2.3 is no longer sufficient for G to be a Čebyšev subspace, as shown by the example at the end of section 2.2.

We shall now give a negative result on best approximation in certain spaces $L^1(T, \nu)$; although in the proof of this result only theorem 3.4 of Chap. I is used, it is presented in this chapter because it concerns best approximation by elements of finite dimensional linear subspaces.

We recall that an *atom* (of a positive measure space (T, ν)) is a measurable set $A \subset T$ with the following properties: a) $\nu(A) > 0$; b) if B is a measurable subset of A, then either $\nu(B) = 0$ or $\nu(A \setminus B) = 0$.

THEOREM 2.6. *Let $E = L^1(T, \nu)$, where (T, ν) is a positive measure space having no atom and such that the conjugate space $L^1(T, \nu)^*$ is canonically equivalent to $L^\infty(T, \nu)$. Then E has no finite dimensional Čebyšev subspace.*

Proof. Let $G = [x_1, \ldots, x_n]$ be an arbitrary n-dimensional linear subspace of E and let $g_0 \in G \setminus \{0\}$ be arbitrary. Put

$$U = \{t \in T \mid g_0(t) \neq 0\}, \tag{2.62}$$

$$y_k(t) = x_k(t) \operatorname{sign} g_0(t) \qquad (t \in T;\ k = 1, \ldots, n). \tag{2.63}$$

We claim that there exists a ν-measurable set $B_0 \subset U$ such that

$$\int_{B_0} y_k(t)\, d\nu(t) = \int_{U \setminus B_0} y_k(t) d\nu(t) \qquad (k = 1, \ldots, n). \tag{2.64}$$

Indeed, put

$$y_k = \operatorname{Re} y_k + i \operatorname{Im} y_k = (\operatorname{Re} y_k)^+ - (\operatorname{Re} y_k)^- + $$
$$+ i\, [(\operatorname{Im} y_k)^+ - (\operatorname{Im} y_k)^-] \qquad (k = 1, \ldots, n), \tag{2.65}$$

where $(\operatorname{Re} y_k)^+, (\operatorname{Re} y_k)^-, (\operatorname{Im} y_k)^+, (\operatorname{Im} y_k)^- \geqslant 0$ $(k = 1, \ldots, n)$, and for every ν-measurable set $B \subset U$ and every $k = 1, \ldots, n$ put

$$m_k^+ (B) = \int_B (\operatorname{Re} y_k)^+(t)\, d\nu(t), \tag{2.66}$$

$$m_k^-(B) = \int_B (\operatorname{Re} y_k)^-(t) d\nu\, (t), \tag{2.67}$$

$$\mu_k^+ (B) = \int_B (\operatorname{Im} y_k)^+\, (t) d\nu\, (t), \tag{2.68}$$

$$\mu_k^-(B) = \int_B (\operatorname{Im} y_k)^-(t)\, d\nu(t). \tag{2.69}$$

Then $m_k^+, m_k^-, \mu_k^+, \mu_k^-$ $(k = 1, \ldots, n)$ are $4n$ non-atomic positive measures on U, hence by virtue of a theorem of A. A. Liapunov *) the set of all $4n$-tuples of the form $\{m_1^+(B), m_1^-(B), \mu_1^+(B), \mu_1^-(B), \ldots, m_n^+(B), m_n^-(B), \mu_n^+(B), \mu_n^-(B)\}$, where B runs over the ν-measurable subsets of U, is convex. Since

---

*) See A. A. Liapunov [135] and P. R. Halmos [81].

$0, \ldots, 0\}$ and $\{m_1^+(U), \ldots, \mu_n^-(U)\}$ belong to this set, it follows that there exists a ν-measurable subset B_0 of U such that

$$m_k^+(B_0) = \frac{m_k^+(U)}{2},\ m_k^-(B_0) = \frac{m_k^-(U)}{2},\ \mu_k^+(B_0) = \frac{\mu_k^+(U)}{2}, \mu_k^-(B_0) = $$
$$= \frac{\mu_k^-(U)}{2} \qquad (k = 1, \ldots, n).$$

Then we have

$$\int_{B_0} y_k(t) \mathrm{d}\nu(t) = \int_{B_0} (\mathrm{Re}\, y_k)^+(t) \mathrm{d}\nu(t) - \int_{B_0} (\mathrm{Re}\, y_k)^-(t) \mathrm{d}\nu(t) +$$
$$+ \mathrm{i}\left[\int_{B_0} (\mathrm{Im}\, y_k)^+(t) \mathrm{d}\nu(t) - \int_{B_0} (\mathrm{Im}\, y_k)^-(t) \mathrm{d}\nu(t)\right] = m_k^+(B_0) -$$
$$- m_k^-(B_0) + \mathrm{i}\, [\mu_k^+(B_0) - \mu_k^-(B_0)] = \frac{m_k^+(U)}{2} - \frac{m_k^-(U)}{2} + \mathrm{i}\left[\frac{\mu_k^+(U)}{2} - \right.$$
$$\left. - \frac{\mu_k^-(U)}{2}\right] = m_k^+(U \setminus B_0) - m_k^-(U \setminus B_0) + \mathrm{i}\left[\mu_k^+(U \setminus B_0) - \right.$$
$$\left. - \mu_k^-(U \setminus B_0)\right] = \int_{U \setminus B_0} y_k(t) \mathrm{d}\nu(t) \qquad (k = 1, \ldots, n),$$

i.e. (2.64).

Consequently, by (2.63) and (2.64) we have

$$\int_{B_0} x_k(t) \operatorname{sign} g_0(t) \mathrm{d}\nu(t) = \int_{U \setminus B_0} x_k(t) \operatorname{sign} g_0(t) \mathrm{d}\nu(t) \ (k = 1, \ldots, n) \tag{2.70}$$

Now put

$$\beta(t) = \begin{cases} \operatorname{sign} g_0(t) & (t \in B_0), \\ -\operatorname{sign} g_0(t) & (t \in U \setminus B_0), \\ 0 & (t \in T \setminus U). \end{cases} \tag{2.71}$$

By virtue of (2.62), (2.70) and (2.71) we then have the relations (3.53)—(3.57) of Chap. I, whence, by virtue of theorem 3.4 of Chap. I, it follows that G is not a Čebyšev subspace, which completes the proof.

In the particular case when T is a Lebesgue measurable bounded set on the real axis, ν is the Lebesgue measure and the scalars are real, theorem 2.6 has been given by M. G. Krein [125]. In the case when (T, ν) is σ-finite and the scalars are real, it has been given by R. R. Phelps [176], theorem 2.5, with the mention that both the result and the proof are due to the referee of the paper and, independently, by R. M. Moroney [152]. Finally, the fact that the result holds also for complex scalars has been observed by B. R. Kripke and T. J. Rivlin ([129], theorem 2.1).

As we have already remarked, in the above proof no use has been made of theorem 2.5 but only of theorem 3.4 of Chap. I; obviously, this latter is more convenient for proving that G *is not* a Čebyšev subspace, whereas theorem 2.5 is useful when we want to prove that a certain subspace *is* a Čebyšev subspace (e.g. theorem 2.5 has been applied to prove corollary 2.3).

In the particular case when the scalars are real, more precise results are known. Namely, we shall give the following sharpening of the real case of theorem 2.6:

THEOREM 2.7. *Let* $E = L^1_R(T, \nu)$, *where* (T, ν) *is a positive measure space having no atom and such that the conjugate space* $L^1_R(T, \nu)^*$ *is canonically equivalent to* $L^\infty_R(T, \nu)$. *Then* E *has no* n*-dimensional* $(n-1)$*-Čebyšev subspace.*

Proof. Let $G = [x_1, \ldots, x_n]$ be an arbitrary n-dimensional linear subspace of E. Then, putting

$$x_k = x_k^+ - x_k^- \qquad (k = 1, \ldots, n), \tag{2.72}$$

where $x_k^+, x_k^- \geqslant 0$ $(k = 1, \ldots, n)$, and putting for every ν-measurable set $B \subset T$ and every $k = 1, \ldots, n$

$$m_k^+(B) = \int_B x_k^+(t)\, d\nu(t), \tag{2.73}$$

$$m_k^-(B) = \int_B x_k^-(t)\, d\nu(t), \tag{2.74}$$

one obtains, in the same way as in the above proof of theorem 2.6, that there exists a ν-measurable subset B_0 of T such that

$$\int_{B_0} x_k(t)\, d\nu(t) = \int_{T \setminus B_0} x_k(t)\, d\nu(t) \qquad (k=1, \ldots, n). \tag{2.75}$$

Put

$$\beta(t) = \begin{cases} 1 & (t \in B_0) \\ -1 & (t \in T \setminus B_0). \end{cases} \tag{2.76}$$

Then we have

$$|\beta(t)| = 1 \qquad (t \in T), \tag{2.77}$$

$$\int_T x_k(t)\, \beta(t)\, d\nu(t) = 0 \qquad (k=1, \ldots, n). \tag{2.78}$$

Now put

$$U = T \setminus \bigcap_{k=1}^{n} Z(x_k) = \bigcup_{k=1}^{n} [T \setminus Z(x_k)]. \tag{2.79}$$

By virtue of (2.77), (2.78) and (2.79) we then have the relations (4.31), (4.32), (4.35) (with x_i instead of g_{i-1} and $n-1$ instead of k) and (4.36) of Chap. I, whence, by virtue of theorem 4.3 of Chap. I, it follows that G is not an $(n-1)$-Čebyšev subspace, which completes the proof.

In the particular case when (T, ν) is σ-finite, theorem 2.7 has been given by A. L. Garkavi ([68], p. 569, lemma 2). The existence of a real function β with the properties (2.77) and (2.78) has been established, in the case when T is a Lebesgue measurable bounded set on the real axis and ν is the Lebesgue measure, by M. G. Krein [125], and in the case when (T, ν) is σ-finite, by the referee of the paper [176] of R. R. Phelps ([176], p. 246); from this result and from corollary 2.3 one obtains again theorem 2.6. For $T = [0, 1]$ and $\nu =$ the Lebesgue measure, C. R. Hobby and J. R. Rice [91] have shown that β may even be choosen as a step function having at most n changes of sign.

THEOREM. 2.8. *Let $E = L_R^1(T, \nu)$, where (T, ν) is a positive measure space such that the conjugate space $L_R^1(T, \nu)^*$ is canonically equivalent to $L_R^\infty(T, \nu)$. In order that E have an n-dimensional Čebyšev subspace it is necessary and sufficient that (T, ν) have at least n atoms.*

Proof. Assume that (T, ν) has only r atoms $A_1, \ldots, A_r$, where $r < n$, and let G be an arbitrary n-dimensional linear subspace of E. Put

$$A = \bigcup_{k=1}^{n} A_k, \qquad N = T \setminus A, \tag{2.80}$$

$$G_0 = \{g|_N \mid g \in G\}. \tag{2.81}$$

Then $(N, \nu|_N)$ has no atom and G_0 is a linear subspace of dimension $m \leqslant n$ of the space $L_R^1(N, \nu|_N)$, hence by virtue of theorem 2.7 G_0 is not an $(m-1)$-Čebyšev subspace, i.e. there exists an $x' \in L_R^1(N, \nu|_N)$ with

$$\dim \mathfrak{P}_{G_0}(x') = m = \dim G_0. \tag{2.82}$$

Let $g_0' \in \operatorname{Int}_l \mathfrak{P}_{G_0}(x')$ be arbitrary and let (by (2.81)) $g_0 \in G$ be arbitrary such that $g_0' = g_0|_N$. Put

$$x(t) = \begin{cases} x'(t) & (t \in N), \\ g_0(t) & (t \in A). \end{cases} \tag{2.83}$$

We have then, by (2.83) and (2.81),

$$\|x - g_0\|_T = \|x' - g_0'\|_N = \inf_{g' \in G_0} \|x' - g'\|_N \leqslant \inf_{g \in G} \|x - g\|_T,$$

whence $g_0 \in \mathfrak{P}_G(x)$. On the other hand, since $r < n$, there exists a $g_1 \in G \setminus \{0\}$ such that

$$g_1(t) = 0 \qquad (t \in A). \tag{2.84}$$

Let $g_1' = g_1|_N$. Then by (2.82) and $g_0' \in \mathrm{Int}_i\, \mathfrak{P}_{G_0}(x')$ we shall have, for a sufficiently small $\varepsilon > 0$,

$$g_0' + \varepsilon g_1' \in \mathfrak{P}_{G_0}(x'), \tag{2.85}$$

whence, taking into account (2.83), (2.84) and (2.81),

$$\|x - g_0 - \varepsilon g_1\| = \|x' - g_0' - \varepsilon g_1'\|_N = \inf_{g' \in G_0} \|x' - g'\|_N \leqslant \inf_{g \in G} \|x - g\|,$$

whence $g_0 + \varepsilon g_1 \in \mathfrak{P}_G(x)$, and thus G is not a Čebyšev subspace. Consequently, the condition is necessary.

Now assume that (T, ν) has at least n atoms $A_1, \ldots, A_n$ and let

$$G = \{g \in E \mid g(t) = 0 \qquad (t \in T \setminus A)\}, \tag{2.86}$$

where $A = \bigcup_{k=1}^{n} A_k$. Then the functions

$$x_k(t) = \begin{cases} 1 & (t \in A_k) \\ 0 & (t \in T \setminus A_k) \end{cases} \qquad (k = 1, \ldots, n) \tag{2.87}$$

form a basis of G, and thus $\dim G = n$. We shall prove that G is a Čebyšev subspace.

Indeed, otherwise, by virtue of theorem 3.4 of Chap. I there would exist a measurable set $U \subset T$ with $\nu(U) > 0$, a $\beta \in L_R^\infty(T, \nu)$ and a $g_0 \in G \setminus \{0\}$ such that we have the relations (3.53), (3.54), (3.57) and (3.58) of Chap. I. Then, applying (3.54) of Chap. I to the functions (2.87), we obtain

$$\beta(t) = 0 \qquad (t \in A = \bigcup_{k=1}^{n} A_k). \tag{2.88}$$

On the other hand, by (3.57) of Chap. I and (2.86) we have

$$g_0(t) = 0 \qquad (t \in (T \setminus U) \cup (T \setminus A) = T \setminus (U \cap A)),$$

whence, by $g_0 \in G \setminus \{0\}$,

$$\nu(U \cap A) \neq 0. \tag{2.89}$$

However, by (3.58) of Chap. I we have

$$|\beta(t)| = 1 \qquad (t \in U \cap A),$$

which, together with (2.89), contradicts (2.88). Consequently, G is a Čebyšev subspace and thus the condition is sufficient, which completes the proof.

In the particular case when (T, ν) is σ-finite, theorem 2.8 has been given by A. L. Garkavi ([68], theorem 11). In a certain sense, it has a character similar to that of theorem 2.3 concerning the spaces $C_R(Q)$. Furthermore, from the necessity part of theorem 2.8 there follows immediately the real case of theorem 2.6.

Let us now pass to the spaces $C^1(Q, \nu)$.

THEOREM 2.9. *Let $E = C^1(Q, \nu)$, where Q is a compact space and ν a positive Radon measure on Q with the carrier $S(\nu) = Q$, and let $G = [x_1, \ldots, x_n]$ be an n-dimensional linear subspace of E. The following statements are equivalent:*

1° G is a Čebyšev subspace.

2° There do not exist a non-void open set $U \subset Q$, h ν-measurable and ν-essentially bounded functions $\beta_1, \ldots, \beta_h$ with

$$|\beta_j(q)| = 1 \quad \nu\text{-a.e. on } Q, \tag{2.90}$$

where $1 \leqslant h \leqslant n$ if the scalars are real and $1 \leqslant h \leqslant 2n - 1$ if the scalars are complex, h numbers $\lambda_1, \ldots, \lambda_h > 0$ with $\sum_{j=1}^{h} \lambda_j = 1$ and such that $\sum_{j=1}^{h} \lambda_j \beta_j|_U$ be continuous, and $g_0 \in G \setminus \{0\}$, with the properties

$$\sum_{j=1}^{h} \lambda_j \int_Q x_k(q) \beta_j(q) \, d\nu(q) = 0 \quad (k = 1, \ldots, n), \tag{2.91}$$

$$g_0(q) \neq 0 \qquad (q \in U), \tag{2.92}$$

$$\sum_{j=1}^{h} \lambda_j \beta_j(q) = \pm \operatorname{sign} g_0(q) \qquad (q \in U), \tag{2.93}$$

$$g_0(q) = 0 \qquad (q \in Q \setminus U). \tag{2.94}$$

In the particular case when the scalars are real (i.e. when $E = C^1_R(Q, \nu)$), these statements are equivalent to the following:

3° There do not exist two disjoint sets U_1 and U_2 open in Q, with $U_1 \cup U_2 \neq \emptyset$, h ν-measurable subsets $M_1, \ldots, M_h$ of Q, where $1 \leqslant h \leqslant n$, h numbers $\lambda_1, \ldots, \lambda_h > 0$ with $\sum_{j=1}^{h} \lambda_j = 1$, and $g_0 \in G \setminus \{0\}$ such that

$$\sum_{j=1}^{h} \lambda_j \int_{Q \setminus (U_1 \cup U_2)} x_k(q) \alpha_{M_j}(q) \, d\nu(q) + \int_{U_1} x_k(q) \, d\nu(q) - \int_{U_2} x_k(q) \, d\nu(q) = 0 \quad (k = 1, \ldots, n), \tag{2.95}$$

$$M_j \supset U_1, \ Q\setminus M_j \supset U_2 \qquad (j = 1, \ldots, h), \tag{2.96}$$

$$g_0(q) = 0 \qquad (q \in Q\setminus(U_1 \cup U_2)), \tag{2.97}$$

where

$$\alpha_{M_j}(q) = \beta_{M_j}|_{Q\setminus(U_1\cup U_2)}(q) = \begin{cases} 1 \ \nu\text{-a.e. on } (Q\setminus U_1)\cap M_j \\ -1 \ \nu\text{-a.e. on } Q\setminus(U_2\cap M_j) \end{cases} \quad (j=1,\ldots,h). \tag{2.98}$$

The proof of theorem 2.9 is similar to the above proof of theorem 2.5, with the following differences : in the part $1°\Rightarrow 2°$, one makes use of theorem 3.5 of Chap. I; in the part $2°\Rightarrow 1°$, the remark preceding theorem 1.10 of Chap. I is also used; finally, the implication *) $3°\Rightarrow 2°$ for real scalars is obtained by applying the argument of the proof of the implication non $2°\Rightarrow$ non $3°$ of Chap. I, theorem 3.5, for $\beta = \sum_{j=1}^{h} \lambda_j \beta_j$, and the argument at the end of the proof of theorem 1.9 (in order to obtain the relations (2.96)).

In the particular case when $Q = [a,b]$, ν is the Lebesgue measure and the scalars are real, the equivalence $1°\Leftrightarrow 3°$ of theorem 2.9 has been given in [227], theorem 2.

COROLLARY 2.4 (D. Jackson [97] — M. G. Krein [125]). *Let $E = C^1_R([0,1], \nu)$, where ν is the Lebesgue measure and let $G = [x_1, \ldots, x_n]$ be an n-dimensional linear subspace of E, such that $x_1, \ldots, x_n$ form a Čebyšev system. Then G is a Čebyšev subspace.*

Proof. Assume that G is not a Čebyšev subspace. Then by theorem 2.9**) there exist two disjoint open subsets U_1 and U_2 of $Q = [0,1]$, h ν-measurable subsets $M_1, \ldots, M_h \subset Q$, where $1 \leqslant h \leqslant n$, h numbers $\lambda_1, \ldots, \lambda_h > 0$ with $\sum_{j=1}^{h} \lambda_j = 1$, and $g_0 \in G\setminus\{0\}$, such that we have (2.95) — (2.97). Now if $[0,1]\setminus(U_1\cup U_2)$ contains at least n distinct points $q_1, \ldots, q_n$, then, by (2.97), $x_1,\ldots,x_n$ do not form a Čebyšev system, which contradicts the hypothesis. Consequently, $[0,1]\setminus(U_1\cup U_2)$ contains l distinct points $q_1, \ldots, q_l$, where $1 \leqslant l \leqslant n-1$. Taking arbitrary distinct points $q_{l+1}, \ldots, q_{n+1} \in [0,1]$, different from $q_1, \ldots, q_l$ and arranging the points $q_1,\ldots, q_{n+1}$ in increasing order $q'_1 < \ldots < q'_{n+1}$, we have then

$$[0,1]\setminus(U_1\cup U_2) = \{q_1,\ldots, q_l\}, \quad U_1\cup U_2 \overset{\nu}{\sim} \bigcup_{j=1}^{n} (q'_j, q'_{j+1}), \tag{2.99}$$

---

*) The implication $3° \Rightarrow 1°$ can be proved also directly, using theorem 1.9.

**) We observe that instead of theorem 2.9 one may also use here theorem 3.5 of Chap. I.

whence (2.95) may be written in the form

$$\sum_{j=1}^{n} \varepsilon_j \int_{q'_j}^{q'_{j+1}} x_k(q)\, \mathrm{d}\nu(q) = 0 \quad (j = 1, \ldots, n), \tag{2.100}$$

where $\varepsilon_j = \pm 1$ (since each open interval (q'_j, q'_{j+1}) is a connected set, and U_1, U_2 are disjoint open sets, whence both U_1 and U_2 are ν-equivalent to finite unions of intervals (q'_j, q'_{j+1})). Consequently, the transposed system of equations

$$\sum_{k=1}^{n} \alpha_k \int_{q'_j}^{q'_{j+1}} x_k(q)\, \mathrm{d}\nu(q) = 0 \qquad (j = 1, \ldots, n) \tag{2.101}$$

also has a non-trivial solution $\alpha_1, \ldots, \alpha_n$, whence for $g'_0 = \sum_{k=1}^{n} \alpha_k x_k \in G \setminus \{0\}$ we have

$$\int_{q'_j}^{q'_{j+1}} g'_0(q)\, \mathrm{d}\nu(q) = 0 \qquad (j = 1, \ldots, n). \tag{2.102}$$

But from these relations it follows that there exist n points $t_1, \ldots, t_n \in [0,1]$, where $t_j \in (q'_j, q'_{j+1})$ $(j = 1, \ldots, n)$, such that

$$g'_0(t_1) = \ldots = g'_0(t_n) = 0,$$

and hence $x_1, \ldots, x_n$ do not form a Čebyšev system, which contradicts the hypothesis. This completes the proof of corollary 2.4.

Other results on polynomial best approximation in the spaces $L^1(T, \nu)$ and $C^1(Q, \nu)$, and, in particular, other proofs of corollary 2.4 have been given by M. G. Krein [125], N. I. Ahiezer [1], S. Ya. Havinson [87], V. Pták [181], [182], V. N. Nikolsky [169], B. R. Kripke and T. J. Rivlin [128], [129], J. L. Walsh and T. S. Motzkin [260] and E. W. Cheney [33].

§ 3. FINITE DIMENSIONAL k-ČEBYŠEV SUBSPACES

An n-dimensional linear subspace of a normed linear space E is obviously a k-Čebyšev subspace for any integer k with $n \leqslant k < \infty$. Using the fundamental lemma of the preceding paragraph (lemma 2.2), we shall now extend the results obtained there on the characterizations of n-dimensional 0-Čebyšev subspaces to the case of n-dimensional k-Čebyšev subspaces, where k is an arbitrary integer with $0 \leqslant k \leqslant n - 1$.

3.1. FINITE DIMENSIONAL k-ČEBYŠEV SUBSPACES IN GENERAL NORMED LINEAR SPACES

THEOREM 3.1 ([225], theorem 4). *Let E be a normed linear space, $G = [x_1, \ldots, x_n]$ an n-dimensional linear subspace of E and k an integer with $0 \leqslant k \leqslant n-1$. The following statements are equivalent:*

1° G is a k-Čebyšev subspace.

2° There do not exist h extremal points $f_1, \ldots, f_h$ of the unit cell S_{E^*}, where $1 \leqslant h \leqslant n-k$ if the scalars are real and $1 \leqslant h \leqslant 2n-2k-1$ if the scalars are complex, h numbers $\lambda_1, \ldots, \lambda_h > 0$ with $\sum_{j=1}^{h} \lambda_j = 1$, an element $x \in E$ and $k+1$ linearly independent polynomials $g_0, g_1, \ldots, g_k \in G$ such that

$$\sum_{j=1}^{h} \lambda_j f_j(g) = 0 \qquad (g \in G), \tag{3.1}$$

$$\sum_{j=1}^{h} \lambda_j f_j(x) = \|x\| = \|x - g_0\| = \ldots = \|x - g_k\|. \tag{3.2}$$

3° There do not exist h extremal points $f_1, \ldots, f_h$ of S_{E^*}, where $1 \leqslant h \leqslant n-k$ if the scalars are real and $1 \leqslant h \leqslant 2n-2k-1$ if the scalars are complex, h numbers $\lambda_1, \ldots, \lambda_h > 0$ with $\sum_{j=1}^{h} \lambda_j = 1$, an element $x \in E$ and $k+1$ linearly independent polynomials $g_0, g_1, \ldots, g_k \in G$ such that we have (3.1) and

$$f_j(x) = \|x\| = \|x - g_0\| = \ldots = \|x - g_k\| \quad (j = 1, \ldots, h). \tag{3.3}$$

Proof. If condition 2° is not satisfied, then the functional $f \in E^*$ defined by

$$f = \sum_{j=1}^{h} \lambda_j f_j \tag{3.4}$$

and the elements $x \in E$, $g_0, g_1, \ldots, g_k \in G$ satisfy the conditions (4.7) — (4.9) of Chap. I, whence, by virtue of theorem 4.1 of Chap. I, it follows that G is not a k-Čebyšev subspace. Thus $1° \Rightarrow 2°$.

The implication $2° \Rightarrow 3°$ is obvious.

Finally, assume that the scalars are real and condition 1° is not satisfied, i.e. there exists a $y \in E$ with $\dim \mathfrak{L}_G(y) > k$. Then, by Chap. I, lemma 4.1, $\mathfrak{L}_G(y)$ contains $k+2$ baricentrically independent polynomials $g_0', g_1', \ldots, g_{k+1}'$. Consequently, for

$$x = y - g_0', \; g_0 = g_1' - g_0', \ldots, g_k = g_{k+1}' - g_0' \tag{3.5}$$

we have

$$\mathfrak{L}_G(x) = \mathfrak{L}_G(y - g_0') = \mathfrak{L}_G(y) - g_0' \ni 0, g_0, g_1, \ldots, g_k, \tag{3.6}$$

and $g_0, g_1, \ldots, g_k$ are linearly independent. But by virtue of corollary 1.2 and lemma 2.2, there exist then h extremal points $f_1, \ldots, f_h$ of S_{E^*}, where $1 \leqslant h \leqslant n+1$, with linearly independent restrictions $f_1|_{E_0}, \ldots, f_h|_{E_0}$, where $E_0 = G \oplus [x]$, and h numbers $\lambda_1, \ldots \lambda_h > 0$ with $\sum_{j=1}^{h} \lambda_j = 1$, such that we have (3.1) and (3.3), whence also

$$f_j(g_i) = 0 \quad (j = 1, \ldots, h;\ i = 0, 1, \ldots, k). \tag{3.7}$$

Now, as in the proof of the implication $3^\circ \Rightarrow 1^\circ$ of theorem 2.1, it follows that we cannot have $h \geqslant n - k + 1$, hence we have $1 \leqslant h \leqslant n - k$. Thus, $3^\circ \Rightarrow 1^\circ$ if the scalars are real.

In the case of complex scalars the proof of the implication $3^\circ \Rightarrow 1^\circ$ is similar (see the proof of the implication $3^\circ \Rightarrow 1^\circ$ of theorem 2.1). This completes the proof of theorem 3.1.

In the particular case when $k = 0$, theorem 3.1 reduces to theorem 2.1.

From theorem 3.1 we shall now deduce

Corollary 3.1 ([225], corollary 1). *Let E be a real normed linear space, $G = [x_1, \ldots, x_n]$ an n-dimensional linear subspace of E, and k an integer with $0 \leqslant k \leqslant n - 1$. Any of the following equivalent conditions is sufficient in order that G be a k-Čebyšev subspace:*

1° *There do not exist $n - k$ linearly independent extremal points $f_1, \ldots, f_{n-k}$ of the unit cell S_{E^*} and $n - k$ numbers $\lambda_1, \ldots, \lambda_{n-k} \geqslant 0$ with $\sum_{j=1}^{n-k} \lambda_j \neq 0$, such that*

$$\sum_{j=1}^{n-k} \lambda_j f_j(x_m) = 0 \qquad (m = 1, \ldots, n). \tag{3.8}$$

2° *There do not exist $n - k$ linearly independent extremal points $f_1, \ldots, f_{n-k}$ of S_{E^*} and $n - k$ numbers $\mu_1, \ldots, \mu_{n-k}$ with $\sum_{j=1}^{n-k} |\mu_j| \neq 0$, such that*

$$\sum_{j=1}^{n-k} \mu_j f_j(x_m) = 0 \qquad (m = 1, \ldots, n). \tag{3.9}$$

3° *For any $n - k$ linearly independent extremal points $f_1, \ldots, f_{n-k}$ of S_{E^*} we have*

$$\operatorname{rank}\, (f_j(x_m))_{\substack{j = 1, \ldots, n-k \\ m = 1, \ldots, n}} = n - k. \tag{3.10}$$

4° *There do not exist $n - k$ linearly independent extremal points $f_1, \ldots, f_{n-k}$ of S_{E^*} and $k+1$ linearly independent polynomials $g_0, g_1, \ldots, g_k \in G$ such that*

$$f_j(g_i) = 0 \quad (j = 1, \ldots, n - k;\ i = 0, 1, \ldots, k). \tag{3.11}$$

The above statement holds also for complex normed liniar spaces, if "linearly independent" is replaced everywhere by "pairwise linearly independent".

Proof. The equivalence of the conditions 1°—4° being immediate (by an elementary argument of linear algebra), we shall show that they are sufficient in order that G be a k-Čebysev subspace.

Assume that G is not a k-Čebyšev subspace. Then by virtue of theorem 3.1 there exist h extremal points of S_{E^*}, where $1 \leqslant h \leqslant n - k$ if the scalars are real and $1 \leqslant h \leqslant 2n - 2k - 1$ if the scalars are complex, h numbers $\lambda_1, \ldots, \lambda_h > 0$ with $\sum_{j=1}^{h} \lambda_j = 1$, an element $x \in E$ and $k+1$ linearly independent polynomials $g_0, g_1, \ldots, g_k \in G$ such that we have (3.1) and (3.3); as we have seen in the above proof of theorem 3.1, $f_1, \ldots, f_h$ may be assumed to be real-linearly independent. If we have $h < n - k$, then by lemma 2.3 we can choose extremal points $f_{h+1}, \ldots, f_{n-k}$ of S_{E^*} such that $f_1, \ldots, f_{n-k}$ be real-linearly independent, whence, putting $\lambda_{h+1} = \ldots = \lambda_{n-k} = 0$, by virtue of (3.1) we shall have (3.8). On the other hand, if $h \geqslant n - k$, then by virtue of (3.1) and (3.3) we shall have (3.11), which completes the proof in the case of real scalars. For complex scalars the proof is similar to that of corollary 2.1.

In the particular case when $k = 0$, corollary 3.1 reduces to corollary 2.1. As has been remarked in §2, section 2.2, the conditions of corollary 3.1 are not necessary in order that $G = [x_1, \ldots, x_n]$ be a 0-Čebyšev subspace; it is easy to see that this remark remains valid also for every integer k with $0 \leqslant k \leqslant n - 1$.

3.2. APPLICATIONS IN THE SPACES $C(Q)$ AND $C_R(Q)$

In the present section we shall deduce from the foregoing general results the following well-known theorem on the characterization of finite-dimensional k-Čebyšev in the spaces $C(Q)$:

THEOREM 3.2. *Let $E = C(Q)$ (Q compact), let $G = [x_1, \ldots, x_n]$ be an n-dimensional linear subspace of E and k an integer with $0 \leqslant k \leqslant n - 1$. In order that G be a k-Čebyšev subspace it is necessary and sufficient that there do not exist $n - k$ distinct* *points* $q_1, \ldots, q_{n-k} \in Q$ *and* $k+1$ *linearly independent polynomials* $g_0, g_1, \ldots, g_k \in G$, *such that*

$$g_i(q_j) = 0 \quad (j = 1, \ldots, n-k;\ i = 0, 1, \ldots, k). \qquad (3.12)$$

In the case when the scalars are real, theorem 3.2 has been given by G. Š. Rubinstein [199] and in the case of complex scalars, by Z. S. Romanova [198]. In the particular case when $k = 0$, theorem 3.2 reduces to the classical theorem of Haar-Kolmogorov (theorem 2.2).

Just as in the particular case $k = 0$, several proofs of theorem 3.2 may be given. We shall give here, following [225], p. 398—399, an extension of the second proof of the necessity part of theorem 2.2 and of the first proof of the sufficiency part of theorem 2.2.

Assume first that the condition of theorem 3.2 is not satisfied, i.e. that there exist linearly independent $g_0, g_1, \ldots, g_k \in G$ and distinct $q_1, \ldots, q_{n-k} \in Q$ such that we have (3.12). Then the system of equations

$$\sum_{m=1}^{n} \alpha_m x_m(q_j) = 0 \qquad (j = 1, \ldots, n-k) \qquad (3.13)$$

has at least $k+1$ linearly independent solutions $\{\alpha_1^l, \ldots, \alpha_n^l\}$ $(l = 0, 1, \ldots, k)$, whence

$$\operatorname{rank}\,(x_m(q_j))_{\substack{j=1,\ldots,n-k\\ m=1,\ldots,n}} \leqslant n - k - 1, \qquad (3.14)$$

and consequently there exist $n-k$ numbers $\mu_1, \ldots, \mu_{n-k}$ such that

$$\sum_{j=1}^{n-k} \mu_j x_m(q_j) = 0 \qquad (m = 1, \ldots, n), \qquad (3.15)$$

$$\sum_{j=1}^{n-k} |\mu_j| \neq 0; \qquad (3.16)$$

obviously, we may assume that $\sum_{j=1}^{n-k} |\mu_j| = 1$. But then the Radon measure μ on Q defined by

$$\int_Q y(q)\,d\mu(q) = \sum_{j=1}^{n-k} \mu_j y(q_j) \qquad (y \in E) \qquad (3.17)$$

and the polynomials $g_0, g_1, \ldots, g_k \in G$ satisfy the conditions (4.24) — (4.27) of Chap. I, whence, by virtue of theorem 4.2 of Chap. I, it follows that G is not a k-Čebyšev subspace.

Conversely, assume that G is not a k-Čebyšev subspace. Then there exist, by virtue of condition 4° of corollary 3.1, $n-k$ distinct points $q_1, \ldots, q_{n-k} \in Q$, $n-k$ scalars $\beta_1, \ldots, \beta_{n-k}$ of modulus 1, and $k+1$ linearly independent polynomials g_0, $g_1, \ldots, g_k \in G$ such that we have

$$\beta_j g_i(q_j) = 0 \quad (j = 1, \ldots, n-k;\ i = 0,1, \ldots, k); \tag{3.18}$$

since $|\beta_j| = 1$ ($j = 1, \ldots, n-k$), it follows that we have (3.12), which completes the proof of theorem 3.2.

The systems $x_1, \ldots, x_n \in C(Q)$ with the property occurring in theorem 3.2 are also called*) "systems of Čebyšev rank $\leqslant k$". There arises naturally the following problem: which are the compact spaces Q which admit systems $x_1, \ldots, x_n$ of Čebyšev rank $\leqslant k$, where $n \geqslant 2$ and $0 \leqslant k \leqslant n-1$? Partial results have been given by V. G. Boltiansky, S. S. Ryškov and Yu. A. Šaškin [19] and also in the paper [240] of Yu. A. Šaškin.

The dimension of the set $\mathfrak{P}_G(x)$ (G an n-dimensional linear subspace) in spaces of differentiable functions has been studied by A. L. Garkavi [60].

We leave to the reader the application of the general results of the preceding section to the spaces $C_E(Q)$, $L^1(T, \nu)$, $L^1_R(T, \nu)$, $C^1(Q, \nu)$ and $C^1_R(Q, \nu)$, in order to obtain the extension of the corresponding results of §2 from the case $k = 0$ to the case when k is an arbitrary integer with $0 \leqslant k \leqslant n-1$. Let us mention that a result of this type in the spaces $L^1_R(T, \nu)$ (namely, theorem 2.7) was given in the preceding paragraph, since it was necessary for proving theorem 2.8.

§ 4. POLYNOMIAL INTERPOLATIVE BEST APPROXIMATION. BEST APPROXIMATION BY ELEMENTS OF FINITE DIMENSIONAL LINEAR MANIFOLDS

4.1. THE CASE OF GENERAL NORMED LINEAR SPACES

Using the remarks made in Chap. I, §5, one immediately obtains, from the preceding results on polynomial best approximation, the solutions of the corresponding problems for polynomial interpolative best approximation (respectively, for best approximation by elements of finite dimensional linear manifolds). Thus, theorem 1.1 yields

THEOREM 4.1 ([222], theorems 2.4 and 3.5). *Let E be a normed linear space, $G = [x_1, \ldots, x_n]$ an n-dimensional linear*

*) See e.g. Yu. A. Šaškin [240]. For the definition of the Čebyšev rank of a linear subspace $G \subset E$ see Chap. I, formula (4.12).

subspace of E, $\varphi_1, \ldots, \varphi_m \in G^*$ *linearly independent (where* $m < n$), $c_1, \ldots, c_m$ m *scalars,* V *the linear manifold in* E *defined by*

$$V = \{g \in G \mid \varphi_i(g) = c_i \ (i = 1, \ldots, m)\} \tag{4.1}$$

(or, what is equivalent, let V *be an arbitrary* $(n - m)$*-dimensional linear manifold in* E, *where* $1 \leqslant n - m < \infty$), *and let* $x \in E \setminus V$, $v_0 \in V$. *The following statements are equivalent:*

1° v_0 *is an interpolatory element of best approximation of* x *(respectively, an element of best approximation of* x *by elements of the linear manifold* V).

2° *There exist* h *extremal points* $f_1, \ldots, f_h$ *of* S_{E^*}, *where* $1 \leqslant h \leqslant n - m + 1$ *if the scalars are real and* $1 \leqslant h \leqslant 2n - 2m + 1$ *if the scalars are complex, and* h *numbers* $\lambda_1, \ldots, \lambda_h > 0$ *with* $\sum_{j=1}^{h} \lambda_j = 1$, *such that*

$$\sum_{j=1}^{h} \lambda_j f_j(v - v_0) = 0 \qquad (v \in V), \tag{4.2}$$

$$\sum_{j=1}^{h} \lambda_j f_j(x - v_0) = \|x - v_0\|. \tag{4.3}$$

3° *There exist* h *extremal points* $f_1, \ldots, f_h$ *of* S_{E^*}, *where* $1 \leqslant h \leqslant n - m + 1$ *if the scalars are real and* $1 \leqslant h \leqslant 2n - 2m + 1$ *if the scalars are complex, and* h *numbers* $\lambda_1, \ldots, \lambda_h > 0$ *with* $\sum_{j=1}^{h} \lambda_j = 1$, *such that we have (4.2) and*

$$f_j(x - v_0) = \|x - v_0\| \qquad (j = 1, \ldots, h). \tag{4.4}$$

Similarly, theorem 2.1 yields

THEOREM 4.2 ([222], theorem 4.2). *Let* E *be a normed linear space,* $G = [x_1, \ldots, x_n]$ *an* n*-dimensional linear subspace of* E, $\varphi_1, \ldots, \varphi_m \in G^*$ *linearly independent (where* $m < n$), $c_1, \ldots, c_m$ m *scalars, and let* V *be the manifold in* E *defined by (4.1) (or, what is equivalent, let* V *be an arbitrary* $(n - m)$*-dimensional linear manifold in* E, *where* $1 \leqslant n - m < \infty$). *The following statements are equivalent:*

1° V *is a Čebyšev linear manifold.*

2° *There do not exist* h *extremal points* $f_1, \ldots, f_h$ *of the unit cell* S_{E^*}, *where* $1 \leqslant h \leqslant n - m$ *if the scalars are real and* $1 \leqslant h \leqslant 2n - 2m - 1$ *if the scalars are complex,* h *numbers* $\lambda_1, \ldots, \lambda_h > 0$ *with* $\sum_{j=1}^{h} \lambda_j = 1$, *and* $x \in E$, $v_0, v_1 \in V$, $v_0 \neq v_1$, *such that we have (4.2) and*

$$\sum_{j=1}^{h} \lambda_j f_j(x - v_0) = \|x - v_0\| = \|x - v_1\|. \tag{4.5}$$

3° *There do not exist* h *extremal points* $f_1, \ldots, f_h$ *of the unit cell* S_{E^*}, *where* $1 \leqslant h \leqslant n - m$ *if the scalars are real and* $1 \leqslant h \leqslant 2n - 2m - 1$ *if the scalars are complex,* h *numbers* $\lambda_1, \ldots, \lambda_h > 0$ *with* $\sum_{j=1}^{h} \lambda_j = 1$, *and* $x \in E$, $v_0, v_1 \in V$, $v_0 \neq v_1$, *such that we have (4.2) and*

$$f_j(x - v_0) = \|x - v_0\| = \|x - v_1\| \qquad (j = 1, \ldots, h). \tag{4.6}$$

4.2. APPLICATIONS IN THE SPACES $C(Q)$ AND $C_R(Q)$

Taking into account the general form of the extremal points of the unit cell $S_{C(Q)^*}$, from theorem 4.1 it follows

THEOREM 4.3 ([222], theorem 5.1). *Let* $E = C(Q)$ *(*Q *compact),* $G = [x_1, \ldots, x_n]$ *an* n*-dimensional linear subspace of* E, *where* $x_1, \ldots, x_n$ *are continuous and linearly independent on a compact space* S *containing* Q, *let* $\varphi_1, \ldots, \varphi_m \in G^*$ *be defined by*

$$\varphi_i(g) = g(s_i) \qquad (g \in G;\ i = 1, \ldots, m), \tag{4.7}$$

where $m < n$ *and the "nodes"* $s_1, \ldots, s_m$ *belong to the set*)* $S \setminus Q$, *let* $c_1, \ldots, c_m$ *be* m *scalars,* V *the linear manifold (4.1) corresponding to these* φ_i, c_i, *and let* $x \in E \setminus V$, $v_0 \in V$. *The following statements are equivalent:*

1° $v_0 \in \mathfrak{P}_V(x)$.

2° *There exist* h *points* $q_1, \ldots, q_h \in Q$, *where* $1 \leqslant h \leqslant n - m + 1$ *if the scalars are real and* $1 \leqslant h \leqslant 2m - 2m + 1$ *if the scalars are complex, and* h *numbers* $\mu_1, \ldots, \mu_h \neq 0$ *with* $\sum_{j=1}^{h} |\mu_j| = 1$, *such that*

$$\sum_{j=1}^{h} \mu_j [v(q_j) - v_0(q_j)] = 0 \qquad (v \in V), \tag{4.8}$$

$$\sum_{i=1}^{h} \mu_j [(x(q_j) - v_0(q_j)] = \max_{t \in Q} |x(t) - v_0(t)|. \tag{4.9}$$

3° *There exist* h *points* $q_1, \ldots, q_h \in Q$, *where* $1 \leqslant h \leqslant n - m + 1$ *if the scalars are real and* $1 \leqslant h \leqslant 2n - 2m + 1$ *if the scalars are complex, and* h *numbers* $\mu_1, \ldots, \mu_h \neq 0$ *with* $\sum_{j=1}^{h} |\mu_j| = 1$, *such that we have (4.8) and*

$$x(q_j) - v_0(q_j) = (\operatorname{sign} \mu_j) \max_{t \in Q} |x(t) - v_0(t)| \quad (j = 1, \ldots, h). \tag{4.10}$$

---

*) We have $\varphi_1, \ldots, \varphi_m \in G^*$ since $\varphi_1, \ldots, \varphi_m$ are linear on G and $\dim G < \infty$.

In the particular case when $x_1, \ldots, x_n$ form a Čebyšev system on S, one obtains

THEOREM 4.4 ([222], theorem 5.2). *Let $E = C(Q)$ (Q compact), let $G = [x_1, \ldots, x_n]$ be an n-dimensional linear subspace of E such that $x_1, \ldots, x_n$ form a Čebyšev system on a compact space S containing Q, and let $s_i, \varphi_i, c_i (i = 1, \ldots, m)$, V, x and v_0 be as in theorem 4.3. The following statements are equivalent:*

1° $v_0 \in \mathfrak{L}_V(x)$.

2° *There exist h points $q_1, \ldots, q_h \in Q$, where $h = n - m + 1$ if the scalars are real and $n - m + 1 \leqslant h \leqslant 2n - 2m + 1$ if the scalars are complex, such that we have (4.8) and (4.9).*

These statements are implied by — and in the case when the scalars are real, equivalent to — the following statement:

3° *There exist $n-m-1$ distinct points $q_1, \ldots, q_{n-m+1} \in Q$ such that*

$$x(q_j) - v_0(q_j) = \left(\operatorname{sign} \frac{\Delta_j}{\Delta}\right) \max_{t \in Q} |x(t) - v_0(t)| \qquad (j = 1, \ldots, n - m + 1), \tag{4.11}$$

where

$$\Delta = \begin{vmatrix} g_1(q_1) & \ldots & g_1(q_{n-m+1}) \\ \cdot & \cdot\cdot\cdot & \cdot \\ g_{n-m}(q_1) & \ldots & g_{n-m}(q_{n-m+1}) \\ x(q_1) - v_0(q_1) & \ldots & x(q_{n-m+1}) - v_0(q_{n-m+1}) \end{vmatrix} \neq 0, \tag{4.12}$$

$g_1, \ldots, g_{n-m}$ being $n - m$ arbitrary linearly independent polynomials in $V - v_0$, and where Δ_j is the cofactor of the element $x(q_j)$ in Δ.

Indeed, we have $V - v_0 = [g_1, \ldots, g_{n-m}]$, where $g_1, \ldots, g_{n-m}$ form a Čebyšev system on Q (since each polynomial $\sum_{i=1}^{n-m} \alpha_i g_i$ vanishes for $s_1, \ldots, s_m \in S \setminus Q$, hence it can have at most $n - m + 1$ zeros on Q), and the relations (4.8) are equivalent to the relations

$$\sum_{j=1}^{h} \mu_j g_k(q_j) = 0 \qquad (k = 1, \ldots, n - m), \tag{4.13}$$

hence one may apply the method used in the proof of theorem 1.4 of §1; moreover, theorem 4.4 can be also deduced directly from theorem 1.4, with the method of translation used in section 4.1.

In the particular case when $Q = [a, b]$ and the scalars are real, from theorem 4.4 there results easily the following classical theorem of S. N. Bernstein ([15], p. 5):

Let $E = C_R([a, b])$, $G = [x_1, \ldots, x_n]$ an n-dimensional linear subspace of E such that $x_1, \ldots, x_n$ form a Čebyšev system on a segment S containing $[a, b]$, let V be the linear manifold

$$V = \{g \in G \mid g(s_i) = c_i \ (i = 1, \ldots, m)\}, \tag{4.14}$$

where $s_1, \ldots, s_m \in S \setminus [a, b]$ and $c_1, \ldots, c_m$ are m real numbers and let $x \in E \setminus V$, $v_0 \in V$. We have $v_0 \in \mathfrak{P}_V(x)$ if and only if there exist $n - m + 1$ points $q_1 < q_2 < \ldots < q_{n-m+1}$ of $[a, b]$ at which the difference $x(q) - v_0(q)$ takes the value $\max_{t \in [a,b]} |x(t) - v_0(t)|$ with alternating signs.

Finally, let us consider the problem of the uniqueness of interpolatory polynomials of best approximation. With the preceding methods one obtains

THEOREM 4.5 ([222], theorem 5.3). *Let $E = C(Q)$ (Q compact), let $G = [x_1, \ldots, x_n]$ be an n-dimensional linear subspace of E, where $x_1, \ldots, x_n$ are continuous and linearly independent on a compact space containing Q, and let s_i, φ_i, $c_i (i = 1, \ldots, m)$ and V be as in theorem 4.3. In order that V be a Čebyšev linear manifold it is necessary and sufficient that there exist a polynomial $v_0 \in V$ such that the linear subspace $V - v_0$ be spanned by $n - m$ polynomials $g_1, \ldots, g_{n-m} \in G$ which form a Čebyšev system on Q.*

As has been seen in the above, this condition is satisfied if, in particular, $x_1, \ldots, x_n$ form a Čebyšev system on S. Hence, we have

COROLLARY 4.1 ([222], theorem 5.4). *Let $E = C(Q)$ (Q compact), let $G = [x_1, \ldots, x_n]$ be an n-dimensional linear subspace of E such that $x_1, \ldots, x_n$ form a Čebyšev system on a compact space S containing Q, and let s_i, φ_i, $c_i (i = 1, \ldots, m)$ and V be as in theorem 4.3. Then V is Čebyšev linear manifold.*

In the particular case when $Q = [a, b]$, $S =$ a segment containing $[a, b]$ and the scalars are real, corollary 4.1 has been given by S. Bernstein ([15], p. 5).

§ 5. THE OPERATORS π_G AND THE FUNCTIONALS e_G FOR LINEAR SUBSPACES G OF FINITE DIMENSION

5.1. THE OPERATORS π_G FOR LINEAR SUBSPACES G OF FINITE DIMENSION

As has been observed in section 6.1 of Chap. I, § 6, in inner product spaces*) $E = \mathcal{H}$ the operators π_G have the following properties, for any linear subspace G of E:

*) We recall (see Chap. I, § 1, section 1.7) that we do not assume the completeness of the space $\mathcal{H}$.

a) we have

$$\|\pi_G(x)\| \leqslant \|x\| \qquad (x \in D(\pi_G)); \tag{5.1}$$

b) π_G is one-valued on $D(\pi_G)$ and we have

$$x+y \in D(\pi_G),\ \pi_G(x+y) = \pi_G(x) + \pi_G(y)\ (x, y \in D(\pi_G)) \tag{5.2}$$

(or, what is equivalent, $\pi_G^{-1}(0)$ is a closed linear subspace of E).

There arises naturally the problem whether in other normed linear spaces E, too, the operators π_G have at least one of these properties. In the present section we shall show that in the case when dim $E \geqslant 3$ the answer is negative even if we assume the property *a*) only for all linear subspaces of a certain fixed finite dimension n, where $1 \leqslant n \leqslant$ dim $E - 1$, or the property *b*) only for all linear subspaces of a certain fixed finite dimension n, where $1 \leqslant n \leqslant$ dim $E - 2$.

THEOREM 5.1. *Let E be a normed linear space of (finite or infinite) dimension $\geqslant 3$ with the property that for every linear subspace G of a certain fixed finite dimension n, where $1 \leqslant n \leqslant$ dim $E - 1$, the mapping π_G satisfies (5.1). Then E is equivalent*) to an inner product space.*

Proof. Let us observe first that it is sufficient to consider the case when $n \geqslant 2$. Indeed, assume that the theorem is valid for linear subspaces G_1 of a certain fixed finite dimension $n \geqslant 2$ and that we have (5.1) for every linear subspace G of dimension 1 of E. Let G_1 be an arbitrary linear subspace of E with dim $G_1 = n$ and let $x \in E$ and $g_1 = \pi_{G_1}(x) \in \mathfrak{P}_{G_1}(x)$ be arbitrary. Then for the 1-dimensional linear subspace $G = [g_1]$ of E we have $g_1 \in G \subset G_1$, hence $g_1 \in \mathfrak{P}_G(x)$, hence $g_1 = \pi_G(x)$, whence, taking into account the hypothesis (5.1), we obtain $\|\pi_{G_1}(x)\| = \|g_1\| = \|\pi_G(x)\| \leqslant \|x\|$. Consequently, for every linear subspace $G_1 \subset E$ with dim $G_1 = n$ we have

$$\|\pi_{G_1}(x)\| \leqslant \|x\| \qquad (x \in E),$$

whence, since the theorem has been assumed to be valid for n-dimensional linear subspaces, it follows that E is equivalent to an inner product space.

Thus, let $n \geqslant 2$. We shall now show that it is sufficient to consider the case when dim $E = n + 1$. Indeed, if the theorem is valid for spaces of dimension $n + 1$, then from the hypothesis (5.1) it follows that all $(n + 1)$-dimensional linear subspaces of E are equivalent to inner product spaces and consequently in these spaces, whence also in E, the "parallelogram law"

$$\|x+y\|^2 + \|x-y\|^2 = 2(\|x\|^2 + \|y\|^2) \qquad (x, y \in E), \tag{5.3}$$

*) We recall that we use the term "equivalent" in the sense of S. Banach [8], i.e.: linearly isometric.

holds, whence it follows, by virtue of a classical theorem of P. Jordan and J. von Neumann*), that the space E itself is equivalent to an inner product space.

We shall now show that it is sufficient to consider the case when $\dim E = 3$ and $n = 2$. Indeed, let $\dim E = n + 1$, where $n \geqslant 2$, and let E_1 be an arbitrary linear subspace of E, with $\dim E_1 = 3$, and G_1 an arbitrary linear subspace of E_1, with $\dim G_1 = 2$. Also, let $x \in E_1 \setminus G_1$ and $g_1 = \pi_{G_1}(x) \in \mathfrak{P}_{G_1}(x)$ be arbitrary. Then, by virtue of theorem 1.1 of Chap. I, there exists an $f \in E^*$ such that

$$\|f\| = 1, \tag{5.4}$$

$$f(g) = 0 \qquad (g \in G_1), \tag{5.5}$$

$$f(x - g_1) = \|x - g_1\|. \tag{5.6}$$

Put

$$G = \{y \in E \mid f(y) = 0\}. \tag{5.7}$$

Then G is a hyperplane in E, whence $\dim G = n$. On the other hand, from (5.4), (5.7), (5.6) and theorem 1.1 of Chap. I if follows that we have $g_1 \in \mathfrak{P}_G(x)$, hence we can write $g_1 = \pi_G(x)$, whence, taking into account the hypothesis (5.1), we obtain $\|\pi_{G_1}(x)\| = \|g_1\| = \|\pi_G(x)\| \leqslant \|x\|$. Since $x \in E_1 \setminus G_1$ and $g_1 = \pi_{G_1}(x) \in \mathfrak{P}_{G_1}(x)$ were arbitrary, it follows that we have

$$\|\pi_{G_1}(x)\| \leqslant \|x\| \qquad (x \in E_1) \tag{5.8}$$

and thus, since $G_1 \subset E_1$ with $\dim G_1 = 2$ was arbitrary, the hypotheses of the theorem are satisfied for E_1 instead of E and for 2 instead of n. Since E_1 was an arbitrary linear subspace of E, with $\dim E_1 = 3$, it follows that if we can prove the theorem for 3-dimensional spaces and for $n = 2$, then all the 3-dimensional linear subspaces E_1 of E will be equivalent to inner product spaces, whence the space E itself will be equivalent to an inner product space.

Thus, let $\dim E = 3$ and $n = 2$ and let G be an arbitrary 2-dimensional linear subspace of E, hence a hyerplane in E. Then, by theorem 6.2 of Chap. I, § 6, π_G admits an additive selection. Taking into account that $D(\pi_G) = E$ (by $\dim G < \infty$) and the hypothesis (5.1), it follows that this selection is a continuous linear projection of norm 1 of E onto G. Since G was an arbitrary 2-dimensional linear subspace of E, by virtue of a classical theorem of S. Kakutani**) it follows that E is equivalent to an inner product space, which completes the proof.

---

*) See e.g. N. Bourbaki [23], Chap. V, p. 138, exercise 2.
**) See e.g. N. Bourbaki [23], Chap. V, p. 144, exercise 13.

In the particular case when $n = 2$, theorem 5.1 has been given by R. A. Hirschfeld ([89], theorem 1).

The hypothesis dim $E \geqslant 3$ of theorem 5.1 is essential; indeed, e.g. in the 2-dimensional real normed linear space E in which the unit cell is the hexagon formed by the convex hull of the points $\{\pm 2, 0\}$ and $\{\pm 1, \pm 1\}$, condition (5.1) is satisfied for every linear subspace G.

COROLLARY 5.1. *Let E be a normed linear space of (finite or infinite) dimension $\geqslant 3$, with the property that for every linear subspace G of a certain fixed finite dimension n, where $1 \leqslant n \leqslant \dim E - 1$, the mapping π_G is "contractive", i.e. satisfies*

$$\|\pi_G(x) - \pi_G(y)\| \leqslant \|x - y\| \qquad (x, y \in D(\pi_G) = E). \quad (5.9)$$

Then E is equivalent to an inner product space.

Proof. Taking in (5.9) $y = 0$, we obtain (5.1) and thus the equivalence of E to an inner product space follows from theorem 5.1.

In the particular case when $n = 1$, corollary 5.1 has been given, essentially, by R. R. Phelps ([74], lemma 5.4). For an arbitrary n and a strictly convex E, it has been given by W. J. Stiles ([238 b], p. 221).

THEOREM 5.2. *Let E be a normed linear space of (finite or infinite) dimension $\geqslant 3$, with the property that for every linear subspace G of a given fixed finite dimension n, where $1 \leqslant n \leqslant \dim E - 2$, the mapping π_G is one-valued on $D(\pi_G) = E$ and satisfies (5.2) (or, what is equivalent, $\pi_G^{-1}(0)$ is a linear subspace of E). Then E is equivalent to an inner product space.*

Proof. Let us first observe that it is sufficient to consider the case when $\dim E = n + 2$. Indeed, if the theorem is valid for spaces of dimension $n + 2$, then from the hypothesis of the theorem it follows that all $(n + 2)$-dimensional linear subspaces of E are equivalent to inner product spaces, whence, by virtue of the theorem of P. Jordan and J. von Neumann used in the foregoing, the space E itself is equivalent to an inner product space.

Thus, let $\dim E = n + 2$ and let G_0 be an arbitrary 2-dimensional linear subspace of E. Let x_1, x_2 be an arbitrary basis of G_0 such that

$$\|x_i\| = 1 \qquad (i = 1, 2), \quad (5.10)$$

and let $f_i \in E^*$ $(i = 1, 2)$ be such that

$$\|f_i\| = 1 \qquad (i = 1, 2), \quad (5.11)$$

$$f_i(x_i) = 1 \qquad (i = 1, 2). \quad (5.12)$$

Since by our hypothesis all n-dimensional linear subspaces of E are Čebyšev subspaces, the space E is, by virtue of corollary 3.3 of Chap. I, strictly convex, and consequently the above functionals $f_1, f_2 \in E^*$ are linearly independent (since if we had $f_1 = \alpha f_2$, then from (5.11) it would follow that $|\alpha| = 1$, whence by (5.12) and (5.10) f_1 would have two linearly independent maximal elements x_1 and $\bar{\alpha} x_2$, contradicting the strict convexity of E). Put

$$G = \{y \in E \mid f_1(y) = f_2(y) = 0\}. \tag{5.13}$$

Since f_1, f_2 are linearly independent and $\dim E = n + 2$, we have $\dim G = n$, and consequently, by virtue of the hypothesis, the mapping π_G is one-valued and additive on $D(\pi_G) = E$. On the other hand, from (5.10) — (5.13) it follows, by theorem 1.1 of Chap. I, that we have $0 \in \mathfrak{P}_G(x_i)$ $(i = 1, 2)$, and thus, since π_G is one-valued, we have

$$\pi_G(x_1) = \pi_G(x_2) = 0. \tag{5.14}$$

Consequently, taking into account the relations

$$\|x - \pi_G(x)\| \leqslant \|x\| \qquad (x \in E), \tag{5.15}$$

it follows that $I - \pi_G$ is a continuous linear projection of norm 1 of E onto $G_0 = [x_1, x_2]$ (where I is the identical mapping of E onto itself). Since G_0 was an arbitrary 2-dimensional linear subspace of E, it follows, by virtue of the theorem of S. Kakutani used also in the foregoing, that E is equivalent to an inner product space, which completes the proof.

In the particular case when $n = 1$, theorem 5.2 has been given by R. A. Hirschfeld ([89], theorem 2), and for $n = 2$ by R. C. James ([100], p. 562). For an arbitrary n and real scalars, it has been given by W. Rudin and K. T. Smith ([201], theorem 1); also, the above proof is due, essentially, to R. C. James [100] and W. Rudin and K. T. Smith ([201], p. 102). A closely related result, based esentially on the same argument, has been given by R. R. Phelps ([174], lemma 4.4); see also Appendix I, §4.

The hypothesis $1 \leqslant n \leqslant \dim E - 2$ (hence also the hypothesis $\dim E \geqslant 3$) in theorem 5.2 is essential; indeed, e. g. in every strictly convex normed linear space of finite dimension $k \geqslant 2$, the condition of theorem 5.2 is satisfied for all linear subspaces of dimension $k - 1$ (by virtue of theorem 6.2 of Chap. I). For 2-dimensional spaces E the following converse statement is also true:

THEOREM 5.3. *Let E be a 2-dimensional normed linear space with the property that for every one-dimensional linear sub-*

space G the set $\pi_G^{-1}(0)$ is a linear subspace of E. Then E is strictly convex.

Proof. If E is not strictly convex, there exist an $f \in E^*$ with $\|f\| = 1$ and $x, y \in E$ with $x \neq y$ such that $f(x) = f(y) = 1 = \|x\| = \|y\|$.

Put

$$G = [x - y]. \tag{5.16}$$

Then dim $G = 1$ and by theorem 1.1 of Chap. I we have

$$x, y \in \pi_G^{-1}(0). \tag{5.17}$$

Since by our hypothesis $\pi_G^{-1}(0)$ is a linear subspace, from (5.17) it follows that we must have also

$$x - y \in \pi_G^{-1}(0),$$

which is absurd since $x - y \in G \setminus \{0\}$. This completes the proof.

Theorem 5.3 has been given, essentially, by R. R. Phelps ([174], lemma 4.6).

Finally, we mention also a continuity property of π_G, which will be used in the next paragraph:

THEOREM 5.4. *Let E be a normed linear space and let G be a finite dimensional Čebyšev subspace of E (hence $D(\pi_G) = E$ and π_G is one-valued on E). Then π_G is continuous on E.*

Proof. Assume that G is a Čebyšev subspace and that there exist $x \in E$, $\{x_n\} \subset E$ such that

$$\lim_{n\to\infty} x_n = x, \quad \lim_{n\to\infty} \pi_G(x_n) \neq \pi_G(x). \tag{5.18}$$

Then $\|x_n\| \leqslant \gamma$ $(n = 1, 2, \ldots)$ for a suitable $\gamma > 0$, whence, by Chap. I, theorem 6.1 b),

$$\|\pi_G(x_n)\| \leqslant 2\,\|x_n\| \leqslant 2\gamma \quad (n = 1, 2, \ldots),$$

and thus, taking into account the hypothesis dim $G < \infty$, we can extract from $\{\pi_G(x_n)\}$ a convergent subsequence $\pi_G(x_{n_k}) \to g_0 \in G$; by virtue of (5.18) we may assume that $g_0 \neq \pi_G(x)$. By Chap. I, theorem 6.1 b), we have then

$$\|x - g_0\| \leqslant \|x - x_{n_k}\| + \|x_{n_k} - \pi_G(x_{n_k})\| + \|\pi_G(x_{n_k}) - g_0\| \leqslant$$

$$\leqslant 2\|x - x_{n_k}\| + \|x - \pi_G(x)\| + \|\pi_G(x_{n_k}) - g_0\|,$$

whence, for $k \to \infty$, we obtain $g_0 \in \mathfrak{P}_G(x)$, which, together with $\pi_G(x) \in \mathfrak{P}_G(x)$ and $g_0 \neq \pi_G(x)$, contradicts the hypothesis that G is a Čebyšev subspace.

Theorem 5.4 is well known (for a more general result and for other continuity properties of π_G, see Appendix II).

5.2. THE OPERATORS π_{G_n} FOR INCREASING SEQUENCES $\{G_n\}$ OF LINEAR SUBSPACES OF FINITE DIMENSION

We have seen in the preceding section (theorem 5.2) that if the normed*) linear space E is not equivalent to an inner product space, then for every integer n with $1 \leqslant n < \infty$ there exist linear subspaces G of E with $\dim G = n$ such that the operator π_G is non-linear. There arises naturally the problem whether there exists a sequence $\{G_n\}$ of linear subspaces of E, with $G_1 \subset G_2 \subset \ldots$ and $\dim G_n = n$ $(n = 1, 2, \ldots)$ such that the operators π_{G_n} $(n = 1,2, \ldots)$ be one-valued and linear on E. We shall now give an important case in which the answer is affirmative and in which the operators π_{G_n} are expressed in a simple form, convenient for applications.

We recall that a sequence $\{x_n\}$ in a Banach space E is called a *basis* of the space E if for every $x \in E$ there exists a unique sequence of scalars $\{\alpha_n\}$ such that we have

$$x = \sum_{i=1}^{\infty} \alpha_i x_i , \tag{5.19}$$

the convergence being taken in the sense of the norm of E (i.e. in the sense $\lim_{n\to\infty} \|x - \sum_{i=1}^{n} \alpha_i x_i\| = 0$). As is well known**), in this case the functionals

$$f_j(x) = \alpha_j \qquad (x = \sum_{i=1}^{\infty} \alpha_i x_i \in E \,;\, j = 1, 2, \ldots) \tag{5.20}$$

are linear and continuous on E, i.e. $f_j \in E^*$ $(j=1, 2, \ldots)$, and

$$\left|\left|\left| \sum_{i=1}^{\infty} \alpha_i x_i \right|\right|\right| = \sup_{1 \leqslant n < \infty} \left\| \sum_{i=1}^{n} \alpha_i x_i \right\| \qquad (x = \sum_{i=1}^{\infty} \alpha_i x_i \in E) \tag{5.21}$$

is a norm on E, equivalent***) to the initial norm of the space E. To every basis $\{x_n\}$ one associates, in a natural way, a sequence $\{s_n\}$ of continuous linear operators (the "partial sum operators") defined by

$$s_n(x) = \sum_{i=1}^{n} f_i(x) x_i \qquad (x \in E,\ n = 1,\ 2,\ \ldots). \tag{5.22}$$

---

*) For simplicity, throughout the sections 5.2 and 5.3 we shall assume, without any special mention, that the space E is infinite dimensional.

**) See S. Banach [8], p. 111.

***) We recall that two norms $\|x\|_1$, $\|x\|_2$, on E are said to be *equivalent*, if the metrics ρ_1 and ρ_2 induced by them on E are equivalent.

If a Banach space E has a basis $\{x_n\}$, then it is separable, since the set of all linear combinations of the form $\sum_{i=1}^{n} r_i x_i$, where the r_i are (complex or real) rational numbers and $n =$ = a positive integer, is a countable dense set in E. However, it is not known, up to the present, whether the converse of this statement is true, i.e. whether in every separable Banach space there exists a basis; this problem has been raised in 1932 by S. Banach ([8], p. 111—112 and p. 245).

Following V. N. Nikolsky [164], [165], the norm in a Banach space E with a basis $\{x_n\}$ is said to be a *T-norm* (with respect to the basis $\{x_n\}$), if:

a) For every $x \in E$ and $n = 1, 2, \ldots$ there exists a unique polynomial $\pi_{G_n}(x)$ of best approximation of x by means of the elements of the n-dimensional linear subspace $G_n = [x_1, \ldots$ $\ldots, x_n]$.

b) This polynomial coincides with the n-th partial sum of the expansion of the element x by means of the basis $\{x_n\}$, i.e. we have

$$\pi_{G_n}(x) = s_n(x) \quad (x \in E, n = 1, 2, \ldots). \tag{5.23}$$

THEOREM 5.5 (V. N. Nikolsky [156], § 4). *Let E be a Banach space with a basis $\{x_n\}$ and let $\{f_n\} \subset E^*$ be the sequence of coefficient functionals associated to the basis $\{x_n\}$, i.e. the sequence (5.20). Then one can introduce on E a T-norm equivalent to the initial norm of the space E, by the formula*

$$((x((= \max_{1 \leqslant n < \infty} \left\{ \frac{1}{n} \sum_{i=1}^{n} \|f_i(x) x_i\| + \left\| \sum_{i=n+1}^{\infty} f_i(x) x_i \right\| \right\}. \tag{5.24}$$

Proof. We observe first that the max in (5.24) is indeed attained, since $\{x_n\}$ being a basis, both summands in (5.24) tend to 0 as $n \to \infty$.

Let us show that $((x(($ is a norm on E. Obviously $((x((\geqslant 0$ and $((0((= 0$. If we have $((x((= 0$, then by

$$((x((\geqslant \|f_1(x) x_1\| + \left\| \sum_{i=2}^{\infty} f_i(x) x_i \right\| \geqslant \left\| \sum_{i=1}^{\infty} f_i(x) x_i \right\| = \|x\| \ (x \in E) \tag{5.25}$$

we have $x = 0$. Finally, the inequality $((x + y((\leqslant ((x((+$ $+ ((y(($ is obvious from $((x + y((= \frac{1}{n_0} \sum_{i=1}^{n_0} \|f_i(x + y) x_i\| +$ $+ \left\| \sum_{i=n_0+1}^{\infty} f_i(x + y) x_i \right\|$ (for a suitable n_0), and $((\alpha x((= |\alpha| ((x(($ is obvious from (5.24). Thus, $((x(($ is a norm on E.

Assume now that $\|y_k\| \to 0$. Then, since we have

$$((y_k((= \frac{1}{n_k} \sum_{i=1}^{n_k} \|f_i(y_k)x_i\| + \left\| \sum_{i=n_k+1}^{\infty} f_i(y_k)x_i \right\| \qquad (k = 1, 2, \ldots)$$

(for suitable integers n_k), and since

$$\left\| \sum_{i=n_k+1}^{\infty} f_i(y_k)x_i \right\| = \left\| y_k - \sum_{i=1}^{n_k} f_i(y_k)x_i \right\| \leqslant \|y_k\| + \left\| \sum_{i=1}^{n_k} f_i(y_k)x_i \right\| \leqslant$$

$$\leqslant \|y_k\| + \sup_{1 \leqslant n < \infty} \left\| \sum_{i=1}^{n} f_i(y_k)x_i \right\| \qquad (k = 1, 2, \ldots),$$

$$\frac{1}{n_0} \sum_{i=1}^{n_k} \|f_i(y_k)x_i\| \leqslant \max_{1 \leqslant i \leqslant n_k} \|f_i(y_k)x_i\| \leqslant \max_{1 \leqslant i \leqslant n_k} \left(\left\| \sum_{j=1}^{i} f_j(y_k)x_j \right\| + \right.$$

$$\left. + \left\| \sum_{j=1}^{i-1} f_j(y_k)x_j \right\| \right) \leqslant 2 \sup_{1 \leqslant n < \infty} \left\| \sum_{i=1}^{n} f_i(y_k)x_i \right\| \qquad (k = 1, 2, \ldots),$$

from the remark that (5.21) is a norm on E, equivalent to the initial norm of the space E, it follows that $((y_k((\to 0$. Conversely, if $((y_k((\to 0$, then by (5.25) we have also $\|y_k\| \to 0$. Consequently, the norm $((x(($ is equivalent to the initial norm of the space E.

Finally, let us show that $((x(($ is a T-norm. If $x \in E$ and $g = \sum_{j=1}^{n} \alpha_j x_j \in G_n$ are arbitrary, we have, by (5.24) and by $f_i\left(\sum_{j=1}^{n} \alpha_j x_j\right) = \alpha_i$ for $1 \leqslant i \leqslant n$ and 0 for $n+1 \leqslant i < \infty$,

$$((x - g((= ((x - \sum_{j=1}^{n} \alpha_j x_j ((= \max_{1 \leqslant k < \infty} \lambda_k,$$

where

$$\lambda_k = \begin{cases} \dfrac{1}{k} \displaystyle\sum_{i=1}^{k} \|(f_i(x) - \alpha_i)x_i\| + \left\| \sum_{i=k+1}^{n} (f_i(x) - \alpha_i)x_i + \sum_{i=n+1}^{\infty} f_i(x)x_i \right\| & \text{for } 1 \leqslant k \leqslant n-1 \\ \dfrac{1}{n} \displaystyle\sum_{i=1}^{n} \|(f_i(x) - \alpha_i)x_i\| + \left\| \sum_{i=n+1}^{\infty} f_i(x)x_i \right\| & \text{for } k = n \\ \dfrac{1}{k} \left[\displaystyle\sum_{i=1}^{n} \|(f_i(x) - \alpha_i)x_i\| + \sum_{i=n+1}^{k} \|f_i(x)x_i\| \right] + \left\| \sum_{i=k+1}^{\infty} f_i(x)x_i \right\| & \text{for } n+1 \leqslant k < \infty. \end{cases}$$

In particular, for $g = s_n(x) = \sum_{i=1}^{n} f_i(x) x_i$ we have

$$((x - s_n(x)((= \max_{1 \leqslant k < \infty} \lambda'_k,$$

where

$$\lambda'_k = \begin{cases} \left\| \sum_{i=n+1}^{\infty} f_i(x) x_i \right\| & \text{for } 1 \leqslant k \leqslant n \\ \frac{1}{k} \sum_{i=n+1}^{k} \|f_i(x) x_i\| + \left\| \sum_{i=k+1}^{\infty} f_i(x) x_i \right\| & \text{for } n + 1 \leqslant k < \infty. \end{cases}$$

Consequently, for any $g \neq s_n(x)$ we have

$$\lambda'_k < \lambda_n \qquad (1 \leqslant k \leqslant n), \tag{5.26}$$

$$\lambda'_k < \lambda_k \qquad (n + 1 \leqslant k < \infty). \tag{5.27}$$

Since $((x - s_n(x)((= \lambda'_{n_0}$ for a suitable n_0 and since by 5.26) and (5.27) there exists an n_1 such that $\lambda'_{n_0} < \lambda_{n_1}$, it ollows that for any $g \neq s_n(x)$ we have

$$((x - s_n(x)((= \lambda'_{n_0} < \lambda_{n_1} \leqslant \max_{1 \leqslant k < \infty} \lambda_k = ((x - g((.$$

Thus, $((x(($ is a T-norm on E, which completes the proof o theorem 5.5.

Let us mention that theorem 5.5 admits, in a certain sense, a converse. Namely, following V. N. Nikolsky [165], the norm in a Banach space E is said to be a *weak T-norm* with respect to a sequence $\{x_n\} \subset E$, if for every polynomial $g = \sum_{i=1}^{k} \alpha_i x_i$ and every positive integer $n \leqslant k$ the polynomial $\sum_{i=1}^{n} \alpha_i x_i \in \in G_n = [x_1, \ldots, x_n]$ is a polynomial of best approximation of g, i.e. $\sum_{i=1}^{n} \alpha_i x_i \in \mathfrak{P}_{G_n}(\sum_{i=1}^{k} \alpha_i x_i)$. It can be shown that if *$\{x_n\}$ is a complete sequence in E (i.e. the closed linear subspace $[x_1, x_2, \ldots]$ spanned by $\{x_n\}$ coincides with E), such that $x_n \neq 0$ $(n = 1, 2, \ldots)$, and if one can introduce in E a weak T-norm with respect to $\{x_n\}$, equivalent to the initial norm of E, then $\{x_n\}$ is a basis of the space E* (V. N. Nikolsky [165], § 6).

Concerning the initial norm of the space E, from theorem 5.5 there follows

COROLLARY 5.2 (V. N. Nikolsky [165], § 7). *Let E be a Banach space with a basis $\{x_n\}$ and let $G_n = [x_1, \ldots, x_n]$ $(n =$*

$= 1, 2, \ldots$). *Then there exists a constant* c *with* $0 < c \leqslant 1$, *depending only on the basis* $\{x_n\}$, *such that*

$$c\|x-s_n(x)\| \leqslant e_{G_n}(x) \leqslant \|x-s_n(x)\| \quad (x \in E,\ n=1, 2, \ldots), \tag{5.28}$$

where s_n *are the partial sum operators* (5.22).

Proof. By virtue of theorem 5.5, one can introduce in E a T-norm $((x(($ equivalent to the initial norm of the space E. Since the mapping $x \rightarrow x$ of E into the space E endowed with the norm $((x(($ is then a topological linear isomorphism, there exist two constants c_1, c_2 with $0 < c_1 \leqslant c_2 < \infty$, such that

$$c_1 \|x\| \leqslant ((x((\leqslant c_2 \|x\| \qquad (x \in E). \tag{5.29}$$

In the initial norm of the space E we have, obviously,

$$e_{G_n}(x) = \|x - \pi_{G_n}(x)\| \leqslant \|x - s_n(x)\| \qquad (x \in E). \tag{5.30}$$

On the other hand, by (5.29) and since $((x(($ is a T-norm, we have

$$e_{G_n}(x) = \|x - \pi_{G_n}(x)\| \geqslant \frac{1}{c_2} ((x - \pi_{G_n}(x)((\geqslant \frac{1}{c_2}((x - s_n(x)((\geqslant$$

$$\geqslant \frac{1}{c_2} c_1 \|x - s_n(x)\| \qquad (x \in E). \tag{5.31}$$

From (5.30) and (5.31) it follows that we have (5.25) with $c = \frac{c_1}{c_2}$, which completes the proof.

Corollary 5.2 shows that if $\{x_n\}$ is a basis of the space E and $G_n = [x_1, \ldots, x_n]$, then $e_{G_n}(x)$ and $\|x - s_n(x)\|$ are of the same order of magnitude. Consequently, if we want to classify the elements $x \in E$ according to the rapidity of the convergence to 0 of the sequences $\{e_{G_n}(x)\}$ (see the next section), we can make use of the sequences $\{\|x - s_n(x)\|\}$, whose computation is much easier.

Now there arises naturally the following problem : Let $\{x_n\}$ be a complete sequence in E such that every finite subsequence of $\{x_n\}$ is linearly independent *) and let $G_n = [x_1, \ldots, x_n]$ ($n = 1, 2, \ldots$). Under what conditions on the sequence $\{x_n\}$ does there exist a sequence of continuous linear mappings $u_n : E \rightarrow E$ with $u_n(E) = G_n$ ($n = 1, 2, \ldots$), such that $u_n(x)$ give, for every $x \in E$ and $n=1, 2, \ldots$, an approximation of x of the same order as $e_{G_n}(x)$ i.e.

$$\|x - u_n(x)\| \leqslant C\, e_{G_n}(x) \quad (x \in E,\ n = 1, 2, \ldots), \tag{5.32}$$

*) Obviously, E has such a sequence $\{x_n\}$ if and only if E is separable.

where $C \geqslant 1$ is a constant independent of x and n? In other words: under what conditions may one "replace" the operators π_{G_n} by continuous linear operators? If such a sequence $\{u_n\}$ exists, then, since by the hypothesis $[x_1, x_2, \dots] = E$ we have $\lim_{n\to\infty} e_{G_n}(x) = 0$ $(x \in E)$, the sequence $\{u_n\}$ satisfies

$$\lim_{n\to\infty} u_n(x) = x \qquad (x \in E), \tag{5.33}$$

whence, by virtue of the principle of uniform boundedness*),

$$\sup_{1\leqslant n<\infty} \|u_n\| < \infty. \tag{5.34}$$

Also, by $e_{G_n}(g)=0$ $(g \in G_n)$ and (5.32) we have in this case

$$u_n(g) = g \qquad (g \in G_n;\ n = 1, 2, \dots), \tag{5.35}$$

i.e. u_n is a continuous linear projection of E onto G_n $(n = 1, 2, \dots \dots)$. Put

$$\lambda_n = \lambda_n(E, G_n) = \inf \|p_n\| \qquad (n = 1, 2, \dots), \tag{5.36}$$

where the inf is taken over all continuous linear projections p_n of E onto G_n. The sequence $\{x_n\}$ is called, following M. I. Kadec [104], a *Lozinsky-Haršiladze system*, or briefly, *a Λ-system*, if

$$\lim_{n\to\infty} \lambda_n = \infty. \tag{5.37}$$

S. M. Lozinsky and F. Haršiladze have shown (see I. P. Natanson [158], Appendix 3), that *the sequence*

$$x_n(t) = t^{n-1} \qquad (t \in [a, b];\ n = 1, 2, \dots) \tag{5.38}$$

is a Λ-system in the space $E = C_R([a, b])$, namely

$$\lambda_n \geqslant \frac{\ln n}{8\sqrt{\pi}} \qquad (n = 1, 2, \dots), \tag{5.39}$$

and consequently, in the space $E = C_R([a, b])$ there exists no sequence of operators $\{u_n\}$ satisfying (5.32) for this sequence $\{x_n\}$; for various extensions of this theorem of S. M. Lozinsky and F. I. Haršiladze see e.g. S. A. Teliakovsky [246] and the papers quoted therein. As has shown M. I. Kadec [104], *in the spaces $L^p_R([a, b])$, where $1 \leqslant p \neq 2$, there also exist Λ-systems.* In $L^2([a, b])$, obviously, there are no Λ-systems, since we have always $\lambda_n = 1$ $(n = 1, 2, \dots)$.

---

*) See e.g. N. Dunford and J. Schwartz [49], p. 66, corollary 2.1. Let us mention that if we have (5.33), then obviously every compact linear operator $v: E \to E$ can be approximated uniformly by linear operators of finite rank; it is still not known whether every separable Banach space E has this approximation property.

We also mention the following generalization of the preceding problems: Let $\{x_n\}$ be a complete sequence in E such that every finite subsequence of $\{x_n\}$ is linearly independent, let $G_n = [x_1, \ldots, x_n]$ $(n = 1, 2, \ldots)$ and let A be a convex set in E. Does there exist a sequence of continuous linear mappings $u_n : E \to E$ with $u_n(E) = G_n$ $(n = 1, 2, \ldots)$, such that

$$\sup_{x \in A} \|x - u_n(x)\| = \sup_{x \in A} e_{G_n}(x) \qquad (n = 1, 2, \ldots), \tag{5.40}$$

or at least such that

$$\sup_{x \in A} \|x - u_n(x)\| \leqslant C \sup_{x \in A} e_{G_n}(x) \qquad (n = 1, 2, \ldots), \tag{5.41}$$

where C is a constant independent of n? The first question has been raised by S. M. Nikolsky [163] (see also Chap. I, formula (6.78) for the case of a single linear subspace G). From the characterization of compactness by means of finite ε-nets it follows immediately that the above problem, for the case when A runs over all sets consisting of a single point $\{x\}$, is equivalent to the problem for the case when A runs over all compact convex sets in E. For other connections between linear operators and best approximation, see the monograph of P. P. Korovkin [121].

Finally, we shall consider the problem of operatorial characterization of a sequence of mappings π_{G_n} $(n = 1, 2, \ldots)$, more precisely, the problem, under what conditions on a sequence $\{v_n\}$ of mappings of E into E with $v_n(E) = G_n$ $(n = 1, 2, \ldots)$ can one introduce on E a norm equivalent to the initial norm, such that in this norm $v_n(x)$ be (for every $x \in E$ and n) polynomials of best approximation of x. The answer to this problem is given by

THEOREM 5.6 (V. N. Nikolsky [166], theorem 1). *Let E be a separable Banach space, $\{x_n\}$ a complete sequence in E such that every finite subsequence of $\{x_n\}$ is linearly independent, $G_0 = \{0\}$, $G_n = [x_1, \ldots, x_n]$ $(n = 1, 2, \ldots)$ and v_n a mapping of E into E, with $v_n(E) = G_n$ $(n = 0, 1, 2, \ldots$; hence, in particular, $v_0(x) \equiv 0)$. In order that there exist on E a norm $|x|$ equivalent to the initial norm, with the property that in this new norm*

$$v_n(x) = \pi_{G_n}(x) \in \mathfrak{P}_{G_n}(x) \qquad (x \in E, n = 0, 1, 2, \ldots), \tag{5.42}$$

it is necessary and sufficient that the following six conditions be satisfied:

a) *We have*

$$v_n(g) = g \qquad (g \in G_n; n = 1, 2, \ldots). \tag{5.43}$$

b) *Each* v_n $(n = 1, 2, \ldots)$ *is continuous at the origin.*

c) *We have*

$$v_n(x+g) = v_n(x) + v_n(g) = v_n(x) + g \quad (x \in E, g \in G_n;\ n = 1, 2, \ldots). \tag{5.44}$$

d) *We have*

$$v_n(\alpha x) = \alpha v_n(x) \qquad (x \in E,\ \alpha = \text{scalar};\ n = 1, 2, \ldots). \tag{5.45}$$

e) *There exists a norm* $\|x\|_1$ *on* E, *equivalent to the initial norm of the space* E, *such that*

$$\|r_{k_1} r_{k_2} \ldots r_{k_m}(x+y)\|_1 \leqslant \|r_{k_1} r_{k_2} \ldots r_{k_m}(x) + r_{k_1} r_{k_2} \ldots r_{k_m}(y)\|_1 \tag{5.46}$$

for every $x, y \in E$ *and every finite system of natural numbers* $0 \leqslant k_1 < k_2 < \ldots < k_m$, *where*

$$r_n(x) = x - v_n(x) \qquad (n = 0, 1, 2, \ldots). \tag{5.47}$$

f) *We have*

$$\sup_{0 \leqslant k_1 < \ldots < k_m} \|r_{k_1} r_{k_2} \ldots r_{k_m}(x)\| < \infty \qquad (x \in E). \tag{5.48}$$

Proof. The necessity of the conditions a) — d) follows from Chap. I, theorem 6.1. In order to show the necessity of condition e) we observe first that in the desired norm $|x|$ we have, for any natural numbers n, m with $n \leqslant m$,

$$|r_m(x)| = |x - v_m(x)| \leqslant |x - v_m(x) - v_n[x - v_m(x)]| = |r_n r_m(x)| =$$
$$= |r_m(x) - v_n[r_m(x)]| \leqslant |r_m(x)| \qquad (x \in E),$$

whence

$$|r_n r_m(x)| = |r_m(x)| = e_{G_m}(x) \qquad (x \in E,\ n \leqslant m). \tag{5.49}$$

The necessity of condition e) follows now from the fact that in the desired norm $|x|$ we have, for every $x, y \in E$ and $0 \leqslant k_1 \leqslant k_2 < \ldots < k_m$,

$$|r_{k_1} r_{k_2} \ldots r_{k_m}(x+y)| = |r_{k_m}(x+y)| = e_{G_{k_m}}(x+y) \leqslant |x+y - v_{k_m}(x) -$$
$$- v_{k_m}(y)| = |r_{k_m}(x) + r_{k_m}(y)| = |r_{k_1} r_{k_2} \ldots r_{k_m}(x) + r_{k_1} r_{k_2} \ldots r_{k_m}(y)|,$$

and thus we may take $\|x\|_1 = |x|$.

Finally, the necessity of condition f) follows from the relations

$$|r_{k_1} r_{k_2} \ldots r_{k_m}(x)| = |r_{k_m}(x)| = e_{G_{k_m}}(x) \to 0 \qquad (m \to \infty)$$

and from the equivalence of the norm $|x|$ to the initial norm of the space E.

Conversely, assume now that the conditions a) — f) are satisfied. Put

$$|x| = \sup \|r_{k_1} r_{k_2} \dots r_{k_m}(x)\|_1 \qquad (x \in E), \tag{5.50}$$

where $\|x\|_1$ is the norm of condition e) and where the sup is taken over all finite systems of natural numbers $0 \leqslant k_1 < < k_2 < \dots < k_m$ $(m = 1, 2, \dots)$; by virtue of condition f), this sup is finite.

We claim that then $|x|$ is a norm on E. Indeed, obviously $|x| \geqslant 0$, and $|x| = 0$ if and only if $x = 0$; on the other hand, the triangle inequality follows from condition e), and the homogeneity of the norm follows from condition d).

We shall show now that the norm $|x|$ is equivalent to the initial norm of the space E. By (5.50) and $r_0(x) = x$ $(x \in E)$ we have

$$\|x\|_1 \leqslant |x| \qquad (x \in E). \tag{5.51}$$

On the other hand, since by e) we have

$$|\, \|r_{k_1} r_{k_2} \dots r_{k_m}(x)\|_1 - \|r_{k_1} r_{k_2} \dots r_{k_m}(y)\|_1 | \leqslant \|r_{k_1} r_{k_2} \dots r_{k_m}(x-y)\|_1$$

and since by b) and the equivalence of the norm $\|x\|_1$ to the initial norm, the (non-linear) functionals

$$\varphi_{k_1 k_2 \dots k_m}(x) = \|r_{k_1} r_{k_2} \dots r_{k_m}(x)\|_1 \qquad (x \in E)$$

are continuous at the origin, it follows that they are continuous on the whole space E with respect to the norm $\|x\|_1$. Consequently, taking into account e) and d), it follows*) that there exists a constant $c > 0$ such that

$$\|r_{k_1} r_{k_2} \dots r_{k_m}(x)\|_1 \leqslant c \|x\|_1 \qquad (x \in E,\ 0 \leqslant k_1 < k_2 < \dots < k_m),$$

hence we have

$$|x| \leqslant c \|x\|_1 \qquad (x \in E). \tag{5.52}$$

From (5.51) and (5.52) it follows that the norm $|x|$ is equivalent to the norm $\|x\|_1$, whence also to the initial norm of the space E.

Finally, in order to show that in the norm $|x|$ we have (5.42), let us observe first that by c) and a) we have, for any positive integers, n, m with $n \leqslant m$,

$$r_m r_n(x) = r_n(x) - v_m[r_n(x)] = x - v_n(x) - v_m(x) + v_n(x) = r_m(x) \quad (x \in E). \tag{5.53}$$

*) See e.g. N. Dunford and J. Schwartz [49], p. 53, lemma 13.

Consequently,

$$|x - v_n(x)| = |r_n(x)| = \sup \|r_{k_1} r_{k_2} \dots r_{k_m}[r_n(x)]\|_1 \leqslant$$

$$\leqslant \sup \|r_{k_1} r_{k_2} \dots r_{k_m}(x)\|_1 = |x| \qquad (x \in E,\ n = 0, 1, 2, \dots),$$

whence, taking into account condition c), we obtain

$$|x - v_n(x)| = |x - g - v_n(x) + g| = |x - g - v_n(x - g)| \leqslant |x - g| \qquad (g \in G_n),$$

hence (5.42), which completes the proof of theorem 5.6.

V. N. Nikolsky has shown, by an example [167], that in theorem 5.6 condition f) cannot be replaced by the following:

f′) *We have*

$$\lim_{n\to\infty} r_n(x) = 0 \qquad (x \in E). \tag{5.54}$$

In the particular case *when the mappings* $v_n : E \to G_n$ ($n = 1, 2, \dots$) *are linear*, conditions c), d), e) are automatically satisfied, whence *conditions* a), b) *and* f) *are necessary and sufficient in order that there exist on* E *an equivalent norm with the property* (*5.42*) (V. N. Nikolsky [166], theorem 2). As has shown V. N. Nikolsky ([166], theorem 3), in this case the space E has a basis consisting of polynomials of all orders*), e.g. such a basis is the sequence

$$y_1 = x_1,\ y_n = r_1 r_2 \dots r_{n-1}(x_n) \qquad (n = 2, 3, \dots). \tag{5.55}$$

Finally, let us mention that V. N. Nikolsky ([166], theorem 4) has given necessary and sufficient conditions in order that there exist on E an equivalent norm $|x|$ with the property that in this new norm π_{G_n} is *one-valued* on E and we have $v_n = \pi_{G_n}$ ($n = 0, 1, 2, \dots$). However, this "norm" does not satisfy, in general, the triangle axiom; a sufficient condition in order that it satisfy this axiom, is the linearity of the mappings v_n ($n = 0, 1, 2, \dots$).

---

*) The *order* of a polynomial $\sum_{i=1}^{n} \alpha_i x_i$ is the number $\max_{\alpha_j \neq 0} j$. The condition of the existence of a basis consisting of polynomials of all orders is more restrictive than that of the existence of a polynomial basis $\sum_{i=1}^{m_n} \alpha_i^{(n)} x_i$; indeed, e.g. the space $E = C_R([a, b])$ has a polynomial basis with respect to the sequence (5.38) (M. G. Krein, D. P. Milman and M. A. Rutman [127]), but by virtue of the theorem of S. M. Lozinsky and F. I. Haršiladze mentioned above, E has no basis consisting of polynomials of all orders with respect to that sequence.

5.3. THE FUNCTIONALS e_{G_n} FOR INCREASING SEQUENCES $\{G_n\}$ OF LINEAR SUBSPACES OF FINITE DIMENSION

Let E be a separable Banach space, $\{x_n\}$ a complete sequence in E such that every finite subsequence of $\{x_n\}$ is linearly independent, and let $G_n = [x_1, \ldots, x_n]$ $(n = 1, 2, \ldots)$. Then we have

$$\lim_{n\to\infty} e_{G_n}(x) = 0 \qquad (x \in E), \tag{5.56}$$

and the elements $x \in E$ can be classified according to the rapidity of the convergence (5.56). Since for concrete functional spaces E this problem belongs to the domain of the constructive theory of functions (see e.g. I. P. Natanson [158], A. F. Timan [250]), here we shall only give, in general Banach spaces, one example. We shall say that an element $x \in E$ is *quasi-analytic* (with respect to $\{x_n\}$), if there exists a constant δ with $0 < \delta < 1$, independent of n, such that

$$\underline{\lim}_{n\to\infty} \frac{e_{G_n}(x)}{\delta^n} = 0. \tag{5.57}$$

For instance, in the space $E = C_R([a, b])$ the quasi-analytic elements with respect to the sequence (5.38) are nothing else than the quasi-analytic functions in the sense of S. N. Bernstein ([15], Appendix I).

THEOREM 5.7. *Let E be a separable Banach space, $\{x_n\}$ a complete sequence in E such that every finite subsequence of $\{x_n\}$ is linearly independent, $G_n = [x_1, \ldots, x_n]$ $(n = 1, 2, \ldots)$, and $\{\varepsilon_n\}$ a non-increasing sequence of positive numbers with* $\lim_{n\to\infty} \varepsilon_n = 0$. *Then for every element $x \in E$ there exist two elements $y_1, y_2 \in E$ such that*

$$y_1 + y_2 = x, \tag{5.58}$$

$$\underline{\lim}_{n\to\infty} \frac{e_{G_n}(y_k)}{\varepsilon_n} = 0 \qquad (k = 1, 2). \tag{5.59}$$

*In particular: α) Every element $x \in E$ is the sum of two elements which are quasi-analytic with respect to $\{x_n\}$. β) If there exists a sequence $\{f_n\} \subset E^*$ such that (x_n, f_n) is a biorthogonal**)

*) I.e. $f_i(x_j) = 1$ for $i = j$ and 0 for $i \neq j$, $(i, j = 1, 2, \ldots)$; obviously, in this case every finite subsequence of $\{x_n\}$ is linearly independent. In every separable space there exists a complete sequence $\{x_n\}$ admitting such a sequence $\{f_n\} \subset E^*$.

system, then for every $x \in E$ *we have* (5.58), *where the series* $\sum_{i=1}^{\infty} f_i(y_k)x_i$ $(k = 1, 2)$ *each have a sequence of partial sums converging to* y_1 *and* y_2 *respectively.*

Proof. Let us construct by recurrence a sequence $\{g_n\}$ of polynomials as follows: Put $g_1 = x_1$; if the order of g_n is m_n, choose g_{n+1} such that $m_{n+1} > m_n$ and $\|x - g_{n+1}\| < \frac{\varepsilon_{m_n}}{2^{n+1}}$. Put $g_0 = 0$ and

$$y_1 = \sum_{i=1}^{\infty} (g_{2i} - g_{2i-1}), \qquad y_2 = \sum_{i=1}^{\infty} (g_{2i-1} - g_{2i-2}). \tag{5.60}$$

Then obviously $y_1 + y_2 = x$. On the other hand, $\sum_{i=1}^{n} (g_{2i} - g_{2i-1}) = g$ is a polynomial of order m_{2n} and we have

$$e_{G_{2n}}(y_1) \leqslant \|y_1 - g\| \leqslant \sum_{i=n+1}^{\infty} \|g_{2i} - g_{2i-1}\| \leqslant$$

$$\leqslant \sum_{i=n+1}^{\infty} (\|x - g_{2i}\| + \|x - g_{2i-1}\|) < \sum_{i=n+1}^{\infty} \left(\frac{\varepsilon_{m_{2i-1}}}{2^{2i}} + \frac{\varepsilon_{m_{2i-2}}}{2^{2i-1}}\right) < \frac{\varepsilon_{m_{2n}}}{2^{2n}},$$

whence $\lim_{n\to\infty} \frac{e_{G_{2n}}(y_1)}{\varepsilon_{m_{2n}}} = 0$ and thus we have (5.59) for $k = 1$. The relation (5.59) for $k = 2$ follows in a similar way.

The particular case α) is obtained by taking $\varepsilon_n = \delta^n$ $(n = 1, 2, \ldots)$, where δ is an arbitrary number such that $0 < \delta < 1$. On the other hand, the statement of the particular case β) is obvious if $\{x_n\}$ is a basis of the space E. If $\{x_n\}$ is not a basis, then for the sequence $\{s_n\}$ of partial sum operators (5.22) associated to the biorthogonal system (x_n, f_n) we have $\sup_{1 \leqslant n \leqslant \infty} \|s_n\| = \infty$. Since

$$\|y_k - s_n(y_k)\| \leqslant \|y_k - \pi_{G_n}(y_k)\| + \|\pi_{G_n}(y_k) - s_n(y_k)\| = e_{G_n}(y_k) +$$

$$+ \left\| \sum_{i=1}^{n} f_i[\pi_{G_n}(y_k) - y_k]\, x_i \right\| \leqslant e_{G_n}(y_k) + (\sup_{1 \leqslant j \leqslant n} \|s_j\|)\, e_{G_n}(y_k)$$

$$(k = 1, 2;\ n = 1, 2, \ldots),$$

from (5.56) and (5.59) for $\varepsilon_n = \frac{1}{\sup_{1 \leqslant j \leqslant n} \|s_j\|}$ $(n = 1, 2, \ldots)$ it follows that we have $\varliminf_{n\to\infty} \|y_k - s_n(y_k)\| = 0$ $(k = 1, 2)$, which completes the proof.

In the particular case when $E = C([-\pi, \pi])$ and $\{x_n\}$ is the trigonometric system (hence $\sum_{i=1}^{\infty} f_i(x) x_i$ is the Fourier series of x), the statement β) of theorem 5.7 has been given by D. E. Menšov ([147], theorem 4). In the general case, theorem 5.7 has been given by A. I. Markuševič ([144], theorems 2 and 3).

We shall consider now, for G_n of the form $[x_1, \ldots, x_n]$ $(n = 1, 2, \ldots)$, $G_0 = \{0\}$, the problem of existence of elements $x \in E$ with prescribed values $e_{G_n}(x)$ $(n = 1, 2, \ldots)$, which has been called by S. N. Bernstein [16] "the inverse problem of the theory of best approximation". The answer to this problem is affirmative, as shown by

THEOREM 5.8.*) *Let E be a separable real Banach space, $\{x_n\}$ a complete sequence in E such that every finite subsequence of $\{x_n\}$ is linearly independent, $G_0 = \{0\}$ and $G_n = [x_1, \ldots, x_n]$ $(n = 1, 2, \ldots)$. Then for every sequence of numbers $\{e_n\}$ with the properties*

$$e_0 \geqslant e_1 \geqslant e_2 \geqslant \ldots \geqslant 0, \lim_{n\to\infty} e_n = 0, \tag{5.61}$$

there exists an element $x \in E$ such that

$$e_{G_k}(x) = e_k \qquad (k = 0, 1, 2, \ldots). \tag{5.62}$$

Proof. Let us first remark that for every $y \in E$, every natural number n and every number $e \geqslant e_{G_{n+1}}(y)$ there exists a number λ such that

$$e_{G_n}(y + \lambda x_{n+1}) = e. \tag{5.63}$$

Indeed, since $x_{n+1} \notin G_n$ and since

$$\min_{\lambda} e_{G_n}(y + \lambda x_{n+1}) = \min_{\lambda} \min_{g \in G_n} \|y + \lambda x_{n+1} - g\| = e_{G_{n+1}}(y) \leqslant e,$$

the statement follows from Chap. I, corollary 6.1.

Assume now that there exists an n such that $e_{n+1} = 0$. We shall show that then there exists a polynomial x of order $n + 1$ satisfying (5.62). Indeed, applying the above remark for $y = 0$ (hence $e_{G_{n+1}}(y) = 0$), it follows that there exists**) a λ_{n+1} such that

$$e_{G_n}(\lambda_{n+1} x_{n+1}) = e_n. \tag{5.64}$$

---

*) From this theorem it follows immediately that every real Banach space E has the property (B) with respect to the family $(\mathfrak{S})$ of all finite dimensional linear subspaces of E (see Chap. I, section 6.3).

**) Besides, by Chap. I, theorem 6.5 d) we have obviously $\lambda_{n+1} = \pm \dfrac{e_n}{e_{G_n}(x_{n+1})}$.

Applying now the above remark for $y = \lambda_{n+1}x_{n+1}$ and $n-1$ instead of n, it follows that there exists a λ_n such that

$$e_{G_{n-1}}(\lambda_{n+1}x_{n+1} + \lambda_n x_n) = e_{n-1}. \tag{5.65}$$

At the same time, by Chap. I, theorem 6.5 c) and by (5.64) we have

$$e_{G_n}(\lambda_{n+1}x_{n+1} + \lambda_n x_n) = e_{G_n}(\lambda_{n+1}x_{n+1}) = e_n. \tag{5.66}$$

Applying again the above remark for $y = \lambda_{n+1}x_{n+1} + \lambda_n x_n$ and $n-2$ instead of n, it follows that there exists a λ_{n-1} such that

$$e_{G_{n-2}}(\lambda_{n+1}x_{n+1} + \lambda_n x_n + \lambda_{n-1}x_{n-1}) = e_{n-2}. \tag{5.67}$$

At the same time, by Chap. I, theorem 6.5 c) and by (5.65), (5.66) we have

$$e_{G_k}(\lambda_{n+1}x_{n+1} + \lambda_n x_n + \lambda_{n-1}x_{n-1}) = e_k \qquad (k = n-1, n). \tag{5.68}$$

Continuing in this way, we thus obtain a polynomial $x = \sum_{i=1}^{n+1} \lambda_i x_i \in G_{n+1}$ satisfying (5.62), which proves the theorem for all almost zero sequences $\{e_n\}$ (i.e. zero except a finite number of terms) with the properties (5.61).

Now let $\{e_n\}$ be an arbitrary sequence with the properties (5.61). Then, by the above, for every natural number n there exists a polynomial $g_{n+1} \in G_{n+1}$ with the properties

$$e_{G_k}(g_{n+1}) = e_k \qquad (k = 0, 1, \ldots, n). \tag{5.69}$$

Let $\pi_{G_k}(g_{n+1}) \in \mathfrak{L}_{G_k}(g_{n+1})$ $(k = 0, 1, \ldots, n;\ n = 0, 1, 2, \ldots)$ be arbitrary. Since by Chap. I, theorem 6.1 b) we have

$$\|\pi_{G_k}(g_{n+1})\| \leqslant 2\|g_{n+1}\| = 2e_{G_0}(g_{n+1}) = 2e_0$$

$$(k = 0, 1, \ldots, n;\ n = 0, 1, 2, \ldots),$$

and since the cells $e_0 S_{G_k} \subset G_k$ are compact (by dim $G_k < \infty$), we can construct, using the diagonal procedure, an increasing sequence of indices $n_1, n_2, \ldots$ such that

$$\lim_{j \to \infty} \pi_{G_k}(g_{n_j}) = g_k^{(0)} \in G_k \qquad (k = 0, 1, 2, \ldots). \tag{5.70}$$

Consequently, there exist natural numbers N_k such that for every $n_j > N_k$ we have

$$\|g_{n_j} - g_k^{(0)}\| \leqslant \|g_{n_j} - \pi_{G_k}(g_{n_j})\| + \|\pi_{G_k}(g_{n_j}) - g_k^{(0)}\| \leqslant 2e_k,$$

and thus, for every $n_i, n_j > N_k$,

$$\|g_{n_i} - g_{n_j}\| \leqslant \|g_{n_i} - g_k^{(0)}\| + \|g_k^{(0)} - g_{n_j}\| \leqslant 4e_k. \tag{5.71}$$

Since $\lim_{k\to\infty} e_k = 0$ and since the space E is complete, it follows that the sequence $\{g_{n_j}\}$ converges to an element $x \in E$. By Chap. I, theorem 6.5 f) and by (5.69) we have then

$$e_{G_k}(x) = e_{G_k}(\lim_{j\to\infty} g_{n_j}) = \lim_{j\to\infty} e_{G_k}(g_{n_j}) = e_k \quad (k = 0, 1, 2, \ldots),$$

whence (5.62), which completes the proof of theorem 5.8.

In the particular case when $E = C_R([a, b])$ and $\{x_n\}$ is the sequence (5.38), theorem 5.8 has been given by S. N. Bernstein [16]. The remark that this theorem remains valid, with the same proof, in the case when $\{x_n\}$ is a "T-system" in an arbitrary separable real Banach space E, has been made by S. M. Nikolsky ([162 a], p. 297); a sequence $\{x_n\} \subset E$ is said to be*) a *T-system* in the space E, if it is complete in E, every finite subsequence of $\{x_n\}$ is linearly independent and all mappings π_{G_n}, where $G_n = [x_1, \ldots, x_n]$ $(n = 1, 2, \ldots)$, are one-valued on E. S. N. Bernstein [16] in the case of the T-system (5.38) in $E = C_R([a, b])$ and M. I. Kadec [102] for T-systems in arbitrary separable real Banach spaces E, have shown that, denoting

$$\pi_{G_n}(x) = \sum_{i=1}^{n} \alpha_i^{(n)} x_i \qquad (n = 1, 2, \ldots), \tag{5.72}$$

the statement of theorem 5.8 remains valid if the functionals e_{G_n} are replaced by the functionals

$$\tilde{e}_{G_n}(x) = \varepsilon_n(x)\, e_{G_n}(x) \qquad (x \in E,\ n = 0, 1, 2, \ldots), \tag{5.73}$$

where

$$\varepsilon_n(x) = \begin{cases} \operatorname{sign} \alpha_{n+1}^{(n+1)}(x) & \text{for } \alpha_{n+1}^{(n+1)} \neq 0 \\ \varepsilon_{n+1}(x) & \text{for } \alpha_{n+1}^{(n+1)} = 0 \end{cases} \qquad (x \in E,\ n = 0, 1, 2, \ldots) \tag{5.74}$$

(in the second case we have obviously $e_{G_n}(x) = e_{G_{n+1}}(x)$; consequently, if $\alpha_{n+1}^{(n+1)} = 0$ for $n = 0, 1, 2, \ldots$, we have $e_{G_n}(x) = 0$ for $n = 0, 1, 2, \ldots$ and in this case we shall put $\tilde{e}_{G_n}(x) = 0$ for $n = 0, 1, 2, \ldots$), and the sequence of non-negative num-

*) This term is used e.g. by M. I. Kadec [102], V. L. Klee and R. G. Long [116]. Some authors use a different terminology, e.g. K. Tatarkiewicz [243] calls such sequences $\{x_n\}$ "(U)-systems". In the particular case when $E = C_R([a, b])$ and the completeness of $\{x_n\}$ is not required, such sequences $\{x_n\}$ are called "Markov systems" (see e.g. N. I. Ahiezer [1]).

bers $\{e_n\}$ by a sequence of real numbers $\{\tilde{e}_n\}$ with the properties

$$|\tilde{e}_0| \geqslant |\tilde{e}_1| \geqslant |\tilde{e}_2| \geqslant \ldots, \lim_{n\to\infty} \tilde{e}_n = 0, \tag{5.75}$$

$$0 \leqslant n < \infty \text{ and } |\tilde{e}_n| = |\tilde{e}_{n+1}| \text{ imply } \tilde{e}_n = \tilde{e}_{n+1}. \tag{5.76}$$

V. L. Klee and R. G. Long [116] have observed that the proofs of Bernstein and Kadec "are convincing" only in the hypothesis of the continuity of the functionals $\tilde{e}_{G_n}$ (see the final part of the above proof of theorem 5.8), and these are obviously discontinuous for $n > 2$; the result of Bernstein-Kadec is however true, using another proof (V. L. Klee and R. G. Long [116]).

A T-system $\{x_n\}$ in a separable real Banach space E is called, following M. I. Kadec [102], a *Bernstein system*, if for every sequence $\{\tilde{e}_n\}$ with the properties (5.75), (5.76) there exists a *unique* element $x \in E$ such that

$$\tilde{e}_{G_k}(x) = \tilde{e}_k \qquad (k = 0,1,2, \ldots). \tag{5.77}$$

For instance, the natural basis*) $\{x_n\}$ in the spaces $l^p_R(1 \leqslant \leqslant p < \infty)$ is a Bernstein system (M. I. Kadec [102]). There exist T-systems which are not Bernstein systems; e.g. in the space $(c_0)_R$ endowed with the equivalent T-norm (5.24) with respect to the natural basis $\{x_n\}$, this basis is not a Bernstein system (R. G. Long [137]). It is not known whether or not the T-system (5.38) in $E = C_R([a, b])$ is a Bernstein system; this problem has been raised by S. N. Bernstein [16]. In any case, it can be shown (M. I. Kadec [102]) that for a T-system in a separable Banach space E the set of all elements $x \in E$ which are uniquely determined by the sequence $\{\tilde{e}_{G_n}(x)\}$ is a dense set in E, of type G_δ, hence of the second category.

M. I. Kadec has proved ([102], theorem 1) that *two infinite dimensional separable real Banach spaces E_1, E_2 having Bernstein systems $\{x_n^{(i)}\} \subset E_i$ $(i = 1, 2)$ are homeomorphic, by the mapping which carries each element $x^{(1)} \in E_1$ into the element $x^{(2)} \in E_2$ with the property $e_{G_k^{(1)}}(x^{(1)}) = e_{G_k^{(2)}}(x^{(2)})$ $(k = 0, 1, 2, \ldots)$, where* $G_0^{(i)} = \{0\}, G_k^{(i)} = [x_1^{(i)}, \ldots, x_k^{(i)}]$ $(i = 1, 2;\ k = 1,2, \ldots)$. This result was the first important step towards the solution of the problem of homeomorphism of infinite dimensional separable Banach spaces, raised by M. Fréchet ([59], p. 95) and S. Banach ([8], p. 242). It is known at present that the answer to this problem is affirmative: *all infinite dimensional separable (real or complex) Banach spaces are homeomorphic* (M. I. Kadec [105]).

---

*) I.e. the basis $x_n = \{\underbrace{0, \ldots, 0}_{n-1}, 1, 0, 0, \ldots\}$ $(n = 1, 2, \ldots)$.

§ 6. n-DIMENSIONAL DIAMETERS. BEST n-DIMENSIONAL SECANTS

Let E be a normed linear space, A a set in E and n an integer with $0 \leqslant n < \infty$. The number

$$d_n(A, E) = \inf_{\dim G = n} \delta(A, G) = \inf_{\dim G = n} \sup_{x \in A} e_G(x) =$$

$$= \inf_{\dim G = n} \sup_{x \in A} \inf_{g \in G} \|x - g\| = \inf_{\dim G = n} \inf_{\substack{\varepsilon > 0 \\ A \subset G + \varepsilon S_E}} \varepsilon, \tag{6.1}$$

where the inf is taken over all n-dimensional linear subspaces G of E, is called*) the *n-dimensional diameter* of the set A (with respect to the space E). The diameter $d_n(A,E)$ may be considered as a measure of the n-dimensional "thickness" of A. In the case when this will lead to no confusion, we shall use the notation $d_n(A)$ for $d_n(A,E)$; it should be observed, however, that although for every linear subspace E_1 of E and for every set $A \subset E_1$ we obviously have

$$d_n(A,E_1) \geqslant d_n(A,E), \tag{6.2}$$

in general we may have in (6.2) also the sign $>$ (see section 6.2, the remark made after theorem 6.3).

An n-dimensional linear subspace $G \subset E$ is called**) *extremal* for A (with respect to E), or a *best n-dimensional secant* of A (with respect to E), if

$$\delta(A, G) = d_n(A). \tag{6.3}$$

In the present paragraph we shall study the n-dimensional diameters and the best n-dimensional secants.

---

*) The numbers $d_n(A,E)$ have been introduced by A. N. Kolmogorov [117]. Some authors use a different terminology, e.g. A. L. Brown [27] proposes for $d_n(A, E)$ the term *n-dimensional radius* of A.

**) This notion was introduced by A. N. Kolmogorov [117]. The term "extremal" is used by V. M. Tihomirov [247], while A. L. Garkavi [65] uses the term "best n-dimensional section", but it seems to be more adequate the term "best n-dimensional secant", since the "section" of A by the "secant" G is at most the set $G \cap A$; moreover, $G \cap A$ may be also void, e.g. if $E =$ the euclidean plane, $A = \{1, 1\} \cup \{1, -1\}$ and $n = 1$.

6.1. PRELIMINARY LEMMAS

We shall give first the following result of M. G. Krein — M. A. Krasnoselsky — D. P. Milman [126] *), which will have a fundamental role in the present paragraph :

LEMMA 6.1. *Let E be a normed linear space and G_1, G_2 two linear subspaces of E such that*

$$\dim G_1 < \infty, \qquad \dim G_1 < \dim G_2. \tag{6.4}$$

Then there exists a $y \in G_2 \setminus \{0\}$ such that

$$y \perp G_1. \tag{6.5}$$

Proof. Obviously we may assume, without loss of generality, that we have

$$\dim G_1 = n, \ \dim G_2 = n + 1. \tag{6.6}$$

Let $E_1 = [G_1, G_2] =$ the linear subspace of E spanned by G_1 and G_2.

Assume first that E_1 is strictly convex, hence $\pi_{G_1}|_{E_1}$ is one-valued on E_1. Then, by theorem 5.5 and Chap. I, theorem 6.1, $\pi_{G_1}|_{E_1}$ is continuous on E_1 and we have

$$\pi_{G_1}(-x) = -\pi_{G_1}(x) \qquad (x \in E_1). \tag{6.7}$$

If we now assume that there exists no $y \in G_2 \setminus \{0\}$ satisfying (6.5), we have

$$\pi_{G_1}(y) \neq 0 \qquad (y \in G_2 \setminus \{0\}), \tag{6.8}$$

whence, taking into account the continuity of $\pi_{G_1}|_{E_1}$ and the compactness of the set

$$\mathbb{S}_2 = \{y \in G_2 \mid \|y\| = 1\}, \tag{6.9}$$

it follows that the operator $\psi : \mathbb{S}_2 \to \mathbb{S}_1 = \{x \in G_1 \mid \|x\| = 1\}$, defined by

$$\psi(y) = \frac{\pi_{G_1}(y)}{\|\pi_{G_1}(y)\|} \qquad (y \in \mathbb{S}_2), \tag{6.10}$$

is continuous. Also, by (6.7) we have

$$\psi(-y) = -\psi(y) \qquad (y \in \mathbb{S}_2). \tag{6.11}$$

---

*) See also I. Ts. Gohberg — M. G. Krein [75], where there are reproduced part of the results of the paper of M. G. Krein — M. A. Krasnoselsky — D. P. Milman [126].

Consider now G_2 as a *real* space $(G_2)_{(r)}$ in the usual way ($(2n+2)$-dimensional or $(n+1)$-dimensional according to whether E is complex or real) and let $y_1, \ldots, y_{2n+2}$ (respectively $y_1, \ldots, y_{n+1}$) be a basis of $(G_2)_{(r)}$. Then the mapping *)

$$\varphi_2 : \sum_{i=1}^{2n+2} \alpha_i y_i \to \left\{ \frac{\alpha_i}{\sqrt{\sum_{k=1}^{2n+2} \alpha_k^2}} \right\}_{i=1}^{2n+2} \tag{6.12}$$

is a homeomorphism of $\mathfrak{S}_2$ onto $\Sigma_2 = \left\{ \{\beta_i\}_{i=1}^{2n+2} \in R^{2n+2} \,\middle|\, \sum_{i=1}^{2n+2} \beta_i^2 = 1 \right\}$, satisfying

$$\varphi_2(-y) = -\varphi_2(y) \qquad (y \in \mathfrak{S}_2). \tag{6.13}$$

Similarly, considering G_1 as a real space $(G_1)_{(r)}$ ($2n$-dimensional or n-dimensional according to whether E is complex or real) and taking a basis $x_1, \ldots, x_{2n}$ (respectively $x_1, \ldots, x_n$) of $(G_1)_{(r)}$, the mapping

$$\varphi_1 : \sum_{i=1}^{2n} \alpha_i x_i \to \left\{ \frac{\alpha_i}{\sqrt{\sum_{k=1}^{2n} \alpha_k^2}} \right\}_{i=1}^{2n} \tag{6.14}$$

will be a homeomorphism of $\mathfrak{S}_1$ onto $\Sigma_1 = \left\{ \{\beta_i\}_{i=1}^{2n} \in R^{2n} \,\middle|\, \sum_{i=1}^{2n} \beta_i^2 = 1 \right\}$, satisfying

$$\varphi_1(-x) = -\varphi_1(x) \qquad (x \in \mathfrak{S}_1). \tag{6.15}$$

Consequently,

$$\chi = \varphi_1 \psi \varphi_2^{-1} \tag{6.16}$$

is a continuous mapping of Σ_2 into Σ_1, satisfying

$$\chi(-z) = -\chi(z) \qquad (z \in \Sigma_2), \tag{6.17}$$

which is impossible by virtue of a well known theorem of K. Borsuk **). This proves lemma 6.1 in the hypothesis that E_1 is strictly convex.

In view of reducing the general case to the case when E_1 is strictly convex, we shall show first that for every $\varepsilon > 0$ one can construct on E_1 a new norm $\|x\|_0$ such that E_1 endowed with this norm be strictly convex and that we have

$$\|x\| \leqslant \|x\|_0 \leqslant (1+\varepsilon)\|x\| \qquad (x \in E_1). \tag{6.18}$$

*) Since the case when E is real is treated in a similar way, we omit it.
**) See K. Borsuk [21], or M. A. Krasnoselsky [123].

Indeed, since $\dim E_1 = \dim [G_1, G_2] < \infty$, there exists on E_1 a strictly convex norm $\|x\|_1$, e.g. $\left\| \sum_{i=1}^{m} \alpha_i z_i \right\|_1 = \sqrt{\sum_{i=1}^{m} |\alpha_i|^2}$, where $z_1, \ldots, z_m$ is an arbitrary basis of E_1. Then for an arbitrary $\delta > 0$ the norm $\|x\|_0$ on E_1 defined by

$$\|x\|_0 = \|x\| + \delta \|x\|_1 \qquad (x \in E_1) \tag{6.19}$$

is also strictly convex, since for every $x_1, x_2, \in E_1$ with $x_1 \neq \alpha x_2$ $(\alpha > 0)$ we have

$$\|x_1 + x_2\|_0 = \|x_1 + x_2\| + \delta \|x_1 + x_2\|_1 < \|x_1\| + \|x_2\| + \\ + \delta (\|x_1\|_1 + \|x_2\|_1) = \|x_1\|_0 + \|x_2\|_0.$$

On the other hand, putting $\gamma = \max_{\substack{z \in E_1 \\ \|z\|=1}} \|z\|_1$, we have $\left\| \frac{x}{\|x\|} \right\|_1 \leqslant \gamma$ $(x \in E)$, hence

$$\|x\|_1 \leqslant \gamma \|x\| \qquad (x \in E_1),$$

whence for $\varepsilon = \delta\gamma$ we obtain, by (6.19),

$$\|x\| \leqslant \|x\|_0 \leqslant \|x\| + \delta\gamma \|x\| = (1 + \varepsilon)\|x\| \quad (x \in E_1),$$

and thus we have (6.18).

Applying this remark for $\varepsilon_n = \frac{1}{n}$ $(n = 1, 2, \ldots)$ and taking into account that the lemma has been already proved for strictly convex E_1 it follows that for every n there exists a $y_n \in G_2$ with $\|y_n\| = 1$ such that

$$\|y_n\| \leqslant \|y_n\|_n \leqslant \|y_n + g\|_n \leqslant \left(1 + \frac{1}{n}\right) \|y_n + g\| \qquad (g \in G_1),$$

(where $\|x\|_n$ is a strictly convex norm on E_1 satisfying (6.18) with $\varepsilon = \frac{1}{n}$). Choosing then a convergent subsequence $y_{n_k} \to y \in G_2$ we obtain $\|y\| = 1$ and

$$\|y\| \leqslant \|y + g\| \qquad (g \in G_1),$$

i.e. (6.5), which completes the proof of lemma 6.1.

The condition $\dim G_1 < \infty$ in lemma 6.1 is essential. Indeed, as has been observed by A. Yu. Levin [134], there exists a Banach space E having two linear subspaces G_1 and G_2 with *)

---

*) For an infinite dimensional normed linear space X the dimension of X is defined as the smallest of the cardinalities of dense subsets of X.

$$\dim G_1 = \aleph_0, \ \dim G_2 = \mathfrak{c}, \tag{6.20}$$

such that no $y \in G_2$ satisfies (6.5). In the case when

$$\dim G_1 = \infty, \ \dim G_1 < \dim G_2, \tag{6.21}$$

M. G. Krein, M. A. Krasnoselsky and D. P. Milman [126] have proved that for every $\varepsilon > 0$ there exists a $y \in G_2 \setminus \{0\}$ such that

$$\|y - g\| \geqslant \left(\frac{1}{2} - \varepsilon\right) \|y\| \qquad (g \in G_1), \tag{6.22}$$

but up to the present it is not known whether there exists also a $y \in G_2 \setminus \{0\}$ such that

$$\|y - g\| \geqslant \frac{1}{2} \|y\| \qquad (g \in G_1). \tag{6.23}$$

From lemma 6.1 it results the following property of the opening $\theta(G_1, G_2)$ (defined in Chap. I, formula (6.81)), given by M. G. Krein, M. A. Krasnoselsky and D. P. Milman [126]:

COROLLARY 6.1. *Let E be a normed linear space and G_1, G_2 two linear subspaces of E such that*

$$\dim G_1 < \infty, \qquad \theta(G_1, G_2) < 1. \tag{6.24}$$

Then we have

$$\dim G_1 = \dim G_2. \tag{6.25}$$

Proof. If $\dim G_1 < \dim G_2$, then by virtue of lemma 6.1 there exists a $y_0 \in G_2$ with $\|y_0\| = 1$ such that

$$1 \geqslant \rho(y_0, G_1) \geqslant 1,$$

hence we have

$$\sup_{\substack{y \in G_2 \\ \|y\| = 1}} \rho(y, G_2) = 1,$$

whence

$$\theta(G_1, G_2) = 1. \tag{6.26}$$

On the other hand, if $\dim G_2 < \dim G_1$, then by (6.24) we have $\dim G_2 < \infty$, whence changing the roles of G_1 and G_2 in the argument above, we obtain again (6.26), which completes the proof.

Let us also mention that to the result (6.22) above there corresponds the following property of the opening $\theta(G_1, G_2)$, given by M. G. Krein, M. A. Krasnoselsky and D. P. Milman [126] : *If G_1, G_2 are linear subspaces of a normed linear space E, such that*

$$\theta(G_1, G_2) < \frac{1}{2}, \tag{6.27}$$

then $\dim G_1 = \dim G_2$. Although for Hilbert spaces the constant $\frac{1}{2}$ of (6.27) may be replaced by 1 (M. G. Krein, M. A. Krasnoselsky and D. P. Milman [126]), in the general case it is not known even whether it can be replaced by a constant $\alpha > \frac{1}{2}$.

Finally, we give the following result of H. Auerbach *), which will be used in section 6.3 :

LEMMA 6.2. *Let E_n be a Banach space of dimension n ($<\infty$). Then there exist n linearly independent elements $x_1, \ldots, x_n \in E_n$ and n functionals $f_1, \ldots, f_n \in E_n^*$ such that*

$$\|x_k\| = \|f_k\| = 1 \qquad (k = 1, \ldots, n), \tag{6.28}$$

$$f_i(x_k) = \begin{cases} 1 \text{ for } i = k \\ 0 \text{ for } i \neq k \end{cases} \qquad (i, k = 1, \ldots, n). \tag{6.29}$$

Consequently, for every $x = \sum_{i=1}^{n} \alpha_i x_i \in E_n$ we have then

$$|\alpha_i| \leqslant \|x\| \qquad (i = 1, \ldots, n). \tag{6.30}$$

Proof. Introduce in E_n a system of cartesian coordinates and let

$$D(y_1, \ldots, y_n) = \det (\eta_i^{(k)})_{i, k = 1, \ldots, n}, \tag{6.31}$$

where

$$y_k = \{\eta_1^{(k)}, \ldots, \eta_n^{(k)}\} \in E_n \qquad (k = 1, \ldots, n).$$

Since $\dim E_n < \infty$, the set $\mathrm{Fr}\, S_E = \{x \in E \mid \|x\| = 1\}$, whence also the set $(\mathrm{Fr}\, S_E)^n = \mathrm{Fr}\, S_E \times \ldots \times \mathrm{Fr}\, S_E$ (n times), is compact. Consequently, since $D(y_1, \ldots, y_n)$ is continuous on $(\mathrm{Fr}\, S_E)^n$, we can choose $x_1, \ldots, x_n \in E_n$ with $\|x_k\| = 1$ ($k = 1, \ldots$

*) In the particular case when the scalars are real, this result has been given in the monograph of S. Banach ([8], p. 238), with the mention that it is due to H. Auerbach; again for real scalars, the result has been rediscovered by A. E. Taylor ([245], theorem 2). The above proof, valid for real or complex scalars, is in A. F. Timan ([250], p. 407) and A. F. Ruston [203].

$\dots, n)$, which maximize $|D(y_1, \dots, y_n)|$. Such a choice being made, put

$$f_i(x) = \frac{D(x_1, \dots, x_{i-1}, x, x_{i+1}, \dots, x_n)}{D(x_1, \dots, x_n)} \quad (x \in E_n;\ i = 1, \dots, n). \tag{6.32}$$

We then have (6.29) and, by the choice of $x_1, \dots, x_n$,

$$\|f_i\| = \sup_{\substack{x \in E_n \\ \|x\|=1}} |f_i(x)| = \sup_{\|x\|=1} \left| \frac{D(x_1, \dots, x_{i-1}, x, x_{i+1}, \dots, x_n)}{D(x_1, \dots, x_n)} \right| =$$

$$= 1 = \|x_i\| \qquad (i = 1, \dots, n),$$

whence (6.28). Finally if $x = \sum_{i=1}^{n} \alpha_i x_i$, then by (6.29) and (6.28) we have

$$|\alpha_i| = |f_i(x)| \leqslant \|x\|,$$

hence (6.30), which completes the proof of lemma 6.2.

6.2. n-DIMENSIONAL DIAMETERS

We shall first give some elementary properties of n-dimensional diameters, collected in

THEOREM 6.1. *Let E be a normed linear space, A a set in E and n an integer with $0 \leqslant n < \infty$. Then*

a) *We have*

$$d_n(\bar{A}) = d_n(A). \tag{6.33}$$

b) *If A_1 is a set in E, we have*

$$d_n(A_1) - \delta(A_1, A) \leqslant d_n(A); \tag{6.34}$$

consequently, if $A_1 \subset A$, we have

$$d_n(A_1) \leqslant d_n(A). \tag{6.35}$$

c) *For every scalar α we have*

$$d_n(\alpha A) = |\alpha| d_n(A). \tag{6.36}$$

d) *For the circled hull $\tau(A)$ of A we have*

$$d_n[\tau(A)] = d_n(A). \tag{6.37}$$

e) *We have*

$$d_n(\text{co}\, A) = d_n(A). \tag{6.38}$$

f) *We have*

$$d_0(A) \geqslant d_1(A) \geqslant \ldots \geqslant d_n(A) \geqslant \ldots \qquad (6.39)$$

g) *If A is compact, we have*

$$\lim_{n\to\infty} d_n(A) = 0. \qquad (6.40)$$

h) *If* $\dim l(A \cup \{0\}) = n$, *we have*

$$d_n(A) = d_{n+1}(A) = \ldots = 0. \qquad (6.41)$$

Proof. a) — f) are immediate consequences of the corresponding results of Chap. I, theorem 6.10.

g) Let A be compact and let $\varepsilon > 0$ be arbitrary. Take an ε-net $\{x_1, \ldots, x_N\}$ for A and let $G = [x_1, \ldots, x_N]$. Then for every $x \in A$ we have

$$e_G(x) \leqslant \min_{1\leqslant i\leqslant N} \|x - x_i\| < \varepsilon,$$

hence

$$\delta(A, G) = \sup_{x\in A} e_G(x) \leqslant \varepsilon,$$

and thus

$$d_N(A) = \inf_{\dim G = N} \delta(A, G) \leqslant \varepsilon,$$

whence, by (6.39), there follows (6.40).

h) If $\dim l(A \cup \{0\}) = n$, we have

$$d_n(A) = \inf_{\dim G = n} \delta(A, G) \leqslant \delta(A, l(A \cup \{0\})) = 0,$$

which, taking into account (6.39), completes the proof.

The properties a)—e) of theorem 6.1 have been given by A. L. Brown ([27], p. 585), and f)—h) by V. M. Tihomirov ([247], p. 82—83).

LEMMA 6.3. *Let E be a normed linear space and A a bounded closed circled convex set in E, such that $0 \in \operatorname{Int} A$. Then we have*

$$\sup_{\substack{\lambda>0\\ \lambda S_E\subset A}} \lambda = \inf_{x\in \operatorname{Fr} A} \|x\|. \qquad (6.42)$$

Proof. Let $\lambda > 0$ be such that $\lambda S_E \subset A$. Then for the Minkowsky functional*) $p = p_{A,0}$ of A we have

$$p(y) \leqslant \frac{1}{\lambda} \|y\| \qquad (y \in E),$$

*) See § 1, section 1.1.

whence, taking into account $p(x) = 1$ $(x \in \operatorname{Fr} A)$, we obtain

$$\lambda \leqslant \|x\| \qquad (x \in \operatorname{Fr} A),$$

and thus

$$\sup_{\substack{\lambda>0 \\ \lambda S_E \subset A}} \lambda \leqslant \inf_{x \in \operatorname{Fr} A} \|x\|. \tag{6.43}$$

On the other hand, for every $y \in S_E$ we have

$$\|(\inf_{x \in \operatorname{Fr} A} \|x\|)y\| \leqslant \inf_{x \in \operatorname{Fr} A} \|x\|,$$

whence, taking into account $\inf_{x \in \operatorname{Fr} A} \|x\| \leqslant \left\| \frac{z}{p(z)} \right\|$ $(z \in E)$, we obtain (for $z = (\inf_{x \in \operatorname{Fr} A} \|x\|)y$) $p[(\inf_{x \in \operatorname{Fr} A} \|x\|)y] \leqslant 1$, hence

$$(\inf_{x \in \operatorname{Fr} A} \|x\|)y \in A \qquad (y \in S_E),$$

and thus

$$\sup_{\substack{\lambda>0 \\ \lambda S_E \subset A}} \lambda \geqslant \inf_{x \in \operatorname{Fr} A} \|x\|. \tag{6.44}$$

From (6.43) and (6.44) there follows (6.42), which completes the proof.

Lemma 6.3 is well known (see e.g. A. L. Brown [27], p. 586); although in the sequel we shall not apply directly this lemma, it presents interest because of theorem 6.2 below.

Denoting (6.42) by $r(A)$, we have

$$r(A)S_E \subset A, \tag{6.45}$$

and in the case when E is of finite dimension, the inf in (6.42) is obviously attained, hence we also have

$$\operatorname{Fr} r(A)S_E \cap \operatorname{Fr} A \neq \emptyset. \tag{6.46}$$

The set $r(A)S_E$ is nothing else than the "cell inscribed" in A, i.e. the largest cell with the center at the origin, contained in the set A, and $r(A)$ is the radius of the cell inscribed in A.

THEOREM 6.2 (A. L. Brown [27], lemma 5). *Let E_{n+1} be an $(n+1)$-dimensional Banach space and A a bounded* *) *closed circled convex set in E_{n+1} with $0 \in \operatorname{Int} A$. Then we have*

$$d_n(A, E_{n+1}) = r(A), \tag{6.47}$$

where $r(A)$ is the number (6.42).

*) Here, as well as in corollary 6.2, theorem 6.5 and theorem 6.6, the hypothesis of the boundedness of A is made for the sake of simplicity; it can be shown that the same results remain also valid under the more general hypothesis $d_n(A, E_{n+1}) < \infty$.

Proof. By (6.46) there exists an $x \in \operatorname{Fr} r(A) S_{E_{n+1}} \cap \operatorname{Fr} A$. Then, since $x \in \operatorname{Fr} A$, there exists a functional $f \in E_{n+1}^* \setminus \{0\}$ such that

$$f(x) = \sup_{y \in A} |f(y)|; \tag{6.48}$$

we may assume $\left(\text{multiplying, if necessary, by } \frac{1}{\|f\|}\right)$ that we have $\|f\| = 1$. Let G be the hyperplane

$$G = \{z \in E_{n+1} \,|\, f(z) = 0\}. \tag{6.49}$$

Then, by (6.48), $\|f\| = 1$ and Chap. I, lemma 1.2, we have

$$\rho(x, G) = \sup_{y \in A} \rho(y, G) = \delta(A, G). \tag{6.50}$$

Also, by $\|f\|=1$, (6.48), (6.45) and $x \in \operatorname{Fr} r(A) S_{E_{n+1}}$ we have

$$\|x\| \geqslant |f(x)| = \sup_{y \in A} |f(y)| \geqslant \sup_{y \in r(A) S_{E_{n+1}}} |f(y)| = r(A) = \|x\|,$$

hence $\rho(x, G) = |f(x)| = r(A)$, whence, by (6.50),

$$r(A) = \delta(A, G), \tag{6.51}$$

and thus

$$r(A) \geqslant d_n(A, E_{n+1}). \tag{6.52}$$

On the other hand, since for every n-dimensional linear subspace G' of E there exists, by Chap. I, corollary 2.2 and theorem 2.1, a $y \in E_{n+1} \setminus \{0\}$ such that $y \perp G'$, we have

$$r(A) = \delta(r(A) S_{E_{n+1}}, G') \leqslant \delta(A, G'),$$

whence, since G' was an arbitrary n-dimensional linear subspace,

$$r(A) \leqslant d_n(A, E_{n+1}). \tag{6.53}$$

From (6.52) and (6.53) there follows (6.47), which completes the proof of theorem 6.2.

The fundamental result of the present section, having numerous applications to the evaluation of n-dimensional diameters of sets in concrete spaces, is

THEOREM 6.3 (V. M. Tihomirov [247], theorem 1). *Let E be a normed linear space, n an integer with $0 \leqslant n < \infty$ and E_{n+1} an $(n+1)$-dimensional linear subspace of E. Then we have*

$$d_n(S_{E_{n+1}}, E) = 1. \tag{6.54}$$

Proof. Let G be an arbitrary n-dimensional linear subspace of E. Then, applying lemma 6.1 for $G_1 = G$, $G_2 = E_{n+1}$, it follows that there exists a $y_0 \in E_{n+1} \setminus \{0\}$ such that $y_0 \perp G$, hence

$$\left\| \frac{y_0}{\|y_0\|} - g \right\| \geqslant 1 \qquad (g \in G).$$

Consequently, we have

$$1 \geqslant \delta(S_{E_{n+1}}, G) \geqslant \rho\left(\frac{y_0}{\|y_0\|}, G\right) = 1,$$

hence $\delta(S_{E_{n+1}}, G) = 1$, whence, since G was an arbitrary n-dimensional linear subspace of E,

$$d_n(S_{E_{n+1}}, E) = \inf_{\dim G = n} \delta(S_{E_{n+1}}, G) = 1,$$

which completes the proof of theorem 6.3.

From theorems 6.1 f) and 6.3 it follows that we have

$$1 \geqslant d_0(S_{E_{n+1}}, E) \geqslant d_1(S_{E_{n+1}}, E) \geqslant \ldots \geqslant d_n(S_{E_{n+1}}, E) = 1,$$

whence

$$d_0(S_{E_{n+1}}, E) = d_1(S_{E_{n+1}}, E) = \ldots = d_n(S_{E_{n+1}}, E) = 1. \quad (6.55)$$

On the other hand, since for every n-dimensional linear subspace G of E_{n+1} there exists, by Chap. I, corollary 2.2 and theorem 2.1, a $y \in E_{n+1} \setminus \{0\}$ such that $y \perp G$, it follows that we have *) $d_n(S_{E_{n+1}}, E_{n+1}) = 1$, whence

$$d_0(S_{E_{n+1}}, E_{n+1}) = d_1(S_{E_{n+1}}, E_{n+1}) = \ldots = d_n(S_{E_{n+1}}, E_{n+1}) = 1. \quad (6.56)$$

Consequently, theorem 6.3 may be interpreted also as follows: *the k-dimensional diameter of an $(n+1)$-dimensional cell $S_{E_{n+1}}$, where $0 \leqslant k \leqslant n$, does not decrease when the cell $S_{E_{n+1}}$ is embedded into a space E of higher dimension.*

For $k = 0$, the statement obviously remains valid when the cell $S_{E_{n+1}}$ is replaced by an arbitrary set $A \subset E_{n+1}$ (since by definition (6.1) we have $d_0(A, E) = \sup_{x \in A} \|x\|$). But in the case when $k > 0$, the situation is different; indeed, as has observed V. M. Tihomirov ([247], p. 118), there exist a Banach space E_4 of dimension 4 and a circled set $A \subset E_4$ such that for a suitable 3-dimensional linear subspace $E_3 \subset E_4$ with $A \subset E_3$ we have $d_1(A, E_3) > d_1(A, E_4)$. However, for $k = n$ the above statement is also valid when the cell $S_{E_{n+1}}$ is replaced by an arbitrary bounded set $A \subset E_{n+1}$, as shown by

*) This follows also from theorem 6.3 applied in the space $E = E_{n+1}$.

COROLLARY 6.2. *Let E be a normed linear space, n an integer with $0 \leqslant n < \infty$ and E_{n+1} an $(n+1)$-dimensional linear subspace of E. Then for every bounded set $A \subset E_{n+1}$ we have*

$$d_n(A, E) = d_n(A, E_{n+1}). \tag{6.57}$$

Proof. If the linear subspace $[A]$ spanned by A does not coincide with E_{n+1}, the statement is obvious, since the diameters in (6.57) are equal to zero; consequently, we may assume that $[A] = E_{n+1}$. Also, by virtue of theorem 6.1 a), d), e) we may assume that A is convex, circled and closed; then, by $[A] = E_{n+1}$, we also have $0 \in \operatorname{Int} A$. By virtue of (6.45) for E_{n+1} and of theorem 6.1 b) we have then

$$d_n(r(A) S_{E_{n+1}}, E) \leqslant d_n(A, E). \tag{6.58}$$

But by theorems 6.3, 6.1 c) and 6.2 we have

$$d_n(r(A) S_{E_{n+1}}, E) = r(A) = d_n(A, E_{n+1}),$$

which, together with (6.58), gives

$$d_n(A, E_{n+1}) \leqslant d_n(A, E).$$

Since the opposite inequality is always true (by (6.3)), it follows that we have (6.57), which completes the proof.

Corollary 6.2 has been given, essentially, by A. L. Brown ([27], p. 587).

Another theorem which is useful for the lower evaluation of n-dimensional diameters of sets in some concrete spaces is

THEOREM 6.4. *Let $E = l_R^\infty(T)$, where T is an arbitrary set and let A be a set in E with the property that there exist an $\alpha > 0$ and elements $t_0, t_1, \ldots, t_n \in T$ such that for every set of indices $J \subset \{0, 1, \ldots, n\}$ we can find an $x \in A$ satisfying*

$$x(t_j) \geqslant \alpha \qquad (j \in J), \tag{6.59}$$

$$x(t_j) \leqslant -\alpha \qquad (j \in \{0, 1, \ldots, n\} \setminus J). \tag{6.60}$$

Then we have

$$d_n(A, E) \geqslant \alpha. \tag{6.61}$$

Proof. The mapping

$$u : x \to \{x(t_0), x(t_1), \ldots, x(t_n)\} \qquad (x \in E)$$

of E into the $(n+1)$-dimensional real Banach space $(l_{n+1}^\infty)_R$ is obviously of norm $\|u\| \leqslant 1$, hence for every n-dimensional linear subspace G of E we have

$$\delta(A, G) = \sup_{x \in A} \rho(x, G) \geqslant \sup_{x \in A} \rho(u(x), u(G)) = \\ = \delta(u(A), u(G)),$$

and thus

$$d_n(A, E) \geqslant d_n(u(A), u(E)).$$

Consequently, it will be sufficient to prove that *if a set* B *in the space* $\widetilde{E} = (l_{n+1}^\infty)_R$ *has the property that there exists a scalar* $\alpha > 0$ *such that for every set of indices* $J \subset \{0, 1, \ldots, n\}$ *we can find* $y = \{\eta_j\} \in B$ *satisfying*

$$\eta_j \geqslant \alpha \qquad (j \in J), \tag{6.62}$$

$$\eta_j \leqslant -\alpha \qquad (j \in \{0, 1, \ldots, n\} \setminus J), \tag{6.63}$$

then

$$d_n(B, \widetilde{E}) \geqslant \alpha. \tag{6.64}$$

It will be sufficient for this purpose to show that for every n-dimensional linear subspace G of $\widetilde{E} = (l_{n+1}^\infty)_R$ we have

$$\delta(B, G) \geqslant \alpha. \tag{6.65}$$

Let D be one of the two closed half-spaces of $\widetilde{E}$ determined by the hyperplane G and let $\{e_0, e_1, \ldots, e_n\}$ be the unit vector basis of $\widetilde{E}$. Let $\mathcal{C}$ be the convex cone with vertex at 0 generated by $\{e_i \mid e_i \in D\} \cup \{-e_i \mid e_i \notin D\}$, and let $J = \{i \mid e_i \in D\}$. For this set of indices J there exists, by our hypothesis, a $y = \{\eta_j\} \in B$ satisfying (6.62) and (6.63). We have then

$$y = \sum_{i=0}^{n} \eta_i e_i = \sum_{e_i \in D} \eta_i e_i + \sum_{e_i \notin D} (-\eta_i)(-e_i),$$

whence, taking into account (6.62) and (6.63), we obtain $y \in \mathcal{C}$. Since $\mathcal{C} \subset D$, we thus have

$$\rho(y, G) \geqslant \rho(y, \operatorname{Fr} \mathcal{C}). \tag{6.66}$$

On the other hand, since for every $z = \{\zeta_j\} \in \operatorname{Fr} \mathcal{C}$ at least one coordinate ζ_{j_0} is equal to 0, we have

$$\|y - z\| = \max_{0 \leqslant j \leqslant n} |\eta_j - \zeta_j| \geqslant |\eta_{j_0}| \geqslant \alpha,$$

whence

$$\rho(y, \operatorname{Fr} \mathcal{C}) \geqslant \alpha. \tag{6.67}$$

From (6.66) and (6.67) we obtain

$$\delta(B, G) \geqslant \rho(y, G) \geqslant \rho(y, \operatorname{Fr} \mathfrak{C}) \geqslant \alpha,$$

whence (6.65), which completes the proof.

Theorem 6.4 has been given by A. N. Kolmogorov (see V. M. Tihomirov [247], theorem 2).

Some authors, e.g. V. M. Tihomirov [247] and A. L. Garkavi [61], [65] use a modified definition of $d_n(A, E)$, where the inf in (6.1) is taken over all the n-dimensional linear manifolds V of E, i.e.

$$d'_n(A, E) = \inf_{\dim V = n} \delta(A, V) = \inf_{\dim V = n} \sup_{x \in A} \inf_{v \in V} \|x - v\|. \qquad (6.68)$$

Since every linear subspace is a linear manifold, we have

$$d'_n(A, E) \leqslant d_n(A, E). \qquad (6.69)$$

Obviously, we may also have in (6.69) the sign $<$, e.g. for $A = \{x\}$, where $x \neq 0$, we have $d'_0(A, E) = 0$, $d_0(A, E) = \|x\|$. Therefore it presents interest the remark (made, in the case of real scalars, by V. M. Tihomirov [247], p. 82—83) that *for circled sets A we have*

$$d'_n(A, E) = d_n(A, E). \qquad (6.70)$$

Indeed, let V be an arbitrary n-dimensional linear manifold in E. Then $V = z + G$, where $z \in E$ and G is an n-dimensional linear subspace of E. If $V \neq G$, then $z \notin G$, hence V is a hyperplane in the linear subspace $F_z = G \oplus [z]$ of E and thus we can write

$$V = \{y \in F_z \mid \varphi(y) = \alpha\},$$

where $\varphi \in (F_z)^*$ and α is a scalar $\neq 0$. Consequently, for $x \in A$ and $\beta = -\dfrac{\operatorname{sign} \varphi(x)}{\operatorname{sign} \alpha}$ (hence $|\beta| = 1$) we have, by Chap. I, lemma 1.2,

$$\rho(\beta x, V) = \frac{|\varphi(\beta x) - \alpha|}{\|\varphi\|} = \frac{|\varphi(x)| + |\alpha|}{\|\varphi\|} > \frac{|\varphi(x)|}{\|\varphi\|} = \rho(x, G),$$

whence, since A is circled, we obtain $\delta(A, V) > \rho(x, G)$, hence

$$\delta(A, V) \geqslant \delta(A, G),$$

and thus, since V was an arbitrary n-dimensional linear manifold,

$$d'_n(A, E) \geqslant d_n(A, E). \qquad (6.71)$$

But, from (6.69) and (6.71) it follows that we have (6.70), which completes the proof.

Sometimes, instead of the n-dimensional diameters $d_n(A, E)$, $d'_n(A, E)$ above, one considers *the n-dimensional linear diameter* $d''_n(A, E)$ defined by

$$d''_n(A, E) = \inf_{\dim G = n} \delta''(A, G), \qquad (6.72)$$

where the inf is taken over all n-dimensional linear subspaces (or sometimes over all n-dimensional linear manifolds) G of E and where δ'' is the linear deviation defined in Chap. I, formula (6.74). From Chap. I, formula (6.76) it follows that we always have

$$d''_n(A, E) \geqslant d_n(A, E). \qquad (6.73)$$

As has shown V. M. Tihomirov ([247], p. 118—119), in general we may also have in (6.73) the sign $>$.

We shall not give here the known results on the exact value or on the evaluations of n-dimensional diameters of various classes of functions in concrete functional spaces (just as we did not give the known results on the values $e_G(x)$ for concrete E, G, x), since this would be beyond the scope of the present monograph. These results are presented in the papers of V. M. Tihomirov [247], [249]; see also G. G. Lorentz [139].

If the set A is convex, the rate of decrease of the diameters $d_n(A)$ may be considered as a natural measure of the size of A. For compact sets A a different measure of their size has been also studied, namely the rate of increase of the "ε-entropy" $H_\varepsilon(A)$, introduced by A. N. Kolmogorov [119]; see e.g. A. F. Timan [250], Chap. III, A. G. Vituškin [258] and the expository paper of A. N. Kolmogorov and V. M. Tihomirov [120]. The relations between diameters and ε-entropy have been studied by Yu. A. Brudnyĭ and A. F. Timan [28]; see also the literature quoted in G. G. Lorentz [140], § 16.

6.3. BEST n-DIMENSIONAL SECANTS

We have the following useful characterization of extremal hyperplanes in finite dimensional spaces:

THEOREM 6.5 (A. L. Brown [27], lemma 5). *Let E_{n+1} be an $(n+1)$-dimensional Banach space and A a bounded closed circled convex set in E_{n+1}, with $0 \in \operatorname{Int} A$. An n-dimensional linear subspace G of E_{n+1} is a best n-dimensional secant of A with respect to E_{n+1} if and only if there exists an $x \in \operatorname{Fr} r(A) S_{E_{n+1}} \cap \cap \operatorname{Fr} A$, where $r(A)$ is the number (6.42), such that $H = x + G$ supports* *) *both A and $r(A) S_{E_{n+1}}$.*

*) See Chap. I, section 1.2 (the definition given there applies also to A, since A is the unit cell in the space E_{n+1} endowed with the norm $p = p_{A,0}$ = the Minkowsky functional of A).

Proof. Let $G = \{z \in E_{n+1} | f(z) = 0\}$, where $\|f\| = 1$, be a best n-dimensional secant of A with respect to E_{n+1}. Then by theorem 6.2 we have

$$\sup_{y \in A} \rho(y, G) = \delta(A, G) = d_n(A, E_{n+1}) = r(A).$$

Consequently, since A is compact and since the functional $e_G(y) = \rho(y, G)$ is continuous (by Chap. I, theorem 6.5 f)), there exists an $x \in A$ such that

$$\rho(x, G) = \sup_{y \in A} \rho(y, G) = r(A). \tag{6.74}$$

By Chap. I, lemma 1.2, we then have

$$\rho(0, x + G) = |f(x)| = \rho(x, G) = r(A),$$

and hence, by Chap. I, lemma 1.3, $x + G$ supports the cell $r(A)S_{E_{n+1}}$.

On the other hand, by (6.74) and Chap. I, lemma 1.2, we have

$$|f(x)| = \sup_{y \in A} |f(y)|,$$

whence, denoting by ρ_1 the distance induced on E_{n+1} by the Minkowsky functional $p = p_{A,0}$ of *) A, we have, again by Chap. I, lemma 1.2,

$$\rho_1(0, x + G) = \frac{|f(x)|}{\sup_{y \in A} |f(y)|} = 1,$$

and thus, by Chap. I, lemma 1.3, $x + G$ supports the "unit cell" $\{y \in E_{n+1} | p(y) = \rho_1(y, 0) \leqslant 1\} = A$.

Conversely, let $G = \{z \in E_{n+1} | f(z) = 0\}$, where $\|f\| = 1$, by an n-dimensional linear subspace of E_{n+1} with the property that there exists an $x \in \operatorname{Fr} r(A)S_{E_{n+1}} \cap \operatorname{Fr} A$ such that $x + G$ supports both A and $r(A)S_{E_{n+1}}$. Since $x + G$ supports $r(A)S_{E_{n+1}}$, we have, by Chap. I, lemmas 1.2 and 1.3,

$$|f(x)| = \rho(0, x + G) = r(A).$$

On the other hand, since $x + G$ supports A, we have, again by Chap. I, lemmas 1.2 and 1.3,

$$\frac{|f(x)|}{\sup_{y \in A} |f(y)|} = \rho_1(0, x + G) = 1,$$

*) I.e. $\rho_1(x, y) = p(x - y)$.

where by ρ_1 we have denoted the distance induced on E_{n+1} by the Minkowsky functional $p = p_{A,0}$ of A. Consequently, taking into account theorem 6.2 and Chap. I, lemma 1.2, we obtain

$$d_n(A, E_{n+1}) = r(A) = |f(x)| = \sup_{y \in A} |f(y)| = \sup_{y \in A} \rho(y, G) = \\ = \delta(A, G),$$

and thus G is a best n-dimensional secant of A with respect to E_{n+1}, which completes the proof of theorem 6.5.

Let us consider now the problem of existence of best n-dimensional secants. The proof of theorem 6.2 (equalities (6.51) and (6.47)) shows that for every bounded closed circled convex set $A \subset E_{n+1}$ with $0 \in \operatorname{Int} A$ there exists a best n-dimensional secant of A with respect to E_{n+1}. However, using corollary 6.2 one can prove more; namely, we have

THEOREM 6.6 (A. L. Brown [27], theorem 6). *Let E be a normed linear space, n an integer with $0 \leqslant n < \infty$ and E_{n+1} an $(n+1)$-dimensional linear subspace of E. Then for every bounded set $A \subset E_{n+1}$ there exists an n-dimensional linear subspace $G \subset E_{n+1}$ which is a best n-dimensional secant of A with respect to E.*

Proof. If the linear subspace $[A]$ spanned by A does not coincide with E_{n+1}, the statement is obvious since every n-dimensional linear subspace $G \subset E_{n+1}$ containing A is a best n-dimensional secant of A with respect to E; consequently, we may assume that $[A] = E_{n+1}$. Also, by theorem 6.1 a), d), e) and Chap. I, theorem 6.10 a), d), e), we may assume that A is convex, circled, and closed; then, by $[A] = E_{n+1}$, we have also $0 \in \operatorname{Int} A$. Consequently, by virtue of the remark made before the statement of the theorem, there exists a best n-dimensional secant $G \subset E_{n+1}$ of A with respect to E_{n+1}. But, by corollary 6.2, G is then also a best n-dimensional secant of A with respect to E, which completes the proof of theorem 6.6.

Obviously, every n-dimensional linear subspace $G \subset E_{n+1}$ is the unique best n-dimensional secant with respect to E of a suitable set $A \subset E_{n+1}$, e.g. of any set $A \subset G$ with $[A] = G$. Also, as has observed A. L. Brown ([27], theorem 7), for any n-dimensional linear subspace $G \subset E_{n+1}$, any $n+1$ points $y_1, \ldots, y_{n+1} \in G$ with the property that 0 is in the interior with respect to G of the set co $\{y_1, \ldots, y_{n+1}\}$ and any $x \in E_{n+1} \setminus G$, there exists a $\mu \neq 0$ such that G is the unique best n-dimensional secant with respect to E_{n+1} of the set of $n+1$ linearly independent points $A = \{\mu x + y_i | i = 1, \ldots, n+1\}$.

THEOREM 6.7 (A. L. Garkavi [65], theorem 7). *Let E be a normed linear space. Then for every set A in the conjugate space E^*, with $d_n(A) < \infty$, there exists a best n-dimensional secant of A with respect to E^*.*

Proof. Let $\{\Gamma_k\}$ be a sequence of n-dimensional linear subspaces of E^* such that

$$\sup_{f\in A} e_{\Gamma_k}(f) \leqslant d_n(A) + \frac{1}{k} \quad (k = 1, 2, \ldots). \tag{6.75}$$

In each Γ_k there exist, by lemma 6.2, n linearly independent elements $\gamma_1^k, \ldots, \gamma_n^k \in \Gamma_k$ with $\|\gamma_i^k\| = 1$ $(i = 1, \ldots, n)$ such that for every $\gamma^k = \sum_{i=0}^{n} a_i^k \gamma_i^k \in \Gamma_k$ we have

$$|a_i^k| \leqslant \|\gamma^k\| \qquad (i = 1, \ldots, n). \tag{6.76}$$

Take now an arbitrary $f \in A$ and let

$$\gamma^k = a_{f,1}^k \gamma_1^k + \ldots + a_{f,n}^k \gamma_n^k \in \mathfrak{P}_{\Gamma_k}(f) \; (k = 1, 2, \ldots) \tag{6.77}$$

be arbitrary. By (6.76) and Chap. I, theorem 6.1 b) we have then

$$|a_{f,j}^k| \leqslant \|\gamma^k\| \leqslant 2\|f\| \quad (j = 1, \ldots, n;\ k = 1, 2, \ldots). \tag{6.78}$$

Let Q be the product topological space

$$Q = S_{E^*,1} \times \ldots \times S_{E^*,n} \times \prod_{\varphi\in A} (I_{\varphi,1} \times \ldots \times I_{\varphi,n}), \tag{6.79}$$

where $S_{E^*,j} = S_{E^*}$ endowed with the weak topology $\sigma(E^*, E)$, and $I_{\varphi,j} = \{\zeta = \text{scalar} \mid |\zeta| \leqslant 2\|\varphi\|\}$ with its natural topology $(j = 1, \ldots, n)$. Since each factor in this product is compact, the space Q is compact. Consider the sequence $\{q_k\} \subset Q$, where

$$q_k = \{\gamma_1^k, \ldots, \gamma_n^k, a_{\varphi,1}^k, \ldots, a_{\varphi,n}^k (\varphi \in A)\} \quad (k = 1, 2, \ldots). \tag{6.80}$$

Since the space Q is compact, the sequence $\{q_k\}$ has at least one limit point, say

$$q = \{\gamma_1, \ldots, \gamma_n, a_{\varphi,1}, \ldots, a_{\varphi,n} (\varphi \in A)\}. \tag{6.81}$$

We shall show that then for every $f \in A$ we have

$$\|f - a_{f,1}\gamma_1 - \ldots - a_{f,n}\gamma_n\| \leqslant d_n(A), \tag{6.82}$$

whence, either the linear subspace $\Gamma = [\gamma_1, \ldots, \gamma_n]$ of E^*, or (if $\dim \Gamma < n$) every n-dimensional linear subspace of E^* containing Γ, is a best n-dimensional secant of A with respect to E^*, which will complete the proof.

Let $f \in A$, $x \in E$ with $\|x\| = 1$ and $\varepsilon > 0$ be arbitrary. By (6.75) and (6.77) we have then, for every $k \geqslant \frac{1}{\varepsilon}$,

$$|f(x) - a_{f,1}^k\gamma_1^k(x) - \ldots - a_{f,n}^k\gamma_n^k(x)| \leqslant \|f - a_{f,1}^k\gamma_1^k - \ldots \\ \ldots - a_{f,n}^k\gamma_n^k\| \leqslant d_n(A) + \varepsilon. \tag{6.83}$$

Consider the neighborhood $V = V(q)$ of the above point $q \in Q$ defined by

$$V = \{\{\gamma_{(1)}, \ldots, \gamma_{(n)}, a_{(\varphi, 1)}, \ldots, a_{(\varphi, n)} (\varphi \in A)\} \in Q \,|\, |\gamma_{(j)}(x) - \\ - \gamma_j(x)| < \varepsilon, |a_{(\varphi, j)} - a_{\varphi, j}| < \varepsilon \; (\varphi \in A; j = 1, \ldots, n)\}.$$

Since q is a limit point of the sequence $\{q_k\}$, there exists an index $k_0 > \frac{1}{\varepsilon}$ such that $q_{k_0} \in V$, hence such that

$$|\gamma_j^{k_0}(x) - \gamma_j(x)| < \varepsilon, |a_{\varphi, j}^{k_0} - a_{\varphi, j}| < \varepsilon \qquad (\varphi \in A; \; j = 1, \ldots, n). \tag{6.84}$$

Consequently, we have, taking into account (6.84), (6.78) and $\|x\| = 1 = \|\gamma_j\| \; (j = 1, \ldots, n)$,

$$|a_{f,1}^{k_0} \gamma_1^{k_0}(x) + \ldots + a_{f,n}^{k_0} \gamma_n^{k_0}(x) - a_{f,1}\gamma_1(x) - \ldots - a_{f,n}\gamma_n(x)| \leqslant$$

$$\leqslant \sum_{j=1}^{n} |a_{f,j}^{k_0} \gamma_j^{k_0}(x) - a_{f,j}\gamma_j(x)| \leqslant \sum_{j=1}^{n} |a_{f,j}^{k_0} \gamma_j^{k_0}(x) - a_{f,j}^{k_0}\gamma_j(x)| +$$

$$+ \sum_{j=1}^{n} |a_{f,j}^{k_0} \gamma_j(x) - a_{f,j}\gamma_j(x)| \leqslant n\varepsilon \,(\max_{1 \leqslant j \leqslant n} |a_{f,j}^{k_0}| + \max_{1 \leqslant j \leqslant n} |\gamma_j(x)|) \leqslant$$

$$\leqslant n\varepsilon(2\|f\| + 1),$$

which, together with (6.83) for $k = k_0$, gives

$$|f(x) - a_{f,1}\gamma_1(x) - \ldots - a_{f,n}\gamma_n(x)| \leqslant d_n(A) + \varepsilon + n\varepsilon\,(2\|f\| + 1).$$

Since $\varepsilon > 0$ was arbitrary, it follows that we have

$$|f(x) - a_{f,1}\gamma_1(x) - \ldots - a_{f,n}\gamma_n(x)| \leqslant d_n(A), \tag{6.85}$$

whence, since $x \in E$ with $\|x\| = 1$ and $f \in A$ were arbitrary, it follows that we have (6.82) for every $f \in A$, which completes the proof of theorem 6.7.

Applying the result of theorem 6.7, one can prove the following extension of this theorem:

THEOREM 6.8 (A. L. Garkavi [65], theorem 8). *Let E be a Banach space with the property that there exists a projection* *) *$p: E^{**} \to E$ of norm $\|p\| = 1$. Then for every set $A \subset E$ with $d_n(A) < \infty$ there exists a best n-dimensional secant of A with respect to E.*

Proof. By virtue of theorem 6.7 there exists a best n-dimensional secant $\Gamma = [\Psi_1, \ldots, \Psi_n] \subset E^{**}$ of A with respect to E^{**}. We claim that then either $G = p(\Gamma) \subset E$, or (if

*) We identify E with its image under the canonical embedding of E into E^{**}.

dim $G<n$) every n-dimensional linear subspace of E containing G, is a best n-dimensional secant of A with respect to E. Indeed, let $x \in A$ and $\Psi = a_1\Psi_1 + \ldots + a_n\Psi_n \in \mathfrak{L}_\Gamma(x)$ be arbitrary. Then we have, taking into account (6.2),

$$\|x - a_1p(\Psi_1) - \ldots - a_np(\Psi_n)\| = \|p(x - a_1\Psi_1 - \ldots - a_n\Psi_n)\| \leqslant$$
$$\leqslant \|x - a_1\Psi_1 - \ldots - a_n\Psi_n\| = e_\Gamma(x) \leqslant d_n(A, E^{**}) \leqslant d_n(A, E),$$

whence $\rho(x, G) \leqslant d_n(A, E)$ $(x \in A)$, which completes the proof.

Theorem 6.8 is effectively an extension of theorem 6.7, since for every normed linear space E there exists *) a projection $p: E^{***} \to E^*$ of norm $\|p\| = 1$ and since there exist normed linear spaces E satisfying the condition of theorem 6.8, which are not equivalent to any conjugate space (e.g. **) $E = L^1([0, 1], \nu)$, where ν is the Lebesgue measure).

In the spaces E which do not satisfy the condition of theorem 6.8, in general there exist bounded sets $A \subset E$ which have no best n-dimensional secant [65].

Furthermore, although for every finite set $A \subset C_R(Q)$ (Q compact) there exists a best n-dimensional secant (A. L. Brown [27], theorem 8), one can show that if Q admits a $y \in C_R(Q)$ such that $\overline{\{q \in Q \mid y(q) > 0\}} \cap \overline{\{q \in Q \mid y(q) < 0\}} \neq \emptyset$, then for every integer n with $1 \leqslant n < \infty$ there exists a subset $A \subset E_{n+3} \subset C_R(Q)$ (where E_{n+3} is a suitable $(n+3)$-dimensional linear subspace of $C_R(Q)$), which has no best n-dimensional secant with respect to $C_R(Q)$ (A. L. Brown [27], theorem 11).

6.4. BEST n-DIMENSIONAL $\mathfrak{V}$-SECANTS. ČEBYŠEV CENTERS. CLOSEST POINTS TO A SET. BEST n-NETS. BEST n-COVERINGS

Let E be a normed linear space, A a set in E and n an integer with $0 \leqslant n < \infty$. We shall say that an n-dimensional linear manifold $V \subset E$ is a *best n-dimensional $\mathfrak{V}$-secant* of A (with respect to E) if

$$\delta(A, V) = d'_n(A, E), \tag{6.86}$$

where $d'_n(A, E)$ is the "$\mathfrak{V}$-diameter" of A defined by (6.68). In the particular case when the set A is circled, the best n-dimensional $\mathfrak{V}$-secants of A coincide, by virtue of (6.70), with the best n-dimensional secants of A.

Theorems 6.7 and 6.8 on the existence of best n-dimensional secants can be extended, by a simple adaptation of the above proofs, to the case of best n-dimensional $\mathfrak{V}$-secants (A. L. Garkavi [65], theorems 7 and 8).

*) See e.g. N. Bourbaki [23], Chap. IV, p. 122, exercise 16 a).

**) See e.g. A. F. Ruston [202].

In the particular case when $n = 0$, the best 0-dimensional $\mathfrak{V}$-secants of a set $A \subset E$, i.e. the elements $y_0 \in E$ with the property

$$\sup_{x \in A} \|x - y_0\| = \inf_{y \in E} \sup_{x \in A} \|x - y\|, \tag{6.87}$$

are called *Čebyšev centers* of the set A. Obviously, every such point $y_0 \in E$ is the center of a "smallest cell" containing the set A, i.e. the center of a cell $S(y_0, r) \supset A$ of minimal radius *).

Theorems 6.7 and 6.8 extended to $\mathfrak{V}$-secants give for $n = 0$ theorems on the existence of Čebyšev centers. Concerning the problem of uniqueness of Čebyšev centers, A. L. Garkavi has proved the following result ([65], theorem 6): *In order that every bounded set A in a normed linear space E have at most one Čebyšev center, it is necessary and sufficient that the space E be uniformly convex in every direction.***)

In particular, from this result and from theorem 6.7 (extended to $\mathfrak{V}$-secants) it follows that in a complete inner product space $\mathcal{H}$ every bounded set $A \subset \mathcal{H}$ has a Čebyšev center y_0 and only one. Also, in complete inner product spaces we have always $y_0 \in \text{co}\, A$ (A. L. Garkavi [70], theorem 2). Moreover, already a part of these properties characterize the complete inner product spaces of dimension $\geqslant 3$: *In order that every bounded set A in a Banach space E have a Čebyšev center y_0 with $y_0 \in \text{co}\, A$, it is necessary and sufficient that E be a complete inner product space or $\dim E = 2$* (A. L. Garkavi [70], theorem 1); in the case when $\dim E \geqslant 3$, it is even sufficient to assume that for any three points $x_1, x_2, x_3 \in E$ there exists a Čebyšev center $y_0 \in l(\{x_1, x_2, x_3\})$ (A. L. Garkavi [70], theorem 5). For other results on Čebyšev centers, see A. L. Garkavi [70] and P. K. Belobrov [11], [12].

A notion related to the notion of Čebyšev center is that of "closest point". Following L. Fejér [56], a $y_0 \in E$ is said to be a *closest point* to a set $A \subset E$, if there exists no $y \in E$ such that

$$\|x - y\| < \|x - y_0\| \qquad (x \in A). \tag{6.88}$$

---

*) In euclidean spaces, such cells have been studied by H. Lebesgue, H. W. E. Jung and others. The term "Čebyšev center" has been introduced by A. L. Garkavi [63]; in the case when the set A consists of a finite number of points $x_1, \ldots, x_m$, sometimes the term "Čebyšev point" (S. I. Zuhovitzky [256]) or "proximity point" (H. Rademacher and I. J. Schoenberg [183]; in fact, H. Rademacher and I. J. Schoenberg [183] consider the more general case when the points $x_1, \ldots, x_m$ are replaced by bounded closed convex sets $D_1, \ldots, D_m$, and the distances $\|x_i - y\|$ of (6.87) are replaced by $\rho(y, D_i)$ $(i = 1, \ldots, m)$) is also used.

**) A normed linear space E is called, following A. L. Garkavi [63], [65], *uniformly convex in every direction*, if for every $\varepsilon > 0$ and every $x \in E$ there exists a $\delta(\varepsilon, x) > 0$ such that the relations $\|y\| = \|z\| = 1$, $y - z = \lambda x$ and $\|y + z\| \geqslant 2 - \delta(\varepsilon, x)$ imply $|\lambda| \leqslant \varepsilon$.

L. Fejér [56] has shown that in the euclidean plane for every set A the set $\varkappa(A)$ of the points $y_0 \in E$ closest to A coincides with the convex hull co A, and R. R. Phelps has observed that the proof of Fejér extends to arbitrary complete inner product spaces ([175], theorem 4). Moreover, already a part of this property characterizes the complete inner product spaces of dimension $\geqslant 3$: *In order that every set $A \subset E$ satisfy $\varkappa(A) \subset$ co A it is necessary and sufficient that E be complete inner product space of dimension $\geqslant 3$ or a strictly convex space of dimension 2* (R. R. Phelps [175], theorems 5 and 7).

The above results on Čebyšev centers are not a consequence of those on closest points, since a Čebyšev center of A need not be a closest point to A, as shown by the following example of A. L. Garkavi ([70], p. 143) : Let $E = C_R([0,1])$ and let A be the sequence

$$x_n(t) = \begin{cases} 0 & \text{for } 0 \leqslant t \leqslant \dfrac{1}{2\pi n} \\ \sin \dfrac{1}{t} & \text{for } \dfrac{1}{2\pi n} \leqslant t \leqslant 1 \end{cases} \quad (n = 1, 2, \ldots); \quad (6.89)$$

then 0 is a Čebyšev center of the set A, but the function $y(t) = t \sin \dfrac{1}{t}$ $(0 \leqslant t \leqslant 1)$ is "closer" to A than 0 (in the sense (6.88)).

We mention also the following extension of the notion of Čebyšev center : a system of n points $y_1^0 \ldots, y_n^0 \in E$ is said to be (A. L. Garkavi [61], [65]) a *best n-net* *) for a set $A \subset E$, if we have

$$\sup_{x \in A} \min_{1 \leqslant k \leqslant n} \|x - y_k^0\| = \inf_{y_1, \ldots, y_n \in E} \sup_{x \in A} \min_{1 \leqslant k \leqslant n} \|x - y_k\| ; \quad (6.90)$$

obviously, the best 1-nets for A are nothing else than the Čebyšev centers of A.

As has shown A. L. Garkavi ([65], theorem 1), *in every normed linear space E, any set $A \subset E$ consisting of $n + 1$ points has a best n-net, but there exist a Banach space E_0 and a set $A \subset E_0$ consisting of $n + 2$ points which has no best n-net.* The second statement shows that from the compactness of A there does not follow the existence of a best n-net for A ; in connection with this A. L. Garkavi ([65], p. 95) has observed that *in the space $E = C_R([0,1])$ for every compact set* (but not for

*) The term "n-net" is used here in another sense than the term "ε-net" used in the preceding. Namely, by *n-net* we mean here an arbitrary system of n elements $y_1, \ldots, y_n \in E$.

every bounded set) $A \subset E$ *there exists a best* n*-net.* Theorems 6.7 and 6.8 for 0-dimensional $\mathcal{V}$-secants can be extended to best n-nets : *If* E *is a normed linear space with the property that there exists a projection* $p : E^{**} \to E$ *of norm* $\|p\| = 1$ (*in particular, if* E *is a conjugate space*), *then every bounded set* $A \subset E$ *has a best* n*-net* (A. L. Garkavi [65], theorem 3). However, up to the present there are not known any characterizations of the spaces E with the property that every bounded set $A \subset E$ has a best n-net.

The problem of finding a best n-net for a set $A \subset E$ is obviously equivalent to that of finding a covering of A by n cells of minimal radius. Replacing the cells by arbitrary sets we arrive at the notion of best n-covering of A : following A. L. Garkavi [61], [65], a system of n sets $e_1^0, \ldots, e_n^0$ of E is said to be a *best* n*-covering* of a set $A \subset E$, if $A \subset \bigcup_{k=1}^{n} e_k^0$ and if we have

$$\max_{1 \leqslant k \leqslant n} d(e_k^0) = \inf_{A \subset \bigcup_{k=1}^{n} e_k} \max_{1 \leqslant k \leqslant n} d(e_k), \tag{6.91}$$

where by $d(e)$ one denotes the diameter *) of the set e, and where the inf is taken over all coverings of A by n sets $e_1, \ldots, e_n$. *In every normed linear space* E *any bounded set* $A \subset E$ *has a best* n*-covering*; for compact sets this result can be found in a paper of A. N. Kolmogorov and V. M. Tihomirov [120], p. 7, and in the general case it has been given by A. L. Garkavi ([65], theorem 4).

---

*) We recall that by definition $d(e) = \sup_{y,z \in e} \|y - z\|$.

CHAPTER III

BEST APPROXIMATION IN NORMED LINEAR SPACES BY ELEMENTS OF CLOSED LINEAR SUBSPACES OF FINITE CODIMENSION

An important particular case of best approximation in normed linear spaces E by elements of linear subspaces G is that when G is closed and the codimension of G (complex or real, according to the space E) is finite : codim $G = \dim E/G = n < \infty$. Naturally, the results of Chap. I are applicable, in particular, also in the case of best approximation by elements of closed linear subspaces G of finite codimension. However, by using the hypothesis codim $G = n$, one can obtain additional results, which we shall present in this Chapter.

Since the theorems of characterization of the elements of best approximation, given in Chap. I, are sufficiently convenient for applications to closed linear subspaces of finite codimension, we shall consider only the other problems of best approximation, especially the problems related to the existence and uniqueness of elements of best approximation.

As regards the applications in concrete spaces of the general theorems of best approximation in normed linear spaces by elements of closed linear subspaces of finite codimension, up to the present in the literature such applications have been given only in real spaces. In the sequel we shall present only these known results (although some of them extend immediately to complex spaces). We observe also that even in real spaces some of the results are still incomplete (e.g. the results on the existence of Čebyšev subspaces of finite codimension in the spaces $C_R(Q)$, the characterization of Čebyšev subspaces of finite codimension in the spaces $L^1_R(T,\nu)$ etc.), the definitive results being so far an open problem.

§ 1. BEST APPROXIMATION BY ELEMENTS OF FACTOR-REFLEXIVE CLOSED LINEAR SUBSPACES

A closed linear subspace G of a normed linear space E is said to be *factor-reflexive* if the space E/G is reflexive (consequently, in particular, every linear subspace G of finite codimension is factor-reflexive).

From Chap. I, theorem 2.1, it follows that *a factor-reflexive linear subspace G of a normed linear space E is proximinal if and only if for every $\Phi \in (G^{\perp})^*$ there exists an element $y \in E$ such that*

$$\Phi(f) = f(y) \qquad (f \in G^{\perp}), \tag{1.1}$$

$$\|\Phi\| = \|y\| \tag{1.2}$$

(A. L. Garkavi [66], theorem 1)*). Taking into account also Chap. I, theorem 3.2, it follows that *a factor-reflexive linear subspace $G \subset E$ is a Čebyšev subspace if and only if for every $\Phi \in (G^{\perp})^*$ there exists one and only one element $y \in E$ satisfying (1.1) and (1.2)* (A. L. Garkavi [66], theorem 2).

A useful property of proximinal factor-reflexive closed linear subspaces is given in

LEMMA 1.1. *Let E be a normed linear space and G a proximinal factor-reflexive closed linear subspace of E. Then for every***) *$f_0 \in G^{\perp} \setminus \{0\}$ the set*

$$\mathfrak{M}_{f_0} = \{x \in E \mid f_0(x) = \|f_0\|, \quad \|x\| = 1\} \tag{1.3}$$

is non-void. In the particular case when codim $G = 1$, *the converse statement is also valid, i.e. if $\mathfrak{M}_{f_0} \neq \emptyset \, (f_0 \in G^{\perp} \setminus \{0\})$, then G is proximinal.*

Proof. Let $f_0 \in G^{\perp} \setminus \{0\}$ be arbitrary. Then there exists a $\Phi \in (G^{\perp})^*$ with $\|\Phi\| = 1$, such that $\Phi(f_0) = \|f_0\|$. If G is proximinal, by virtue of the remark made at the beginning of this paragraph, there exists a $y \in E$ satisfying (1.1) and (1.2), whence $\|f_0\| = \Phi(f_0) = f_0(y)$, $1 = \|\Phi\| = \|y\|$, and thus $y \in \mathfrak{M}_{f_0}$. For codim $G = 1$ the converse statement follows immediately from Chap. I, theorem 2.1 (implication $3° \Rightarrow 1°$), which completes the proof.

---

*) We mention that by virtue of a classical theorem of E. Helly, for every $G, \Phi \in (G^{\perp})^*$ and $\varepsilon > 0$ there exists a $y \in E$ satisfying (1.1) and $\|y\| \leqslant \|\Phi\| + \varepsilon$ (see e.g. N. Dunford and J. Schwartz [49], p. 86—87).

**) Here, as well as in the next theorems involving the functionals $f_0 \in G^{\perp} \setminus \{0\}$, one obtains obviously the same result if the condition $f_0 \neq 0$ is replaced by the condition $\|f_0\| = 1$, which has been used in Chap. I and II (since $\mathfrak{M}_{f_0} = \mathfrak{M}_{\frac{f_0}{\|f_0\|}}$). However, in the sequel the condition $f_0 \in G^{\perp} \setminus \{0\}$ will be preferred, since it will simplify the exposition.

The first part of lemma 1.1 has been given by A. L. Garkavi ([66], p. 106, corollary of theorem 1) and, independently, by R. R. Phelps ([177], p. 649, remark to proposition 2).

We shall give now some properties of Čebyšev factor-reflexive closed linear subspaces.

For $\{x_n\}\subset E$, $x\in E$ and a linear subspace Γ of the conjugate space E^* we shall write $x_n \xrightarrow{\Gamma} x$ if $\|x_n\|\to\|x\|$ and *) $\|x_n - x\|_\Gamma \to 0$; we shall say that a set $A\subset E$ is Γ-*closed* if the relations $\{x_n\}\subset A$, $x\in E$, $x_n \xrightarrow{\Gamma} x$ imply $x\in A$.

THEOREM 1.1. (A. L. Garkavi [66], theorems 4 and 5). *Let E be a normed linear space and let G be a factor-reflexive Čebyšev subspace of E. Then*

a) *For every $f_0\in G^\perp\setminus\{0\}$ the set $\mathfrak{M}_{f_0}$ defined by (1.3) is non-void and coincides with the $G^\perp$-closed convex hull of a set of extremal points of the unit cell $S_E=\{x\in E \mid \|x\|\leqslant 1\}$.*

b) *E is spanned by G and the $G^\perp$-closed convex hull of the set $\mathcal{E}(S_E)$.*

c) *The cardinality of the set $\mathcal{E}(S_E)$ is $\geqslant$ the cardinality of the set $\mathcal{E}(S_{E/G})$.*

Proof. Let $f_0\in G^\perp\setminus\{0\}$ be arbitrary. Then by lemma 1.1 the set $\mathfrak{M}_{f_0}$ is non-void.

Let u be the canonical mapping $E\to E/G$. Then, since u is linear, continuous and open, and $\mathfrak{M}_{f_0}$ is convex, bounded and closed, the set $u(\mathfrak{M}_{f_0})\subset E/G$ is convex, bounded and closed. Since E/G is reflexive, it follows that $u(\mathfrak{M}_{f_0})$ is also $\sigma(E/G, (E/G)^*)$-compact, hence by virtue of the Krein-Milman theorem the set $\mathcal{E}[u(\mathfrak{M}_{f_0})]$ is non-void and $u(\mathfrak{M}_{f_0})$ coincides with the $\sigma(E/G, (E/G)^*)$-closed convex hull, whence also with the strongly convex hull of the set $\mathcal{E}[u(\mathfrak{M}_{f_0})]$.

We shall show now that for every $u(x)\in\mathcal{E}[u(\mathfrak{M}_{f_0})]$ there exists a $y\in u(x)$ such that $y\in\mathcal{E}(\mathfrak{M}_{f_0})$ (hence, in particular, $\mathcal{E}(\mathfrak{M}_{f_0})\neq\emptyset$). For this purpose it is sufficient to show that $u|_{\mathfrak{M}_{f_0}}$ is a one-to-one linear mapping of $\mathfrak{M}_{f_0}$ onto $u(\mathfrak{M}_{f_0})\subset E/G$, since then for every $u(x)\in\mathcal{E}[u(\mathfrak{M}_{f_0})]$ the element $y\in u(x)$ with $y\equiv\mathfrak{M}_{f_0}$ will be in $\mathcal{E}(\mathfrak{M}_{f_0})$. Assume that $u|_{\mathfrak{M}_{f_0}}$ is not one-to-one, i.e. there exist $x, y\in\mathfrak{M}_{f_0}$ with $x\neq y$, such that $u(x)=u(y)$. Since $x, y\in\mathfrak{M}_{f_0}$, we have then

$$\frac{f_0}{\|f_0\|}(x)=\frac{f_0}{\|f_0\|}(y)=1,\quad \|x\|=\|y\|=1, \tag{1.4}$$

and on the other hand, since $u(x)=u(y)$ in E/G, we have

$$x-y\in G\setminus\{0\}. \tag{1.5}$$

---

*) For the definition of the semi-norm $\|x\|_\Gamma$ see Chap. I, formula (1.11).

But from (1.4) and (1.5) it follows, by virtue of the implication $1° \Rightarrow 2°$ of Chap. I, theorem 3.2, that G is not a Čebyšev subspace, which contradicts the hypothesis. This proves that $u|_{\mathfrak{M}_{f_0}}$ is one-to-one.

Now let $x \in \mathfrak{M}_{f_0}$ be arbitrary. Then by the above there exists a sequence $\{u(x_n)\} \subset \operatorname{co} \mathcal{E}[u(\mathfrak{M}_{f_0})]$ such that $u(x_n) \to u(x)$. Again by the above, there exists then a sequence $y_n \in u(x_n)$ $(n = 1, 2, \ldots)$ with $\{y_n\} \subset \operatorname{co} \mathcal{E}(\mathfrak{M}_{f_0})$. Since $\|x\| = 1 = \|y_n\|$ $(n = 1, 2, \ldots)$ and since by the canonical equivalence $G^\perp \equiv (E/G)^*$ we have

$$\|y_n - x\|_{G^\perp} = \sup_{\substack{f \in G^\perp \\ \|f\| \leqslant 1}} |f(y_n - x)| = \|u(y_n - x)\| = \|u(y_n) - u(x)\| \to 0,$$

it follows that x, whence also $\mathfrak{M}_{f_0}$, are contained in the $G^\perp$-closed convex hull $\Omega[\mathcal{E}(\mathfrak{M}_{f_0})]$ of $\mathcal{E}(\mathfrak{M}_{f_0})$. Since $\mathfrak{M}_{f_0}$ is convex and $G^\perp$-closed (because the relations $x_n \in \mathfrak{M}_{f_0}, x \in E, x_n \xrightarrow{G^\perp} x$ imply $|f_0(x) - \|f_0\|| = |f_0(x) - f_0(x_n)| \leqslant \|f_0\| \, \|x - x_n\|_{G^\perp} \to 0$ and $\|x\| = \lim_{n \to \infty} \|x_n\| = 1$, whence $x \in \mathfrak{M}_{f_0}$), it follows that we have $\mathfrak{M}_{f_0} = \Omega[\mathcal{E}(\mathfrak{M}_{f_0})]$. But by virtue of an argument dual to that of Chap. I, the proof of corollary 1.8, $\mathfrak{M}_{f_0}$ is an extremal subset of S_E ([211], theorem 1.2), hence $\mathcal{E}(\mathfrak{M}_{f_0}) \subset \mathcal{E}(S_E)$, which completes the proof of the statement a).

b) is a consequence of a), since by Chap. I, theorem 1.1, for every $x \in E$ and $g_0 \in \mathfrak{P}_G(x)$ there exists an $f_0 \in G^\perp$ with $\|f_0\| = 1$ such that $\frac{x - g_0}{\|x - g_0\|} \in \mathfrak{M}_{f_0}$ and since we have obviously $x = g_0 + \|x - g_0\| \frac{x - g_0}{\|x - g_0\|}$.

c) Let $\dot{x} \in \mathcal{E}(S_{E/G})$ be arbitrary and let $\varphi \in (E/G)^*$ be such that $\|\varphi\| = 1 = \varphi(\dot{x})$. Then by the canonical equivalence $(E/G)^* \equiv G^\perp$ there exists an $f_0 \in G^\perp$ such that $\|f_0\| = \|\varphi\| = 1$, $f_0(x) = \varphi(\dot{x}) = 1 (x \in \dot{x})$, hence $x \in \mathfrak{M}_{f_0} (x \in \dot{x})$. Consequently, by the above proof of part a), there exists a $y \in \dot{x}$ such that $y \in \mathcal{E}(\mathfrak{M}_{f_0}) \subset \mathcal{E}(S_E)$. Thus we have obtained in each equivalence class $\dot{x} \in \mathcal{E}(S_{E/G})$ an element $y \in \dot{x}$ with $y \in \mathcal{E}(S_E)$, which completes the proof of theorem 1.1.

Corollary 1.1 (A. L. Garkavi [66], corollary of theorem 5). *Let E be a normed linear space.*

a) *If $\mathcal{E}(S_E) = \emptyset$, then E has no factor-reflexive Čebyšev subspace.*

b) *If $\mathcal{E}(S_E)$ consists of a finite number of points, then E has no factor-reflexive Čebyšev subspace of infinite codimension.*

Proof. Both statements follow from theorem 1.1 c), since if E/G is reflexive, then $\mathcal{E}(S_{E/G}) \neq \emptyset$, and if E/G is reflexive and $\dim E/G = \infty$, then the set $\mathcal{E}(S_{E/G})$ is infinite (by virtue of the weak compactness of $S_{E/G}$ and of the theorem of Krein-Milman).

Examples of spaces satisfying the condition of statement a) are $E = c_0$ and $E = L^1(T, \nu)$, where (T, ν) is a positive measure space which has no atom and is such that $L^1(T, \nu)^* \equiv \equiv L^\infty(T, \nu)$ (the fact that the spaces $E = c_0$ and $E = L^1(T, \nu)$ for a σ-finite (T, ν) without atoms, have no Čebyšev subspace of finite codimension, has been observed by R. R. Phelps [176], p. 250 and 245) and examples of spaces satisfying the condition of statement b) are the infinite dimensional spaces $E = C(Q)$ with $Q =$ a compact space having a finite number of connected components *).

§2. BEST APPROXIMATION BY ELEMENTS OF CLOSED LINEAR SUBSPACES OF FINITE CODIMENSION

If G is a closed linear subspace of finite codimension n of a normed linear space E, then by virtue of the canonical isometry $G^\perp \equiv (E/G)^*$ we have

$$\dim G^\perp = \dim (E/G)^* = \dim E/G = \operatorname{codim} G = n, \tag{2.1}$$

hence

$$G^\perp = [f_1, \ldots, f_n], \tag{2.2}$$

where $f_1, \ldots, f_n \in E^*$ are linearly independent. Consequently,

$$G = \overline{G} = (G^\perp)_\perp = \{x \in E \mid f_k(x) = 0 \ (k = 1, \ldots, n)\}. \tag{2.3}$$

In the present paragraph we shall study proximinal, semi-Čebyšev and Čebyšev subspaces of finite codimension of normed linear spaces.

2.1. BEST APPROXIMATION BY ELEMENTS OF CLOSED LINEAR SUBSPACES OF FINITE CODIMENSION IN GENERAL NORMED LINEAR SPACES

LEMMA 2.1. *Let E be a normed linear space and let $\mathfrak{M}$ be an extremal subset of the unit cell S_E, with*

$$0 \leqslant \dim \mathfrak{M} = r < \infty. \tag{2.4}$$

---

*) See e.g. G. Köthe [122], p. 336–337. Concerning $E = c_0$ we have seen in Chap. I, § 3, section 3.1, that it has no Čebyšev subspace of infinite dimension.

Then $\mathfrak{M}$ *contains at least* $r+1$ *linearly independent extremal points* $x_0, x_1, \ldots, x_r \in \mathcal{E}(S_E) \cap \mathfrak{M} = \mathcal{E}(\mathfrak{M})$.

Proof. Let $E_1 = [\mathfrak{M}]$ = the linear subspace of E spanned by $\mathfrak{M}$. Then, taking*) an arbitrary $y \in \operatorname{Int}_l \mathfrak{M}$, $\mathfrak{M}$ is the smallest extremal subset of S_{E_1} containing y, whence, by Chap. II, the proof of lemma 1.1 (applied in $(E_1^*)^* \equiv E_1$), we have $r = \dim \mathfrak{M} \leqslant \dim E_1 - 1$, whence $\dim E_1 \geqslant r+1$, and every $x \in \mathfrak{M}$ can be expressed as a convex combination of a finite number of suitable extremal points $z_1, \ldots, z_h$ of the smallest extremal subset $\Gamma \subset S_{E_1}$ containing x, whence also of $\mathfrak{M}$. Now assume that $\mathfrak{M}$ has altogether m linearly independent extremal points $x_0, x_1, \ldots, x_{m-1}$, where $m < r+1$. Then each z_j $(j=1,\ldots,h)$ is a linear combination of $x_0, x_1, \ldots, x_{m-1}$ and thus every $x \in \mathfrak{M}$, whence (by $E_1 = [\mathfrak{M}]$) also every $x \in E_1$, can be expressed as a linear combination of $x_0, x_1, \ldots, x_{m-1}$. Consequently, $\dim E_1 = m < r+1$, in contradiction with the above. Thus, there exist at least $r+1$ linearly independent extremal points $x_0, x_1, \ldots, x_r \in \mathcal{E}(\mathfrak{M})$ $(= \mathcal{E}(S_E) \cap \mathfrak{M}$ by Chap. I, lemma 1.7), which completes the proof.

In the particular case when the scalars are real, E is a conjugate space B^* and $\mathfrak{M} = \{\varphi \in B^* \mid \varphi(b_0) = \|b_0\|, \|\varphi\| = 1\}$ (where $b_0 \in B \setminus \{0\}$), lemma 2.1 has been given, with a different proof, by R. R. Phelps ([176], lemma 1.6).

THEOREM 2.1. *Let* E *be a normed linear space and* G *a closed linear subspace of codimension* n *of* E, *hence* $G^\perp = [f_1, \ldots, f_n]$. *The following statements are equivalent*:

1° G *is a semi-Čebyšev subspace.*

2° *For every* $f_0 \in G^\perp \setminus \{0\}$ *the set* $\mathfrak{M}_{f_0}$ *defined by (1.3) is of dimension* $r = r(f_0) \leqslant n-1$ *and contains* $r+1$ *elements* $x_0, x_1, \ldots, x_r \in \mathfrak{M}_{f_0}$ *such that*

$$\operatorname{rank}(f_k(x_i))_{\substack{k=1,\ldots,n\\ i=0,1,\ldots,r}} = r+1. \tag{2.5}$$

3° *For every* $f_0 \in G^\perp \setminus \{0\}$ *the set* $\mathfrak{M}_{f_0}$ *is of dimension* $r = r(f_0) \leqslant n-1$ *and contains* $r+1$ *extremal points* $x_0, x_1, \ldots, x_r \in \mathcal{E}(S_E) \cap \mathfrak{M}_{f_0} = \mathcal{E}(\mathfrak{M}_{f_0})$ *satisfying* (2.5).

4° *For every* $f_0 \in G^\perp \setminus \{0\}$ *the set* $\mathfrak{M}_{f_0}$ *is of dimension* $r = r(f_0) \leqslant n-1$ *and for any* $r+1$ *linearly independent extremal points* $x_0, x_1, \ldots, x_r \in \mathcal{E}(S_E) \cap \mathfrak{M}_{f_0} = \mathcal{E}(\mathfrak{M}_{f_0})$ *we have* (2.5).

5° *For every* $f_0 \in G^\perp \setminus \{0\}$ *the set* $\mathfrak{M}_{f_0}$ *is of dimension* $r = r(f_0) \leqslant n-1$ *and for any* $r+1$ *linearly independent elements* $x_0, x_1, \ldots, x_r \in \mathfrak{M}_{f_0}$ *we have* (2.5).

Proof. Assume that there exists an $f_0 \in G^\perp \setminus \{0\}$ such that $\dim \mathfrak{M}_{f_0} \geqslant n$ and let $x \in \operatorname{Int}_l \mathfrak{M}_{f_0}$ be arbitrary. Then $l(\mathfrak{M}_{f_0}) - x = l(\mathfrak{M}_{f_0} - x)$ contains an n-dimensional linear subspace G_1,

*) For the definitions of $\operatorname{Int}_l \mathfrak{M}$, $l(\mathfrak{M})$, etc. see Chap. I, § 4.

and since $0 \in \operatorname{Int}_l(\mathfrak{M}_{f_0} - x)$, it follows that 0 is an interior point of the set $G_1 \cap (\mathfrak{M}_{f_0} - x)$ in the linear subspace G_1, i.e. every $y \in G_1$ is a positive multiple of an element of $G_1 \cap (\mathfrak{M}_{f_0} - x)$. If we had $G_1 \cap G = \{0\}$, then from $\dim G_1 = n = \operatorname{codim} G$ it would follow that $E = G_1 \oplus G$, whence, taking into account $f_0|_{G_1} = 0$ (by $f_0|_{G_1 \cap (\mathfrak{M}_{f_0} - x)} = 0$) and $f_0|_G = 0$ (by the hypothesis $f_0 \in G^\perp$), it would follow that $f_0 = 0$ on E, which contradicts the hypothesis $f_0 \neq 0$. Consequently, there exists a $g_0 \in G_1 \cap G \setminus \{0\}$; taking a sufficiently small positive multiple of $-g_0$, we may assume that $-g_0 \in G_1 \cap (\mathfrak{M}_{f_0} - x)$, hence $x - g_0 \in \mathfrak{M}_{f_0}$. By $g_0 \in G \setminus \{0\}$, $f_0 \in G^\perp \setminus \{0\}$ and $x, x - g_0 \in \mathfrak{M}_{f_0}$ we then have the relations (3.9), (3.10) and (3.12) of Chap. I $\left(\text{with } f = \frac{f_0}{\|f_0\|}\right)$, whence, by virtue of theorem 3.2 of Chap. I, G is not a semi-Čebyšev subspace. Thus, the condition $\dim \mathfrak{M}_{f_0} \leqslant n - 1$ ($f_0 \in G^\perp \setminus \{0\}$) is necessary in order that G be a semi-Čebyšev subspace.

Now let $f_0 \in G^\perp \setminus \{0\}$ be arbitrary and $\dim \mathfrak{M}_{f_0} = r \leqslant n - 1$. If $\mathfrak{M}_{f_0} = \emptyset$, then $r = -1$, whence the condition (2.5) is trivial. Assume that $\mathfrak{M}_{f_0} \neq \emptyset$ and that $\mathfrak{M}_{f_0}$ contains $r + 1$ linearly independent elements $x_0, x_1, \ldots, x_r$ such that $\operatorname{rank} (f_k(x_i))_{\substack{k=1,\ldots,n \\ i=0,1,\ldots,r}} < r + 1$. Then, taking into account that $r + 1 \leqslant n$, the system of n homogeneous linear equations with $r + 1$ unknowns

$$\sum_{i=0}^{r} \gamma_i f_k(x_i) = 0 \qquad (k = 1, \ldots, n) \tag{2.6}$$

has a non-trivial solution $\gamma_0^{(0)}, \gamma_1^{(0)}, \ldots, \gamma_r^{(0)}$, hence $g_0 = \sum_{i=0}^{r} \gamma_i^{(0)} x_i \in G$; since $x_0, x_1, \ldots, x_r$ are linearly independent, we have also $g_0 \neq 0$, hence $g_0 \in G \setminus \{0\}$. We claim that, in this case, by taking a sufficiently small positive multiple of $-g_0$, we may assume that $-g_0 \in \mathfrak{M}_{f_0} - x$, where

$$x = \frac{1}{r+1} \sum_{i=0}^{r} x_i \in \operatorname{Int}_l \mathfrak{M}_{f_0}. \tag{2.7}$$

Indeed, by $x_i \in \mathfrak{M}_{f_0} (i = 0, 1, \ldots, r)$ and $f_0 \in G^\perp \setminus \{0\}$ we have

$$\sum_{i=0}^{r} \gamma_i^{(0)} = \frac{f_0}{\|f_0\|} \left(\sum_{i=0}^{r} \gamma_i^{(0)} x_i \right) = \frac{f_0}{\|f_0\|} (g_0) = 0,$$

whence

$$-g_0 = -\sum_{i=0}^{r} \gamma_i^{(0)} (x_i - x) \in l\,(\mathfrak{M}_{f_0} - x),$$

whence, taking into account that $0 \in \operatorname{Int}_t (\mathfrak{M}_{f_0} - x)$ (by (2.7)), the statement follows. But, by $g_0 \in G \setminus \{0\}$, $f_0 \in G^\perp \setminus \{0\}$ and $x, x - g_0 \in \mathfrak{M}_{f_0}$ we have then the relations (3.9), (3.10) and (3.12) of Chap. I $\left(\text{with } f = \frac{f_0}{\|f_0\|}\right)$, whence, by virtue of theorem 3.2 of Chap. I, G is not a semi-Čebyšev subspace. Thus, we have proved that $1° \Rightarrow 5°$.

The implications $5° \Rightarrow 4°$ and $3° \Rightarrow 2°$ are obvious. The implication $4° \Rightarrow 3°$ is also immediate, taking into account the fact that $\mathfrak{M}_{f_0}$ is an extremal subset of S_E (see the proof of theorem 1.1 a)) and lemma 2.1.

Finally, assume that condition 2° is satisfied but G is not a semi-Čebyšev subspace. Then by theorem 3.2 of Chap. I there exist an $f_0 \in G^\perp$ with $\|f_0\| = 1$ and $x \in E$, $g_0 \in G \setminus \{0\}$ such that $\frac{x}{\|x\|} \in \mathfrak{M}_{f_0}$, $\frac{x}{\|x\|} - \frac{g_0}{\|x\|} \in \mathfrak{M}_{f_0}$. For this f_0 we have, by virtue of 2°,

$$\dim \mathfrak{M}_{f_0} = r \leqslant n - 1, \tag{2.8}$$

and there exist $x_0, x_1, \ldots, x_r \in \mathfrak{M}_{f_0}$ satisfying (2.5); by $r+1 \leqslant n$ and (2.5), $x_0, x_1, \ldots, x_r$ are linearly independent. Since $\dim \mathfrak{M}_{f_0} = r$, it follows that there exist scalars α_i, β_i $(i = 0, 1, \ldots, r)$ such that

$$\frac{x}{\|x\|} = \sum_{i=0}^{r} \alpha_i x_i, \quad \frac{x}{\|x\|} - \frac{g_0}{\|x\|} = \sum_{i=0}^{r} \beta_i x_i, \tag{2.9}$$

whence, for $\gamma_i = (\alpha_i - \beta_i)\|x\|$ $(i = 0, 1, \ldots, r)$, we obtain

$$g_0 = \|x\| \left(\frac{x}{\|x\|} - \frac{x - g_0}{\|x\|} \right) = \sum_{i=0}^{r} \gamma_i x_i,$$

whence

$$\sum_{i=0}^{r} \gamma_i f_k(x_i) = f_k(g_0) = 0 \qquad (k = 1, \ldots, n). \tag{2.10}$$

Taking into account that $g_0 \neq 0$, we have $\max_{0 \leqslant i \leqslant r} |\gamma_i| \neq 0$, hence $\gamma_0, \gamma_1, \ldots, \gamma_r$ is a non-trivial solution of the system (2.10) of n homogeneous linear equations with $r + 1$ unknowns. Since $r + 1 \leqslant n$, it follows that rank $(f_k(x_i))_{\substack{k=1,\ldots,n \\ i=0,1,\ldots,r}} < r + 1$, which contradicts (2.5). Thus, $2° \Rightarrow 1°$, which completes the proof.

In the case of real scalars the necessary condition $\dim \mathfrak{M}_{f_0} \leqslant n-1$ $(f_0 \in G^\perp \setminus \{0\})$ for G to be a semi-Čebyšev subspace has been given, essentially with the above proof, by R. R. Phelps ([176], theorem 1.4). Again for real scalars, the equivalence $1° \Leftrightarrow 3°$ of theorem 2.1 above has been given, with a different proof, by A. L. Garkavi ([66], theorem 6), and the equivalence $1° \Leftrightarrow 2°$ has been also stated by A. L. Garkavi ([72], proposition B).

Let us mention also the following alternative proof of the necessity of condition $\dim \mathfrak{M}_{f_0} \leqslant n - 1$ ($f_0 \in G^{\perp} \setminus \{0\}$) in order that G be a semi-Čebyšev subspace: Assume that for an $f_0 \in G^{\perp} \setminus \{0\}$ we have $\dim \mathfrak{M}_{f_0} \geqslant n$ and let E_1 be the linear subspace of E spanned by $\mathfrak{M}_{f_0}$. Then, since $l(\mathfrak{M}_{f_0}) \not\ni 0$, we have $\dim E_1 = \dim \mathfrak{M}_{f_0} + 1 \geqslant n + 1$. Taking $n + 1$ arbitrary linearly independent elements $y_1, \ldots, y_{n+1} \in E_1$, the system

$$\sum_{i=1}^{n+1} \alpha_i f_k(y_i) = 0 \qquad (k = 1, \ldots, n) \tag{2.11}$$

of n homogeneous linear equations with $n + 1$ unknowns has a non-trivial solution $\alpha_1^{(0)}, \ldots, \alpha_{n+1}^{(0)}$, hence we have

$$g_0 = \sum_{i=1}^{n+1} \alpha_i^{(0)} y_i \in (E_1 \cap G) \setminus \{0\}. \tag{2.12}$$

Let $\mathcal{C}_{f_0} \subset E_1$ be the closed convex cone

$$\mathcal{C}_{f_0} = \{x \in E \mid f_0(x) = \|f_0\| \, \|x\|\}, \tag{2.13}$$

and let $x \in \operatorname{Int}_l \mathfrak{M}_{f_0} \subset \operatorname{Int}_l \mathcal{C}_{f_0}$. Then by $g_0 \in E_1 = l(\mathcal{C}_{f_0})$ we have, for $\lambda > 0$ sufficiently small,

$$x + \lambda(g_0 - x) \in \mathcal{C}_{f_0},$$

whence, taking into account $f_0(g_0) = 0$ and $f_0(x) = \|f_0\|$, we obtain

$$\|f_0\| \, \|(1 - \lambda)x + \lambda g_0\| = \|f_0\| \|x + \lambda(g_0 - x)\| =$$
$$= f_0[x + \lambda(g_0 - x)] = \|f_0\|(1 - \lambda),$$

whence, since $\lambda \neq 1$ (otherwise we would have $g_0 \in \mathcal{C}_{f_0}$, which contradicts $f_0(g_0) = 0$), we have

$$x + \frac{\lambda}{1 - \lambda} g_0 \in \mathfrak{M}_{f_0}.$$

By $g_0 \in G \setminus \{0\}$, $\lambda > 0$, $f_0 \in G^{\perp} \setminus \{0\}$ and x, $x + \dfrac{\lambda}{1 - \lambda} g_0 \in \mathfrak{M}_{f_0}$ we then have the relations (3.9), (3.10) and (3.12) of Chap. I

$\left(\text{with } \frac{-\lambda}{1-\lambda} g_0 \text{ instead of } g_0 \text{ and } f = \frac{f_0}{\|f_0\|}\right)$, whence, by virtue of theorem 3.2 of Chap. I, G is not a semi-Čebyšev subspace, which completes the proof.

COROLLARY 2.1 (A. L. Garkavi [66], theorem 7 and 8). *Let E be a normed linear space and let G be a Čebyšev subspace of E of codimension n. Then*

a) *E is spanned by G and $\mathcal{E}(S_E)$.*

b) *We have*

$$\dim l[\mathcal{E}(S_E)] \geqslant n. \tag{2.14}$$

Proof. a) Let $x \in E$ and $g_0 \in \mathfrak{P}_G(x)$ be arbitrary. Then by Chap. I, theorem 1.1, there exists an $f_0 \in G^{\perp}$ with $\|f_0\| = 1$ such that $\frac{x - g_0}{\|x - g_0\|} \in \mathfrak{M}_{f_0}$. Since by theorem 2.1 we have $\dim E_1 < \infty$, where E_1 is the linear subspace of E spanned by $\mathfrak{M}_{f_0}$ and since $\mathfrak{M}_{f_0}$ is an extremal subset of S_E (see the proof of theorem 1.1 a)), by applying lemma 1.1 of Chap. II for $(E_1^*)^* \equiv E_1$ we obtain that $\frac{x - g_0}{\|x - g_0\|}$ is a finite convex combination $\sum_{i=0}^{r} \lambda_i x_i$, where $x_0, x_1, \ldots, x_r \in \mathcal{E}(\mathfrak{M}_{f_0}) = \mathcal{E}(S_E) \cap \mathfrak{M}_{f_0}$. Consequently,

$$x = g_0 + \|x - g_0\| \frac{x - g_0}{\|x - g_0\|} = g_0 + \sum_{i=0}^{r} \lambda_i \|x - g_0\| x_i,$$

with $x_0, x_1, \ldots, x_r \in \mathcal{E}(S_E)$.

b) is an immediate consequence of the statement a) proved above and of the hypothesis $\operatorname{codim} G = n$, which completes the proof.

Denote by s the least upper bound of the codimensions of Čebyšev subspaces of finite codimension of a normed linear space E. Then from corollary 2.1 b) it follows that we have

$$\dim l[\mathcal{E}(S_E)] \geqslant s. \tag{2.15}$$

As has observed A. L. Garkavi ([66], theorem 8), for every positive integer m, and also for $m = \infty$, there exists a Banach space E such that

$$s = \dim l[\mathcal{E}(S_E)] = m; \tag{2.16}$$

obviously, in the case $m = \infty$ one can take for E any infinite dimensional Hilbert space. On the other hand, as has been observed by A. L. Garkavi ([66], theorem 9), strict inequality in (2.15) is also possible, since there exist normed linear spaces for which we have even

$$\dim l[\mathcal{E}(S_E)] = \infty, \qquad s = 0. \tag{2.17}$$

Indeed, e.g. let $E = l^\infty([0,1])$ = the space of all bounded functions on $[0,1]$, with the norm $\|x\| = \sup_{t\in[0,1]} |x(t)|$. Then $\dim E = \infty$ and E is the conjugate space of the space $l^1([0,1])$ of all summable families of scalars $\zeta = \{\zeta_t\}_{t\in[0,1]}$ with the norm $\|\zeta\| = \sum_{t\in[0,1]} |\zeta_t|$, hence (by the Krein-Milman theorem) we have $\dim l[\mathcal{E}(S_E)] = \infty$*). On the other hand, if E had a Čebyšev subspace G of finite codimension, say n, then by theorem 2.1 we should have $\dim \mathfrak{M}_{f_0} \leqslant n-1 < \infty$ for every $f_0 \in G^\perp \setminus \{0\}$. However, we shall show that this does not occur for any $f_0 \in E^* \setminus \{0\}$ with $\mathfrak{M}_{f_0} = \emptyset$, hence we have $s = 0$. Let $f_0 \in E^* \setminus \{0\}$ be arbitrary with $\mathfrak{M}_{f_0} \neq \emptyset$, and let

$$T_0 = \{\tau \in [0,1] \mid f_0(x_\tau) = 0\}, \tag{2.18}$$

where the elements $x_\tau \in E$ are defined by

$$x_\tau(t) = \begin{cases} 1 & \text{for } t = \tau \\ 0 & \text{for } t \in [0,1] \setminus \{\tau\}. \end{cases} \tag{2.19}$$

The set T_0 is infinite, since otherwise there would exist $\left(\text{by } [0,1]\setminus T_0 = \bigcup_{n=1}^{\infty} \left\{\tau \in [0,1] \,\middle|\, |f_0(x_\tau)| \geqslant \frac{1}{n}\right\}\right)$ an $\varepsilon > 0$ and an infinite sequence $\{\tau_k\} \subset [0,1]$ such that

$$|f_0(x_{\tau_k})| \geqslant \varepsilon \qquad (k = 1, 2, \ldots),$$

hence we would have, putting $\alpha_k = \operatorname{sign} f_0(x_{\tau_k})$ $(k = 1, 2, \ldots)$, the relations

$$\left\|\sum_{k=1}^m \alpha_k x_{\tau_k}\right\| = 1, \left|f_0\left(\sum_{k=1}^m \alpha_k x_{\tau_k}\right)\right| = \sum_{k=1}^m |f_0(x_{\tau_k})| \geqslant m\varepsilon \ (m = 1,2,\ldots),$$

and thus the functional f_0 would not be bounded, which contradicts the hypothesis $f_0 \in E^*$. Consequently, fixing a $z \in \mathfrak{M}_{f_0}$ and putting

$$\beta_\tau = \begin{cases} \operatorname{sign} \overline{z(\tau)} & \text{if } \tau \in T_0,\ z(\tau) \neq 0, \\ 1 & \text{if } \tau \in T_0,\ z(\tau) = 0, \end{cases} \tag{2.20}$$

the set of all $y_\tau \in E$ $(\tau \in T_0)$ of the form

$$y_\tau(t) = z(t) - \beta_\tau x_\tau(t) \qquad (t \in [0,1]) \tag{2.21}$$

spans an infinite dimensional linear manifold. But, by $z \in \mathfrak{M}_{f_0}$ and (2.18) – (2.21), we have $f_0(y_\tau) = \|f_0\|$, $\|y_\tau\| = 1$ $(\tau \in T_0)$, i.e.

*) We mention that although $\dim l[\mathcal{E}(S_E)] = \infty$, yet S_E has no exposed point, since in Chap. I, §3, section 3.1 we have observed that if S_{E_0} has an exposed point, then E_0 has a Čebyšev hyperplane (hence $s \geqslant 1$).

$y_\tau \in \mathfrak{M}_{f_0}$ $(\tau \in T_0)$, and thus $\dim \mathfrak{M}_{f_0} = \infty$, which completes the proof of (2.17) for the space $E = l^\infty([0,1])$.

In the next section we shall also give examples of normed linear spaces E for which we have

$$\dim l[\mathcal{E}(S_E)] > s > 0. \tag{2.22}$$

2.2. APPLICATIONS IN THE SPACES $C_R(Q)$

THEOREM 2.2 (A. L. Garkavi [72], theorem 1). *Let $E = C_R(Q)$ (Q compact) and let G be a closed linear subspace of E of codimension n, hence* *) $G^\perp = [\mu_1, \ldots, \mu_n]$. *In order that G be proximinal it is necessary and sufficient that the following three conditions be satisfied:*

α) *For every $\mu \in G^\perp \setminus \{0\}$ the carrier $S(\mu)$ admits a Hahn-decomposition into two closed sets $S(\mu)^+$ and $S(\mu)^- = S(\mu) \setminus S(\mu)^+$.*

β) *For every pair of measures $\mu, \bar{\mu} \in G^\perp \setminus \{0\}$ the set $S(\bar{\mu}) \setminus S(\mu)$ is closed.*

γ) *For every pair of measures $\mu, \bar{\mu} \in G^\perp \setminus \{0\}$ the measure μ is absolutely continuous with respect to $\bar{\mu}$ on the set* **) $S(\bar{\mu})$.

Proof. Assume that G is proximinal and let $\mu \in G^\perp \setminus \{0\}$ be arbitrary. Then by lemma 1.1 μ has at least one maximal element $x \in E$ with $\|x\| = 1$, whence, by Chap. I, lemma 1.6, $S(\mu)$ admits a Hahn decomposition into two closed sets $S(\mu)^+$ and $S(\mu)^-$. Thus we have α).

Now assume that condition β) is not satisfied, i.e. that there exist $\mu, \bar{\mu} \in G^\perp \setminus \{0\}$ for which the set $S(\bar{\mu}) \setminus S(\mu)$ is not closed. Then for $q_0 \in \overline{S(\bar{\mu}) \setminus S(\mu)} \setminus [S(\bar{\mu}) \setminus S(\mu)]$ we have

$$q_0 \in S(\mu) \cap \overline{S(\bar{\mu}) \setminus S(\mu)}, \tag{2.23}$$

since otherwise we would have $q_0 \notin S(\mu)$, whence $q_0 \notin S(\bar{\mu})$ (by $q_0 \notin S(\bar{\mu}) \setminus S(\mu)$), which contradicts $q_0 \in \overline{S(\bar{\mu}) \setminus S(\mu)} \subset \overline{S(\bar{\mu})} = S(\bar{\mu})$. Since $S(\mu)^- = S(-\mu)^+$ and $S(\mu) = S(-\mu)$, we may assume, replacing, if necessary, μ by $-\mu$ or $\bar{\mu}$ by $-\bar{\mu}$ (or both), that we have

$$q_0 \in S(\mu)^+ \cap S(\bar{\mu})^+. \tag{2.24}$$

Define now a function α on $S(\mu) \cup S(\bar{\mu})$ by

$$\alpha(q) = \begin{cases} 1 \text{ for } q \in S(\mu)^+ \cup [S(\bar{\mu})^- \setminus S(\mu)] \\ -1 \text{ for } q \in S(\mu)^- \cup [S(\bar{\mu})^+ \setminus S(\mu)]. \end{cases} \tag{2.25}$$

---

*) For the sake of simplicity of expression, we shall identify in the sequel the functionals $f \in E^*$ with their kernels $\mu \in \mathfrak{M}_R^1(Q)$ (see Chap. I, §1, section 1.4).

**) This condition is equivalent to the condition that μ be absolutely continuous with respect to $\bar{\mu}$ on the set $S(\bar{\mu})'$ of all limit points of $S(\bar{\mu})$, since at the isolated points q of $S(\bar{\mu})$ we have $|\bar{\mu}|(\{q\}) \neq 0$ (by the definition of $S(\bar{\mu})$).

Then α is discontinuous at the point q_0. Indeed, we have

$$q_0 \in \overline{S(\bar{\mu}) \setminus S(\mu)} = \overline{[S(\bar{\mu})^+ \cup S(\bar{\mu})^-] \setminus S(\mu)} = \\ = \overline{S(\bar{\mu})^+ \setminus S(\mu)} \cup \overline{S(\bar{\mu})^- \setminus S(\mu)},$$

and also $q_0 \notin \overline{S(\bar{\mu})^- \setminus S(\mu)}$ (since otherwise we would have, by α) proved above, $q_0 \in \overline{S(\bar{\mu})^-} = S(\bar{\mu})^-$, which contradicts (2.24)), whence $q_0 \in \overline{S(\bar{\mu})^+ \setminus S(\mu)}$. Consequently, in every neighborhood V of q_0 there exist points $q \in S(\bar{\mu})^+ \setminus S(\mu)$, hence points at which $\alpha(q) = -1$; on the other hand, by $q_0 \in S(\mu)^+$ we have $\alpha(q_0) = 1$, hence α is discontinuous at q_0.

Define now $\Phi \in (G^{\perp})^*$ by

$$\Phi(\mu') = \int_{S(\mu) \cup S(\bar{\mu})} \alpha(q) \mathrm{d}\mu'(q) \qquad (\mu' \in G^{\perp}). \tag{2.26}$$

We shall show that for this functional Φ there exists no $y \in E$ satisfying (1.1) and (1.2), which contradicts the hypothesis that G is proximinal, and this contradiction will prove the necessity of condition β). Assume that a $y \in E$ satisfies (1.1) and (1.2). Then for $\mu' = \mu$ we obtain

$$\int_Q y(q)\, \mathrm{d}\mu(q) = \mu(y) = \Phi(\mu) = \int_{S(\mu)} \alpha(q)\, \mathrm{d}\mu(q) = \\ = \int_{S(\mu)^+} \mathrm{d}\mu(q) - \int_{S(\mu)^-} \mathrm{d}\mu(q) = \|\mu\|, \tag{2.27}$$

and on the other hand we have

$$\|y\| = \|\Phi\| = \max_{q \in S(\mu) \cup S(\bar{\mu})} |\alpha(q)| = 1, \tag{2.28}$$

whence, by Chap. I, corollary 1.6 (implication $1° \Rightarrow 3°$),

$$y(q) = \begin{cases} 1 & \text{for } q \in S(\mu)^+ \\ -1 & \text{for } q \in S(\mu)^-. \end{cases} \tag{2.29}$$

Consequently, for $\mu' = \bar{\mu}$ we obtain

$$\int_{S(\bar{\mu}) \setminus S(\mu)} y(q)\, \mathrm{d}\bar{\mu}(q) + \int_{S(\bar{\mu}) \cap S(\mu)} y(q)\, \mathrm{d}\bar{\mu}(q) = \\ = \int_{S(\bar{\mu})} y(q)\, \mathrm{d}\bar{\mu}(q) = \bar{\mu}(y) = \Phi(\bar{\mu}) = \int_{S(\bar{\mu})} \alpha(q)\, \mathrm{d}\bar{\mu}(q) = \\ = -\int_{S(\bar{\mu})^+ \setminus S(\mu)} \mathrm{d}\bar{\mu}(q) + \int_{S(\bar{\mu})^- \setminus S(\mu)} \mathrm{d}\bar{\mu}(q) + \int_{S(\bar{\mu}) \cap S(\mu)} \alpha(q)\, \mathrm{d}\bar{\mu}(q) = \\ = -\int_{S(\bar{\mu}) \setminus S(\mu)} \mathrm{d}|\bar{\mu}|(q) + \int_{S(\bar{\mu}) \cap S(\mu)} y(q)\, \mathrm{d}\bar{\mu}(q),$$

whence

$$\int_{S(\bar{\mu})\setminus S(\mu)} y(q)\, d\bar{\mu}(q) = -|\bar{\mu}|(S(\bar{\mu})\setminus S(\mu)), \tag{2.30}$$

which, together with (2.28), implies

$$y(q) = \begin{cases} -1 \text{ for } q \in S(\bar{\mu})^{+}\setminus S(\mu) \\ \ \ 1 \text{ for } q \in S(\bar{\mu})^{-}\setminus S(\mu). \end{cases} \tag{2.31}$$

But, from (2.29), (2.31) and (2.25) there follows

$$y(q) = \alpha(q) \qquad (q \in S(\mu) \cup S(\bar{\mu})),$$

which contradicts the discontinuity of α in q_0, observed above. Thus, condition β) is necessary in order that G be proximinal.

Assume now that condition γ) is not satisfied, i.e. that there exist $\mu, \bar{\mu} \in G^{\perp}\setminus\{0\}$ such that μ is not absolutely continuous with respect to $\bar{\mu}$ on the set $S(\bar{\mu})$. Then there exists a set $e \subset S(\bar{\mu})$ such that

$$|\bar{\mu}|(e) = 0, \qquad \mu(e) \neq 0, \tag{2.32}$$

hence for the sets

$$e_1 = e \cap [S(\mu)^{+} \cap S(\bar{\mu})^{+}], \qquad e_2 = e \cap [S(\mu)^{+} \cap S(\bar{\mu})^{-}],$$

$$e_3 = e \cap [S(\mu)^{-} \cap S(\bar{\mu})^{+}], \qquad e_4 = e \cap [S(\mu)^{-} \cap S(\bar{\mu})^{-}]$$

we have $e_i \subset S(\mu) \cap S(\bar{\mu})$, $|\bar{\mu}|(e_i) = 0$ $(i = 1, 2, 3, 4)$ and for at least one of them, say e_{i_0}, we have $\mu(e_{i_0}) \neq 0$. Consequently, taking this set e_{i_0} as e and replacing μ by $-\mu$ or $\bar{\mu}$ by $-\bar{\mu}$ (or both), we may assume that we have

$$e \subset S(\mu)^{+} \cap S(\bar{\mu})^{+}. \tag{2.33}$$

Define now a function α on $S(\mu) \cup S(\bar{\mu})$ by *)

$$\alpha(q) = \begin{cases} \ \ 1 \text{ for } q \in [S(\bar{\mu})^{+}\setminus e] \cup [S(\mu)^{-}\setminus S(\bar{\mu})] \\ -1 \text{ for } q \in S(\bar{\mu})^{-} \cup [S(\mu)^{+}\setminus S(\bar{\mu})] \cup e, \end{cases} \tag{2.34}$$

and $\Phi \in (G^{\perp})^{*}$ by

$$\Phi(\mu') = \int_{S(\mu)\cup S(\bar{\mu})} \alpha(q)\, d\mu'(q) \qquad (\mu' \in G^{\perp}). \tag{2.35}$$

We shall show that for this functional Φ there exists no $y \in E$ satisfying (1.1) and (1.2), which contradicts the hypothesis that G is proximinal, and this contradiction will prove the

---

*) Using the hypotheses $e \subset S(\bar{\mu})$, $|\bar{\mu}|(e) = 0$ and $e \neq \varnothing$ (by $\mu(e) \neq 0$), one can show that there exists a $q_0 \in e$ at which α is discontinuous, but we shall not make use of this remark.

necessity of condition γ). Assume that a $y \in E$ satisfies (1.1) a n (1.2). Then for $\mu' = \bar{\mu}$ we obtain, taking into account $e \subset S(\bar{\mu})^+$ and $|\bar{\mu}|(e) = 0$,

$$\int_Q y(q)\, d\bar{\mu}(q) = \bar{\mu}(y) = \Phi(\bar{\mu}) = \int_{S(\bar{\mu})} \alpha(q)\, d\bar{\mu}(q) = -\int_{S(\bar{\mu})^- \cup e} d\bar{\mu}(q) +$$

$$+\int_{S(\bar{\mu})^+ \setminus e} d\bar{\mu}(q) = -\int_{S(\bar{\mu})^-} d\bar{\mu}(q) + \int_{S(\bar{\mu})^+} d\bar{\mu}(q) = \|\bar{\mu}\|, \tag{2.36}$$

and on the other hand we have

$$\|y\| = \|\Phi\| = \max_{q \in S(\mu) \cup S(\bar{\mu})} |\alpha(q)| = 1, \tag{2.37}$$

whence, by Chap. I, corollary 1.6 (implication $1^\circ \Rightarrow 3^\circ$),

$$y(q) = \begin{cases} 1 & \text{for } q \in S(\bar{\mu})^+ \\ -1 & \text{for } q \in S(\bar{\mu})^-. \end{cases} \tag{2.38}$$

Consequently, for $\mu' = \mu$ we obtain

$$\int_{[S(\mu) \setminus S(\bar{\mu})] \cup e} y(q)\, d\mu(q) + \int_{S(\mu) \cap [S(\bar{\mu}) \setminus e]} y(q) d\mu(q) = \int_{S(\mu)} y(q)\, d\mu(q) =$$

$$= \mu(y) = \Phi(\mu) = \int_{S(\mu)} \alpha(q)\, d\mu(q) = \int_{S(\mu)^- \setminus S(\bar{\mu})} d\mu(q) -$$

$$-\int_{[S(\mu)^+ \setminus S(\bar{\mu})] \cup e} d\mu(q) + \int_{S(\mu) \cap [S(\bar{\mu}) \setminus e]} \alpha(q)\, d\mu(q) =$$

$$= -\int_{[S(\mu) \setminus S(\bar{\mu})] \cup e} d\,|\mu|\,(q) + \int_{S(\mu) \cap [S(\bar{\mu}) \setminus e]} y(q)\, d\mu(q),$$

whence

$$\int_{[S(\mu) \setminus S(\bar{\mu})] \cap e} y(q)\, d\mu(q) = -\,|\mu|\,([S(\mu) \setminus S(\bar{\mu})] \cup e),$$

which, together with (2.37) and $e \subset S(\mu)^+$, implies

$$y(q) = \begin{cases} -1 & \text{for } q \in [S(\mu)^+ \setminus S(\bar{\mu})] \cup e \\ 1 & \text{for } q \in S(\mu)^- \setminus S(\bar{\mu}). \end{cases} \tag{2.39}$$

But, by $e \subset S(\bar{\mu})^+$ and (2.38) we have $y(q) = 1$ $(q \in e)$, and by (2.39) we have $y(q) = -1$ $(q \in e)$, which is impossible, since $e \neq \emptyset$ (by $\mu(e) \neq 0$). Thus, condition γ) is necessary in order that G be proximinal.

Conversely, assume now that conditions α), β) and γ) are satisfied and let $\Phi \in (G^{\perp})^*$ be arbitrary with $\|\Phi\|=1$*). We shall show that there exists then a $y \in E$ satisfying (1.1) and (1.2), hence G is proximinal, which will complete the proof. Put

$$\nu = \sum_{i=1}^{n} |\mu_i|, \tag{2.40}$$

and let Γ be the linear subspace of $E^* = \mathfrak{M}^1(Q)$ consisting of all measures μ absolutely continuous with respect to ν. Obviously, $G^{\perp}$ is a linear subspace of Γ, hence there exists an extension $\Psi \in \Gamma^*$, with the same norm $\|\Psi\|=1$, of $\Phi \in (G^{\perp})^*$. By virtue of the canonical equivalence **) $\Gamma \equiv L^1(Q, \nu)$ we then have $\Gamma^* \equiv L^1(Q, \nu)^* = L^{\infty}(Q,\nu)$ canonically, whence there exists an $\alpha \in L^{\infty}(Q, \nu)$ such that

$$\Psi(\mu) = \int_Q \alpha(q) z_{\mu}(q) \mathrm{d}\nu(q) = \int_Q \alpha(q) \mathrm{d}\mu(q) \qquad (\mu \in \Gamma) \tag{2.41}$$

(where z_{μ} is the image of μ under the canonical equivalence $\Gamma \to L^1(Q, \nu)$, i.e. $\mu = z_{\mu} \nu$), and

$$1 = \|\Psi\| = \underset{(\nu)\ q \in Q}{\text{ess sup}} |\alpha(q)|. \tag{2.42}$$

Since dim $G^{\perp} < \infty$ (by (2.2)) and $\|\Phi\| = 1$, there exists a $\mu_0 \in G^{\perp} \setminus \{0\}$ such that $\Phi(\mu_0) = \|\mu_0\|$, hence we have, taking into account (2.41),

$$\int_Q \alpha(q) \mathrm{d}\mu_0(q) = \Psi(\mu_0) = \Phi(\mu_0) = \|\mu_0\|. \tag{2.43}$$

On the other hand, since μ_0 is absolutely continuous with respect to ν, we have, taking into account (2.42),

$$\underset{(\mu_0)\ q \in Q}{\text{ess sup}} |\alpha(q)| \leqslant \underset{(\nu)\ q \in Q}{\text{ess sup}} |\alpha(q)| = 1. \tag{2.44}$$

From the relations (2.43) and (2.44) it follows

$$\alpha(q) = \begin{cases} 1 & \mu_0\text{-a.e. on } S(\mu_0)^+ \\ -1 & \mu_0\text{-a.e. on } S(\mu_0)^-; \end{cases} \tag{2.45}$$

indeed, if there existed a set $e \subset S(\mu_0)^+$ with $\mu_0(e) > 0$ such that $\alpha(q) < 1$ on e, by (2.44) we would have $\int_e \alpha(q) \mathrm{d}\mu_0(q) < \mu_0(e)$,

*) The condition $\|\Phi\| = 1$ does not restrict the generality and allows a simplification of the formulas.

**) See e.g. N Bourbaki [22], Chap. V, p. 54.

whence $\int_Q \alpha(q)\, \mathrm{d}\mu_0(q) < \|\mu_0\|$, and if there existed a set $e \subset S(\mu_0)^-$ with $\mu_0(e) < 0$ such that $\alpha(q) > -1$ on e, by (2.44) we would have $\int_e \alpha(q)\, \mathrm{d}\mu_0(q) = \int_e [-\alpha(q)]\mathrm{d}[-\mu_0(q)] < -\mu_0(e)$, whence $\int_Q \alpha(q)\,\mathrm{d}\mu_0(q) < \mu_0(S(\mu_0)^+) - \mu_0(S(\mu_0)^-) = \|\mu_0\|$, which contradicts (2.43).

Put

$$\alpha_0(q) = \begin{cases} 1 & \text{for} \quad q \in S(\mu_0)^+ \\ -1 & \text{for} \quad q \in S(\mu_0)^- \\ \alpha(q) & \text{for} \quad q \in Q \setminus S(\mu_0). \end{cases} \tag{2.46}$$

Then by condition α) the function $\alpha_0|_{S(\mu_0)}$ is continuous, and by (2.45) and (2.46) we have $\alpha(q) = \alpha_0(q)$ on $Q \setminus e$, where $e \subset S(\mu_0)$, $|\mu_0|(e) = 0$. Since by condition γ) every $\mu \in G^\perp$ is absolutely continuous with respect to μ_0 on the set $S(\mu_0)$, it follows that we have $\mu(e) = 0$ $(\mu \in G^\perp)$, whence

$$\Phi(\mu) = \Psi(\mu) = \int_Q \alpha(q)\,\mathrm{d}\mu(q) = \int_Q \alpha_0(q)\,\mathrm{d}\mu(q) \quad (\mu \in G^\perp). \tag{2.47}$$

Now put

$$Q_1 = S(\nu) \setminus S(\mu_0). \tag{2.48}$$

Since by condition β) the sets $S(\mu_i) \setminus S(\mu_0)$ $(i = 1, \ldots, n)$ are closed and since we have

$$Q_1 = S(\nu) \setminus S(\mu_0) = S\left(\sum_{i=1}^{n} |\mu_i|\right) \setminus S(\mu_0) = \left[\bigcup_{i=1}^{n} S(|\mu_i|)\right] \setminus S(\mu_0) =$$

$$= \bigcup_{i=1}^{n} [S(\mu_i) \setminus S(\mu_0)],$$

the set Q_1 is closed. On the other hand, since all $\mu \in G^\perp$ are absolutely continuous with respect to ν (hence $S(\mu) \subset S(\nu)$), by (2.47) and (2.48) we have

$$\Phi(\mu) = \int_Q \alpha_0(q)\,\mathrm{d}\mu(q) = \int_{S(\nu)} \alpha_0(q)\,\mathrm{d}\mu(q) =$$

$$= \int_{S(\mu_0)} \alpha_0(q)\,\mathrm{d}\mu(q) + \int_Q \alpha_0(q)\,\mathrm{d}\mu(q) \qquad (\mu \in G^\perp). \tag{2.49}$$

Consider the linear subspace Δ_1 of $\mathfrak{M}^1(Q_1) \equiv C(Q_1)^*$ defined by

$$\Delta_1 = \{\mu|_{Q_1} | \mu \in G^{\perp}\} \tag{2.50}$$

and the functional $\Phi_1 \in \Delta_1^*$ defined by

$$\Phi_1(\mu|_{Q_1}) = \int_{Q_1} \alpha_0(q) \, d\mu|_{Q_1}(q) \qquad (\mu|_{Q_1} \in \Delta_1). \tag{2.51}$$

Assume that we have

$$\|\Phi_1\| = 1. \tag{2.52}$$

Then, since $\dim \Delta_1 \leqslant \dim G^{\perp} < \infty$, there exists a $\mu_0^{(1)} \in G^{\perp}$ such that $\Phi_1(\mu_0^{(1)}|_{Q_1}) = \|\mu_0^{(1)}|_{Q_1}\| \neq 0$. Consequently, proceeding as in the case of the measure μ_0, we obtain that the function α_0 in (2.47) can be replaced by a function α_1 having the restriction to $S(\mu_0) \cup S(\mu_0^{(1)}|_{Q_1})$ continuous. Furthermore, since $\emptyset \neq S(\mu_0^{(1)}|_{Q_1}) \subset Q_1 = S(\nu) \setminus S(\mu_0)$, the measures $\mu_0, \mu_0^{(1)} \in G^{\perp}$ are) linearly independent.

Now put

$$Q_2 = Q_1 \setminus S(\mu_0^{(1)}|_{Q_1}) \tag{2.53}$$

and consider the linear subspace Δ_2 of $\mathfrak{M}^1(Q_2) \equiv C(Q_2)^*$ consisting of the restrictions of the measures $\mu \in G^{\perp}$ to Q_2, and the unctional $\Phi_2 \in \Delta_1^*$ defined by

$$\Phi_2(\mu|_{Q_2}) = \int_{Q_2} \alpha_1(q) \, d\mu|_{Q_2}(q) \qquad (\mu|_{Q_2} \in \Delta_2). \tag{2.54}$$

If $\|\Phi_2\| = 1$, taking a $\mu_0^{(2)} \in G^{\perp}$ such that $\Phi_2(\mu_0^{(2)}|_{Q_2}) = \|\mu_0^{(2)}|_{Q_2}\| \neq 0$, we obtain, as previously, that the above function α_1, whence also the function α_0 in (2.47), can be replaced by a function α_2 having the restriction to $S(\mu_0) \cup S(\mu_0^{(1)}|_{Q_1}) \cup S(\mu_0^{(2)}|_{Q_2})$ continuous. In addition, the measures $\mu_0, \mu_{0m}^{(1)}, \mu_0^{(2)} \in G^{\perp}$ are linearly independent, since if $c_0\mu_0 + c_1\mu_0^{(1)} + c_2\mu_0^{(2)} = 0$ and $\sum_{i=0}^{2} |c_i| \neq 0$, then from the linear independence of $\mu_0, \mu_0^{(1)}$ observed above, it follows that $c_2 \neq 0$, whence $\mu_0^{(2)} = -\frac{c_0}{c_2}\mu_0 - \frac{c_1}{c_2}\mu_0^{(1)}$, whence $S(\mu_0^{(2)}|_{Q_1}) \subset S(\mu_0|_{Q_1}) \cup S(\mu_0^{(1)}|_{Q_1}) = S(\mu_0^{(1)}|_{Q_1})$, which contradicts $\emptyset \neq S(\mu_0^{(2)}|_{Q_2}) \subset Q_2 = Q_1 \setminus S(\mu_0^{(1)}|_{Q_1})$. Continuing in this way and taking into account $\dim G = n$, it follows that after m steps, where $1 \leqslant m \leqslant n$, we arrive to a functional $\Phi_m \in \Delta_m^*$ for which

$$\|\Phi_m\| < 1. \tag{2.55}$$

Then, by virtue of the theorem of E. Helly mentioned at the beginning of §1 (footnote), for $0 < \varepsilon < 1 - \|\Phi_m\|$ there exists a $\beta \in C(Q_m)$ such that

$$\|\beta\| \leqslant \|\Phi_m\| + \varepsilon < 1, \quad \Phi_m(\mu|_{Q_m}) = \int_{Q_m} \beta(q)\, d\mu|_{Q_m}(q)$$

$$(\mu|_{Q_m} \in \Delta_m). \qquad (2.56)$$

On the other hand, by (2.48) and $S(\mu_0) \subset S(\nu)$ (since μ_0 is absolutely continuous with respect to ν), we have

$$S(\mu_0) = S(\nu) \setminus Q_1.$$

Also, by $Q_{k+1} = Q_k \setminus S(\mu_0^{(k)}|_{Q_k})$ and $S(\mu_0^{(k)}|_{Q_k}) \subset Q_k$ $(k = 1, \ldots, m-1)$, we have

$$S(\mu_0^{(k)}|_{Q_k}) = Q_k \setminus Q_{k+1} \qquad (k = 1, \ldots, m-1),$$

whence

$$S(\mu_0) \cup S(\mu_0^{(1)}|_{Q_1}) \cup \ldots \cup S(\mu_0^{(m-1)}|_{Q_{m-1}}) = [S(\nu) \setminus Q_1] \cup$$
$$\cup (Q_1 \setminus Q_2) \cup \ldots \cup (Q_{m-1} \setminus Q_m) = S(\nu) \setminus Q_m. \qquad (2.57)$$

Consequently, putting

$$\gamma(q) = \begin{cases} \alpha_{m-1}(q) & \text{for } q \in S(\nu) \setminus Q_m \\ \beta(q) & \text{for } q \in Q_m \end{cases} \qquad (2.58)$$

and extending this continuous function on*) $S(\nu)$ to a continuous function $y \in C(Q)$ with $\|y\| = 1$, we shall have (1.1) and (1.2), which completes the proof of theorem 2.2.

In the particular case when $n = 1$, conditions β) and γ) are automatically satisfied, hence *condition α) is necessary and sufficient in order that G be proximinal* (R. R. Phelps [177], p. 649; besides, this follows immediately also from lemma 1.1, taking into account Chap. I, corollary 1.6). However, in the case when $n \geqslant 2$, the condition α) is no longer sufficient in order that G be proximinal, as shown by the following example of R. R. Phelps ([177], p. 654, example 3): Let $Q = \{1, 2, 3, \ldots \ldots\} \cup \{\infty\}$ with the natural topology (hence $E = C_R(Q) \equiv c_R$), and let $\mu_1, \mu_2 \in E^*$ and $G \subset E$ be defined by

$$\mu_1(\{\infty\}) = 1, \ \mu_1(\{n\}) = \frac{1}{2^n} \qquad (n = 1, 2, \ldots), \qquad (2.59)$$

$$\mu_2(\{\infty\}) = 0, \ \mu_2(\{n\}) = \frac{1}{4^n} \qquad (n = 1, 2, \ldots), \qquad (2.60)$$

$$G = \{x \in E \mid \mu_1(x) = \mu_2(x) = 0\}. \qquad (2.61)$$

---

*) γ is continuous on $S(\nu)$ since we have $S(\nu) = [S(\nu) \setminus Q_m] \cup Q_m$ (by $Q_m \subset S(\nu)$), where $S(\nu) \setminus Q_m$ and Q_m are closed and disjoint, and since $\alpha_{m-1}|_{S(\nu) \setminus Q_m}$ and β are continuous.

Then for any real number b we have $(\mu_1 + b\mu_2)(\{n\}) = \frac{1}{2^n} + \frac{b}{4^n} > 0$ $(2^n > -b)$, hence $S(\mu_1 + b\mu_2)$ admits a Hahn decomposition into two closed sets. Consequently, for any real numbers a, b, the measure $a\mu_1 + b\mu_2$ has the same property and thus G satisfies condition α). On the other hand, we have $|\mu_2|(\{\infty\}) = \mu_2(\{\infty\}) = 0$, $\mu_1(\{\infty\}) = 1$, hence μ_1 is not absolutely continuous with respect to μ_2 on $S(\mu_2)$ and thus G does not satisfy condition γ).

For any infinite compact space Q and any integer n with $1 \leqslant n < \infty$, *the space* $C_R(Q)$ *contains proximinal subspaces* G *of codimension* n (A. L. Garkavi [66], theorem 12). Indeed, e.g. the closed linear subspace

$$G = \{x \in C_R(Q) \,|\, x(q_k) = 0 \ (k = 1, \ldots, n)\}, \tag{2.62}$$

where $q_1, \ldots, q_n \in Q$, obviously satisfies conditions α), β), γ) of theorem 2.2.

THEOREM 2.3 (A. L. Garkavi [72], theorem 2). *Let* $E = C_R(Q)$ *(Q compact) and let* G *be a closed linear subspace of* E *of codimension* n, *hence* $G^{\perp} = [\mu_1, \ldots, \mu_n]$. *In order that* G *be a semi-Čebyšev subspace it is necessary and sufficient that for every* $\mu \in G^{\perp} \setminus \{0\}$ *satisfying**) *condition* α) *of theorem 2.2, the number of points* $q_1, \ldots, q_r$ *of the set* $Q \setminus S(\mu)$ *be* $\leqslant n-1$, *and that we have*

$$\operatorname{rank}\,(\mu_k(\{q_i\}))_{\substack{k=1,\ldots,n\\ i=1,\ldots,r}} = r. \tag{2.63}$$

Proof. Assume that G is a semi-Čebyšev subspace. Let $\mu \in G^{\perp} \setminus \{0\}$ be arbitrary satisfying condition α) of theorem 2.2 and let $q_1, \ldots, q_r \in Q \setminus S(\mu)$ be arbitrary. Then there exist, by the theorem of Tietze, continuous function $x_0, x_1, \ldots, x_r \in C_R(Q)$ with $\|x_i\| = 1$ $(i = 0, 1, \ldots, r)$ such that

$$x_0(q) = \begin{cases} 1 & \text{for } q \in S(\mu)^+ \\ -1 & \text{for } q \in S(\mu)^- \\ 0 & \text{for } q = q_1, \ldots, q_r, \end{cases} \tag{2.64}$$

$$x_i(q) = \begin{cases} 1 & \text{for } q \in S(\mu)^+ \text{ and } q = q_i \\ -1 & \text{for } q \in S(\mu)^- \\ 0 & \text{for } q = q_1, \ldots, q_{i-1}, q_{i+1}, \ldots, q_r \end{cases} \quad (i = 1, \ldots, r). \tag{2.65}$$

---

*) Of course, in the case when no such $\mu \in G^{\perp} \setminus \{0\}$ exists, the condition of theorem 2.3 is satisfied, hence G is a semi-Čebyšev subspace.

Since the functions x_i $(i = 0, 1, \ldots, r)$ are linearly independent and $x_i \in \mathfrak{M}_\mu$ $(i = 0, 1, \ldots, r)$, we have $\dim \mathfrak{M}_\mu \geqslant r$. Consequently, since G is a semi-Čebyšev subspace, by theorem 2.1 we have $n - 1 \geqslant \dim \mathfrak{M}_\mu \geqslant r$, hence the set $Q \setminus S(\mu)$ consists of at most $n - 1$ elements; for simplicity, we may assume that we have

$$Q \setminus S(\mu) = \{q_1, \ldots, q_r\}, \tag{2.66}$$

hence $x_0, x_1, \ldots, x_r$ are even defined on the whole Q by (2.64) and (2.65)*).

We shall show that in this case we have

$$\dim \mathfrak{M}_\mu = r. \tag{2.67}$$

Indeed, let $x \in \mathfrak{M}_\mu$ be arbitrary. Then by (2.64)—(2.66) there exist constants $\alpha_0, \alpha_1, \ldots, \alpha_r$ such that

$$(x - x_0)|_{Q \setminus S(\mu)} = \sum_{i=1}^{r} \alpha_i (x_i - x_0)|_{Q \setminus S(\mu)}. \tag{2.68}$$

On the other hand, taking into account (2.64), (2.65) and

$$x(q) = \begin{cases} 1 & \text{for } q \in S(\mu)^+ \\ -1 & \text{for } q \in S(\mu)^- \end{cases}$$

(by $x \in \mathfrak{M}_\mu$ and Chap. I, corollary 1.6), it follows that we have also

$$(x - x_0)|_{S(\mu)} = \sum_{i=1}^{r} \alpha_i (x_i - x_0)|_{S(\mu)}, \tag{2.69}$$

which, together with (2.68), gives

$$x - x_0 = \sum_{i=1}^{r} \alpha_i (x_i - x_0).$$

Since $x \in \mathfrak{M}_\mu$ has been arbitrary, it follows that $\dim \mathfrak{M}_\mu = \dim (\mathfrak{M}_\mu - x_0) \leqslant r$, which, together with the inequality $\dim \mathfrak{M}_\mu \geqslant r$ proved above, gives (2.67).

Consequently, since G is a semi-Čebyšev subspace, by theorem 2.1 (implication $1^\circ \Rightarrow 5^\circ$) we have

$$\operatorname{rank} \left(\mu_k(x_i)\right)_{\substack{k=1,\ldots,n \\ i=0,1,\ldots,r}} = r + 1. \tag{2.70}$$

---

*) Replacing in (2.64) and (2.65) 0 by -1, one obtains even linearly independent extremal points of S, but we shall not use this remark in the sequel.

Renumbering, if necessary, the measures $\mu_1, \ldots, \mu_n$ we may then assume that we have

$$0 \neq D = \begin{vmatrix} \mu_1(x_0) & \mu_1(x_1) & \ldots & \mu_1(x_r) \\ \ldots & \ldots & \ldots & \ldots \\ \mu_{r+1}(x_0) & \mu_{r+1}(x_1) & \ldots & \mu_{r+1}(x_r) \end{vmatrix}.$$

Subtracting the first colum from the others, we obtain

$$0 \neq D = \begin{vmatrix} \mu_1(x_0) & \mu_1(x_1 - x_0) & \ldots & \mu_1(x_r - x_0) \\ \ldots & \ldots & \ldots & \ldots \\ \mu_{r+1}(x_0) & \mu_{r+1}(x_1 - x_0) & \ldots & \mu_{r+1}(x_r - x_0) \end{vmatrix}.$$

Expanding this determinant by the elements of the first column it follows that at least one of the corresponding subdeterminants must be $\neq 0$. Since by (2.64)—(2.66) we have

$$\mu_k(x_i - x_0) = \int_Q [x_i(q) - x_0(q)]\, d\mu_k(q_i) = \mu_k(\{q_i\})$$

$$(k = 1, \ldots, n;\ i = 1, \ldots, r), \qquad (2.71)$$

it follows that the respective minor is a determinant of order r of the matrix $(\mu_k(\{q_i\}))_{\substack{k=1,\ldots,n \\ i=1,\ldots,r}}$, whence we have (2.63).

Conversely, assume now that the conditions of theorem 2.3 are satisfied and let $\mu \in G^\perp \setminus \{0\}$ be arbitrary. If $\mathfrak{M}_\mu = \emptyset$, then $\dim \mathfrak{M}_\mu = -1 < n - 1$, and condition (2.5) is trivial. Assume that $\mathfrak{M}_\mu \neq \emptyset$. Then, by Chap. I, corollary 1.6, μ satisfies condition α) of theorem 2.2, hence by virtue of what has been proved above we have (2.67). Since by our hypothesis $r \leqslant n - 1$, it follows that we have $\dim \mathfrak{M}_\mu \leqslant n - 1$. On the other hand, since by our hypothesis we have (2.63), renumbering, if necessary, the measures $\mu_1, \ldots, \mu_n$ we may assume that we have

$$0 \neq D' = \begin{vmatrix} \mu_1(\{q_1\}) & \ldots & \mu_1(\{q_r\}) \\ \ldots & \ldots & \ldots \\ \mu_r(\{q_1\}) & \ldots & \mu_r(\{q_r\}) \end{vmatrix}.$$

Consequently, taking into account (2.71), $x_0, x_1, \ldots, x_r \in \mathfrak{M}_\mu$ (for the functions $x_0, x_1, \ldots, x_r$ above), and also $\mu = \sum_{k=1}^{n} \beta_k \mu_k$, we have

$$0 \neq D' = \begin{vmatrix} \mu_1(x_1 - x_0) & \ldots & \mu_1(x_r - x_0) \\ \ldots & \ldots & \ldots \\ \mu_r(x_1 - x_0) & \ldots & \mu_r(x_r - x_0) \end{vmatrix} =$$

$$
= \frac{1}{\|\mu\|} \begin{vmatrix} \|\mu\| & 0 & \dots & 0 \\ \mu_1(x_0) & \mu_1(x_1 - x_0) & \dots & \mu_1(x_r - x_0) \\ \dots & \dots & \dots & \dots \\ \mu_r(x_0) & \mu_r(x_1 - x_0) & \dots & \mu_r(x_r - x_0) \end{vmatrix} =
$$

$$
= \frac{1}{\|\mu\|} \begin{vmatrix} \|\mu\| & \|\mu\| & \dots & \|\mu\| \\ \mu_1(x_0) & \mu_1(x_1) & \dots & \mu_1(x_r) \\ \dots & \dots & \dots & \dots \\ \mu_r(x_0) & \mu_r(x_1) & \dots & \mu_r(x_r) \end{vmatrix} =
$$

$$
= \frac{1}{\|\mu\|} \begin{vmatrix} \mu(x_0) & \mu(x_1) & \dots & \mu(x_r) \\ \mu_1(x_0) & \mu_1(x_1) & \dots & \mu_1(x_r) \\ \dots & \dots & \dots & \dots \\ \mu_r(x_0) & \mu_r(x_1) & \dots & \mu_r(x_r) \end{vmatrix} =
$$

$$
= \frac{1}{\|\mu\|} \sum_{k=r+1}^{n} \beta_k \begin{vmatrix} \mu_k(x_0) & \mu_k(x_1) & \dots & \mu_k(x_r) \\ \mu_1(x_0) & \mu_1(x_1) & \dots & \mu_1(x_r) \\ \dots & \dots & \dots & \dots \\ \mu_r(x_0) & \mu_r(x_1) & \dots & \mu_r(x_r) \end{vmatrix},
$$

whence at least one of the determinants which appear in this sum is $\neq 0$, and thus we have (2.70). Hence, by virtue of theorem 2.1 (implication $2° \Rightarrow 1°$), G is a semi-Čebyšev subspace, which completes the proof of theorem 2.3.

In the particular case when $n = 1$, hence $G = \{x \in E \mid \mu_1(x) = 0\}$, for every $\mu \in G^{\perp} \setminus \{0\}$ we have $S(\mu) = S(\mu_1)$ and therefore *G is a semi-Čebyšev subspace if and only if either a) μ_1 does not satisfy condition α) of theorem 2.2, or b) μ_1 satisfies condition α) of theorem 2.2 and* $S(\mu_1) = Q$. From the foregoing it follows that we have *a*) if and only if the hyperplane G is non-proximinal and *b*) if and only if G is a Čebyšev subspace.

From theorems 2.2, 2.3 (or from theorem 2.2 and Chap. I, corollary 3.7) it follows immediately

COROLLARY 2.2. *Let $E = C_R(Q)$ (Q compact) and let G be a closed linear subspace of E of codimension n, hence $G^{\perp} = [\mu_1, \dots, \mu_n]$. If G is proximinal and if we have $S(\mu) = Q$ for every $\mu \in G^{\perp} \setminus \{0\}$, then G is a Čebyšev subspace.*

In corollary 2.5 we shall see that if Q has no isolated point, then the condition of corollary 2.2 is necessary and sufficient in order that a proximinal linear subspace G of codimension n be a Čebyšev subspace.

Concerning the existence of semi-Čebyšev hyperplanes, we have seen in Chap. I, § 3, that every non-reflexive Banach space E, hence in particular every space $C_R(Q)$ (Q *infinite* compact) has at least one non-proximinal, hence semi-Čebyšev hyperplane. Also, we have

THEOREM 2.4 (A. L. Garkavi [66], theorem 12). *For every infinite separable compact space (hence, in particular, for every infinite compact metric space) Q and every integer n with $1 \leqslant n < \infty$, the space $C_R(Q)$ contains semi-Čebyšev subspaces G of codimension n.*

Proof. Let $A = \{q_1, q_2, \ldots\}$ be a countable dense set in Q. Put

$$\mu_k(\{q_i\}) = \frac{1}{i^{k+1}} \qquad (k = 1, \ldots, n;\ n = 1, 2, \ldots), \tag{2.72}$$

$$\mu_k(e) = \sum_{q_i \in e} \frac{1}{i^{k+1}} \qquad (k = 1, \ldots, n;\ e \subset Q), \tag{2.73}$$

$$G = \{x \in C_R(Q) \mid \mu_k(x) = 0\ (k = 1, \ldots, n)\}. \tag{2.74}$$

Let $\mu = \sum_{k=1}^{n} a_k \mu_k \in G^{\perp} \setminus \{0\}$ be arbitrary. We have then

$$\mu(\{q_i\}) = \sum_{k=1}^{n} a_k \mu_k(\{q_i\}) = \sum_{k=1}^{n} \frac{a_k}{i^{k+1}} = \frac{1}{i^{n+1}} (a_1 i^{n-1} + a_2 i^{n-2} + \ldots + a_n) \qquad (i = 1, 2, \ldots), \tag{2.75}$$

hence there exist at most $n - 1$ values of i for which $\mu(\{q_i\}) = 0$. Now *) let $q_0 \in Q'$ be arbitrary. Then, since A is dense in Q and $\mu(\{q_i\}) = 0$ for at most $n-1$ values of i, it follows that for every open neighborhood V of q_0 there exists a $q_i \in A \cap V$ such that $\mu(\{q_i\}) \neq 0$. Consequently, we have

$$|\mu|(V) \geqslant |\mu|(\{q_i\}) = |\mu(\{q_i\})| > 0,$$

whence, since V has been an arbitrary open neighborhood of q_0, we obtain $q_0 \in S(\mu)$, and thus $Q' \subset S(\mu)$, hence

$$Q \setminus S(\mu) \subset Q \setminus Q' \qquad (\mu \in G^{\perp} \setminus \{0\}). \tag{2.76}$$

Thus for every $q \in Q \setminus S(\mu)$ ($\mu \in G^{\perp} \setminus \{0\}$) the set $\{q\}$ is open in Q, hence it satisfies $\mu(\{q\}) = 0$. Since A is dense in Q, we have $Q \setminus Q' \subset A$, whence $Q \setminus S(\mu) \subset A$, whence, taking into account that $\mu(\{q_i\}) = 0$ for at most $n - 1$ values of i, it follows that for every $\mu \in G^{\perp} \setminus \{0\}$ the set $Q \setminus S(\mu)$ consists of at most $n - 1$ points, i.e. G satisfies the first condition of theorem 2.3.

---

*) We recall that we denote by Q' the set of all limit points of Q.

Now let $Q\setminus S(\mu) = [Q\setminus S(\mu)]\cap A = \{q_{i_1}, \ldots, q_{i_r}\}$ $(r \leqslant n-1)$. Then we have

$$(\mu_k(\{q_{i_l}\}))_{\substack{k=1,\ldots,n\\ l=1,\ldots,r}} = \begin{pmatrix} \frac{1}{i_1^2} & \frac{1}{i_2^2} & \cdots & \frac{1}{i_r^2} \\ \frac{1}{i_1^3} & \frac{1}{i_2^3} & \cdots & \frac{1}{i_r^3} \\ \cdots & \cdots & \cdots & \cdots \\ \frac{1}{i_1^{n+1}} & \frac{1}{i_2^{n+1}} & \cdots & \frac{1}{i_r^{n+1}} \end{pmatrix}.$$

Expanding the determinant $\det(\mu_k(\{q_{i_l}\}))_{k,l=1,\ldots,r}$ by the elements of the first column, we obtain

$$\det(\mu_k(\{q_{i_l}\}))_{k,l=1,\ldots,r} = \sum_{j=1}^{r} b_j \frac{1}{i_1^{j+1}} = \sum_{j=1}^{r} b_j \left(\frac{1}{i_1}\right)^{j+1},$$

where the polynomial $\sum_{j=1}^{r} b_j t^{j+1}$ has zeros at the points $\frac{1}{i_l}$ $(l=2, \ldots, r)$; since this polynomial has at most $r-1$ positive zeros, it follows that at $\frac{1}{i_1}$ it does not vanish, whence $\det(\mu_k(\{q_{i_l}\}))_{k,l=1,\ldots,r} \neq 0$, and thus G also satisfies the second condition of theorem 2.3, which completes the proof of theorem 2.4.

THEOREM 2.5 (A. L. Garkavi [72], theorem 3*)). *Let $E = C_R(Q)$ (Q compact) and let G be a closed linear subspace of E of codimension n, hence $G^{\perp} = [\mu_1, \ldots, \mu_n]$. In order that G be a Čebyšev subspace it is necessary and sufficient that the following four conditions be satisfied:*

a) *For every $\mu \in G^{\perp}\setminus\{0\}$ the space Q admits a Hahn decomposition with respect to μ into two closed sets Q^+ and $Q^- = Q\setminus Q^+$.*

b) *Any two measures $\mu, \bar{\mu} \in G^{\perp}\setminus\{0\}$ are equivalent (i.e. each of them is absolutely continuous with respect to the other) on the set Q' of all limit points of Q.*

c) *For every $\mu \in G^{\perp}\setminus\{0\}$ the set $Q\setminus S(\mu)$ consists of at most $n-1$ points.*

d) *For any r isolated points $q_1, \ldots, q_r$ of Q, where $1 \leqslant r \leqslant n-1$, we have*

$$\operatorname{rank}\,(\mu_k(\{q_i\}))_{\substack{k=1,\ldots,n\\ i=1,\ldots,r}} = r. \tag{2.77}$$

*) The necessity of condition c) has been observed also by R. R. Phelps ([177], proposition 6).

Proof. Assume that G is a Čebyšev subspace. Then G satisfies the conditions of theorems 2.2 and 2.3, whence in particular we have c). From condition α) of theorem 2.2 and from c) it follows that we have a). Furthermore, from condition γ) of theorem 2.2 and from c) it follows that we have b) (since by c) we have $Q' \subset S(\mu)$ for every $\mu \in G^{\perp} \setminus \{0\}$). Finally, let $q_1, \ldots, q_r$ be isolated points of Q, where $1 \leqslant r \leqslant n-1$. Then the system of r homogeneous linear equations with n unknowns

$$\sum_{k=1}^{n} \alpha_k \mu_k(\{q_i\}) = 0 \qquad (i = 1, \ldots, r) \tag{2.78}$$

has a non-trivial solution $\alpha_1^{(0)}, \ldots, \alpha_n^{(0)}$, hence for the measure $\mu = \sum_{k=1}^{n} \alpha_k^{(0)} \mu_k \in G^{\perp} \setminus \{0\}$ we have $q_1, \ldots, q_r \in Q \setminus S(\mu)$ (since by hypothesis $\{q_1, \ldots, q_r\}$ is an open subset of Q), whence, by condition (2.63) of theorem 2.3 we obtain (2.77); indeed, if we had rank $(\mu_k(\{q_i\}))_{\substack{k=1,\ldots,n\\ i=1,\ldots,r}} < r$ and $Q \setminus S(\mu) = \{q_1, \ldots, q_{r_0}\}$, where $r \leqslant r_0 \leqslant n-1$, it would follow that rank $(\mu_k(\{q_i\}))_{\substack{k=1,\ldots,n\\ i=1,\ldots,r_0}} < r_0$ which contradicts (2.63).

Conversely, assume that conditions a)—d) are satisfied. Then by a) we have α) of theorem 2.2 (putting $S(\mu)^+ = S(\mu) \cap Q^+$, $S(\mu)^- = S(\mu) \cap Q^-$). Furthermore, by c) and $S(\bar{\mu}) \setminus S(\mu) \subset Q \setminus S(\mu)$ ($\mu, \bar{\mu} \in G^{\perp} \setminus \{0\}$) we have β) of theorem 2.2, and b) and c) obviously imply γ) of theorem 2.2, and thus G is proximinal. On the other hand, from c) and d) there follow obviously the conditions of theorem 2.3, and thus G is a semi-Čebyšev subspace, whence also a Čebyšev subspace, which completes the proof of theorem 2.5.

In the particular case when $n = 1$, hence $G = \{x \in E \mid \mu_1(x) = 0\}$, for every $\mu \in G^{\perp} \setminus \{0\}$ we have $S(\mu) = S(\mu_1)$, and hence *G is a Čebyšev subspace if and only if the space Q admits a Hahn decomposition with respect to μ_1 into two closed sets Q^+ and Q^- and* $S(\mu_1) = Q$. This result has been given by A. L. Garkavi, ([66], theorem 10) and, independently, by R. R. Phelps ([177], corollary 5). In particular, *when $n = 1$ and the compact space Q is connected, G is a Čebyšev subspace if and only if the measure μ_1 is either strictly positive or strictly negative* (A. L. Garkavi *) [66], p. 116, corollary of theorem 10); indeed, Q being connected, and Q^+ and Q^- being disjoint closed sets with $Q = Q^+ \cup Q^-$, one of these two sets must be void, whence $S(\mu_1) = Q^+$ or Q^-.

*) The fact that for a connected Q every Čebyšev hyperplane G can be written the form $G = \{x \in E \mid \mu_1(x) = 0\}$, where μ_1 is a positive measure, has been established by R. R. Phelps ([176], theorem 3.4).

COROLLARY 2.3. *Let $E = C_R(Q)$, where Q is a compact space containing at least n isolated points and let G be a closed linear subspace of E of codimension n, hence $G^\perp = [\mu_1, \ldots, \mu_n]$. In order that G be a Čebyšev subspace it is necessary and sufficient that conditions a), b), c) of theorem 2.5 be satisfied. In the particular case when G is proximinal, condition c) is necessary and sufficient in order that G be a Čebyšev subspace.*

Proof. For the first statement it will be sufficient to show that if Q contains at least n isolated points, condition d) of theorem 2.5 is a consequence of conditions a), b), c) of that theorem. Assume that there exist r isolated points $q_1, \ldots, q_r$ of Q, where $1 \leqslant r \leqslant n-1$, such that

$$\operatorname{rank} (\mu_k(\{q_i\}))_{\substack{k=1,\ldots,n \\ i=1,\ldots,r}} < r. \tag{2.79}$$

Taking then isolated points $q_{r+1}, \ldots, q_n$ of Q, different from $q_1, \ldots, q_r$ (such points exist since Q has by hypothesis at least n isolated points), we have

$$\det (\mu_k(\{q_i\}))_{k, i=1,\ldots,n} = 0, \tag{2.80}$$

hence the system of n homogeneous linear equations with n unknowns

$$\sum_{k=1}^{n} \alpha_k \mu_k(\{q_i\}) = 0 \qquad (i = 1, \ldots, n) \tag{2.81}$$

has a non-trivial solution $\alpha_1^{(0)}, \ldots, \alpha_n^{(0)}$. Consequently, for the measure $\mu = \sum_{k=1}^{n} \alpha_k^{(0)} \mu_k \in G^\perp \setminus \{0\}$ we have $q_1, \ldots, q_n \in Q \setminus S(\mu)$ (since by our hypothesis $\{q_1, \ldots, q_n\}$ is an open subset of Q), in contradiction with c) and thus the first statement is proved.

For the second statement it is sufficient to observe that if G is proximinal and satisfies condition c), then, as has been shown in the proof of theorem 2.5, G also satisfies conditions a), b), which completes the proof of corollary 2.3.

The first statement of corollary 2.3 has been given by A. L. Garkavi ([72], p. 515), and the second statement has been given by R. R. Phelps ([177], theorem 9).

Corollary 2.3 is no longer true when the number of isolated points of Q is $\leqslant n - 1 = \operatorname{codim} G - 1$, as shown by the following example of R. R. Phelps ([177], p. 654, example 2): Let $T = [0, 1] \cup \{2\}$, and let λ be the measure equal to the Lebesgue measure on the Borel subsets of [0,1] and equal to 1 on $\{2\}$. There exists then*) a compact space $Q = Q_{T,\lambda}$ and an isometry $\varkappa$ of $L_R^\infty(T, \lambda)$ onto $C_R(Q)$, which is linear, multi-

*) See e.g. N. Dunford and J. Schwartz [49], p.445, theorem 11.

plicative and positive, hence it carries the characteristic function χ_e of a λ-measurable set $e \subset T$ into a characteristic function $\chi_{\psi(e)} \in C_R(Q)$. Thus one obtains a mapping ψ of the σ-algebra of all λ-measurable sets $e \subset T$ (modulo the sets of λ-measure zero) onto the family of all simultaneously closed and open subsets of Q, which maps the atoms of (T, λ) in a one-to-one fashion onto the isolated points of Q; since the only atom of (T, λ) is $\{2\}$, it follows that Q has a single isolated point, say q_0. Now let $y_1, \ldots, y_n \in L_R^1(T, \lambda) \subset L_R^\infty(T, \lambda)^*$, where $n \geqslant 2$, be defined by

$$y_k(2) = 0, \quad y_k(t) = t^{k-1} \quad (t \in [0, 1]; \; k = 1, \ldots, n), \tag{2.82}$$

let $\mu_k \in C_R(Q)^*$ be the image of y_k $(k = 1, \ldots, n)$ by the isometry $L_R^\infty(T, \lambda)^* \equiv C_R(Q)^*$, hence

$$\int_T z(t) y_k(t) \, \mathrm{d}\lambda(t) = \int_Q (\varkappa z)(q) \, \mathrm{d}\mu_k(q) \qquad (z \in L_R^\infty(T, \lambda); k = 1, \ldots, n), \tag{2.83}$$

and let

$$G = \{x \in C_R(Q) \mid \mu_k(x) = 0 \; (k = 1, \ldots, n)\}. \tag{2.84}$$

Then, taking into account that for $\mu = \sum_{k=1}^n \alpha_k \mu_k \in G^\perp$ and $y = \sum_{k=1}^n \alpha_k y_k$ we have $S(\mu) = \psi[S(y)]$, where $S(y) = T \setminus \{t \in T \mid y(t) = 0\}$ (modulo a set of λ-measure zero), and that y has at most $n-1$ zero son $[0, 1]$, whence $\lambda(\{t \in [0, 1] \mid y(t) = 0\}) = 0$, it follows that G satisfies the conditions α), β), γ) of theorem 2.2, hence it is proximinal*). Also, we have $Q \setminus S(\mu) = \{q_0\}$ $(\mu \in G^\perp \setminus \{0\})$, and thus, since $n \geqslant 2$, G satisfies condition c) of theorem 2.5, hence also conditions a), b). At the same time, the only isolated point of Q is q_0, and by (2.83) for $z = \chi_{\{2\}}$ (hence $\varkappa(z) = \chi_{\psi(\{2\})} = \chi_{\{q_0\}}$) and (2.82) we have

$$\mu_k(\{q_0\}) = \int_Q \chi_{\{q_0\}}(q) \mathrm{d}\mu_k(q) = \int_T \chi_{\{2\}}(t) \, y_k(t) \mathrm{d}\lambda(t) = 0 \qquad (k = 1, \ldots, n),$$

---

*) This also follows from the fact that G is the image by the above isometry of the $\sigma(L_R^\infty(T, \lambda), L_R^1(T, \lambda))$-closed linear subspace $\left\{z \in L_R^\infty(T, \lambda) \,\middle|\, \int_T z(t) y_k(t) \, \mathrm{d}\lambda(t) = 0 \; (k = 1, \ldots, n)\right\}$ of $L_R^\infty(T, \lambda)$ (see Chap. I, corollary 2.5).

whence rank $(\mu_k(\{q_0\}))_{k=1,\ldots,n} = 0$, and thus G does not satisfy condition d) of theorem 2.5.

COROLLARY 2.4 (A. L. Garkavi [68], theorem 9 and [72], p. 515). *Let $E = C_R(Q)$, where Q is a countable infinite compact space and let G be a closed linear subspace of E of codimension $n \geqslant 2$, hence $G^\perp = [\mu_1, \ldots, \mu_n]$. In order that G be a Čebyšev subspace it is necessary and sufficient that there be satisfied conditions* a), c) *of theorem 2.5 and condition*

b′) *We have*

$$\mu(\{q\}) = 0 \qquad (q \in Q', \ \mu \in G^\perp \setminus \{0\}). \tag{2.85}$$

Proof. Since Q contains an infinity of isolated points, it will be sufficient, taking into account corollary 2.3, to show that *in the case when Q is a countable infinite compact space and $n \geqslant 2$, condition* b) *of theorem 2.5 is equivalent to the condition* b′) *above*; obviously, for a countable Q we have the implication b′) $\Rightarrow$ b), hence it will be sufficient to show that b) $\Rightarrow$ b′). Assume that we do not have b′), i.e. that there exist $q_0 \in Q'$ and $\mu_0 \in G^\perp \setminus \{0\}$ such that

$$\mu_0(\{q_0\}) \neq 0.$$

Take, by $n \geqslant 2$, a measure $\mu \in G^\perp \setminus \{0\}$ linearly independent of μ_0 and put

$$\bar{\mu}_0 = \mu - \frac{\mu(\{q_0\})}{\mu_0(\{q_0\})} \mu_0.$$

We have then $\bar{\mu}_0 \in G^\perp \setminus \{0\}$ and

$$|\bar{\mu}_0| (\{q_0\}) = |\bar{\mu}_0(\{q_0\})| = 0,$$

whence μ_0 is not absolutely continuous with respect to $\bar{\mu}_0$ on the set Q', and thus condition b) of theorem 2.5 is not satisfied. Consequently b) $\Rightarrow$ b′), which completes the proof.

In particular, for $Q = \{1, 2, 3, \ldots\} \cup \{\infty\}$ with the natural topology, we obtain that *in the space $E = C_R(Q) \equiv c_R$ a linear subspace G of codimension $n \geqslant 2$ is a Čebyšev subspace if and only if it satisfies conditions a) and c) of theorem 2.5 and condition*

$$\mu(\{\infty\}) = 0 \qquad (\mu \in G^\perp \setminus \{0\}). \tag{2.86}$$

Indeed, we have $Q' = \{\infty\}$, hence (2.86) is nothing else than condition b′) of corollary 2.4.

We observe that the above proof of the implication b) $\Rightarrow$ b′) is valid for *arbitrary* compact spaces Q, in the hypothesis $n \geqslant 2$. Consequently, if $G \subset C_R(Q)$ (Q compact), with $2 \leqslant \operatorname{codim} G = n < \infty$, is a Čebyšev subspace, then we have (2.85); in particular, if $Q' = Q$ (i.e. if Q has no isolated point), it follows that every $\mu \in G^\perp \setminus \{0\}$ must be diffuse.

COROLLARY 2.5. *Let $E = C_R(Q)$, where Q is a compact space which has no isolated point and let G be a closed linear subspace of E of codimension n, hence $G^\perp = [\mu_1, \ldots, \mu_n]$. In order that G be a Čebyšev subspace, it is necesary and sufficient that there be satisfied conditions* a), b) *of theorem 2.5 and condition*

c′) *We have*

$$S(\mu) = Q \qquad (\mu \in G^\perp \setminus \{0\}). \tag{2.87}$$

In the particular case when G is proximinal, condition c′) *is necessary and sufficient in order that G be a Čebyšev subspace.*

Proof. For the first statement it is sufficient to observe that if Q has no isolated point, conditions a), b), c), d) of theorem 2.5 reduce to conditions a), b), c′) (since by c) every point of $Q \setminus S(\mu)$ would be an isolated point of Q). For the second statement it is sufficient to observe that if G is proximinal and satisfies condition c′) then by theorem 2.2 it also satisfies conditions a), b), which completes the proof of corollary 2.5.

The second statement of corollary 2.5 has been given by R. R. Phelps ([177], theorem 10).

Corollary 2.5 is no longer valid when Q contains an isolated point, as shown by the following example of R. R. Phelps ([177], p. 653, example 1) : Let $Q = Q_{T,\lambda}$ be the compact space of the example given after corollary 2.3, let $y_k \in L^1_R(T, \lambda) \subset L^\infty_R(T, \lambda)^*$ $(k = 1, 2)$ be defined by

$$y_1(t) = 1\ (t \in T),\ y_2(2) = 0,\ \ y_2(t) = t\ (t \in [0, 1]), \tag{2.88}$$

let $\mu_k \in C_R(Q)^*$ be the image of y_k $(k = 1,2)$ under the isometry $L^\infty_R(T, \lambda)^* \equiv C_R(Q)^*$, hence

$$\int_T z(t) y_k(t) \mathrm{d}\lambda(t) = \int_Q (\varkappa z)(q) \mathrm{d}\mu_k(q) \qquad (z \in L^\infty_R(T, \lambda),\ k = 1, 2), \tag{2.89}$$

and let

$$G = \{x \in C_R(Q) \mid \mu_k(x) = 0 \quad (k = 1, 2)\}. \tag{2.90}$$

Then for the isolated point q_0 of Q (corresponding to the atom $\{2\}$ of (T, λ)) we have $q_0 \notin S(\mu_2) = \psi[S(y_2)]$ (since $2 \notin S(y_2)$), hence condition c′) is not satisfied. At the same time, G satisfies conditions a), b), c), d) of theorem 2.5 (since by (2.89) for $z = \chi_{\{2\}}$, hence $\varkappa(z) = \chi_{\psi(\{2\})} = \chi_{\{q_0\}}$, we have $\mu_1(\{q_0\}) =$

$$= \int_Q \chi_{\{q_0\}}(q) \mathrm{d}\mu_1(q) = \int_T \chi_{\{2\}}(t) y_1(t) \mathrm{d}\lambda(t) = \lambda(\{2\}) = 1, \quad \text{whence}$$

rank $(\mu_k(\{q_0\}))_{k=1,2} = 1$), and thus G is a Čebyšev subspace.

Let us pass now to the problem of existence of Čebyšev subspaces of finite codimension in the space $C_R(Q)$. We shall first give conditions on the compact space Q, sufficient in order that $C_R(Q)$ have Čebyšev subspaces of finite codimension.

THEOREM 2.6 (A. L. Garkavi [66], theorem 11). *For every separable compact space Q (hence, in particular, for every compact metric space) the space $C_R(Q)$ contains Čebyšev subspaces G of codimension 1.*

Proof. Let $\{q_1, q_2, \ldots\}$ be a countable dense set in Q. Put

$$\mu_1(\{q_i\}) = \frac{1}{2^i} \qquad (i = 1, 2, \ldots), \tag{2.91}$$

$$\mu_1(e) = \sum_{q_i \in e} \frac{1}{2^i} \qquad (e \subset Q), \tag{2.92}$$

$$G = \{x \in E \mid \mu_1(x) = 0\}. \tag{2.93}$$

Then the space Q admits a Hahn decomposition with respect to μ_1 into two closed sets, namely $Q^+ = Q$, $Q^- = \emptyset$, and we have $S(\mu_1) = Q$, whence, by virtue of the remark made after the proof of theorem 2.5, G is a Čebyšev subspace, which completes the proof.

THEOREM 2.7 (A. L. Garkavi [68], corollary of theorem 9 and [72], p. 516). *For every separable compact space Q satisfying*

$$Q = \overline{Q \setminus Q'}, \tag{2.94}$$

i.e. coinciding with the closure of the set of its isolated points (hence, in particular, for every countable compact space Q) and every integer n with $1 \leqslant n < \infty$, the space $C_R(Q)$ contains Čebyšev subspaces G of codimension n.

Proof. Obviously, it is sufficient to consider the case when Q is infinite. Let $A = \{q_1, q_2, \ldots\}$ be a countable dense set in Q. We have then $Q \setminus Q' \subset A$, hence by (2.94) we may assume (considering eventually $(Q \setminus Q') \cap A$ instead of A) that we have

$$Q \setminus Q' = A = \{q_1, q_2, \ldots\}. \tag{2.95}$$

Now let $\mu_k \in C_R(Q)^*$ $(k = 1, \ldots, n)$ and $G \subset C_R(Q)$ be defined by (2.72)—(2.74) and let $\mu \in G^\perp \setminus \{0\}$ be arbitrary. Then by (2.75) one of the sets $\{q_i \in A \mid \mu(\{q_i\}) \leqslant 0\}$, $\{q_i \in A \mid \mu(\{q_i\}) \geqslant 0\}$, say the first, is finite. Put

$$Q^+ = \overline{\{q_i \in A \mid \mu(\{q_i\}) > 0\}}, \quad Q^- = \{q_i \in A \mid \mu(\{q_i\}) \leqslant 0\}. \tag{2.96}$$

Then the sets Q^+, Q^- are closed (Q^- is finite) and we have

$$Q^+ \cup Q^- = \overline{\{q_i \in A \mid \mu(\{q_i\}) > 0\} \cup \{q_i \in A \mid \mu(\{q_i\}) \leqslant 0\}} = \overline{A} = Q.$$

Furthermore, the sets Q^+, Q^- are disjoint. Indeed, if $q \in Q^+ \cap Q^-$, then $q = q_{i_0} \in A$, $\mu(\{q_{i_0}\}) \leqslant 0$ for a suitable i_0 and on the other hand, $q_{i_0} \in \overline{\{q_i \in A \mid \mu(\{q_i\}) > 0\}}$, hence in every neighborhood of q_{i_0} there exists a $q_i \in A$ with $\mu(\{q_i\}) > 0$, whence, since by (2.95) q_{i_0} is an isolated point of Q, it follows that $\mu(\{q_{i_0}\}) > 0$; this contradiction proves that $Q^+ \cap Q^- = \emptyset$. Consequently, taking into account (2.96) and (2.73), it follows that the closed sets Q^+, Q^- constitute a Hahn decomposition of Q with respect to μ, and thus G satisfies condition a) of theorem 2.5.

On the other hand, since by (2.95) we have $Q' = Q \setminus A = Q \setminus \{q_1, q_2, \ldots\}$ and since (by (2.73)) for every set $B \subset Q \setminus \{q_1, q_2, \ldots\}$ we have $\mu_k(B) = 0$, whence $\mu(B) = 0$ $(\mu \in G^{\perp} \setminus \{0\})$, condition b) of theorem 2.5 is also satisfied.

Finally, the proof of theorem 2.3 shows that G satisfies condition c) of theorem 2.5. Since by our hypothesis Q is infinite and satisfies (2.94), it follows that Q contains at least n isolated points and thus the conditions *) of corollary 2.3 are satisfied, whence G is a Čebyšev subspace, which completes the proof.

The compact spaces Q which satisfy the condition of theorem 2.7 may contain also infinite connected closed subsets (continuums); thus, e.g. the space $\overline{Q = \left\{\left\{\frac{n}{2^i}, \frac{1}{2^i}\right\} \in R^2 \,|\, n = 0, 1, \ldots, 2^i;\ i = 0, 1, 2, \ldots\right\}}$ contains the segment $\{\{\lambda, 0\} \,|\, 0 \leqslant \lambda \leqslant 1\}$.

The conditions of theorem 2.6 and 2.7 are not necessary in order that $C_R(Q)$ contain Čebyšev subspaces of codimension 1, respectively of every finite codimension n. Indeed, as we shall see in § 3, corollary 3.3, the space $L_R^\infty([0, 1], \nu) \equiv C_R(Q)$, where ν = the Lebesgue measure, contains Čebyšev subspaces of every finite codimension n; at the same time, the corresponding space Q is non-separable (since in Q every set of the first category, whence in particular every countable set, is **) nowhere dense), and contains no isolated point (since $([0, 1], \nu)$ has no atom).

There exist extremally disconnected compact spaces Q such that $C_R(Q)$ has no Čebyšev subspace of finite codimension. Indeed, as has been observed at the end of section 2.1, if $E = l_R^\infty([0, 1])$, then for every $f_0 \in E^* \setminus \{0\}$ with $\mathfrak{M}_{f_0} \neq \emptyset$ we have $\dim \mathfrak{M}_{f_0} = \infty$, whence no subspace $G \subset E$ of finite codimension satisfies the condition of theorem 2.1, whence no such G is a Čebyšev subspace; at the same time, there exists ***) an extremally disconnected compact space Q such that $E = l_R^\infty([0, 1]) = C_R(Q)$.

---

*) Besides, from (2.76) and the second part of the proof of theorem 2.3 it follows directly that G satisfies condition d) of theorem 2.5.

**) See e.g. N. Bourbaki [22], Chap. V, p. 65, exercise 14.

***) See e.g. N. Dunford and J. Schwartz [49], p. 276, theorem 20, and p. 398.

There exist also *connected* compact spaces Q for which $C_R(Q)$ has no Čebyšev subspace of finite codimension, as shown by the following example of A. L. Garkavi ([66], p. 118): Let $W(\omega_1)$ be the well ordered set of all ordinal numbers $< \omega_1$. Let us order the set $W(\omega_1) \times [0, 1)$ putting $(\alpha, t) < (\alpha', t')$ if $\alpha < \alpha'$ or if $\alpha = \alpha', t < t'$, and let us add to this totally ordered set the point ω_1, putting $(\alpha, t) < \omega_1$ $(\alpha \in W(\omega_1),\ 0 \leqslant t < 1)$. Introducing in the totally ordered set Q thus obtained the topology having as a base for the open sets the finite unions of sets (of "open intervals") of the form $\{q \in Q \,|\, q_1 < q < q_2\}$, we obtain a connected *) compact space $Q = \Delta(\omega_1 + 1)$. Taking then (by the lemma of Uryson) $x_\beta \in C_R(Q)$ $(\beta \in W(\omega_1))$ such that

$$x_\beta(\alpha, t) = \begin{cases} 1 & \text{for} \quad (\alpha, t) = \left(\beta, \dfrac{1}{2}\right) \\ 0 & \text{for} \quad (\alpha, t) \leqslant (\beta, 0) \text{ or } (\alpha, t) \geqslant (\beta + 1, 0), \end{cases} \tag{2.97}$$

$$\|x_\beta\| = 1, \tag{2.98}$$

one can show, with the same argument as that used at the end of section 2.1 (for the space $l^\infty([0, 1])$) that $\dim \mathfrak{M}_\mu = \infty$ for every $\mu \in C_R(Q)^* \setminus \{0\}$ with $\mathfrak{M}_\mu \neq \emptyset$, hence $C_R(Q)$ has no Čebyšev subspace. This statement remains also valid, with a similar proof, for every compact space Q containing an uncountable number of disjoint open subsets; in other words, we have the following *necessary* condition in order that $C_R(Q)$ have Čebyšev subspaces of finite codimension:

THEOREM 2.8 (A. L. Garkavi [66], p. 118; R. R. Phelps [177], theorem 8). *If the space $E = C_R(Q)$ (Q compact) contains a Čebyšev subspace of finite codimension, then Q has at most a countable number of disjoint open subsets (hence, in particular, at most a countable number of isolated points).*

We shall give now one more proof of theorem 2.8, using theorem 2.5. Let G be a Čebyšev subspace of codimension n of $E = C_R(Q)$, and let $\mu \in G^\perp \setminus \{0\}$ be arbitrary. Then by theorem 2.5 $Q \setminus S(\mu)$ contains at most $n - 1$ points. On the other hand, $|\mu|(U) > 0$ for every open set $U \subset S(\mu)$. Since $S(\mu)$ contains at most a countable number of subsets of positive $|\mu|$-measure (by the complete additivity of $|\mu|$), it follows that $S(\mu)$, whence also $Q = [Q \setminus S(\mu)] \cup S(\mu)$, contains at most a countable number of disjoint open subsets, which completes the proof of theorem 2.8.

THEOREM 2.9. *If the space $E = C_R(Q)$ (Q compact) contains a Čebyšev subspace of finite codimension $n \geqslant 2$, then Q contains no infinite connected open subset.*

---

*) See P. S. Alexandrov — P. S. Uryson [4].

Proof. Let G be a Čebyšev subspace of E of finite codimension $n \geqslant 2$, hence $G^\perp = [\mu_1, \ldots, \mu_n]$, and let U be an infinite connected open subset of Q. Since by theorem 2.5 every set $Q \setminus S(\mu)$ $(\mu \in G^\perp \setminus \{0\})$ consists of at most $n-1$ points which are necessarily isolated points of Q and since U is infinite and connected, we have then

$$U \subset S(\mu) \qquad (\mu \in G^\perp \setminus \{0\}), \tag{2.99}$$

whence, since U is open,

$$|\mu|(U) > 0 \qquad (\mu \in G^\perp \setminus \{0\}). \tag{2.100}$$

But, since G is proximinal, for every $\mu \in G^\perp \setminus \{0\}$ the carrier $S(\mu)$ admits, by virtue of theorem 2.2, a Hahn decomposition with respect to μ into two closed sets $S(\mu)^+$ and $S(\mu)^-$. Since U is connected and the sets $S(\mu)^+ \cap U$, $S(\mu)^- \cap U$ are open and disjoint, one of these sets must be void, whence we have either $U \subset S(\mu)^+$, or $U \subset S(\mu)^-$. Consequently, we have either $\mu(U) = |\mu|(U)$, or $\mu(U) = -|\mu|(U)$, whence, by (2.100), we obtain

$$\mu(U) \neq 0 \qquad (\mu \in G^\perp \setminus \{0\}). \tag{2.101}$$

On the other hand, since $\dim G^\perp = n \geqslant 2$, there exist linearly independent $\mu_1, \mu_2 \in G^\perp$. But then for the measure $\mu = \mu_1(U)\mu_2 - \mu_2(U)\mu_1 \in G^\perp \setminus \{0\}$ we have $\mu(U) = 0$, contradicting (2.101), which completes the proof of theorem 2.9.

The fact that under the hypotheses of theorem 2.9 Q is not connected, has been proved by R. R. Phelps ([176], theorem 3.4). Theorem 2.9 has been given by R. R. Phelps ([177], theorem 7) and, independently, by A. L. Garkavi ([66], theorem 11 and [72], p. 516).

Corollary 2.6. *If the space $E = C_R(Q)$ (Q compact) contains a Čebyšev subspace of finite codimension $n \geqslant 2$, then either Q is finite or Q has an infinity of connected components.*

Proof. If Q has a finite number of connected components, then each of them is open in Q, whence by virtue of theorem 2.9 it reduces to an isolated point and thus Q is finite, which completes the proof.

Yu. I. Makovoz ([143], theorem 1) has observed that if Q has a finite number, say m, of connected components and if G is a Čebyšev subspace of finite codimension n of $C_R(Q)$, then $n \leqslant m$. Corollary 2.6 above says more, namely that under the same hypotheses, in the case when Q is infinite we have even $n = 1$; on the other hand, when Q is finite, we have $\dim C_R(Q) = m$, whence obviously $n \leqslant m$.

Let us also observe that if Q is an infinite compact metric space with a finite number, say m, of connected components, and $E=C_R(Q)$, then $\mathcal{E}(S_E)$ consists of 2^m points and $\dim l[\mathcal{E}(S_E)]=$ $=m$ (since for every $x\in\mathcal{E}(S_E)$ we have $x(q)\equiv 1$ $(q\in Q)$), and by the above the least upper bound s of the codimensions of the Čebyšev subspaces of finite codimension of E is 1. We have thus, as has been announced at the end of section 2.1, an example of a normed linear space E satisfying relation (2.22).

The problem of the *characterization* of compact spaces Q for which $C_R(Q)$ admits Čebyšev subspaces of finite codimension, is open*).

2.3. APPLICATIONS IN THE SPACES $L_R^1(T, \nu)$

THEOREM 2.10. *Let $E = L_R^1(T, \nu)$, where (T, ν) is a positive measure space with the property that the dual $L_R^1(T, \nu)^*$ is canonically equivalent to $L_R^\infty(T, \nu)$ and let G be a closed linear subspace of codimension n of E, hence**) $G^\perp = [\beta_1, \ldots, \beta_n]$. In order that G be proximinal, it is necessary, and in the particular case when $n = 1$, it is necessary and sufficient, that for every $\beta\in G^\perp\setminus\{0\}$ we have*

$$\nu(\{t\in T \mid |\beta(t)| = \|\beta\|\}) > 0. \tag{2.102}$$

Proof. Let G be proximinal and let $\beta\in G^\perp\setminus\{0\}$ be arbitrary. Then by lemma 1.1 we have $\mathfrak{M}_\beta \neq \emptyset$, whence (2.102) (since otherwise for every $x\in L_R^1(T, \nu)\setminus\{0\}$ we would have $\int_T x(t)\beta(t)\mathrm{d}\nu(t) < \|\beta\| \int_T |x(t)|\mathrm{d}\nu(t)$, which contradicts $\mathfrak{M}_\beta \neq \emptyset$).

Conversely, let $n=1$, and let $\beta\in G^\perp\setminus\{0\}$, satisfying (2.102), be arbitrary. Then for $x_e\in L_R^1(T, \nu)$ defined by

$$x_e(t) = \begin{cases} \dfrac{\operatorname{sign}\beta(t)}{\nu(e)} & \text{for} \quad t\in e \\ 0 & \text{for} \quad t\in T\setminus e, \end{cases}$$

---

*) After the present monograph has gone to print, A. L. Garkavi has communicated at the International Congress of Mathematicians (Moscow, 1966) that he solved the problem in the case when Q is a compact *metric* space, proving that for such a Q condition (2.94) of theorem 2.7 is not only sufficient but also necessary in order that $C_R(Q)$ have Čebyšev subspaces G of codimension n.

**) For the sake of simplicity, in the sequel we shall identify the functionals $f\in E^*$ with their kernels $\beta\in L_R^\infty(T, \nu)$, i.e. with their image under the canonical equivalence $L_R^1(T, \nu)^* \equiv L_R^\infty(T, \nu)$ (see Chap. I, §1, section 1.5).

where $e \subset \{t \in T \mid |\beta(t)| = \|\beta\|\}$, $0 < \nu(e) < \infty$, we have

$$\int_T |x_e(t)| \,\mathrm{d}\nu(t) = \int_e \frac{1}{\nu(e)} \,\mathrm{d}\nu(t) = 1,$$

$$\int_T x_e(t)\, \beta(t) \,\mathrm{d}\nu(t) = \int_e \frac{|\beta(t)|}{\nu(e)} \,\mathrm{d}\nu(t) = \|\beta\|,$$

whence $x_e \in \mathfrak{M}_\beta$. Consequently, by virtue of lemma 1.1, G is proximinal, which completes the proof.

THEOREM 2.11. *Let $E = L^1_R(T, \nu)$ where (T, ν) is a positive measure space with the property that the dual $L^1_R(T, \nu)^*$ is canonically equivalent to $L^\infty_R(T, \nu)$ and let G be a closed linear subspace of codimension n of E, hence $G^\perp = [\beta_1, \ldots, \beta_n]$. In order that G be a semi-Čebyšev subspace it is necessary and sufficient that for every $\beta \in G^\perp \setminus \{0\}$ with the property*) (2.102) the following two conditions be satisfied:*

a) *We have*

$$|\beta(t)| < \|\beta\| \quad \nu\text{-a.e. on } N = T \setminus \bigcup_{i \in I} A_i, \tag{2.103}$$

where $\{A_i\}_{i \in I}$ is the set of all atoms of (T, ν).

b) *The number p of the atoms $A_{i_1}, \ldots, A_{i_p}$ for which*

$$|\beta(A_{i_j})| = \|\beta\| \qquad (j = 1, \ldots, p), \tag{2.104}$$

satisfies $1 \leqslant p \leqslant n$, and we have

$$\operatorname{rank}\, (\beta_k(A_{i_j}))_{\substack{k=1,\ldots,n \\ j=1,\ldots,p}} = p. \tag{2.105}$$

Proof. Let G be a semi-Čebyšev subspace and let $\beta \in G^\perp \setminus \{0\}$ be arbitrary with the property (2.102); we may assume, without loss of generality, that we have $\|\beta\| = 1$.

Assume that condition a) is not satisfied, i.e. there exists a set $e_\beta \subset N$ with $\nu(e_\beta) > 0$, such that

$$|\beta(t)| = \|\beta\| = 1 \qquad \nu\text{-a.e. on } e_\beta. \tag{2.106}$$

Since the set e_β is atomless, we have $\dim L^1_R(e_\beta, \nu|_{e_\beta}) = \infty$, hence there exists a $\tilde{g}_0 \in L^1_R(e_\beta, \nu|_{e_\beta}) \setminus \{0\}$ such that

$$\int_{e_\beta} \tilde{g}_0(t)\, \beta_k(t) \,\mathrm{d}\nu(t) = 0 \qquad (k = 1, \ldots, n). \tag{2.107}$$

---

*) Of course, when there exists no such $\beta \in G^\perp \setminus \{0\}$, the condition of theorem 2.11 is satisfied, hence G is a semi-Čebyšev subspace.

Put

$$g_0(t) = \begin{cases} \tilde{g}_0(t) & \text{for } t \in e \\ 0 & \text{for } t \in T \setminus e. \end{cases} \tag{2.108}$$

Then the function $\beta \in L^\infty_R(T, \nu)$, the set $U = e_\beta \subset T$ and the element $g_0 \in G \setminus \{0\}$ defined by (2.108), satisfy the conditions (3.53), (3.54), (3.57) and (3.58) of Chap. I, whence by virtue of theorem 3.4 of Chap. I, G is not a semi-Čebyšev subspace, which contradicts the hypothesis. Consequently, we have a).

Now let $A_{i_1}, \ldots, A_{i_p}$ be arbitrary atoms of (T, ν) satisfying (2.104). Then by (2.102) and a) we have $p \geqslant 1$. Put

$$x_j(t) = \begin{cases} \dfrac{\operatorname{sign} \beta(A_{i_j})}{\nu(A_{i_j})} & \text{for } t \in A_{i_j} \\ 0 & \text{for } t \in T \setminus A_{i_j} \end{cases} \qquad (j = 1, \ldots, p). \tag{2.109}$$

Then, as we have seen in the proof of theorem 2.10, $x_j \in \mathfrak{M}_\beta$ $(j = 1, \ldots, p)$, whence, $x_1, \ldots, x_p$ being linearly independent, $\dim \mathfrak{M}_\beta \geqslant p - 1$. Consequently, since G is a semi-Čebyšev subspace, by theorem 2.1 we have $n - 1 \geqslant \dim \mathfrak{M}_\beta \geqslant p - 1$, whence $p \leqslant n$; for simplicity, we may assume that $A_{i_1}, \ldots, A_{i_p}$ are the only atoms of (T, ν) which satisfy (2.104). We shall show that in this case we have

$$\dim \mathfrak{M}_\beta = p - 1. \tag{2.110}$$

Indeed, let $x \in \mathfrak{M}_\beta$ be arbitrary. Then by (2.109) there exist constants $\alpha_1, \ldots, \alpha_p$ such that

$$x \Big|_{\bigcup_{j=1}^p A_{i_j}} = \sum_{l=1}^p \alpha_l x_l \Big|_{\bigcup_{j=1}^p A_{i_j}}. \tag{2.111}$$

On the other hand, since by a) and the hypothesis on $A_{i_1}, \ldots, A_{i_p}$ we have $|\beta(t)| < \|\beta\|$ ν-a.e on $T \setminus \bigcup_{j=1}^p A_{i_j}$ and since $x \in \mathfrak{M}_p$, it follows that we have

$$x \Big|_{T \setminus \bigcup_{j=1}^p A_{i_j}} = 0 = \sum_{l=1}^p \alpha_l x_l \Big|_{T \setminus \bigcup_{j=1}^p A_{i_j}} \tag{2.112}$$

(the relation $x \neq 0$ on $e \subset T \setminus \bigcup_{l=1}^p A_{i_j}$ with $\nu(e) > 0$ would imply $\int_e x(t)\, \beta(t)\, d\nu(t) < \|\beta\| \int_e |x(t)|\, d\nu(t)$, whence $\int_T x(t) \beta(t)\, d\nu(t) <$

$< \|\beta\| \, \|x\| = \|\beta\|$, contradicting $x \in \mathfrak{M}_\beta$), which, together with (2.111), gives

$$x = \sum_{j=1}^{p} \alpha_j x_j.$$

Since $x \in \mathfrak{M}_\beta$ has been arbitrary and $l(\mathfrak{M}_\beta) \not\ni 0$, it follows that $\dim \mathfrak{M}_\beta = \dim l(\mathfrak{M}_\beta) \leqslant p - 1$, which, together with the inequality $\dim \mathfrak{M}_\beta \geqslant p - 1$ proved above, gives (2.110).

Consequently, since G is a semi-Čebyšev subspace, we have by theorem 2.1 (implication $1° \Rightarrow 5°$)

$$\begin{aligned} p &= \operatorname{rank} \left(\int_T x_j(t)\, \beta_k(t)\, \mathrm{d}\nu(t) \right)_{\substack{k=1,\ldots,n \\ j=1,\ldots,p}} = \\ &= \operatorname{rank} \left(\int_{A_{i_j}} \frac{\operatorname{sign} \beta(A_{i_j})}{\nu(A_{i_j})}\, \beta_k(t)\, \mathrm{d}\nu(t) \right)_{\substack{k=1,\ldots,n \\ j=1,\ldots,p}} = \\ &= \operatorname{rank} \left([\operatorname{sign} \beta(A_{i_j})]\, \beta_k(A_{i_j}) \right)_{\substack{k=1,\ldots,n \\ j=1,\ldots,p}} = \\ &= \operatorname{rank} \left(\beta_k(A_{i_j}) \right)_{\substack{k=1,\ldots,n \\ j=1,\ldots,p}}, \end{aligned} \tag{2.113}$$

whence (2.105).

Conversely, assume now that the conditions of theorem 2.11 are satisfied and let $\beta \in G^\perp \setminus \{0\}$ be arbitrary. If $\mathfrak{M}_\beta = \emptyset$, then $\dim \mathfrak{M}_\beta = -1 < n - 1$ and condition (2.5) is trivial. Assume that $\mathfrak{M}_\beta \neq \emptyset$. Then β satisfies (2.102) (see the proof of theorem 2.10), whence, by virtue of what has been proved above, we have (2.110). Since by hypothesis $p \leqslant n$, it follows that we have $\dim \mathfrak{M}_\beta \leqslant n - 1$. On the other hand, since by our hypothesis we have (2.105), taking into account (2.113) we obtain $\operatorname{rank} \left(\int_T x_j(t)\, \beta_k(t)\, \mathrm{d}\nu(t) \right)_{\substack{k=1,\ldots,n \\ j=1,\ldots,p}} = p = \dim \mathfrak{M}_\beta + 1$. Thus, by virtue of theorem 2.1 (implication $2° \Rightarrow 1°$), G is a semi-Čebyšev subspace, which completes the proof of theorem 2.11.

In the particular case when $n = 1$, hence $G = \{x \in E \mid \int_T x(t) \beta_1(t)\, \mathrm{d}\nu(t) = 0\}$, for any $\beta \in G^\perp \setminus \{0\}$ we have $\beta = \lambda \beta_1$ and consequently *G is a semi-Čebyšev subspace if and only if we have either* 1) *β_1 does not satisfy (2.102), or* 2) *β_1 satisfies (2.102) and (2.103), and there exists an atom A_{i_1} and only one such that $|\beta_1(A_{i_1})| = \|\beta_1\|$.* From the foregoing it follows that we have 1) if and only if the hyperplane G is non-proximinal, and 2) if and only if the hyperplane G is a Čebyšev subspace.

Theorem 2.12. *Let $E = L^1_R(T, \nu)$, where (T, ν) is a positive measure space with the property that the dual $L^1_R(T, \nu)^*$ is cano-*

nically equivalent to $L_R^\infty(T, \nu)$ *and let* G *be a closed linear subspace of codimension* n *of* E, *hence* $G^\perp = [\beta_1, \ldots, \beta_n]$. *In order that* G *be a Čebyšev subspace it is necessary that for every* $\beta \in G^\perp \setminus \{0\}$ *the conditions* a), b) *of theorem 2.11 be satisfied, and it is sufficient that for every* $\beta \in G^\perp \setminus \{0\}$ *these conditions be satisfied together with the condition*

c) *There exists an* $\varepsilon > 0$ *such that for every atom* A_i *which does not satisfy (2.104) we have*

$$|\beta(A_i)| \leqslant \|\beta\| - \varepsilon. \tag{2.114}$$

Proof. Let G be a Čebyšev subspace and let $\beta \in G^\perp \setminus \{0\}$ be arbitrary. Then G is proximinal, hence by theorem 2.10 we have (2.102). Since G is a semi-Čebyšev subspace, by virtue of theorem 2.11 β satisfies then conditions a) and b).

Conversely, assume that every $\beta \in G^\perp \setminus \{0\}$ satisfies conditions a), b) and c). Then by theorem 2.11 G is a semi-Čebyšev subspace, hence there remains to show that G is also proximinal. Let $\Phi \in (G^\perp)^*$ be arbitrary with $\|\Phi\| = 1$. We shall show that there exists then a $y \in E$ satisfying (1.1) and (1.2), which will complete the proof.

By virtue of condition a) we have

$$\|\beta\| = \operatorname*{ess\,sup}_{t \in T} |\beta(t)| = \max\left(\operatorname*{ess\,sup}_{t \in N} |\beta(t)|, \sup_{i \in I} |\beta(A_i)|\right) =$$

$$= \sup_{i \in I} |\beta(A_i)| \qquad (\beta \in G^\perp \setminus \{0\}), \tag{2.115}$$

hence the mapping $\beta \to \{\beta(A_i)\}_{i \in I}$ is a linear isometry of $G^\perp$ into a closed linear subspace Γ of $l_R^\infty(I)$; thus, identifying $G^\perp$ with Γ by this mapping, we may consider that $\Phi \in \Gamma^*$. Let $\Psi \in l_R^\infty(I)^*$ be an extension with the same norm $\|\Psi\| = 1$, of $\Phi \in \Gamma^*$. Then there exists*) a finitely additive set function $\lambda(e)$ $(e \subset I)$ such that

$$\Phi(\beta) = \Psi(\beta) = \int_I \beta(A_i)\, d\lambda(e) \qquad (\beta \in G^\perp), \tag{2.116}$$

$$1 = \|\Psi\| = \operatorname*{Var}_{e \subset I} \lambda(e). \tag{2.117}$$

Since $\dim G^\perp < \infty$ (by (2.2)) and $\|\Phi\| = 1$, there exists a $\beta_0 \in G^\perp \setminus \{0\}$ such that $\Phi(\beta_0) = \|\beta_0\| = 1$, hence we have,

*) See e.g. N. Dunford and J. Schwartz [49], p. 296, theorem 16.

putting $I_p = \{i_1, \ldots, i_p\}$ and taking into account (2.104) and (2.114) for $\beta = \beta_0$,

$$\operatorname*{Var}_{e\subset I_p} \lambda(e) + \operatorname*{Var}_{e\subset I\setminus I_p} \lambda(e) = \operatorname*{Var}_{e\subset I} \lambda(e) = 1 = \Phi(\beta_0) = \int_I \beta_0(A_i)\, d\lambda(e) =$$

$$= \int_{I_p} \beta_0(A_i)\, d\lambda(e) + \int_{I\setminus I_p} \beta_0(A_i)\, d\lambda(e) \leqslant \operatorname*{Var}_{e\subset I_p} \lambda(e) + (1-\varepsilon) \operatorname*{Var}_{e\subset I\setminus I_p} \lambda(e),$$

whence

$$\operatorname*{Var}_{e\subset I\setminus I_p} \lambda(e) = 0, \tag{2.118}$$

and consequently (2.116) becomes

$$\Phi(\beta) = \sum_{j=1}^{p} \beta(A_{i_j})\, \lambda(A_{i_j}) \qquad (\beta \in G^{\perp}). \tag{2.119}$$

Define $y \in L^1(T, \nu)$ by

$$y(t) = \begin{cases} \dfrac{\lambda(A_{i_j})}{\nu(A_{i_j})} & \text{for } t \in A_{i_j} \quad (j = 1, \ldots, p), \\ 0 & \text{for } t \in T \setminus \bigcup_{j=1}^{p} A_{i_j}. \end{cases} \tag{2.120}$$

Then we have, taking into account (2.119) and (2.118),

$$\int_T y(t)\, \beta(t)\, d\nu(t) = \sum_{j=1}^{p} \beta(A_{i_j}) \lambda(A_{i_j}) = \Phi(\beta) \qquad (\beta \in G^{\perp}),$$

$$\|y\| = \int_T |y(t)|\, d\nu(t) = \sum_{j=1}^{p} |\lambda(A_{i_j})| = \operatorname*{Var}_{e\subset I_p} \lambda(e) = \operatorname*{Var}_{e\subset I} \lambda(e) =$$

$$= 1 = \|\Phi\|,$$

whence (1.1) and (1.2), which completes the proof of theorem 2.12.

In the particular case when (T, ν) is σ-finite (whence the set of all atoms of (T, ν) is at most countable), theorem 2.12 has been given by A. L. Garkavi ([73], theorem 1). As has been observed by A. L. Garkavi ([73], p. 10), condition c) of theorem 2.12 is not necessary in order that G be a Čebyšev subspace, it may be replaced e.g. by the following condition:

c') *There exists a finite set of atoms, say* $\{A_{l_1}, \ldots, A_{l_m}\}$, *such that every* $\beta \in G^{\perp} \setminus \{0\}$ *takes the value* $\pm \|\beta\|$ *on at least one of the atoms of this set.*

Indeed, in this case it follows, similarly to (2.115), that the mapping $\beta \to \{\beta(A_{l_i})\}_{i=1}^m$ is a linear isometry of $G^\perp$ into the m-dimensional space $l_R^\infty(I_m)$, where $I_m = \{1, \ldots, m\}$.

An example of a closed linear subspace G of codimension 2 which satisfies a), b), c') (whence G is a Čebyšev subspace), but does not satisfy c), is obtained (A. L. Garkavi [73], p. 10) taking $E = l_R^1$ and

$$\beta_1 = \left\{1, 1, 1 - \frac{1}{3}, 0, 1 - \frac{1}{5}, 0, \ldots\right\}, \beta_2 = \left\{1, -1, 0, 1 - \right.$$

$$\left. - \frac{1}{4}, 0, 1 - \frac{1}{6}, \ldots\right\} \in l_R^\infty = E^*, \tag{2.121}$$

$$G = \{x \in E \mid \beta_k(x) = 0 \ (k = 1, 2)\}. \tag{2.122}$$

From the remark made after theorem 2.11 it follows that in the particular case when $n = 1$ conditions c) and c') may be omitted, the conditions a) and b) (for every $\beta \in G^\perp \setminus \{0\}$) being necessary and sufficient in order that G be a Čebyšev subspace.

THEOREM 2.13. *Let $E = L_R^1(T, \nu)$, where (T, ν) is a positive measure space such that the dual $L_R^1(T, \nu)^*$ is canonically equivalent to $L_R^\infty(T, \nu)$. In order that E have a Čebyšev subspace of codimension n it is necessary and sufficient that (T, ν) have at least n atoms.*

Proof. Assume that (T, ν) has no atom or that it has r atoms $A_1, \ldots, A_r$, where $r < n$, and let G be an arbitrary linear subspace of codimension n of the space E. Then, since $r < n = \dim G^\perp$, there exists a $\beta \in G^\perp \setminus \{0\}$ such that

$$\beta(A_i) = 0 \quad (i = 1, \ldots, r), \tag{2.123}$$

hence condition b) (the statement $p \geqslant 1$) of theorem 2.11 is not satisfied, and thus, by virtue of theorem 2.12, G is not a Čebyšev subspace.

Conversely, assume that (T, ν) has at least n atoms $A_1, \ldots$ $\ldots, A_n$ and let

$$\beta_k(t) = \begin{cases} 1 & \text{for} \quad t \in A_k \\ 0 & \text{for} \quad t \in T \setminus A_k \end{cases} \quad (k = 1, \ldots, n), \tag{2.124}$$

$$G = \{x \in E \mid \int_T x(t)\, \beta_k(t)\, \mathrm{d}\nu(t) = 0 \ \ (k = 1, \ldots, n)\}. \tag{2.125}$$

Then obviously G satisfies conditions a), b) and c) of theorem 2.12 (besides, G also satisfies condition c′) above), whence it is a Čebyšev subspace, which completes the proof.

In the particular case when (T, ν) is σ-finite, theorem 2.13 has been given by A. L. Garkavi ([68], theorem 10, and [73], theorem 2). Comparing this theorem with theorem 2.7 of Chap. II, we see that if (T, ν) is a positive measure space satisfying $L^1_R(T, \nu)^* \equiv L^\infty_R(T, \nu)$, then the condition for the existence of a Čebyšev subspace of codimension n coincides with the condition for the existence of an n-dimensional Čebyšev subspace.

From theorem 2.13 it follows in particular that *the space* $E = l^1_R$ *has Čebyšev subspaces of every finite codimension* (A. L. Garkavi [68], corollary of theorem 10), and also the following result, obtained in a different way at the end of §1 (see the remark made after corollary 1.1) : *If* (T, ν) *is a positive measure space which has no atom and is such that* $L^1_R(T,\nu)^* \equiv L^\infty_R(T,\nu)$, *then the space* $E = L^1_R(T, \nu)$ *has no Čebyšev subspace of finite codimension.* Taking into account that such an $L^1_R(T, \nu)$ does not have finite-dimensional Čebyšev subspaces either (by Chap. II, theorem 2.6), there arises naturally the problem whether such an $L^1_R(T, \nu)$ has no Čebyšev subspaces at all. The answer is negative, as shown by the following example of R. R. Phelps ([176], p. 245) : *If* $e \subset T$ *is* ν*-measurable and* $\nu(e) > 0$, $\nu(T \setminus e) > 0$, *then the closed linear subspace* G *of* $L^1_R(T,\nu)$ *defined by*

$$G = \{g \in L^1_R(T, \nu) \mid g(t) = 0 \ \nu\text{-a.e. on } e\} \tag{2.126}$$

is a Čebyšev subspace. Indeed, let $x \in L^1_R(T, \nu)$ be arbitrary. Define $g_0 \in G$ by

$$g_0(t) = \begin{cases} 0 & \text{for } t \in e \\ x(t) & \text{for } t \in T \setminus e. \end{cases} \tag{2.127}$$

Then for every $g \in G \setminus \{g_0\}$ we have

$$\|x - g_0\| = \int_T |x(t) - g_0(t)| \, d\nu(t) = \int_e |x(t)| \, d\nu(t) <$$

$$< \int_e |x(t)| \, d\nu(t) + \int_{T \setminus e} |g_0(t) - g(t)| \, d\nu(t) = \int_T |x(t) - g(t)| \, d\nu(t) =$$

$$= \|x - g\|,$$

whence $\mathfrak{P}_G(x) = \{g_0\}$, and thus G is a Čebyšev subspace.

§3. BEST APPROXIMATION IN CONJUGATE SPACES BY ELEMENTS OF WEAKLY* CLOSED LINEAR SUBSPACES OF FINITE CODIMENSION

If E is a normed linear space and Γ a $\sigma(E^*, E)$-closed linear subspace of finite codimension n of the conjugate space E^*, then by virtue of the canonical isometry $(\Gamma_\perp)^* \equiv E^*/(\Gamma_\perp)^\perp = E^*/\Gamma$ we have

$$\dim (\Gamma_\perp)^* = \dim E^*/\Gamma = \operatorname{codim} \Gamma = n,$$

whence

$$\dim \Gamma_\perp = n, \tag{3.1}$$

and thus

$$\Gamma_\perp = [x_1, \ldots, x_n], \tag{3.2}$$

where $x_1, \ldots, x_n \in E$ are linearly independent. Consequently

$$\Gamma = (\Gamma_\perp)^\perp = \{f \in E^* \mid f(x_k) = 0 \quad (k = 1, \ldots, n)\}. \tag{3.3}$$

Naturally, the results of §1 and §2 are applicable, in particular, also in the case of best approximation in conjugate spaces E^* by elements of $\sigma(E^*, E)$-closed linear subspaces Γ of finite codimension. However, by using effectively the hypothesis that Γ is $\sigma(E^*, E)$-closed in E^*, one can obtain additional results, which will be presented in this paragraph. At the same time, since by Chap. I, corollary 3.2, Γ is a Čebyšev subspace if and only if Γ has the property (U) (i.e. every functional $\varphi \in (\Gamma_\perp)^*$ has a unique extension with the same norm to the whole space E), the results on $\sigma(E^*, E)$-closed Čebyšev subspaces $\Gamma \subset E^*$ of finite codimension present interest also by the fact that *they may be considered as results on finite dimensional subspaces $G \subset E$ with the property (U)*.

3.1. WEAKLY* CLOSED ČEBYŠEV SUBSPACES OF FINITE CODIMENSION IN GENERAL CONJUGATE SPACES

By Chap. I, corollary 2.5, every $\sigma(E^*, E)$-closed linear subspace Γ of a conjugate space E^* is proximinal. Consequently, such a Γ *is a Čebyšev subspace if and only if it is a semi-Čebyšev subspace*, and thus the characterizations of §1 and §2 of semi-

Čebyšev subspaces of finite codimension give, taking into account the canonical embedding $E \subset E^{**}$ and (3.3), (3.2), corresponding characterizations of $\sigma(E^*, E)$-closed Čebyšev subspaces Γ of conjugate spaces E^*. Using these observations, from theorem 2.1 it follows

THEOREM 3.1. *Let E be a normed linear space and Γ a $\sigma(E^*, E)$-closed linear subspace of codimension n of the conjugate space E^*, hence $\Gamma_\perp = [x_1, \ldots, x_n]$. The following statements are equivalent:*

1° *Γ is a Čebyšev subspace.*

2° *For every $x_0 \in \Gamma_\perp \setminus \{0\}$ the non-void set*

$$\mathfrak{M}_{x_0} = \{f \in E^* \mid f(x_0) = \|x_0\|, \ \|f\| = 1\} \tag{3.4}$$

is of dimension $r = r(x_0) \leqslant n - 1$ and contains $r + 1$ elements $f_0, f_1, \ldots, f_r \in \mathfrak{M}_{x_0}$ such that

$$\operatorname{rank} \left(f_i(x_k)\right)_{\substack{k=1,\ldots,n \\ i=0,1,\ldots,r}} = r + 1. \tag{3.5}$$

3° *For every $x_0 \in \Gamma_\perp \setminus \{0\}$ the non-void set $\mathfrak{M}_{x_0}$ defined by (3.4) is of dimension $r = r(x_0) \leqslant n - 1$ and contains $r + 1$ extremal points $f_0, f_1, \ldots, f_r \in \mathcal{E}(S_E) \cap \mathfrak{M}_{x_0} = \mathcal{E}(\mathfrak{M}_{x_0})$ satisfying (3.5).*

4° *For every $x_0 \in \Gamma_\perp \setminus \{0\}$ the non-void set $\mathfrak{M}_{x_0}$ is of dimension $r = r(x_0) \leqslant n - 1$ and for any $r + 1$ linearly independent extremal points $f_0, f_1, \ldots, f_r \in \mathcal{E}(S_E) \cap \mathfrak{M}_{x_0} = \mathcal{E}(\mathfrak{M}_{x_0})$ we have (3.5).*

5° *For every $x_0 \in \Gamma_\perp \setminus \{0\}$ the non-void set $\mathfrak{M}_{x_0}$ is of dimension $r = r(x_0) \leqslant n - 1$ and for any $r + 1$ linearly independent elements $f_0, f_1, \ldots, f_r \in \mathfrak{M}_{x_0}$ we have (3.5).*

In the case of real scalars, the necessary condition $\dim \mathfrak{M}_{x_0} \leqslant n - 1$ $(x_0 \in \Gamma_\perp \setminus \{0\})$ in order that Γ be a Čebyšev subspace, has been given by R. R. Phelps([176], theorem 1.5). Again for real scalars, the equivalence 1°$\Leftrightarrow$4° of theorem 3.1 above has been given by A. L. Garkavi ([68], theorem 1); A. L. Garkavi [68] has given also another characterization, of a geometrical nature, of $\sigma(E^*, E)$-closed Čebyšev subspaces Γ of codimension n.

Let us observe that *if E is an arbitrary separable normed linear space, the conjugate space E^* has at least one $\sigma(E^*, E)$-closed Čebyšev hyperplane*, since S_{E^*} has (D. P. Milman [150]) at least one "$\sigma(E^*, E)$-exposed" point (i.e. a point f with the property that there exists a $\sigma(E^*, E)$-closed support hyperplane H of S_{E^*} such that $H \cap S_{E^*} = \{f\}$); however, it is not known whether E^* also has $\sigma(E^*, E)$-closed Čebyšev subspaces of finite codimension $n > 1$. On the other hand, we have seen in §2, section 2.1, that *the conjugate space $E^* = l^\infty([0, 1])$* of the (non-separable) space $E = l^1([0, 1])$ *has no Čebyšev subspace (hence also no $\sigma(E^*, E)$-closed Čebyšev subspace) of finite codimension.*

3.2. APPLICATIONS IN THE SPACES $L_R^1(T, \nu)^*$

THEOREM 3.2. *Let $E = L_R^1(T, \nu)$ where (T, ν) is a positive measure space such that the dual $L_R^1(T, \nu)^*$ is canonically equivalent to $L_R^\infty(T, \nu)$ and let Γ be a $\sigma(E^*, E)$-closed linear subspace of codimension n of E^*, hence $\Gamma_\perp = [x_1, \ldots, x_n]$. In order that Γ be a Čebyšev subspace, it is necessary and sufficient that for every $x \in \Gamma_\perp \setminus \{0\}$ the set*) $Z(x)$ be either of ν-measure zero, or the union of a set of ν-measure zero with r atoms $A_1, \ldots, A_r$ of (T, ν), where $1 \leqslant r \leqslant n-1$, and in this latter case to have*

$$\operatorname{rank}\, (x_k(A_i))_{\substack{k=1,\ldots,n \\ i=1,\ldots,r}} = r. \tag{3.6}$$

Proof. As we have seen in the example given after corollary 2.3, there exist a compact space $Q = Q_{T,\nu}$, and a multiplicative and positive linear isometry $\varkappa$ of $L_R^\infty(T, \nu)$ onto $C_R(Q)$, which induces a mapping ψ of the σ-algebra of all ν-measurable sets $e \subset T$ (modulo the sets of measure zero) onto the family of all simultaneously open and closed subsets of Q, defined by $\chi_{\psi(e)} = \varkappa(\chi_e)$ (where $\chi_B =$ the characteristic function of B). The mapping ψ is one-to-one and monotone, carries the atoms of (T, ν) onto the isolated points of Q and if we denote by μ_k the image of $x_k \in L_R^1(T, \nu) \subset L_R^\infty(T, \nu)^*$ $(k = 1, \ldots, n)$ under the isometry $L_R^\infty(T, \nu)^* \equiv C_R(Q)^*$, hence

$$\int_T z(t)\, x_k(t)\, d\nu(t) = \int_Q (\varkappa z)(q)\, d\mu_k(q) \quad (z \in L_R^\infty(T, \nu),\ k = 1, \ldots, n), \tag{3.7}$$

then for $x = \sum_{k=1}^n \alpha_k x_k \in \Gamma_\perp \setminus \{0\}$ and $\mu = \sum_{k=1}^n \alpha_k \mu_k$ we have $\psi[T \setminus Z(x)] = S(\mu)$. Consequently, we have

$$\varkappa(\chi_{Z(x)}) = \varkappa(1) - \varkappa(1 - \chi_{Z(x)}) = 1 - \varkappa(\chi_{T \setminus Z(x)}) =$$

$$= 1 - \chi_{\psi[T \setminus Z(x)]} = 1 - \chi_{S(\mu)} = \chi_{Q \setminus S(\mu)},$$

whence

$$\psi[Z(x)] = Q \setminus S(\mu), \tag{3.8}$$

*) We recall (see Chap. I, § 1, section 1.5) that we use the same notation for a function in $\mathfrak{L}^p$ and for its equivalence class in L^p $(1 \leqslant p \leqslant \infty)$, and that by $Z(x)$ we denote the set $\{t \in T \mid x(t) = 0\}$.

which shows that we have $S(\mu) = Q$ if and only if $\nu[Z(x)] = 0$. On the other hand, if $A_i \in Z(x)$ is an atom of (T, ν) and $\psi(A_i) = \{q_i\} \subset Q \setminus S(\mu)$ is the set consisting of the corresponding isolated point, then by (3.7) we have

$$x_k(A_i) = \frac{1}{\nu(A_i)} \int_T \chi_{A_i}(t)\, x_k(t)\, \mathrm{d}\nu(t) = \frac{1}{\nu(A_i)} \int_Q (\varkappa\, \chi_{A_i})\,(q)\, \mathrm{d}\mu_k(q) =$$

$$= \frac{1}{\nu(A_i)} \int_Q \chi_{\psi(A_i)}(q)\, \mathrm{d}\mu_k(q) = \frac{1}{\nu(A_i)} \int_Q \chi_{\{q_i\}}(q) \mathrm{d}\mu_k(q) = \frac{1}{\nu(A_i)}\, \mu_k(\{q_i\}).$$

Consequently, taking into account that Γ is a Čebyšev subspace if and only if it is a semi-Čebyšev subspace (see the remark made before theorem 3.1), theorem 3.2 follows now from theorem 2.3. This completes the proof of theorem 3.2.

In the particular case when (T, ν) is σ-finite, theorem 3.2 has been given by A. L. Garkavi ([68], theorem 6), who has deduced it directly from theorem 3.1 (see A. L. Garkavi [68], p. 563). The remark that the theorems on best approximation in the spaces $L^1_R(T, \nu)^* \equiv L^\infty_R(T, \nu)$ can be deduced from those known in the spaces $C_R(Q)$ by using the isometry of $L^\infty_R(T, \nu)$ to a suitable $C_R(Q)$ has been made by R. R. Phelps ([177], p. 655).

Corollary 3.1. *Let $E = L^1_R(T, \nu)$, where (T, ν) is a positive measure space containing at least n atoms and such that the dual $L^1_R(T, \nu)^*$ is canonically equivalent to $L^\infty_R(T, \nu)$, and let Γ be a $\sigma(E^*, E)$-closed linear subspace of codimension n of E^*, hence $\Gamma_\perp = [x_1, \ldots, x_n]$. In order that Γ be a Čebyšev subspace, it is necessary and sufficient that for every $x \in \Gamma_\perp \setminus \{0\}$ the set $Z(x)$ be either of ν-measure zero or the union of a set of ν-measure zero with r atoms of (T, ν), where $1 \leqslant r \leqslant n - 1$.*

Corollary 3.1 can be deduced from theorem 3.2 above, but it follows also from the second part of corollary 2.3, using the isometry of $L^\infty_R(T, \nu)$ to a suitable space $C_R(Q)$ (see the proof of theorem 3.2).

In the particular case when $E = l^1_R$ (hence all points of $T = \{1,2,3, \ldots\}$ are atoms of ν-measure 1 of (T, ν) and the condition $\nu[Z(x)] = 0$ is equivalent to $Z(x) = \emptyset$), corollary 3.1 has been given by R. R. Phelps ([176], theorem 2.3), and in the case when (T, ν) is σ-finite it has been given by R. R. Phelps ([177], p. 655) and, independently, by A. L. Garkavi ([68], p. 564).

Corollary 3.2. *Let $E = L^1_R(T, \nu)$, where (T, ν) is a positive measure space such that the dual $L^1_R(T, \nu)^*$ is canonically equivalent to $L^\infty_R(T, \nu)$, and let Γ be a $\sigma(E^*, E)$-closed linear subspace of codimension n of E^*, hence $\Gamma_\perp = [x_1, \ldots, x_n]$. If we have*

$$\nu[Z(x)] = 0 \qquad (x \in \Gamma_\perp \setminus \{0\}), \tag{3.9}$$

then Γ is a Čebyšev subspace. In the case when (T, ν) has no atom, condition (3.9) is necessary and sufficient in order that Γ be a Čebyšev subspace.

Corollary 3.2 can be deduced from theorem 3.2 above, but it follows also from corollary 2.2 and the second part of corollary 2.5, using the isometry of $L_R^\infty(T, \nu)$ to a suitable space $C_R(Q)$ (see the proof of theorem 3.2).

In the particular case when (T, ν) is σ-finite, corollary 3.2 has been given by R. R. Phelps ([176], theorem 2.2).

Let us pass now to problem of existence of $\sigma(E^*, E)$-closed Čebyšev subspaces of finite codimension in the spaces $E^* = L_R^1(T, \nu)^*$. We shall give conditions on the positive measure space (T, ν), sufficient for $E^* = L_R^1(T, \nu)^*$ to have $\sigma(E^*, E)$-closed Čebyšev subspaces of finite codimension.

THEOREM 3.3 (A. L. Garkavi [68], p. 564). *Let $E = L_R^1(T, \nu)$, where (T, ν) is a σ-finite positive measure space. Then the space $E^* = L_R^1(T, \nu)^*$ contains $\sigma(E^*, E)$-closed Čebyšev subspaces of codimension 1.*

Proof. Since (T, ν) is σ-finite, there exists a sequence of ν-measurable sets $\{e_n\} \subset T$ with $\nu(e_n) < \infty$ $(n = 1,2, \ldots)$, such that $T = \bigcup_{n=1}^{\infty} e_n$. Put

$$y_n(t) = \begin{cases} \dfrac{1}{2^n} \text{ for } t \in e_n & (n = 1,2, \ldots), \\ 0 \text{ for } t \in T \setminus e_n \end{cases} \tag{3.10}$$

$$x_1(t) = \sum_{n=1}^{\infty} y_n(t) \; (\in L_R^1(T, \nu)), \tag{3.11}$$

$$\Gamma = \{f \in E^* \mid f(x_1) = 0\}. \tag{3.12}$$

Then we have $\Gamma_\perp = [x_1]$ and $\nu[Z(x)] = 0$ for every $x \in \Gamma_\perp \setminus \{0\}$, whence, by virtue of theorem 3.2, it follows that Γ is a Čebyšev subspace, which completes the proof.

If (T, ν) is not σ-finite, the statement of theorem 3.3 is in general no longer valid. Indeed, as has been observed at the end of section 3.1, for the space $E = l_R^1([0, 1])$ (here all points of $T = [0, 1]$ are atoms of ν-measure 1 of (T, ν), whence (T, ν) is not σ-finite) the conjugate space $E^* = l_R^1([0, 1])^* \equiv l_R^\infty([0, 1])$ has no $\sigma(E^*, E)$-closed Čebyšev subspace of finite codimension.

THEOREM 3.4 (A. L. Garkavi [68], p. 564).Let $E = L_R^1(T, \nu)$, *where (T, ν) is a σ-finite positive measure space. If there exists a bounded function $x_0 \in L_R^1(T, \nu)$ which is not constant on any atomless*) ν-measurable subset $e \subset T$ with $\nu(e) > 0$, then for*

*) I.e. not containing any atom of (T, ν).

every integer n with $1 \leqslant n < \infty$ the space $E^* = L^1_R(T, \nu)^*$ contains $\sigma(E^*, E)$-closed Čebyšev subspaces of codimension n.

Proof. Since (T, ν) is σ-finite, it has at most a countable number of atoms, say $A_1, A_2, \dots$ Put $A = \bigcup_{i=1}^{\infty} A_i$, $N = T \setminus A$, and let

$$y(t) = \begin{cases} x_0(t) & \text{for } t \in N \\ \dfrac{1}{j^2} & \text{for } t \in A_j \end{cases} \quad (j = 1,2,\dots). \tag{3.13}$$

Then, since $\sum_{j=1}^{\infty} \frac{1}{j^2} < \infty$, we have $y \in L^1_R(T, \nu)$. Consequently, since x_0, whence also y, is bounded, we have $y^k \in L^1_R(T, \nu)$ $(k = 1, \dots, n)$. Let Γ be the $\sigma(E^*, E)$-closed linear subspace of E^* defined by

$$\Gamma = \{f \in E^* \mid f(y^k) = 0 \ (k = 1, \dots, n)\}, \tag{3.14}$$

hence $\Gamma_{\perp} = [y, y^2, \dots, y^n]$. We shall show that Γ satisfies the conditions of theorem 3.2, whence it is a Čebyšev subspace, which will complete the proof.

Let $x = \sum_{k=1}^{n} a_k y^k \in \Gamma_{\perp} \setminus \{0\}$ be arbitrary. Since by our hypothesis x_0 is not constant on any ν-measurable set $e \subset N$, with $\nu(e) > 0$, every real number λ we have $\nu(\{t \in N \mid x_0(t) = \lambda\}) = 0$, whence, taking into account that the polynomial $\sum_{k=1}^{n} a_k \lambda^k$ has at most n real zeros, say $\lambda_1, \dots, \lambda_p$ $(p \leqslant n)$, we obtain

$$\nu[Z(x) \cap N] = \nu\left[Z\left(\sum_{k=1}^{n} a_k y^k |_N\right)\right] = \nu\left[Z\left(\sum_{k=1}^{n} a_k x_0^k |_N\right)\right] =$$

$$= \nu\left[\bigcup_{i=1}^{p} \{t \in N \mid x_0(t) = \lambda_i\}\right] = 0.$$

On the other hand, since for every atom A_j we have

$$x(A_j) = \sum_{k=1}^{n} a_k y^k(A_j) = \sum_{k=1}^{n} \frac{a_k}{j^{2k}} = \frac{1}{j^{2n}} \sum_{k=1}^{n} a_k j^{2n-2k}$$

and since the polynomial $\sum_{k=1}^{n} a_k \lambda^{2n-2k}$ has at most $n-1$ positive zeros, the set $Z(x)$ may contain at most $n-1$ atoms of (T, ν). Consequently, $Z(x)$ is either of ν-measure zero, or the union of a set of ν-measure zero with r atoms $A_{j_1}, \ldots, A_{j_r}$ of (T, ν), where $1 \leqslant r \leqslant n-1$. In this latter case we have

$$\operatorname{rank}\,(y^k(A_{j_i}))_{\substack{k=1,\ldots,n\\ i=1,\ldots,r}} = \operatorname{rank}\left(\frac{1}{j_i^{2k}}\right)_{\substack{k=1,\ldots,n\\ i=1,\ldots,r}} = r$$

(the last equality follows by a computation similar to that in the proof of theorem 2.4), and thus Γ satisfies the conditions of theorem 3.2, which completes the proof.

COROLLARY 3.3 (R. R. Phelps [177], p. 655 and A. L. Garkavi [68], p. 568). *Let* $E = L_R^1([0, 1], \nu)$, *where* $\nu =$ *the Lebesgue measure. Then for every integer* n *with* $1 \leqslant n < \infty$ *the space* $E^* = L_R^1([0,1], \nu)^* \equiv L_R^\infty([0,1], \nu)$ *contains* $\sigma(E^*, E)$*-closed Čebyšev subspaces of codimension* n.

Proof. The function $x_0(t) = t$ $(t \in [0, 1])$ satisfies the conditions of theorem 3.4, whence the statement follows. Besides, corollary 3.3 can be deduced also from corollary 3.2, by taking there $x_k(t) = t^{k-1}$ $(k = 1, \ldots, n)$.

3.3. APPLICATIONS IN THE SPACES $C_R(Q)^*$, $L_R^\infty(T,\nu)^*$ AND $((c_0)_R)^*$

From theorem 3.1 and lemma 2.1 it results, taking into account the general form of the extremal points of $S_{C(Q)^*}$ (Chap. 1, §1, section 1.10, formula (1.135)), the following theorem:

THEOREM 3.5. *Let* $E = C_R(Q)$ (Q *compact*) *and let* Γ *be a* $\sigma(E^*, E)$*-closed linear subspace of codimension* n *of* E^*, *hence* $\Gamma_\perp = [x_1, \ldots, x_n]$. *In order that* Γ *be a Čebyšev subspace, it is necessary and sufficient that for every* $x \in \Gamma_\perp \setminus \{0\}$ *the number* $p = p(x)$ *of the points* $q_1, \ldots, q_p$ *of the set* $\{q \in Q \mid |x(q)| = \|x\|\}$ *satisfy* $1 \leqslant p \leqslant n$ *and that we have*

$$\operatorname{rank}\,(x_k(q_i))_{\substack{k=1,\ldots,n\\ i=1,\ldots,p}} = p. \tag{3.15}$$

The necessity of the condition $1 \leqslant p = p(x) \leqslant n$ $(x \in \Gamma_\perp \setminus \{0\})$ in order that Γ be a Čebyšev subspace, as well as its sufficiency in the case when $n = 1$, have been proved by R. R. Phelps ([176], theorem 3.2). In the above general form, theorem 3.5 has been given by A. L. Garkavi ([68], theorem 5).

By virtue of a classical theorem of S. Kakutani*) there exist a locally compact space T and a positive Radon measure ν on T such that $C_R(Q)^* \equiv \mathfrak{M}_R^1(Q)$ be isometric and latticially isomorphic to $L^1(T, \nu)$. Taking into account that Γ is a Čebyšev subspace if and only if it is a semi-Čebyšev subspace (see the

*) See e.g. A. Grothendieck [79], p. 407, exercise 2 b), or M. M. Day [43], Chap. VI, §4, theorem 2.

remark preceding theorem 3.1), there arises naturally the problem whether it would not be more convenient to deduce theorem 3.5 above from theorem 2.11 (just as in section 3.2 we have seen that it is convenient to deduce theorem 3.2 from theorem 2.3, corollary 3.1 from corollary 2.3 and corollary 3.2 from corollaries 2.2 and 2.5, by using the isometric, algebraic and latticial isomorphism $L_R^\infty(T, \nu) \equiv C_R(Q)$). The answer is *negative*. Indeed, both the evaluation measures $\psi = \varepsilon_q$ in the normed linear lattice $\mathfrak{M}_R^1(Q)$, and the normalized characteristic functions $\psi = \frac{1}{\nu(A)} \chi_A \in L_R^1(T, \nu)$ of the atoms A of (T, ν) are characterized*) by the fact that $\|\psi\| = 1$, $\psi > 0$ and that the relation $|\varphi| \leqslant \psi$ implies $\varphi = \lambda\psi$, where λ is a suitable scalar (the verification is immediate). Consequently, the latticial isometric isomorphism $u : \mathfrak{M}_R^1(Q) \to L_R^1(T, \nu)$ carries in a one-to-one fashion the evaluation measures ε_q onto the normalized characteristic functions $\frac{1}{\nu(A)} \chi_A$ of the atoms A of (T, ν), and thus the condition of theorem 3.5 is nothing else than that obtained by applying condition b) of theorem 2.11 to the closed linear subspace $u(\Gamma) \subset L^1(T, \nu)$. Thus, condition a) of theorem 2.11 is in this case automatically satisfied. On the other hand, it can be translated into a condition in the space $C_R(Q)^* \equiv \equiv \mathfrak{M}_R^1(Q)$, necessary in order that Γ be a Čebyšev subspace; namely, we have

THEOREM 3.6. *Let $E = C_R(Q)$ (Q compact) and let Γ be a $\sigma(E^*, E)$-closed Čebyšev subspace of finite codimension n of E. Then for every $x \in \Gamma_\perp \setminus \{0\}$ and every diffuse**) *positive measure $\mu \in E^* \setminus \{0\}$ we have*

$$|\mu(x)| < \|\mu\| \, \|x\|. \tag{3.16}$$

Proof. Assume that there exist an $x \in \Gamma_\perp \setminus \{0\}$ and a diffuse positive measure $\mu \in E^* \setminus \{0\}$ such that

$$|\mu(x)| = \|\mu\| \, \|x\|. \tag{3.17}$$

Let $y = u(\mu) \in L_R^1(T, \nu)$ where $u : \mathfrak{M}_R^1(Q) \to L_R^1(T, \nu)$ is the above latticial isometric isomorphism. Then $y > 0$ and we have

$$\bigcup_{i \in I} A_i \subset Z(y). \tag{3.18}$$

---

*) Another characterization is that they (and only they) are the positive extremal points of the unit cell of the normed linear lattice $\mathfrak{M}_R^1(Q)$ and $L_R^1(T, \nu)$ respectively.

**) See N. Bourbaki [22], Chap. V, p. 61, definition 5.

Indeed, if there existed an atom $A_i \in T \setminus Z(y)$, then by $y \geqslant y(A_i)\chi_{A_i}$ and by the fact that $\frac{1}{\nu(A_i)}\chi_{A_i} = u(\varepsilon_q)$ for a suitable $q \in Q$ (see the remarks following theorem 3.5), there would result

$$\mu = u^{-1}(y) \geqslant u^{-1}(y(A_i)\chi_{A_i}) = y(A_i)u^{-1}(\chi_{A_i}) = y(A_i)\,\nu(A_i)\,\varepsilon_q,$$

whence $\mu(\{q\}) \geqslant y(A_i)\,\nu(A_i) > 0$, which contradicts the hypothesis that μ is diffuse. Thus, we have (3.18), whence for the set $e = T \setminus Z(y)$ we have

$$e \subset T \setminus \bigcup_{i \in I} A_i, \quad \nu(e) > 0. \tag{3.19}$$

Let $\nu_1 = y\nu$. Then the mapping v defined by

$$[v(z)](t) = \begin{cases} \frac{z(t)}{y(t)} & \text{for } t \in e \\ 0 & \text{for } t \in Z(y) \end{cases} \tag{3.20}$$

is a latticial isometric isomorphism of $L^1_R(T, \nu)$ onto $L^1_R(T, \nu_1)$, hence the mapping

$$w = vu \tag{3.21}$$

is a latticial isometric isomorphism of $\mathfrak{M}^1_R(Q)$ onto $L^1_R(T, \nu_1)$. Since

$$w(\mu) = v[u(\mu)] = v(y) = \chi_e,$$

we have, denoting by $\beta \in L^\infty_R(T, \nu_1) \equiv L^1_R(T, \nu_1)^*$ the image of $x \in E \subset E^{**}$ under the isometric isomorphism $({}^t w)^{-1} : E^{**} \to L^1_R(T, \nu_1)^*$ and taking into account the hypothesis (3.17),

$$\left|\int_e \beta(t)\,d\nu(t)\right| = \left|\int_T \beta(t)\,\chi_e(t)\,d\nu(t)\right| = |\beta(\chi_e)| = |\mu(x)| = \|\mu\|\,\|x\| =$$

$$= \|\chi_e\|\,\|\beta\| = \nu(e)\,\|\beta\|,$$

whence, taking into account $\nu(e) > 0$, we obtain

$$|\beta(t)| = \|\beta\| \qquad \nu\text{-a.e. on } e,$$

which, together with (3.19), means that $w(\Gamma) \subset L^1_R(T, \nu_1)$ does not satisfy condition a) of theorem 2.11. This completes the proof of theorem 3.6*).

---

*) Naturally, theorem 3.6 can be proved also directly (i.e. without using the representation $\mathfrak{M}^1_R(Q) \equiv L^1_R(T, \nu_1)$), by a method similar to that used in the proof of the necessity of condition a) of theorem 2.11.

As we have seen in the preceding section, the spaces $L_R^\infty(T, \nu)$ may be considered as spaces $C_R(Q)$ hence the spaces $L_R^\infty(T, \nu)^*$ as spaces $C_R(Q)^*$. Since in this case $Q = Q_{T,\nu}$ is nothing else but the set of the positive extremal points of $S_{L_R^\infty(T,\nu)^*}$, from theorem 3.5 we thus obtain again theorem 3.1 with $E = L_R^\infty(T, \nu)$. On the other hand, as we have seen in the above (for the spaces $C_R(Q)^*$), neither does the representation $L_R^\infty(T, \nu)^* \equiv L_R^1(T_1, \nu_1)$ lead to a convenient characterization of the $\sigma(E^*, E)$-closed Čebyšev subspaces of finite codimension of $E^* = L_R^\infty(T, \nu)^*$. Such a characterization can be obtained by a direct use of theorem 3.1; namely, we shall prove

THEOREM 3.7. *Let $E = L_R^\infty(T, \nu)$ where (T, ν) is a positive measure space, and let Γ be a $\sigma(E^*, E)$-closed linear subspace of codimension n of the conjugate space E^*, hence $\Gamma_\perp = [x_1, \dots, x_n]$. In order that Γ be a Čebyšev subspace, it is necessary and sufficient that for every $x \in \Gamma_\perp \setminus \{0\}$ the following three conditions be satisfied:*

a) *We have*

$$\operatorname*{ess\,sup}_{t \in N} |x(t)| < \|x\|, \tag{3.22}$$

where $N = T \setminus \bigcup_{i \in I} A_i$, and $\{A_i\}_{i \in I}$ is the set of all atoms of (T, ν).

b) *The number r of the atoms $A_{i_1}, \dots, A_{i_r}$ for which*

$$|x(A_{i_j})| = \|x\| \qquad (j = 1 \dots, r), \tag{3.23}$$

satisfies $1 \leqslant r \leqslant n$, and we have

$$\operatorname{rank}\,(x_k(A_{i_j}))_{\substack{k=1,\dots,n \\ j=1,\dots,r}} = r. \tag{3.24}$$

c) *There exists an $\varepsilon > 0$ such that for every atom A_i which does not satisfy (3.23) we have*

$$|x(A_i)| < \|x\| - \varepsilon. \tag{3.25}$$

Proof. Assume that there exists an $x \in \Gamma_\perp \setminus \{0\}$ which does not satisfy condition a), hence such that

$$\operatorname*{ess\,sup}_{t \in N} |x(t)| = \|x\|. \tag{3.26}$$

Then we have either $\operatorname*{ess\,sup}_{t \in N} x(t) = \|x\|$, or $\operatorname*{ess\,inf}_{t \in N} x(t) = -\|x\|$ and we may assume (considering, if necessary, $-x$ instead of x) that we are in the first case. Since $N = T \setminus \bigcup_{i \in I} A_i$, there exist then $n + 1$ infinite sequences of disjoint sets $e_j^i \subset N$

($e_j^i \cap e_k^l = \emptyset$ for $j \neq k$ or $i \neq l$) with $0 < \nu(e_j^i) < \infty$ ($i=1, \ldots, n+1$; $j = 1,2,\ldots$) such that

$$\lim_{j\to\infty} \frac{1}{\nu(e_j^i)} \int_{e_j^i} x(t)\, \mathrm{d}\nu(t) = \|x\| \qquad (i = 1, \ldots, n+1). \tag{3.27}$$

Let Lim be an arbitrary Banach limit*) on the space l_R^∞ and let

$$f_i(y) = \operatorname*{Lim}_{j\to\infty} \frac{1}{\nu(e_j^i)} \int_{e_j^i} y(t)\, \mathrm{d}\nu(t) \qquad (y \in L_R^\infty(T, \nu);\ i=1, \ldots, n+1). \tag{3.28}$$

Then $f_1, \ldots, f_n \in \mathfrak{M}_x$ and they are linearly independent, whence $\dim \mathfrak{M}_x \geqslant n$, and thus, by virtue of theorem 3.1, Γ is not a Čebyšev subspace. Consequently, condition a) is necessary in order that Γ be a Čebyšev subspace.

The necessity of condition c) is proved similarly. Namely, if there exists an $x \in \Gamma_\perp \setminus \{0\}$ which does not satisfy condition c), then there exist $n+1$ disjoint infinite sequences of atoms $\{A_{i_j}^l\}_{j=1}^\infty$ ($l = 1, \ldots, n+1$) such that

$$\lim_{j\to\infty} x(A_{i_j}^l) = \varepsilon\|x\| \qquad (l = 1, \ldots, n+1), \tag{3.29}$$

where $\varepsilon = \pm 1$ does not depend on l; we may assume (considering, if necessary, $-x$ instead of x) that $\varepsilon = 1$. Taking then an arbitrary Banach limit Lim on l_R^∞ and putting

$$f_l(y) = \operatorname*{Lim}_{j\to\infty} y(A_{i_j}^l) \qquad (l = 1, \ldots, n+1), \tag{3.30}$$

we have $f_1, \ldots, f_{n+1} \in \mathfrak{M}_x$ and $f_1, \ldots, f_{n+1}$ are linearly independent, whence $\dim \mathfrak{M}_x \geqslant n$, and thus, by virtue of theorem 3.1, Γ is not a Čebyšev subspace.

Finally, the necessity of condition b) follows directly from theorem 3.1 (implication $1° \Rightarrow 5°$), taking into account that the functionals of the form

$$f_i(y) = \varepsilon y(A_i) \qquad (y \in L_R^\infty(T, \nu)), \tag{3.31}$$

where A_i is an atom and $\varepsilon = \pm 1$, are in $L_R^\infty(T, \nu)^*$ (moreover, it can be shown**) that they are even in $\mathcal{E}(S_{L_R^\infty(T,\nu)^*})$, but do not exhaust the whole set $\mathcal{E}(S_{L_R^\infty(T,\nu)^*})$).

*) See S. Banach [8], p. 34.

**) This follows e.g. from the equivalence $L_R^\infty(T, \nu) \equiv C_R(Q)$ and the general form of the extremal points of $S_{C_R(Q)^*}$.

Conversely, assume now that every $x \in \Gamma_{\perp} \setminus \{0\}$ satisfies the conditions a), b), c). Since every functional $f \in L_R^\infty(T, \nu)^*$ can be written*) in the form

$$f(y) = \int_T y(s)\, \mathrm{d}\lambda(s) \qquad (y \in L_R^\infty(T, \nu)), \tag{3.32}$$

where $\lambda(e)$ is an additive function defined on the ν-measurable sets $e \subset T$, such that $\lambda(e) = 0$ whenever $\nu(e) = 0$ and that

$$\|f\| = \operatorname*{Var}_{e \subset T} \lambda(e), \tag{3.33}$$

from a) and c) if follows that for $x \in \Gamma_{\perp} \setminus \{0\}$ and $f \in \mathcal{E}(S_{L_R^\infty(T,\nu)*})$ we may have $f(x) = \|x\|$ only if f is of the form (3.31). Consequently, from b), lemma 2.1 and theorem 3.1, it follows that Γ is a Čebyšev subspace, which completes the proof of theorem 3.7.

In the particular case when (T, ν) is σ-finite (hence the set of all atoms of (T, ν) is at most countable), as well as in the particular case when $E = l_R^\infty$, theorem 3.7 has been given by A. L. Garkavi ([68], theorems 8 and 7).

Let us pass now to the problem of existence of $\sigma(E^*, E)$-closed Čebyšev subspaces of finite codimension in the spaces $E^* = C_R(Q)^*$ and, in particular, in the spaces $E^* = L_R^\infty(T, \nu)^*$.

The space $E^* = C_R([0, 1])^*$ has $\sigma(E^*, E)$-closed Čebyšev subspaces of every finite codimension n, e.g.

$$\Gamma = \{f \in E^* \mid f(x_k) = 0 \quad (k = 1, \ldots, n)\}, \tag{3.34}$$

where

$$x_k(t) = t^k \qquad (t \in [0, 1]; k = 1, \ldots, n). \tag{3.35}$$

Indeed, for every $x \in \Gamma_{\perp} \setminus \{0\}$ the number of points $t_1, \ldots, t_p \in [0, 1]$ with the property $|x(t_i)| = \|x\|$ $(i = 1, \ldots, p)$ is $\leqslant n$ and we have

$$\operatorname{rank} (x_k(t_i))_{\substack{k=1,\ldots,n\\ i=1,\ldots,p}} = \operatorname{rank} \begin{pmatrix} t_1 & \ldots & t_p \\ t_1^2 & \ldots & t_p^2 \\ \cdots & \cdots & \cdots \\ t_1^n & \ldots & t_p^n \end{pmatrix} = p,$$

whence, by virtue of theorem 3.5, Γ is a Čebyšev subspace.

Similarly, the space $E^* = c_R^*$ has $\sigma(E^*, E)$-closed Čebyšev subspaces of every finite codimension n, e.g.

$$\Gamma = \{f \in E^* \mid f(x_k) = 0 \qquad (k = 1, \ldots, n)\},$$

*) See e.g. N. Dunford and J. Schwartz [49], p. 296, theorem 16.

where

$$x_k = \left\{1, \frac{1}{2^k}, \ldots, \frac{1}{n^k}, 0, 0, \ldots\right\} \quad (k = 1, \ldots, n). \qquad (3.36)$$

Indeed, for every $x \in \Gamma_\perp \setminus \{0\}$ the number of points $q_1, \ldots$ $\ldots, q_p \in Q = \{1, 2, 3, \ldots\} \cup \{\infty\}$ with the property*) $|x(q_i)| =$ $= \|x\|$ $(i = 1, \ldots, p)$ is $\leqslant n$ and we have

$$\operatorname{rank} (x_k(q_i))_{\substack{k=1,\ldots,n \\ i=1,\ldots,p}} = \operatorname{rank} \begin{pmatrix} \frac{1}{q_1} & \cdots & \frac{1}{q_p} \\ \frac{1}{q_1^2} & \cdots & \frac{1}{q_p^2} \\ \cdot & \cdot\cdot\cdot & \cdot \\ \frac{1}{q_1^n} & \cdots & \frac{1}{q_p^n} \end{pmatrix} = p,$$

whence, by virtue of theorem 3.5, Γ is a Čebyšev subspace.

If Q is a metrizable compact space, the space $E = C_R(Q)$ is separable**), hence by the remark made at the end of section 3.1, *the conjugate space $E^* = C_R(Q)^*$ has at least one $\sigma(E^*, E)$-closed Čebyšev hyperplane*. Moreover, as has observed R. R. Phelps, in this case, *for every $n \geqslant 1$ the conjugate space E^* has a $\sigma(E^*, E)$-closed Čebyšev subspace Γ of codimension n*. Indeed, taking arbitrary points $q_1, \ldots, q_n \in Q$ and disjoint neighbourhoods $U_1, \ldots, U_n$ of them ($q_k \in U_k$, $U_k \cap U_l = \varnothing$ for $k \neq l$), and constructing functions $x_1, \ldots, x_n \in C(Q)$ with the properties

$$x_k(q) = \begin{cases} 1 \text{ for } q = q_k \\ 0 \text{ for } q \in Q \setminus U_k \end{cases} \qquad (k = 1, \ldots, n),$$

$$0 \leqslant x_k(q) \leqslant 1 \text{ for } q \in U_k \setminus \{q\} \qquad (k = 1, \ldots, n),$$

the $\sigma(E^*, E)$-closed linear subspace

$$\Gamma = \{f \in E^* \mid f(x_k) = 0 \ (k = 1, \ldots, n)\}$$

will obviously satisfy the conditions of theorem 3.5.

On the other hand, there exist compact spaces Q for which the space $E^* = C_R(Q)^*$ has no $\sigma(E^*, E)$-closed Čebyšev subspace of finite codimension. Indeed, from theorem 3.8 below it follows that the space $E^* = L_R^\infty([0, 1], \nu)^*$, where ν is

---

*) For the sake of simplicity, we use the same notation for an element $x \in c_R$ and its image under the equivalence $c_R \equiv C_R(Q)$, where $Q = \{1, 2, 3, \ldots\} \cup \{\infty\}$. Consequently, for $x = \{\xi_n\} \in c_R$ and $q \in Q = \{1, 2, 3, \ldots\} \cup \{\infty\}$ by $x(q)$ we mean $x(q) = \begin{cases} \xi_n \text{ if } q = n \ (n = 1, 2, \ldots). \\ \lim_{n\to\infty} \xi_n \text{ if } q = \infty. \end{cases}$

**) See e. g. A. Grothendieck [79], p. 50, proposition 16.

the Lebesgue measure, has no such subspace, and in the foregoing we have seen that there exists a compact space Q such that $E = L_R^\infty([0, 1], \nu) \equiv C_R(Q)$. In this connection it may be interesting to remark that *every space* $E^* = C_R(Q)^*$ (Q *compact*) *has* $\sigma(E^*, E)$-*closed Čebyšev subspaces of infinite dimension*, as shown by the following example of R. R. Phelps ([176], lemma 3.1): *If* $A \subset Q$ *is closed, then the* $\sigma(E^*, E)$-*closed linear subspace* Γ *of* $E^* = C_R(Q)^*$, *defined by**)

$$\Gamma = \{f \in E^* \mid f(x) = 0 \ (x \in I_A)\} = I_A^\perp \tag{3.37}$$

is a Čebyšev subspace. Indeed, let $\varphi \in I_A^*$ be arbitrary with $\|\varphi\| = 1$ and let $\mu_1, \mu_2 \in E^*$ be two arbitrary extensions of φ, with $\|\mu_1\| = \|\mu_2\| = 1$. Then we have

$$|\mu_i|(A) = 0 \qquad (i = 1, 2), \tag{3.38}$$

since otherwise for $i=1$ or $i=2$ we would have $1 - |\mu_i|(A) < 1$, whence, taking into account

$$|\varphi(x)| = |\mu_i(x)| = \left| \int_{Q \setminus A} x(q) \, \mathrm{d}\mu_i(q) \right| \leqslant \|x\| \, |\mu_i|(Q \setminus A) = \\ = \|x\| \, (1 - |\mu_i|(A)) \qquad (x \in I_A),$$

there would follow $\|\varphi\| \leqslant 1 - |\mu_i|(A) < 1$, which contradicts the hypothesis $\|\varphi\| = 1$. Since the measures $|\mu_i|$ are regular, there exists an open set $U \supset A$ such that

$$|\mu_i|(U \setminus A) < \frac{\varepsilon}{4} \qquad (i = 1, 2). \tag{3.39}$$

Now let $x \in E = C_R(Q)$ be arbitrary and let $y \in E$ be such that $y = x$ on $T \setminus U$, $y = 0$ on A and $\|y\| \leqslant \|x\|$. Then we have, taking into account (3.38) and (3.39),

$$|\mu_i(x - y)| = \left| \int_{U \setminus A} [x(q) - y(q)] \, \mathrm{d}\mu_i(q) \right| \leqslant \|x - y\| \frac{\varepsilon}{4} \leqslant \frac{\varepsilon}{2} \|x\|,$$

whence, taking into account $y \in I_A$ and $\mu_1 |_{I_A} = \mu_2 |_{I_A}$,

$$|\mu_1(x) - \mu_2(x)| \leqslant |\mu_1(x) - \mu_1(y)| + |\mu_2(y) - \mu_2(x)| \leqslant \varepsilon \|x\|.$$

Since $x \in E$ was arbitrary, it follows that $\|\mu_1 - \mu_2\| \leqslant \varepsilon$, whence, since $\varepsilon > 0$ was arbitrary, we obtain $\mu_1 = \mu_2$. Thus $\Gamma_\perp = I_A \subset C_R(Q)$ has the property (U), whence $\Gamma = I_A^\perp \subset E^*$ is a Čebyšev subspace, which completes the proof.

For the spaces $L_R^\infty(T, \nu)^*$ we have

---

*) We recall (see Chap. II, §1, section 1.3) that by definition

$$I_A = \{x \in C_R(Q) \mid x(q) = 0 \quad (q \in A)\}.$$

THEOREM 3.8. *Let $E = L_R^\infty(T, \nu)$, where (T, ν) is a positive measure space. In order that E^* have a $\sigma(E^*, E)$-closed Čebyšev subspace of codimension n it is necessary and sufficient that (T, ν) have at least n atoms.*

The proof is entirely similar to the proof of theorem 2.13 (replacing $G^\perp$ by $\Gamma_\perp$ and applying theorem 3.7 instead of theorem 2.12), and therefore we omit it.

The fact that for (T, ν) without atoms the conjugate space $E^* = L_R^\infty(T, \nu)^*$ has no $\sigma(E^*, E)$-closed Čebyšev subspace of finite codimension has been observed by R. R. Phelps ([176], p. 252). In the particular case when (T, ν) is σ-finite (hence the set of all atoms of (T, ν) is at most countable), theorem 3.8 has been given by A. L. Garkavi ([68], theorem 8).

From the example (3.37) and the equivalence $L_R^\infty(T, \nu) \equiv C_R(Q)$ it follows that *every space $E^* = L_R^\infty(T, \nu)^*$ (where (T, ν) is a positive measure space), has $\sigma(E^*, E)$-closed Čebyšev subspaces of infinite dimension* (R. R. Phelps [176], p. 252).

Finally, for the space $((c_0)_R)^*$ we have

THEOREM 3.9 (R. R. Phelps [176], theorem 3.3). *Let $E = (c_0)_R$ and let Γ be a $\sigma(E^*, E)$-closed linear subspace of codimension n of $E^* \equiv l_R^1$, hence $\Gamma_\perp = [x_1, \ldots, x_n]$, where $x_k = \{\xi_i^{(k)}\} \in E$ $(k = 1, \ldots, n)$. In order that Γ be a Čebyšev subspace it is necessary and sufficient that for every $x = \{\xi_i\} \in \Gamma_\perp \setminus \{0\}$ the number $p = p(x)$ of the indices $i_1, \ldots, i_p$ with the property $|\xi_{i_j}| = \|x\|$ $(j = 1, \ldots, p)$ satisfy $1 \leqslant p \leqslant n$.*

Proof. Since a functional $f \in E^*$ is in $\mathcal{E}(S_{E^*})$ if and only if it is of the form

$$f_i(x) = \varepsilon \xi_i \qquad (x = \{\xi_i\} \in E), \tag{3.40}$$

where $\varepsilon = \pm 1$, the necessity of the condition is an immediate consequence of theorem 3.1. Also, by virtue of (3.40), lemma 2.1 and theorem 3.1, in order to prove the sufficiency of the condition it is sufficient to show that if $x = \{\xi_i\} \in \Gamma_\perp \setminus \{0\}$ and if

$$\{i \mid |\xi_i| = \|x\|\} = \{i_1, \ldots, i_p\}, \tag{3.41}$$

(where $1 \leqslant p \leqslant n$ by hypothesis), then

$$\operatorname{rank} \begin{pmatrix} \xi_{i_1}^{(1)} \ldots \xi_{i_p}^{(1)} \\ \ldots\ldots\ldots \\ \xi_{i_1}^{(n)} \ldots \xi_{i_p}^{(n)} \end{pmatrix} = p. \tag{3.42}$$

Assume the contrary, i.e. that there exists an $x = \{\xi_i\} \in \Gamma_\perp \setminus \{0\}$ satisfying (3.41) (with $1 \leqslant p \leqslant n$), such that

$$\operatorname{rank} \begin{pmatrix} \xi_{i_1}^{(1)} \ldots \xi_{i_p}^{(1)} \\ \ldots\ldots\ldots \\ \xi_{i_1}^{(n)} \ldots \xi_{i_p}^{(n)} \end{pmatrix} < p. \tag{3.43}$$

We shall show that there exists then an $x' = \{\xi'_i\} \in \Gamma_\perp \setminus \{0\}$ such that

$$\{i \mid |\xi'_i| = \|x'\|\} \supset \{i_1, \ldots, i_p\} \cup \{i_{p+1}\}, \tag{3.44}$$

where $i_{p+1} \notin \{i_1, \ldots, i_p\}$. Since in this case by (3.43) we have

$$\operatorname{rank} \begin{pmatrix} \xi^{(1)}_{i_1} \ldots \xi^{(1)}_{i_{p+1}} \\ \ldots\ldots\ldots \\ \xi^{(n)}_{i_p} \ldots \xi^{(n)}_{i_{p+1}} \end{pmatrix} < p+1,$$

by induction it will follow that there exists an element $x_0 = \{\xi^{(0)}_i\} \in \Gamma_\perp \setminus \{0\}$ such that the cardinality of the set $\{i \mid |\xi^{(0)}_i| = \|x\|\}$ be $> n$, contradicting the hypothesis, and this contradiction will prove the sufficiency of the condition of the theorem.

By (3.43) and $p \leqslant n$ the system of p homogeneous linear equations with n unknowns

$$\alpha_1 \xi^{(1)}_{i_j} + \ldots + \alpha_n \xi^{(n)}_{i_j} = 0 \qquad (j = 1, \ldots, p)$$

has a non trivial solution, whence there exists a $y = \sum_{i=1}^{n} \alpha_i x_i = \{\eta_i\} \in \Gamma_\perp \setminus \{0\}$ such that

$$\eta_{i_1} = \ldots = \eta_{i_p} = 0; \tag{3.45}$$

we may assume, without loss of generality, that we have $\|y\| \leqslant 1$.

Define $z = \{\zeta_n\} \in E$ by

$$\zeta_i = \begin{cases} \xi_i \text{ for } i \in \{i_1, \ldots, i_p\} \\ 0 \ \text{ for } i \notin \{i_1, \ldots, i_p\}, \end{cases} \tag{3.46}$$

and let $0 < \alpha < \|x\| - \sup_{i \notin \{i_1, \ldots, i_p\}} |\xi_i|$. By (3.45) and (3.46) we have then

$$\xi_i + \alpha\eta_i - \zeta_i = \begin{cases} 0 & \text{for } i \in \{i_1, \ldots, i_p\} \\ \xi_i + \alpha\eta_i & \text{for } i \notin \{i_1, \ldots, i_p\}, \end{cases} \tag{3.47}$$

whence

$$\|x + \alpha y - z\| = \sup_{i \notin \{i_1, \ldots, i_p\}} |\xi_i + \alpha\eta_i| \leqslant \sup_{i \notin \{i_1, \ldots, i_p\}} |\xi_i| + \alpha < \|x\|.$$

Since the function $\varphi(\lambda) = \|x + \lambda y - z\|$ is continuous and $\lim_{\lambda \to +\infty} \varphi(\lambda) = +\infty$, it follows that there exists a $\beta > 0$ such that

$$\|x + \beta y - z\| = \|x\|. \tag{3.48}$$

We claim that then the element $x' = \{\xi_i'\} \in \Gamma_\perp \setminus \{0\}$ defined by

$$x' = x + \beta y \tag{3.49}$$

has the property (3.44). Indeed, by (3.45), (3.46) and (3.49) we have

$$\xi_i' = \begin{cases} \xi_i & \text{for } i \in \{i_1, \ldots, i_p\} \\ \xi_i + \beta\eta_i - \zeta_i & \text{for } i \notin \{i_1, \ldots, i_p\}, \end{cases} \tag{3.50}$$

whence taking into account (3.41) and (3.48),

$$\|x'\| = \sup_{1 \leqslant i < \infty} |\xi_i'| = \max\left(\sup_{i \in \{i_1, \ldots, i_p\}} |\xi_i|, \sup_{i \notin \{i_1, \ldots, i_p\}} |\xi_i + \beta\eta_i - \xi_i| \right) = \|x\|. \tag{3.51}$$

Consequently, taking again into account (3.41) and (3.50), we have

$$\{i \mid |\xi_i'| = \|x'\|\} \supset \{i_1, \ldots, i_p\}. \tag{3.52}$$

On the other hand, since by (3.45) and (3.46) we have $\xi_i + \beta\eta_i - \zeta_i = 0$ $(i \in \{i_1, \ldots, i_p\})$, from (3.48) and $E = (c_0)_R$ it follows that there exists an $i_{p+1} \in \{i_1, \ldots, i_p\}$ such that $|\xi_{i_{p+1}} + \beta\, \eta_{i_{p+1}} - \zeta_{i_{p+1}}| = \|x\|$, whence by (3.50), $|\xi'_{i_{p+1}}| = \|x\|$, and thus, by (3.51),

$$i_{p+1} \in \{i \mid |\xi_i'| = \|x'\|\}. \tag{3.53}$$

The relations (3.52) and (3.53) show that we have (3.44), which completes the proof of theorem 3.9.

For the space $E = c_R$ a statement similar to that of theorem 3.9 is no longer valid; an example of a $\sigma(E^*, E)$-closed linear subspace Γ of codimension 2 of $E^* = (c_R)^*$, which satisfies the conditions of theorem 3.9 but is not a Čebyšev subspace, is obtained (R. R. Phelps [176], p. 249) by taking

$$x_1 = \left\{1, 1 - \frac{1}{4}, 1 - \frac{1}{9}, 1 - \frac{1}{16}, \ldots\right\},$$

$$x_2 = \left\{0, \frac{1}{2}, -\frac{1}{3}, \frac{1}{4}, -\frac{1}{5}, \ldots\right\} \in c_R = E, \tag{3.54}$$

$$\Gamma = \{f \in E^* \mid f(x_k) = 0 \ (k = 1, 2)\}. \tag{3.55}$$

§ 4. THE OPERATORS π_G AND THE FUNCTIONALS e_G FOR CLOSED LINEAR SUBSPACES G OF FINITE CODIMENSION. DIAMETERS OF ORDER n

4.1. THE OPERATORS π_G FOR CLOSED LINEAR SUBSPACES G OF FINITE CODIMENSION

We shall now prove, for closed linear subspaces of finite codimension of normed linear spaces, results similar to those given in Chap. II, § 5 (for linear subspaces of finite dimension).

THEOREM 4.1. *Let E be a normed linear space of (finite or infinite) dimension $\geqslant 3$, with the property that for every closed linear subspace G of a certain fixed finite codimension n, where $1 \leqslant n \leqslant \dim E - 1$, the mapping π_G satisfies*

$$\|\pi_G(x)\| \leqslant \|x\| \qquad (x \in D(\pi_G)). \tag{4.1}$$

Then E is equivalent to an inner product space.

Proof. Let G_1 be an arbitrary linear subspace of E with $\dim G_1 = 1$, and let $x \in E \setminus G_1$ and $g_1 = \pi_{G_1}(x) \in \mathfrak{P}_{G_1}(x)$ be arbitrary. Then by virtue of theorem 1.1 of Chap. I, there exists an $f_1 \in E^*$ such that

$$\|f_1\| = 1, \tag{4.2}$$

$$f_1(g_1) = 0, \tag{4.3}$$

$$f_1(x - g_1) = \|x - g_1\| \neq 0. \tag{4.4}$$

Since by hypothesis we have $\dim \{f \in E^* \mid f(g_1) = 0\} \geqslant \geqslant \dim E^* - 1 \geqslant n$, there exist $n - 1$ functionals $f_2, \ldots, f_n \in E^*$ such that $f_1, f_2, \ldots, f_n$ be linearly independent and that

$$f_2(g_1) = \ldots = f_n(g_1) = 0. \tag{4.5}$$

Put

$$G = \{y \in E \mid f_k(y) = 0 \ (k = 1, \ldots, n)\}. \tag{4.6}$$

We have then $\operatorname{codim} G = n$, $x \in E \setminus G$ (since $f_1(x) \neq 0$ by (4.4) and (4.3)) and $g_1 \in G$. From (4.2), (4.6), (4.4) and theorem 1.1 of Chap. I, it follows that we have $g_1 \in \mathfrak{P}_G(x)$, hence $x \in D(\pi_G)$ and we can write $g_1 = \pi_G(x)$, whence, taking into account the hypothesis (4.1), we obtain $\|\pi_{G_1}(x)\| = \|g_1\| = \|\pi_G(x)\| \leqslant \leqslant \|x\|$. Consequently, for every linear subspace $G_1 \subset E$ with $\dim G_1 = 1$ we have

$$\|\pi_{G_1}(x)\| \leqslant \|x\| \qquad (x \in E), \tag{4.7}$$

i.e. E satisfies the condition of Chap. II, theorem 5.1, whence there follows the equivalence of E to an inner product space, and thus theorem 4.1 is proved.

COROLLARY 4.1. *Let E be a normed linear space of (finite or infinite) dimension $\geqslant 3$, with the property that for every closed linear subspace G of a certain fixed finite codimension n, where $1 \leqslant n \leqslant \dim E - 1$, the mapping π_G is "contractive", i.e. satisfies*

$$\|\pi_G(x) - \pi_G(y)\| \leqslant \|x - y\| \qquad (x, y \in D(\pi_G)). \qquad (4.8)$$

Then E is equivalent to an inner product space.

Proof. Taking in (4.8) $y = 0$, we obtain (4.1) and thus the equivalence of E to an inner product space follows from theorem 4.1.

THEOREM 4.2. *Let E be a normed linear space of (finite or infinite) dimension $\geqslant 3$, with the property that for every closed linear subspace G of a certain fixed finite codimension n, where $2 \leqslant n \leqslant \dim E - 1$, the mapping π_G is one-valued on $D(\pi_G) = E$ and satisfies*

$$\pi_G(x + y) = \pi_G(x) + \pi_G(y) \qquad (x, y \in D(\pi_G) = E). \qquad (4.9)$$

Then E is equivalent to an inner product space.

Proof. By virtue of theorem 5.2 of Chap. II it is sufficient to consider the case when $\dim E = \infty$. Since by our hypothesis all closed linear subspaces of codimension n of E are Čebyšev subspaces, the space E is, by virtue of corollary 3.3 of Chap. I, strictly convex, whence the mapping π_G is one-valued on $D(\pi_G)$, whatever be the linear subspace G of E.

We shall show now that it is sufficient to consider the case when $n = 2$. Indeed, assume that the theorem is valid for closed linear subspaces of codimension 2 and let E_1 be an arbitrary closed linear subspace of E with codim $E_1 = n - 2$. Then for every closed linear subspace G of E_1 with

$$\operatorname{codim}_{E_1} G = \dim E_1/G = 2, \qquad (4.10)$$

we have $\operatorname{codim}_E G = n$, whence (4.9), and consequently (4.9) also in E_1. Since the theorem has been assumed to be valid for closed linear subspaces of codimension 2, it follows that E_1 is equivalent to an inner product space. Consequently, since E_1 was an arbitrary closed linear subspace of E with codim $E_1 = 2$ (hence $\dim E_1 = \infty$), by virtue of the theorem of P. Jordan and J. von Neumann used also in Chap. II, § 5, it follows that E is equivalent to an inner product space.

Thus, let $n = 2$ and let G_0 be an arbitrary two-dimensional linear subspace of E. Then, just as in the proof of theorem 5.2

of Chap. II, it follows that there exists a closed linear subspace G of E with codim $G = 2$ such that $I - \pi_G$ be a continuous linear projection of norm 1 of E onto G_0 (here is used the hypothesis $D(\pi_G) = E$), whence, by virtue of the theorem of S. Kakutani, E is equivalent to an inner product space, which completes the proof.

The hypothesis $2 \leqslant n \leqslant \dim E - 1$ (hence also the hypothesis $\dim E \geqslant 3$) in theorem 4.2 is essential; indeed, e.g. in any strictly convex normed linear space E with the property that all hyperplanes of E are proximinal, the condition of theorem 4.2 is satisfied for every hyperplane G of E (by virtue of theorem 6.2 of Chap. I).

For $n \geqslant 3$ one can prove more, namely

THEOREM 4.3. *Let E be a normed linear space of (finite or infinite) dimension $\geqslant 4$, with the property that for every closed linear subspace G of a certain fixed finite codimension n, where $3 \leqslant n \leqslant \dim E - 1$, the mapping π_G is one-valued on $D(\pi_G)$ and satisfies*

$$\pi_G(x + y) = \pi_G(x) + \pi_G(y) \qquad (x, y, x + y \in D(\pi_G)). \tag{4.11}$$

Then E is equivalent to an inner product space.

Proof. Since by hypothesis all closed linear subspaces of codimension n of E are semi-Čebyšev subspaces, the space E is, by virtue of corollary 3.3 of Chap. I, strictly convex, hence the mapping π_G is one-valued on $D(\pi_G)$, for every linear subspace G of E.

Now let G_1 be an arbitrary linear subspace of E, with $\dim G_1 = 1$, let $x_1, x_2 \in E \setminus G_1$ be arbitrary with $x_1 + x_2 \in E \setminus G_1$ and let $g_1 = \pi_{G_1}(x_1)$, $g_2 = \pi_{G_1}(x_2)$, $g_3 = \pi_{G_1}(x_1 + x_2)$. Then, by virtue of theorem 1.1 of Chap. I, there exist $f_1', f_2', f_3' \in E^*$ such that

$$\|f_i'\| = 1 \qquad (i = 1, 2, 3), \tag{4.12}$$

$$f_i'(g) = 0 \qquad (g \in G_1,\ i = 1, 2, 3), \tag{4.13}$$

$$f_i'(x_i - g_i) = \|x_i - g_i\| \neq 0 \quad (i = 1, 2, 3), \tag{4.14}$$

where we have put $x_3 = x_1 + x_2$. Since by our hypothesis $n \geqslant 3$ and $\dim \{f \in E^* | f(g_1) = 0\} \geqslant \dim E^* - 1 \geqslant n$, there exist n linearly independent functionals $f_1, \ldots, f_n \in E^*$ containing among them a maximal linearly independent subsystem of $\{f_1', f_2', f_3'\}$, such that

$$f_1(g_i) = \ldots = f_n(g_i) = 0 \qquad (i = 1, 2, 3). \tag{4.15}$$

Put

$$G = \{y \in E \,|\, f_k(y) = 0 \quad (k = 1, \ldots, n)\}. \tag{4.16}$$

Then we have codim $G = n$, $x_1, x_2, x_3 \in E \setminus G$ (since $f_i'(x_i) \neq 0$ for $i = 1, 2, 3$ by (4.14) and (4.13)), and $g_1, g_2, g_3 \in G$. From (4.12), (4.16), (4.14) and theorem 1.1 of Chap. I it follows that we have $g_i = \pi_G(x_i)$ $(i = 1, 2, 3)$, hence $x_i \in D(\pi_G)$ $(i = 1, 2, 3)$, whence, taking into account the hypothesis (4.11), we obtain

$$\pi_{G_1}(x_1 + x_2) = g_3 = \pi_G(x_1 + x_2) = \pi_G(x_1) + \pi_G(x_2) = g_1 + g_2 = \\ = \pi_{G_1}(x_1) + \pi_{G_1}(x_2).$$

Consequently, since $x_1, x_2 \in E \setminus G_1$ with $x_1 + x_2 \in E \setminus G_1$ were arbitrary, it follows that also G_1 satisfies (4.11), with $D(\pi_{G_1}) = E$ (the additivity of π_{G_1} for the case when one of x_1, x_2 or $x_1 + x_2$ is in G_1, follows from Chap. I, theorem 6.1 e)). Since G_1 was an arbitrary linear subspace of E with dim $G_1 = 1$, E satisfies the condition of Chap. II, theorem 5.2, whence it follows the equivalence of E to an inner product space, and thus theorem 4.3 is proved.

In particular, the condition of theorem 4.3 is satisfied if for every closed linear subspace G with codim $G = n (\geqslant 3)$ the mapping π_G is one-valued on $D(\pi_G)$ and satisfies

$$x + y \in D(\pi_G),\ \pi_G(x + y) = \pi_G(x) + \pi_G(y) \quad (x, y \in D(\pi_G)) \tag{4.17}$$

(or, what is equivalent, $\pi_G^{-1}(0)$ is a linear subspace of E).

4.2. THE OPERATORS π_{G^n} FOR DECREASING SEQUENCES $\{G^n\}$ OF CLOSED LINEAR SUBSPACES OF FINITE CODIMENSION

In the preceding section (theorems 4.2 and 4.3) we have seen that if a normed linear space *) E is not equivalent to an inner product space, then for every integer n with $2 \leqslant n < \infty$ there exist closed linear subspaces of E with codim $G = n$ such that the operator π_G is non-linear. It arises naturally the problem whether there exists a sequence $\{G^n\}$ of closed linear subspaces of E, with $G^1 \supset G^2 \supset \ldots$ and codim $G^n = n$ $(n = 1, 2, \ldots)$ such that the operators π_{G^n} be one-valued and linear on $D(\pi_{G^n}) = E$ $(n = 1, 2, \ldots)$. We shall give now an important case where the answer is affirmative and where the operators π_{G^n} are expressed in a simple form, convenient for applications.

Following V. N. Nikolsky [164], [165], the norm in a Banach space E with a basis $\{x_n\}$ is said to be a *K-norm* (with respect to the basis $\{x_n\}$), if:

---

*) For the sake of simplicity, throughout the sections 4.2 and 4.3 we shall assume, without any special mention, that the space E is infinite dimensional.

a) For every $x \in E$ and $n = 1, 2, \ldots$ there exists a unique element $\pi_{G^n}(x)$ of best approximation of x by the elements *) of the closed linear subspace $G^n = [x_{n+1}, x_{n+2}, \ldots]$ of codimension n.

b) This element coincides with the nth remainder $r_n(x)$ of the expansion of the element x with respect of the basis $\{x_n\}$, i.e.**)

$$\pi_{G^n}(x) = r_n(x) = x - s_n(x) \qquad (x \in E, n = 1, 2, \ldots). \quad (4.18)$$

THEOREM 4.4. (V. N. Nikolsky [165], §4). *Let E be a Banach space with a basis $\{x_n\}$ and let $\{f_n\} \subset E^*$ be the sequence of coefficient functionals associated to the basis $\{x_n\}$. Then one can introduce on E a K-norm equivalent to the initial norm of the space E, by the formula*

$$))x)) = \sup_{1 \leqslant n < \infty} \left\| \sum_{i=1}^{\infty} f_i(x) x_i \right\| + \sum_{i=1}^{\infty} \frac{1}{2^i} \| f_i(x) x_i \|. \quad (4.19)$$

Proof. Obviously, $))x))$ is a norm on E; it is equivalent to the initial norm since for every $x \in E$ we have

$$\|x\| = \left\| \sum_{i=1}^{\infty} f_i(x) x_i \right\| \leqslant \sup_{1 \leqslant n < \infty} \left\| \sum_{i=1}^{n} f_i(x)\, x_i \right\| \leqslant))x)) \leqslant$$

$$\leqslant \sup_{1 \leqslant n < \infty} \left\| \sum_{i=1}^{n} f_i(x) x_i \right\| + \max_{1 \leqslant i < \infty} \|f_i(x) x_i\| \leqslant \sup_{1 \leqslant n < \infty} \left\| \sum_{i=1}^{n} f_i(x) x_i \right\| +$$

$$+ \max_{1 \leqslant i < \infty} \left(\left\| \sum_{j=1}^{i} f_j(x) x_j \right\| + \left\| \sum_{j=1}^{i-1} f_j(x) x_j \right\| \right) \leqslant 3 \sup_{1 \leqslant n < \infty} \left\| \sum_{i=1}^{n} f_i(x) x_i \right\|,$$

and since $|||x||| = \sup_{1 \leqslant n < \infty} \left\| \sum_{i=1}^{n} f_i(x) x_i \right\|$ is***) a norm on E, equivalent to the initial norm of the space E. Consequently, it remains to show that $))x))$ is a K-norm.

---

*) Some authors, e.g. V. N. Nikolsky [165], call the elements of G^n "complementary polynomials (of the order n)".

**) For the definition of the partial sum operators s_n associated to the basis $\{x_n\}$ see Chap. II, §5, section 5.2.

***) See Chap. II, §5, section 5.2.

Let $x \in E$ and $g = \sum_{j=n+1}^{\infty} \alpha_j x_j \in G^n$ be arbitrary. Putting $\alpha_1 = \ldots = \alpha_n = 0$, we have then

$$))x - g)) =)) \sum_{i=1}^{\infty} [f_i(x) - \alpha_i] x_i)) = \sup_{1 \leqslant k < \infty} \left\| \sum_{i=1}^{k} [f_i(x) - \alpha_i] x_i \right\| + \sum_{i=1}^{\infty} \frac{1}{2^i} \| [f_i(x) - \alpha_i] x_i \|.$$

In particular, for $g = r_n(x) = \sum_{i=n+1}^{\infty} f_i(x) x_i$ we have

$$))x - r_n(x))) = \sup_{1 \leqslant k \leqslant n} \left\| \sum_{i=1}^{k} f_i(x)\, x_i \right\| + \sum_{i=1}^{n} \frac{1}{2^i} \left\| f_i(x) x_i \right\|.$$

Since for $g \neq r_n(x)$ we have obviously

$$\sum_{i=1}^{\infty} \frac{1}{2^i} \| [f_i(x) - \alpha_i] x_i \| > \sum_{i=1}^{n} \frac{1}{2^i} \| f_i(x) x_i \|,$$

it follows that for $g \neq r_n(x)$ we have $))x - g)) >))x - r_n(x)))$. Thus, $))x))$ is a K-norm on E, which completes the proof of theorem 4.4.

As has shown V. N. Nikolsky ([165], §5), *one can introduce on E a norm equivalent to the initial norm of the space E, which is simultaneously a T-norm and a K-norm with respect to the basis $\{x_n\}$, by the formula*

$$((x)) = \sup_{1 \leqslant n < \infty} \left\{ \left(\left(\sum_{i=1}^{n} f_i(x) x_i \left(\left(+ \right) \right) \sum_{i=n+1}^{\infty} f_i(x) x_i \right) \right) \right\}, \tag{4.20}$$

where $((x(($ is the T-norm of Chap. II, formula (5.24), and $))x))$ is the K-norm (4.19) above.

4.3. THE FUNCTIONALS e_{G^n} FOR DECREASING SEQUENCES $\{G^n\}$ OF CLOSED LINEAR SUBSPACES OF FINITE CODIMENSION

From the remark made in Chap. I, the final part of section 6.3, it follows in particular that *if $E = B^*$, where B is a Banach space, then for every decreasing sequence $G^1 \supset G^2 \supset \ldots$ of $\sigma(B^*, B)$-closed linear subspaces of B^* with codim $G^n = n$ ($n = 1, 2, \ldots$) and every sequence of numbers $\{e_n\}$ with the properties*

$$0 \leqslant e_1 \leqslant e_2 \leqslant \ldots \leqslant c < \infty \tag{4.21}$$

there exists an element $x \in E$ satisfying

$$\|x\| = \lim_{n \to \infty} e_n, \quad e_{G^k}(x) = e_k \qquad (k = 1, 2, \ldots). \qquad (4.22)$$

As has shown M. I. Kadec ([103], theorem 2), in separable real spaces $E = B^*$ and for G^n $(n = 1, 2, \ldots)$ being Čebyšev subspaces, this statement remains valid if the functionals e_{G^n} are replaced by the functionals

$$\tilde{e}_{G^n}(x) = \varepsilon_n(x)\, e_{G^k}(x) \qquad (x \in E,\ n = 1, 2, \ldots), \qquad (4.23)$$

where

$$\varepsilon_n(x) = \begin{cases} 1 \text{ for } \pi_{G^{n-1}}(x) \in G_+^{n-1} \\ -1 \text{ for } \pi_{G^{n-1}}(x) \in G_-^{n-1} \\ \varepsilon_{n-1}(x) \text{ for } \pi_{G^{n-1}}(x) \in G^n, \text{ i.e. } e_{G^{n-1}}(x) = e_{G^n}(x) \end{cases} \qquad (4.24)$$

(G_+^{n-1} and G_-^{n-1} being the open half-spaces of G^{n-1} determined by G^n), and the sequence of non-negative numbers $\{e_n\}$ is replaced by a sequence of real numbers $\{\tilde{e}_n\}$ with the properties

$$|\tilde{e}_1| \leqslant |\tilde{e}_2| \leqslant \ldots \leqslant c < \infty, \qquad (4.25)$$

$$1 \leqslant n < \infty \text{ and } |\tilde{e}_n| = |\tilde{e}_{n+1}| \text{ imply } \tilde{e}_n = \tilde{e}_{n+1}; \qquad (4.26)$$

moreover, in this case the element $x \in E$ with the properties

$$\|x\| = \lim_{n \to \infty} |\tilde{e}_n|, \qquad e_{G^k}(x) = e_k \qquad (k = 1, 2, \ldots) \quad (4.27)$$

is unique. Using this result and a theorem on the existence of an equivalent strictly convex norm $|||f|||$ on $E = B^*$ with the property that the relations $\lim_{n \to \infty} (f_n - f)(x) = 0$ $(x \in E)$, $\lim_{n \to \infty} |||f_n||| = |||f|||$ imply $\lim_{n \to \infty} |||f_n - f||| = 0$ (M. I. Kadec [103], theorem 1 and V. Klee [112], corollary 1.5), it has been shown that *all real infinite dimensional separable conjugate Banach spaces* are homeomorphic (M. I. Kadec [103], theorem 3 and V. Klee [112], theorem 1.6); as has been mentioned in Chap. II, the final part of section 5.3, it is known at present that this statement remains also valid for all infinite dimensional separable Banach spaces.

4.4. DIAMETERS OF ORDER n

In Chap. II, §6, we have seen that the n-dimensional diameter $d_n(A)$ of a set A in a normed linear space E may be considered as a measure of the n-dimensional "thickness" of A. I. M. Gelfand and, independently, S. Smolyak (see V. M. Tihomirov [248]) have considered a different measure $d^n(A)$ of this "thickness", which we shall call *the diameter of order n of the set A*, defined by

$$d^n(A) = \inf_{\operatorname{codim} G = n} \inf_{\substack{\varepsilon > 0 \\ A \cap G \subset \varepsilon S_E}} \varepsilon, \tag{4.28}$$

where the last inf is taken over all closed linear subspaces G of codimension n of E. We shall denote

$$\tilde{\delta}(A, G) = \inf_{\substack{\varepsilon > 0 \\ A \cap G \subset \varepsilon S_E}} \varepsilon. \tag{4.29}$$

An important case when the diameters $d_n(A)$ and $d^n(A)$ coincide, is obtained by taking $A = S_{E_{n+1}}$, where E_{n+1} is an $(n+1)$-dimensional linear subspace of E. Indeed, since for every closed linear subspace $G \subset E$ of codimension n we have $G \cap E_{n+1} \neq \{0\}$, we obtain from (4.28)

$$d^n(S_{E_{n+1}}) = 1, \tag{4.30}$$

and on the other hand, by Chap. II, theorem 6.3, we have $d_n(S_{E_{n+1}}) = 1$.

There exist sets A with $d_1(A) < d^1(A)$, as shown by the following example of V. M. Tihomirov [248]: Let $E = (l_3^\infty)_R =$ = the space of all triplets of real numbers $x = \{\xi_1, \xi_2, \xi_3\}$ endowed with the usual vector operations and with the norm $\|x\| = \max_{1 \leqslant i \leqslant 3} |\xi_i|$, and let $A = \bigcup_{-\infty < \lambda < \infty} S(\lambda x_0, 1)$, where $x_0 =$ $= \{1, 1, 1\} \in E$. Then obviously

$$d_1(A) = \inf_{\dim G_1 = 1} \delta(A, G_1) \leqslant \delta(A, [x_0]) = 1.$$

On the other hand, for every plane $G_f^2 = \{y \in E \mid f(y) = 0\}$, where $f \in E^*$, $\|f\| = 1$, we have $A \cap G_f^2 \not\subset S_E$, whence $\tilde{\delta}(A, G_f^2) < 1$. Since $\tilde{\delta}(A, G_f^2)$ is a continuous function of f on the compact set Fr S_{E^*}, it follows that $d^1(A) > 1$.

Such an example can be constructed in every space which is not an inner product space and only in such spaces (V. M. Tihomirov [248], remark 2). Moreover, we have also the following result of V. M. Tihomirov ([249], theorem 3): *Among all Banach spaces E, only the complete inner product spaces have the properties*

$$d_n(A, E) = \tilde{d}_n''(A, E) \geqslant \tilde{d}^n(A), \tag{4.31}$$

$$d_n(A, E) = d_n(A, E_1) \qquad (A \subset E \subset E_1), \tag{4.32}$$

where $\tilde{d}_n''(A, E)$ is the n-dimensional linear diameter of Chap. II, formula (6.72).

There also exist sets A with $\tilde{d}^1(A) < d_1(A)$, as shown by the following example of V. M. Tihomirov [248]: Let $E = (l_3^2)_R$ = the real three-dimensional euclidean space, and let A be the closed convex hull of the set consisting of the unit circle in the plane $\{x \in E \mid \xi_3 = 0\}$ and two equilateral triangles symmetric with respect to 0, situated in the planes $\{x \in E \mid \xi_3 = \pm 1\}$, with the centers at the points $\{0, 0, \pm 1\}$ and with the radii of the circumscribed circles equal to $1 + \varepsilon$. If we choose $\varepsilon > 0$ sufficiently small, then the intersection of A with the plane $G^2 = \{x \in E \mid \xi_3 = 0\}$ coincides with the unit circle, whence

$$\tilde{d}^1(A) \leqslant \tilde{\delta}(A, G^2) = 1.$$

On the other hand, the cells of radius $1 + \varepsilon$ with the centers at the vertices of the upper equilateral triangle intersect only in the point $\{0, 0, 1\}$, hence every line G_1 passing through the origin, except the ξ_3 axis, does not intersect at least one of these cells, hence $G_1 + (1 + \varepsilon)S_E$ does not contain the center of the respective cell, whence

$$d_1(A) = \inf_{\dim G_1 = 1} \inf_{\substack{\eta > 0 \\ A \subset G_1 + \eta S_E}} \eta = 1 + \varepsilon.$$

Such an example can be constructed in every Banach space E (V. M. Tihomirov [248], remark 2).

For the computation of the diameters $\tilde{d}^n(A)$ of certain classes A of functions in concrete functional spaces see V. M. Tihomirov [248].

APPENDIX I

BEST APPROXIMATION IN NORMED LINEAR SPACES BY ELEMENTS OF NON-LINEAR SETS

Practical necessities have required not only the study of best approximation in normed linear spaces E by elements of linear subspaces $G \subset E$ (or of linear manifolds $V \subset E$), but also the study of the more general problem of best approximation by elements of non-linear sets*) $G \subset E$. Beginning with P. L. Čebyšev, who considered the problem in the space $E = C_R([a, b])$ for $G =$ the set of all functions of the form

$$g(t) = z(t) \frac{a_1 + a_2 t + \ldots + a_n t^{n-1}}{b_1 + b_2 t + \ldots + b_m t^{m-1}}, \tag{1.1}$$

where**) m, n are given positive integers and z is a given function in $C_R([a, b])$ with $z(t) > 0$ on $[a, b]$ (see N. I. Ahiezer [1], Chap. II), the problem of best approximation by non-linear sets $G \subset E$ has been studied by many mathematicians. However, the results obtained do not constitute a unified theory (as is the theory of best approximation by elements of linear subspaces), the construction of such a theory being, up to the present, an open problem. In this appendix we shall present, briefly, some results and problems concerning best approximation by non-linear sets $G \subset E$. The notations $\mathfrak{P}_G(x)$, $\pi_G(x)$, $e_G(x)$, ... as well as the notions of proximinal set, Če-

*) By non-linear set we mean any set $G \subset E$ which is not a linear manifold. For the notion of linear manifold see Chap. I, §1, section 1.2 and for best approximation by elements of linear manifolds see Chap. I, § 5, and Chap. II, § 4.

**) In the particular case when $m = 1$ and $z(t) \equiv 1$, this problem reduces to the problem of best approximation by elements of the n-dimensional linear subspace $G_n = [1, t, \ldots, t^{n-1}]$ of $E = C_R([a,b])$.

byšev set etc. are introduced as the obvious extension of the corresponding notations and notions for linear subspaces and linear manifolds.

§ 1. BEST APPROXIMATION BY ELEMENTS OF CONVEX SETS

The first natural step when passing from best approximation in normed linear spaces E by elements of linear subspaces $G \subset E$ (or linear manifolds $V \subset E$) to best approximation by elements of non-linear sets $G \subset E$ is to take as G a convex set in E.

Several results of Chap. I on best approximation in general normed linear spaces E by elements of linear subspaces $G \subset E$ remain valid, with the same proof, for the case when $G =$ a convex set in E (e.g., theorem 1.13 on characterization of elements of best approximation*), corollary 2.1 on existence of elements of best approximation, where S_G is replaced by "all bounded subsets of G", theorem 3.1 on uniqueness of elements of best approximation, properties of the mappings π_G, of the sets $\pi_G^{-1}(g_0)$, and of the functionals e_G not involving the linearity of G, given in § 6), part of them being stated by their authors directly for this more general case.

There arises naturally the problem of extending the other results of Chap. I to the case when $G =$ a convex set in E. This extension has not been realized systematically so far. As an example of such results, we shall give here the following extension of theorem 1.1 of Chap. I on characterization of elements of best approximation:

THEOREM 1.1. *Let E be a normed linear space, G a convex set in E, $x \in E \setminus \overline{G}$ and $g_0 \in G$. We have $g_0 \in \mathfrak{L}_G(x)$ if and only if there exists an $f \in E^*$ with the following properties:*

$$\|f\| = 1, \tag{1.2}$$

$$\operatorname{Re} f(x - g) \geqslant \|x - g_0\| \qquad (g \in G). \tag{1.3}$$

Proof. Assume that $g_0 \in \mathfrak{L}_G(x)$. Then, since $x \in E \setminus \overline{G}$, we have $G \cap \operatorname{Int} S(x, \|x - g_0\|) = \emptyset$. Consequently, by virtue of the well known separation theorem of J. W. Tukey**) there exist

*) See A. L. Garkavi [71].

**) See e.g. N. Dunford and J. Schwartz [49], p. 417, theorem 8.

an $f_0 \in E^* \setminus \{0\}$ and a real number c such that

$$\operatorname{Re} f_0(x-g) \geqslant c \qquad (g \in G), \tag{1.4}$$

$$\operatorname{Re} f_0(x-y) \leqslant c \qquad (y \in S(x, \|x-g_0\|)). \tag{1.5}$$

Then $c = \operatorname{Re} f_0(x-g_0) \neq 0$, whence for $f = \frac{1}{c}\|x-g_0\| f_0 \in E^*$ we have (1.3) and

$$\operatorname{Re} f(x-y) \leqslant \|x-g_0\| \qquad (y \in S(x, \|x-g_0\|)). \tag{1.6}$$

But by (1.3) applied to $g = g_0$ we have

$$\|x-g_0\| \leqslant \operatorname{Re} f(x-g_0) \leqslant |f(x-g_0)| \leqslant \|f\| \, \|x-g_0\|,$$

whence $\|f\| \geqslant 1$, and if we had $\|f\| > 1$, then there would exist a $z \in E$ such that $\|z\| = 1$, $f(z) > 1$, which contradicts (1.6) for $y = x - \|x-g_0\| z \in S(x, \|x-g_0\|)$. Consequently, we have (1.2).

Conversely, assume that there exists an $f \in E^*$ satisfying (1.2) and (1.3). Then for every $g \in G$ we have

$$\|x-g_0\| \leqslant \operatorname{Re} f(x-g) \leqslant |f(x-g)| \leqslant \|f\| \, \|x-g\| = \|x-g\|,$$

whence $g_0 \in \mathfrak{P}_G(x)$, which completes the proof of theorem 1.1.

Let us mention that in the particular case when the scalars are real, the relation "of duality"

$$\inf_{g \in G} \|x-g\| = \max_{f \in \mathfrak{A}_x} \frac{1}{\|f\|}$$

(where G = a convex set in E and $\mathfrak{A}_x = \{f \in E^* \mid \operatorname{Re} f(x-g) \geqslant 1 \ (g \in G)\}$), which implies obviously the necessity part of theorem 1.1, has been given by A. L. Garkavi ([64], theorem 1). However, as has been remarked also in Chap. I, § 1, section 1.1, it is the theorem of characterization of elements of best approximation which presents interest for applications in concrete spaces.

We observe that an $f \in E^*$ satisfies (1.2) and (1.3) if and only if it satisfies (1.2) and the relations

$$\operatorname{Re} f(g_0 - g) \geqslant 0 \qquad (g \in G), \tag{1.7}$$

$$f(x-g_0) = \|x-g_0\|. \tag{1.8}$$

Indeed, by virtue of lemma 1.1 a) of Chap. I, the relations (1.2) and (1.3) for $g = g_0$ are equivalent to (1.2) $\cap$ (1.8). On the other hand, relations (1.7) may be written in the equivalent form

$$\operatorname{Re} f(x-g) \geqslant \operatorname{Re} f(x-g_0) \qquad (g \in G),$$

which, taking into account (1.8), amounts to (1.3). This remark shows that in theorem 1.13 of Chap. I (which remains valid also for G = a convex set in E, as has been remarked above) one may take an $f = f^g$ common for all $g \in G$, if the condition $f^g \in \mathcal{E}(S_{E^*})$ is replaced by the weaker condition $\|f^g\| = 1$. The characterization (1.2) $\cap$ (1.7) $\cap$ (1.8) of elements of best approximation has been given, in the particular case when the scalars are real, by G. Š. Rubinštein [200]*).

Theorem 1.1 above obviously admits the following geometric interpretation: *we have* $g_0 \in \mathfrak{P}_G(x)$ *if and only if there exists a real hyperplane* H *which separates* G *from* $S(x, \|x - g_0\|)$. Since we have $g_0 \in G \cap S(x, \|x - g_0\|)$, such a hyperplane H must pass through g_0, and hence support the cell $S(x, \|x - g_0\|)$. In the particular case when the scalars are real, this geometric characterization of elements of best approximation has been given by V. N. Burov ([31], theorem 4).

In the particular case when G is a linear subspace of E, from theorem 1.1 above one can deduce corollary 1.1 of Chap. I (the equivalence $1° \Leftrightarrow 5°$), hence also theorem 1.1 of Chap. I. Indeed, in this case from (1.3) it follows that we have

$$\operatorname{Re} f(g) = 0 \qquad (g \in G), \tag{1.9}$$

since if there existed a $g \in G$ with $\operatorname{Re} f(g) \neq 0$, then for a suitable real number λ we would have $\lambda g \in G$ and $\operatorname{Re} f(x - \lambda g) < 0$, which contradicts (1.3); on the other hand, from (1.2) and (1.3) for $g = g_0$ we obtain

$$\operatorname{Re} f(x - g_0) = \|x - g_0\|. \tag{1.10}$$

Conversely, from (1.9) and (1.10) it follows

$$\operatorname{Re} f(x - g) = \|x - g_0\| \qquad (g \in G),$$

whence (1.3), which completes the proof.

Another important particular case is that considered in

COROLLARY 1.1. *Let* E *be a normed linear space,* G *a convex cone***) *with vertex in* 0, $x \in E \setminus G$ *and* $g_0 \in G$. *We have* $g_0 \in \mathfrak{P}_G(x)$

*) Actually, in [200] G. Š. Rubinštein has considered the more general problem of characterization of the elements $g_0 \in G$ which minimize the functional $\Phi(g) = \sum_{i=1}^{n} m_i \|x_i - g\|$, where $x_1, \ldots, x_n \in E$ and $m_1, \ldots, m_n > 0$ are fixed. In the particular case when $G = E$ and $m_i = 1$, this problem is, in a certain sense, dual to the problem of characterization of the Čebyšev centers (See Chap. II, §6, section 6.4) of the set $\{x_1, \ldots, x_n\}$.

**) See Chap. I, §4, section 4.2.

if and only if there exists an $f \in E^*$ satisfying (1.2) and the relations

$$\operatorname{Re} f(g) \leqslant 0 \qquad (g \in G), \tag{1.11}$$

$$\operatorname{Re} f(x) = \|x - g_0\|. \tag{1.12}$$

Proof. Assume that $g_0 \in \mathfrak{P}_G(x)$, whence we have (1.2) and (1.3). If there existed a $g \in G$ with $\operatorname{Re} f(g) > 0$, then for a sufficiently great $\lambda > 0$ we would have $\lambda g \in G$ and

$$\operatorname{Re} f(\lambda g) > \operatorname{Re} f(x) - \|x - g_0\|,$$

which contradicts (1.3). Thus, we have (1.11).

But, by (1.2) and (1.3) for $g = g_0$ we have then

$$0 \geqslant \operatorname{Re} f(g_0) = \operatorname{Re} f(x) - \|x - g_0\|,$$

which, together with (1.3) applied to $g = 0$, gives (1.12).

Conversely, conditions (1.11) and (1.12) imply obviously (1.3), which completes the proof of corollary 1.1.

As shown by the above proof, condition (1.12) of corollary 1.1 may be replaced by the conditions

$$\operatorname{Re} f(x - g_0) = \|x - g_0\|, \tag{1.13}$$

$$\operatorname{Re} f(g_0) = 0. \tag{1.14}$$

In this form, corollary 1.1 has been given, in the particular case when the scalars are real, by G. Š. Rubinštein [200].

If G is, in particular, a linear subspace of E, condition (1.11) is obviously equivalent to (1.9), hence we find again the equivalence $1° \Leftrightarrow 5°$ of Chap. I, corollary 1.1.

Concerning the extension of the results of Chap. II to the case when G is a finite dimensional convex set in E, new difficulties arise, at least as regards certain applications in concrete spaces. For instance, if $x_1, \ldots, x_n$ (where $n \geqslant 2$) is a Čebyšev system in the space $E = C_R([0, 1])$ and if G is the convex cone with vertex at 0 defined by

$$G = \left\{ \sum_{i=1}^{n} \alpha_i x_i \,\middle|\, \sum_{i=1}^{n} \alpha_i x_i(t) \geqslant 0 \; (t \in [0,1]) \right\}, \tag{1.15}$$

then, as has shown J. R. Rice [190], the characterization of the elements of best approximation is complicated, and the problem of their uniqueness is still unsolved.

Finally, let us mention that there exist some results on best approximation by elements of convex sets, which are not extensions of the results given in the foregoing chapters. For instance, E. W. Cheney and A. A. Goldstein [34] have studied the operator $u = \pi_{G_1} \pi_{G_2}$, where G_1, G_2 are two disjoint closed convex sets in an inner product space $\mathcal{H}$.

§ 2. THE PROBLEM OF CONVEXITY OF ČEBYŠEV SETS

The implication $3° \Rightarrow 1°$ of Chap. I, corollary 3.4, remains valid, as may be easily verified, also for closed convex sets instead of closed linear subspaces, that is, *if E is a reflexive and strictly convex Banach space, then all closed convex sets in E are Čebyšev sets*; by virtue of the implication $1° \Rightarrow 3°$ of the same corollary, *this property characterizes the strictly convex reflexive Banach spaces.*

There arises naturally *the problem of characterizing the Banach spaces in which every Čebyšev set is convex.*

As have shown L. N. H. Bunt [30], T. S. Motzkin [154] and others (see the bibliography in the paper of N. V. Efimov and S. B. Stečkin [52]), *in a smooth Banach space E of finite dimension every Čebyšev set is convex* *). It is not known whether this result remains valid for Banach spaces of infinite dimension, being unknown even the answer to the following problem: *In a Hilbert space $\mathcal{H}$ is a Čebyšev set necessarily convex?*

We shall now present, briefly, some partial results obtained in connection with these infinite dimensional problems and other open problems related to them.

LEMMA 2.1 (M. Nicolescu [161], proposition 6). *Let E be a normed linear space and G a Čebyšev set in E. We have then* **)

$$\pi_G[\lambda x + (1-\lambda)\pi_G(x)] = \pi_G(x) \qquad (x \in E,\ 0 \leqslant \lambda \leqslant 1). \quad (2.1)$$

Proof. Putting $z = \lambda x + (1-\lambda)\pi_G(x)$ (where $x \in E$ and $0 \leqslant \lambda \leqslant 1$), we have

$$\|x - z\| + \|z - \pi_G(x)\| = (1-\lambda)\,\|x - \pi_G(x)\| + \lambda\,\|x - \pi_G(x)\| = \|x - \pi_G(x)\|,$$

*) As has shown A. Brøndsted [26 a], [26 b], the converse is not true; namely, for every integer $n \geqslant 3$ there exists a non-smooth n-dimensional Banach space E with the property that every Čebyšev set in E is convex. The problem of characterization of n-dimensional Banach spaces in which every Čebyšev set is convex, remains open for $n \geqslant 4$; for $n = 3$ it has been solved by A. Brøndsted [26 b], who proved that a three-dimensional Banach space E has this property if and only if every exposed point of S_E (see Chap. I, § 3, section 3.1) is a normal element (see Chap. I, § 1, section 1.1), while in the case when $n = 2$ N. V. Efimov and S. B. Stečkin ([52], § 3) have observed that the smooth spaces and only such spaces have this property.

**) In the terminology of Chap. I, theorem 6.3, this result is expressed as follows: for $g_0 \in G$ we have the implication

$$x \in \pi_G^{-1}(g_0) \Rightarrow \lambda x + (1-\lambda)\, g_0 \in \pi_G^{-1}(g_0) \qquad (0 \leqslant \lambda \leqslant 1).$$

whence

$$\|z - g\| \geqslant \|x - g\| - \|x - z\| \geqslant \|x - \pi_G(x)\| - \\ - \|x - z\| = \|z - \pi_G(x)\| \qquad (g \in G),$$

whence $\pi_G(x) \in \mathfrak{L}_G(z)$. Since G is a Čebyšev set, it follows that we have (2.1), which completes the proof of lemma 2.1.

A set G in a normed linear space E is said to be *boundedly compact*, if its intersection with every cell $S(x, r) = \{y \in E \mid \|x - y\| \leqslant r\}$ is conditionally compact *). Following N. V. Efimov and S. B. Stečkin ([50], § 7), a Čebyšev set $G \subset E$ is called a *sun* if we have

$$\pi_G[\lambda x + (1 - \lambda)\pi_G(x)] = \pi_G(x) \qquad (x \in E, \lambda \geqslant 0). \qquad (2.2)$$

THEOREM 2.1 (L. P. Vlasov [259]). *Let E be a Banach space. Then every boundedly compact Čebyšev set G in E is a sun.*

Proof. Assume the contrary, i.e. that $G \subset E$ is a boundedly compact Čebyšev set, but not a sun, hence there exist an $x_0 \in E$ and a λ_0 with $0 < \lambda_0 < \infty$ for which we do not have (2.2). From lemma 2.1 it follows that for the least upper bound λ_1 of the numbers $\lambda \geqslant 0$ with the property

$$\pi_G[\lambda x_0 + (1 - \lambda)\pi_G(x_0)] = \pi_G(x_0) \qquad (2.3)$$

we have $\lambda_1 \leqslant \lambda_0 < \infty$; since the set $\pi_G^{-1}[\pi_G(x_0)]$ is closed (by Chap. I, the first part of theorem 6.3, valid for arbitrary sets G, with the same proof), λ_1 is *the greatest number* $\lambda \geqslant 0$ satisfying (2.3) (whence $\lambda_1 < \lambda_0$). Put

$$y = \lambda_1 x_0 + (1 - \lambda_1)\pi_G(x_0). \qquad (2.4)$$

By (2.3) for $\lambda = \lambda_1$ we have then $\pi_G(y) = \pi_G(x_0)$, whence

$$\lambda y + (1 - \lambda)\, \pi_G(y) = \lambda[\lambda_1 x_0 + (1 - \lambda_1)\pi_G(x_0)] + (1 - \lambda)\pi_G(x_0) = \\ = \lambda\lambda_1 x_0 + (1 - \lambda\lambda_1)\pi_G(x_0) \qquad (0 \leqslant \lambda < \infty),$$

hence $\lambda_2 = 1$ is the greatest $\lambda \geqslant 0$ with the property

$$\pi_G[\lambda y + (1 - \lambda)\pi_G(y)] = \pi_G(x_0) = \pi_G(y). \qquad (2.5)$$

Let now $r > 0$ be arbitrary such that $S(y, r) \cap G = \emptyset$ and let

$$\lambda(z) = \frac{r}{\|y - \pi_G(z)\|} + 1 \qquad (z \in S(y, r)), \qquad (2.6)$$

$$u(z) = \lambda(z)y + [1 - \lambda(z)]\pi_G(z) \qquad (z \in S(y, r)). \qquad (2.7)$$

*) I. e. "relatively compact" in the sense of N. Bourbaki [25], that is, having a compact closure.

Since by the boundedly compactness of G the mapping π_G is continuous (this is shown just as theorem 5.4 of Chap. II), and since $\|y - u(z)\| = [\lambda(z) - 1]\,\|y - \pi_G(z)\| = r$ $(z \in S(y, r))$, it follows that u is a continuous mapping of the closed convex set $S(y, r)$ into itself. But, since G is boundedly compact, the set $\pi_G[S(y, r)] = \{\pi_G(z) \mid z \in S(y, r)\} \subset G$, whence also the set $u[S(y, r)]$ is conditionally compact. Consequently, by virtue of a theorem of J. Schauder*), u has at least one fixed point, i.e. there exists a $z_0 \in S(y, r)$ such that

$$\lambda(z_0)y + [1 - \lambda(z_0)]\pi_G(z_0) = u(z_0) = z_0. \tag{2.8}$$

Since we have then

$$y = \frac{1}{\lambda(z_0)} z_0 + \left[1 - \frac{1}{\lambda(z_0)}\right] \pi_G(z_0),$$

by (2.1) applied for $x = z_0$, $\lambda = \dfrac{1}{\lambda(z_0)}$ and by (2.8) we obtain

$$\pi_G(y) = \pi_G(z_0) = \pi_G\{\lambda(z_0)y + [1 - \lambda(z_0)]\pi_G(y)\}. \tag{2.9}$$

But this relation, together with $\lambda(z_0) > 1$, contradicts the hypothesis that $\lambda_2 = 1$ is the greatest $\lambda \geqslant 0$ satisfying (2.5), which completes the proof of theorem 2.1.

LEMMA 2.2. *Let E be a smooth normed linear space. Then every sun G in E is convex.*

Proof. Obviously we may assume, without loss of generality, that the space E is real. Assume that G is a non-convex Čebyšev set in E, whence there exist $g_1, g_2 \in G$ and λ_0 with $0 < \lambda_0 < 1$ such that

$$x = \lambda_0 g_1 + (1 - \lambda_0) g_2 \notin G. \tag{2.10}$$

We claim that then

$$\bigcup_{i=1}^{2} \{\langle g_i, \pi_G(x)\rangle \cap \operatorname{Int} S(x, \|x - \pi_G(x)\|)\} \neq \emptyset, \tag{2.11}$$

where $\langle g_i, \pi_G(x)\rangle = \{\lambda g_i + (1 - \lambda)\pi_G(x) \mid 0 < \lambda < 1\}$ $(i = 1, 2)$. Indeed, if there exists a $\lambda_1 < 0$ such that $z = \lambda_1 g_1 + (1 - \lambda_1)\pi_G(x) \in \operatorname{Int} S(x, \|x - \pi_G(x)\|)$, then for $\lambda_2 = -\dfrac{\lambda_1(1-\lambda_0)}{\lambda_0 - \lambda_1}$ we have $0 < \lambda_2 < 1$ and

$$\|x - \lambda_2 g_2 - (1 - \lambda_2)\pi_G(x)\| = \frac{\lambda_0}{\lambda_0 - \lambda_1} \|x - z\| < \|x - \pi_G(x)\|,$$

*) See e.g. L. V. Kantorovič and G. P. Akilov [106], p. 578, theorem 2.

whence $\lambda_2 g_2 + (1-\lambda_2)\pi_G(x) \in \langle g_2, \pi_G(x)\rangle \cap \operatorname{Int} S(x, \|x-\pi_G(x)\|)$. Similarly, if there exists a $\lambda_2 < 0$ such that $\lambda_2 g_2 + (1-\lambda_2)\pi_G(x) \in \operatorname{Int} S(x, \|x-\pi_G(x)\|)$, we have again (2.11). Assume now that

$$\lambda g_i + (1-\lambda)\pi_G(x) \notin \operatorname{Int} S(x, \|x-\pi_G(x)\|) \quad (-\infty < \lambda < 1;\ i = 1, 2). \tag{2.12}$$

Since for every $\lambda \geqslant 1$ we have

$$\|x - \lambda g_i - (1-\lambda)\pi_G(x)\| \geqslant |\lambda \|x - g_i\| - |1-\lambda|\, \|x - \pi_G(x)\|| \geqslant$$
$$\geqslant \|x - \pi_G(x)\| \qquad (i = 1, 2),$$

it follows that

$$\lambda g_i + (1-\lambda)\pi_G(x) \notin \operatorname{Int} S(x, \|x-\pi_G(x)\|) \quad (-\infty < \lambda < \infty;\ i = 1, 2). \tag{2.13}$$

Consequently, by virtue of the separation theorem of S. Mazur *) there exist hyperplanes H_i with $H_i \cap \operatorname{Int} S(x, \|x - \pi_G(x)\|) = \emptyset$, such that $H_i \supset \{\lambda g_i + (1-\lambda)\pi_G(x) \mid -\infty < \lambda < \infty\}$ $(i = 1, 2)$. Since by $\pi_G(x) \in H_i \cap S(x, \|x - \pi_G(x)\|)$ these hyperplanes support the cell $S(x, \|x - \pi_G(x)\|)$, by virtue of lemma 1.4 of Chap. I there exist $f_i \in E^*$ with $\|f_i\| = 1$ $(i = 1, 2)$ such that

$$f_i[x - \lambda g_i - (1-\lambda)\pi_G(x)] = \|x - \pi_G(x)\| \quad (-\infty < \lambda < \infty;\ i = 1, 2) \tag{2.14}$$

We have then $f_1 \neq f_2$ (since otherwise writing (2.14) for $i = 1$, $\lambda = \lambda_0$ and $i = 2$, $\lambda = 1 - \lambda_0$, and adding, we would obtain $f_1[x - \pi_G(x)] = 2\|x - \pi_G(x)\|$, whence $2\|x - \pi_G(x)\| \leqslant \|f_1\|\, \|x - \pi_G(x)\| = \|x - \pi_G(x)\|$, which is contrary to (2.10)), hence, by $\|f_i\| = 1$ and (2.14) with $\lambda = 0$, the space E is not smooth, which contradicts the hypothesis. Thus we have (2.11), hence for $i = 1$ or $i = 2$ there exists a λ_i with $0 < \lambda_i < 1$ such that $\lambda_i g_i + (1-\lambda_i)\pi_G(x) \in \operatorname{Int} S(x, \|x - \pi_G(x)\|)$. But then for

$$y = \frac{1}{\lambda_i} x + \left(1 - \frac{1}{\lambda_i}\right)\pi_G(x)$$

we have

$$\|y - g_i\| = \left\| \frac{1}{\lambda_i} x + \left(1 - \frac{1}{\lambda_i}\right)\pi_G(x) - g_i \right\| =$$

$$= \frac{1}{\lambda_i}\|x - \lambda_i g_i - (1-\lambda_i)\pi_G(x)\| < \frac{1}{\lambda_i}\|x - \pi_G(x)\| = \|y - \pi_G(x)\|,$$

*) See e.g. G. Köthe [122], p. 191, theorem (2).

whence $\pi_G(y) \neq \pi_G(x)$, and thus G is not a sun, which completes the proof.

Lemma 2.2 is given in the paper of L. P. Vlasov [259] with the mention that it is due to N. V. Efimov and S. B. Stečkin.

From theorem 2.1 and lemma 2.2 it follows

THEOREM 2.2. *Let E be a smooth Banach space. Then every boundedly compact Čebyšev set in E is convex.*

In the particular case when the space E is smooth and has a "uniformly small curvature", theorem 2.2 for compact Čebyšev sets has been given by N. V. Efimov and S. B. Stečkin [51]. In the case when the space E is smooth and uniformly convex*), theorem 2.2 has been given by N. V. Efimov and S. B. Stečkin ([52], theorem 3), and for reflexive smooth Banach spaces satisfying another additional condition, it has been given by V. Klee ([113], theorem 2.6). In the general case theorem 2.2 has been given by L. P. Vlasov [259], who has also proved [259 a] the following generalization of theorem 2.2 : *Let E be a smooth Banach space and G a boundedly compact set in E with the property that for every $x \in E$ the set $\mathfrak{L}_G(x)$ is nonvoid and convex. Then G is convex.*

We have given, in the above, the proof of theorem 2.2, in order to present one of the techniques frequently used in the problem of convexity of Čebyšev sets. We shall give now, without proof, other results obtained in this problem.

Since the condition of boundedly compactness is too restrictive, being not satisfied in various important concrete cases, N. V. Efimov and S. B. Stečkin have introduced [53] the notion of "approximative compactness" : a set $G \subset E$ is called *approximatively compact* if for every $x \in E$ and every sequence $\{g_n\} \subset G$ with $\lim_{n\to\infty} \|x - g_n\| = \rho(x, G)$ there exists a subsequence $\{g_{n_k}\}$ converging to an element of G. For example (N. V. Efimov and S. B. Stečkin [53], lemma 1), in a uniformly convex Banach space E every *sequentially weakly closed* set (hence, in particular, every *weakly closed* set) $G \subset E$ is approximatively compact; characterizations of Banach spaces E with this property have been given in the paper [231]. As have shown N. V. Efimov and S. B. Stečkin ([53], theorem 3), *in a smooth and uniformly convex Banach space E a Čebyšev set G is convex if and only if it is approximatively compact*; in the particular

*) Let us recall that, following J. A. Clarkson [38], a normed linear space E is said to be *uniformly convex,* if for every ε with $0 < \varepsilon \leqslant 2$ there exists a $\delta(\varepsilon) > 0$ such that the relations $\|x\|,\ \|y\| \leqslant 1$ and $\|x - y\| \geqslant \varepsilon$ imply $\left\|\frac{x+y}{2}\right\| \leqslant 1-\delta(\varepsilon)$. Every uniformly convex Banach space is reflexive (see e.g. G. Köthe [122], p. 348, theorem (4)), and uniformly convex in each direction (see Chap. II, § 6, section 6.4).

case when G is a weakly closed convex set in a Banach space E which is "uniformly smooth in each direction" and uniformly convex, this result has been obtained by V. Klee ([113], corollary 4.2). Since for every fixed $n \geqslant 1$ and $m \geqslant 2$ the set $G = G_{n,m}$ of all functions of the form (1.1), where $z(t) \equiv 1$, is weakly closed (hence approximatively compact) in $E = L^p_R([0,1])$ $(1 < p < \infty)$, but not convex, it follows that it is not a Čebyšev set in these spaces (N. V. Efimov and S. B. Stečkin [53], § 3); it is interesting to compare this result with the classical theorem of P. L. Čebyšev (see e. g. N. I. Ahiezer [1], § 34), according to which $G_{n,m}$ is a Čebyšev set in the space $E = C_R([0,1])$.

From the foregoing and from the fact that every closed convex set $G \subset E$ is weakly closed *), it follows that *for a set G in a smooth and uniformly convex Banach space E the following statements are equivalent: 1°. G is convex and closed; 2°. G is a weakly closed Čebyšev set; 3°. G is an approximatively compact Čebyšev set*; indeed, 1° ⇒ 2° ⇒ 3° ⇒ 1°. We thus have *infinite dimensional characterizations* of closed convex sets in terms of the Čebyšev property, which show, among other things, that the problem of convexity of Čebyšev sets in smooth and uniformly convex Banach spaces E is equivalent to each of the following problems: a) *is a Čebyšev set weakly closed?* b) *is a Čebyšev set approximatively compact?*

Instead of prescribing additional conditions directly on the Čebyšev set G in order to be able to conclude that it is convex, one may impose, for the same purpose, conditions on the mapping π_G. Thus, V. Klee has proved ([113], theorem 2.2) that *if G is a Čebyšev set in a smooth and reflexive Banach space E and if every $x \in E \setminus G$ admits a neighbourhood $V(x)$ such that $\pi_G|_{V(x)}$ is both continuous and weakly continuous, then G is convex*; in the particular case when G is boundedly compact and π_G is both continuous and weakly continuous, this result has been given previously by V. Klee ([110], theorem (A 2.5)). However, it is not known, even in Hilbert**) spaces, whether for a Čebyšev set $G \subset E$ the mapping π_G is continuous; if G is an approximatively compact Čebyšev set, then the answer is affirmative (see Appendix II). The condition of continuity of the mapping π_G, where G is a Čebyšev set, does not imply the weak continuity of the mapping π_G, as shown by the example E = a Hilbert space, $G = S_E$; on the other hand, it is not known whether from the weak continuity of the mapping π_G, where G is a Čebyšev set, there follows the strong continuity of this mapping. Various conditions equivalent to the strong continuity

---

*) See e.g. N. Dunford and J. Schwartz [49], p. 418, corollary 14.
**) I.e. complete inner product spaces.

of π_G, for G = a locally compact Čebyšev set, have been given by D. E. Wulbert ([263], theorem 1.53).

We recall that for given $x \in E$ and $r > 0$ *the inversion of* E *in* Fr $S(x, r)$ is the mapping

$$y \to x + \frac{r^2}{\|y - x\|^2}(y - x) \qquad (y \in E \setminus \{x\}) \tag{2.15}$$

of $E \setminus \{x\}$ into itself. Using these inversions, F. A. Ficken and V. Klee [113] have shown that there is a close connection between the convexity of Čebyšev sets in Hilbert spaces and the following problem on farthest points *) : *If a closed convex set* $A \subset \mathcal{H}$ *has the property that every* $y \in \mathcal{H}$ *admits a unique farthest point in* A, *does* A *necessarily consist of one single point?* If the answer to this problem were affirmative, then every Čebyšev set in $\mathcal{H}$ would be convex (V. Klee [113]); from the fact that the answer for compact convex sets $A \subset \mathcal{H}$ is affirmative (V. Klee [113], proposition 1.3), it follows again that every boundedly compact Čebyšev set $G \subset \mathcal{H}$ is convex (V. Klee [113]).

Since the problem of convexity of Čebyšev sets $G \subset \mathcal{H}$ is open, there arises naturally the problem of convexity of sets $G \subset \mathcal{H}$ belonging to wider classes of sets. V. Klee [115] has considered two such classes, C_1 and C_2. Following V. Klee [115], a set G in a normed linear space E is said to be a C_1*-set* (respectively a C_2*-set*), if G is a semi-Čebyšev and closed set (respectively if G is proximinal and for every $x \in E$ the set $\mathfrak{L}_G(x)$ is contractible in itself **)). Since every boundedly compact set is proximinal (see Appendix II), *every boundedly compact* C_1*-set in a smooth Banach space is* a Čebyšev set, whence *convex* by virtue of theorem 2.2. On the other hand, V. Klee has shown [115] that *in every infinite dimensional Hilbert space* $\mathcal{H}$ *there exists a* C_1*-set* G *whose complement* $\mathcal{H} \setminus G$ *is non-void, bounded and convex* (hence G is non-convex). Namely, without loss of generality, it may be assumed that $\mathcal{H} = l^2(I)$ for an infinite set I. Let $\varphi : I \to [1, \infty)$ be such that $\sup_{\iota \in I} \varphi(\iota) = \infty$, and let

$$K = \left\{ x \in l^2(I) \,\middle|\, \sum_{\iota \in I} [\varphi(\iota) x(\iota)]^2 \leqslant 1 \right\}, \tag{2.16}$$

$$G = E \setminus (K + \operatorname{Int} S_E); \tag{2.17}$$

then G is a C_1-set with the required properties. Thus, the answer to the problem of convexity of C_1-sets in Hilbert spaces is

*) See Chap. I, § 6, section 6.4.

**) See e.g. C. Kuratowski [133], vol. II, Chap. VII, § 49.

negative. V. Klee has also shown [115] that *the boundedly compact C_2-sets in strictly convex smooth Banach spaces, as well as the C_2-sets G with* $\operatorname{Int} G = \emptyset$ *in finite dimensional smooth Banach spaces, are convex, but in every infinite dimensional Hilbert space $\mathcal{H}$ there exists a non-convex C_2-set G with $\mathcal{H}\setminus G$ non-void, bounded and convex.*

A particular class of C_1-sets is that of the closed sets which are *very non-proximinal* in the sense of Chap. I, § 2, i.e. of the closed sets $G \subset E$ with the property

$$\mathfrak{Q}_G(x) = \emptyset \qquad (x \in E\setminus G). \tag{2.18}$$

Following V. Klee [115], let $\mathfrak{N}_0$ be the class of all normed linear spaces which contain a very non-proximinal closed set G, and for $i = 1, 2, 3, 4$, let $\mathfrak{N}_i$ be the class of all normed linear spaces E which contain a very non-proximinal closed set G having respectively the properties: (1) G is convex; (2) G is bounded and convex (3) $E\setminus G$ is convex; (4) $E\setminus G$ is bounded and convex. We have obviously $\mathfrak{N}_0 \supset \mathfrak{N}_1 \supset \mathfrak{N}_2$ and $\mathfrak{N}_0 \supset \mathfrak{N}_3 \supset \mathfrak{N}_4$. There arises naturally the problem of characterizing these classes of normed linear spaces in terms of known properties. In connection with this problem, V. Klee [115] has made the following remarks: a) $\mathfrak{N}_1$ is the class of all non-reflexive spaces; b) $\mathfrak{N}_3 \supset \mathfrak{N}_1$; c) no Banach space is in $\mathfrak{N}_2$, but $\mathfrak{N}_2 \neq \emptyset$; d) $\mathfrak{N}_4 \neq \emptyset$; e) it is possible that $\mathfrak{N}_4$ (whence also $\mathfrak{N}_3$, $\mathfrak{N}_0$) coincide with the class of all normed linear spaces.

§ 3. BEST APPROXIMATION BY ELEMENTS OF FINITE DIMENSIONAL SURFACES

Let $g(\alpha_1, \ldots, \alpha_N)$ be a function of N scalar variables, defined on a subset $\mathfrak{A}$ of an N-dimensional (real or complex) Banach space B_N, with values in a normed linear space E. We shall say then that the set $G = \{g(\alpha_1, \ldots, \alpha_N) \in E \mid \{\alpha_1, \ldots, \alpha_N\} \in \mathfrak{A}\}$ is an N-dimensional surface*) in E; in the particular case when $N = 1$, G is said to be a *curve* in E (M. G. Krein [124]). We shall consider now the problem of best approximation by elements of N-dimensional surfaces, raised in [211], p. 137.

In the particular case when $\mathfrak{A} = B_N$ and there exist N linearly independent elements $x_1, \ldots, x_N \in E$ such that

$$g(\alpha_1, \ldots, \alpha_N) = \sum_{i=1}^{N} \alpha_i x_i \qquad (\{\alpha_1, \ldots, \alpha_N\} \in B_N), \tag{3.1}$$

*) One may also use the term *N-parametric set*

G is nothing else than the linear subspace $[x_1, \ldots, x_N]$ of E and thus we find again the problem of polynomial best approximation, studied in Chap. II. On the other hand, in the particular case when E is a space of real functions on $[a, b]$, $N = n + m$, $\alpha_i = a_i$ $(i = 1, \ldots, n)$, $\alpha_{n+i} = b_i$ $(i = 1, \ldots, m)$ and g is of the form (1.1), one finds again the problem of best approximation in $E = C_R([a, b])$, $L_R^p([a, b])$ etc. by elements of the set $G = G_{n,m}$ of all functions of the form (1.1).

In the particular case when E is a space of real functions, the problem of best approximation by elements of N-dimensional surfaces has been considered for the first time by J. W. Young [264], who obtained some results in $E = C_R([a, b])$ on the characterization, existence and uniqueness of elements of best approximation. These investigations have been continued by T. S. Motzkin [155], L. Tornheim [254], M. I. Morozov [153] and others, who have considered "unisolvent" functions $g(\alpha_1, \ldots, \alpha_N)$ and J. R. Rice [188] has introduced "varisolvent" functions $g(\alpha_1, \ldots, \alpha_N)$ even on arbitrary compact spaces Q; also, J. R. Rice has considered the problem of best approximation in general normed linear spaces by elements of finite dimensional surfaces. We shall now present, briefly, some of these results. In the sequel we shall denote $\{\alpha_1, \ldots, \alpha_N\}$ by α, and $[g(\alpha)](q)$ by $g(\alpha, q)$, and we shall assume that $g(\alpha, q)$ is continuous in α and q.

The function $g(\alpha, q)$ is said to be *locally solvent* if for every $q_1, \ldots, q_N \in Q$, $\alpha_0 \in \mathfrak{A}$ and $\varepsilon > 0$ there exists a $\delta = \delta(\alpha_0, \varepsilon, q_i) > 0$ such that the relations

$$|g(\alpha_0, q_i) - \gamma_i| < \delta \qquad (i = 1, \ldots, N) \tag{3.2}$$

imply the existence of an $\alpha \in \mathfrak{A}$ such that

$$g(\alpha, q_i) = \gamma_i \qquad (i = 1, \ldots, N), \tag{3.3}$$

$$\max_{t \in Q} |g(\alpha, t) - g(\alpha_0, t)| < \varepsilon. \tag{3.4}$$

The function $g(\alpha, q)$ is said to be *locally unisolvent* if it is locally solvent and if for every $\alpha_1 \neq \alpha_2$ in $\mathfrak{A}$ the function $g(\alpha_1, q) - g(\alpha_2, q)$ has at most $N - 1$ zeros on Q. The function $g(\alpha, q)$ is said to be *unisolvent* if for every $q_1, \ldots, q_N \in Q$ and $\gamma_1, \ldots, \gamma_N$ the system of equations (3.3) has a unique solution $\alpha \in \mathfrak{A}$.

As has shown J. R. Rice ([191], theorem 1 and its corollary), *if $g(\alpha, q)$ is locally unisolvent, then the surface G is an N-dimensional manifold* (in the usual topological sense), and *if $g(\alpha, q)$ is unisolvent then G is homeomorphic to B_N*. The converse of the second statement is not true, as shown by the following example of J. R. Rice ([191], p. 115): Let $Q = [-1,1]$, $\mathfrak{A} = (-1,1) \subset B_1$

and $g(\alpha, q) = \frac{\alpha}{1 + \alpha q}$; then G is homeomorphic to B_1, but $g(\alpha, q)$ is not unisolvent. Also, the condition that $g(\alpha, q)$ be locally unisolvent, is not sufficient in order that G be homeomorphic to B_N, as shown by the following example of J. R. Rice ([191], p. 115): $Q = [-1,1]$, $\mathfrak{A} = \left(-1, -\frac{1}{2}\right) \cup \left(\frac{1}{2}, 1\right) \subset B_1$ and $g(\alpha, q) = \frac{\alpha}{1 + \alpha q}$.

In the particular case when Q is a finite set $\{q_1, \ldots, q_k\}$, the space $E = C_R(Q)$ can be identified with the finite dimensional space $E_k = (l_k^\infty)_R$. J. R. Rice has observed that the converse of the first statement above is not true in E_k and he has given a characterization of the closed manifolds G in E_k for which $g(\alpha, q)$ is locally unisolvent ([191], theorem 2).

If $G \subset E_k$ is a differentiable manifold and if Fr $S(x, r) \subset E_k$ is a differentiable manifold of dimension $k - 1$, then one can apply the methods of differential geometry to obtain characterizations of elements of best approximation. For example, *in this case the following statements are equivalent*: *1°* $g_0 \in \mathfrak{L}_G(x)$; *2°* g_0 *is a critical point of the function* $\Phi(g) = \|x - g\|$; *3° the normal of* Fr $S(x, \|x - g_0\|)$ *is normal to* G *at* g_0 (J. R. Rice [191], p. 117, corollaries 1 and 2).

Concerning uniqueness of elements of best approximation, we have the following result (J. R. Rice (191], theorem 6): *If* G *is a closed submanifold of a k-dimensional smooth and strictly convex Banach space* E_k, *then* G *is a Čebyšev set if and only if it is a linear manifold*; indeed, if G is a Čebyšev set, then by theorem 2.2 it is convex and one can show that every closed convex manifold is a linear manifold. In particular, from the above it follows that for a non-linear unisolvent $g(\alpha, q)$ the corresponding manifold $G \subset L_R^p([0,1])$ $(1 < p < \infty)$ is not a Čebyšev set (J. R. Rice [191], p. 119, corollary); on the other hand, it is known (T. S. Motzkin [155] and others) that in $C_R([0,1])$ the respective manifold is a Čebyšev set. This result, as well as the similar one of N.V. Efimov and S. B. Stečkin given in § 2, shows that, as concerns uniqueness of elements of best approximation, the situations in the spaces $L_R^p([0,1])$ and $C_R([0,1])$ are reversed when we pass from best approximation by elements of linear manifolds to best approximation by elements of non-linear sets.

For best approximation by elements of sufficiently smooth non-linear manifolds in Euclidean spaces, we have however a result of local uniqueness: *If* G *is a closed submanifold of a k-dimensional Euclidean space, which has its curvature bounded at each point, then there exists a neighbourhood* $V(G)$ *of* G *such that for every* $x \in V(G)$ *the set* $\mathfrak{L}_G(x)$ *contains exactly one element* (J. R. Rice [191], theorem 7); the difficulty of extending this result to other k-dimensional Banach spaces E_k is due to the fact that we do not have a suitable concept of curvature in E_k.

Since there exist important (non-unisolvent) approximating functions, for example of type (1.1) or of type $g(\alpha, t) = \alpha_1\alpha_2^t + \alpha_3$, for which the corresponding surface G is not a manifold, it is necessary to study best approximation by elements of finite dimensional surfaces which are not manifolds. A notion useful for applications *), which includes the above approximating functions, is that of a varisolvent function, introduced by J. R. Rice [188]: the function $g(\alpha, q)$ is said to be *varisolvent*, if there exists a *non-constant* bounded function $p(\alpha)$ such that $g(\alpha, q)$ satisfies the conditions of the definition of locally unisolvent functions, for $p(\alpha)$ $(\alpha \in \mathfrak{A})$ instead of N. *If* $g(\alpha, q)$ *is varisolvent on a finite space* Q, *and* G *is connected, then* G *is not a manifold and is not convex* (J. R. Rice [191], theorem 9 and 10); also, J. R. Rice has conjectured ([191], p. 125) that *in this case* G *is not closed.*

Finally, let us mention that some results have been also obtained in the problem of continuity of the mappings π_G, especially for $E = C_R([0,1])$ and the set $G = G_{n,m}$ of all functions of the form (1.1) (see J. R. Rice [191] and the papers quoted in its bibliography; the literature on best approximation in $E = C_R([0,1])$ by elements of $G_{n,m}$ is extremely rich and its presentation would be outside the scope of the present monograph).

§ 4. BEST APPROXIMATION BY ELEMENTS OF ARBITRARY SETS

In the present paragraph we shall present, briefly, some results which have been obtained in the general problem of best approximation in normed linear spaces by elements of arbitrary sets $G \subset E$.

As concerns the existence of elements of best approximation, from Chap. I, formula (1.1) (obviously valid for arbitrary sets $G \subset E$), it follows that *every proximinal set* $G \subset E$ *is closed.* It is also known that every approximatively compact set (hence, in particular, every boundedly compact closed set) $G \subset E$ is proximinal (N. V. Efimov and S. B. Stečkin [53]; see also Appendix II); in particular, J. R. Rice has observed ([191], theorem 12

*) The applications given by J. R. Rice in $E = C_R([0, 1])$ are obtained by using arguments of the theory of functions of a real variable.

and its corollary) that *every set $G \subset E$ which is "boundedly embeddable"* (i.e. embeddable by means of a homeomorphism which carries bounded sets into bounded sets) *in a space B_N of finite dimension N is boundedly compact, hence proximinal.*

S. B. Stečkin [237] has studied the set

$$\mathfrak{U}_G = \{x \in E \mid g_1, g_2 \in \mathfrak{L}_G(x) \Rightarrow g_1 = g_2\}, \tag{4.1}$$

obtaining, among other things, the following results: *If E is a strictly convex normed linear space, then for every set $G \subset E$ the set $\mathfrak{U}_G$ is dense in E* (S. B. Stečkin [237], theorem 1). *If E is a strictly convex Banach space and $G \subset E$ is a boundedly compact set, then $\mathfrak{U}_G$ is of the second category in E, and if G is also closed, then $\mathfrak{U}_G$ is a set of type G_δ* (S. B. Stečkin [237], theorem 2); however, it is not known whether for every $G \subset E$ the set $\mathfrak{U}_G$ is of the second category. Since by the criterion of strict convexity of M. G. Krein used in Chap. I, § 3, section 3.1, in every space which is not strictly convex there exists a hyperplane H such that $\mathfrak{U}_H = H$, there results the following constructive characterization of strictly convex Banach spaces: *A Banach space is strictly convex if and only if for every set $G \subset E$ we have $\overline{\mathfrak{U}}_G = E$* (S. B. Stečkin [237], theorem 3); however, it is not known whether from the fact that for every *compact* $G \subset E$ the set $\mathfrak{U}_G$ is either dense in E or of the second category in E, there follows that E is strictly convex. Finally, *if E is a locally uniformly convex Banach space* *), *then for every $G \subset E$ the set $\mathfrak{U}_G$ is of the second category and if E is a uniformly convex Banach space, then for every closed set G in E the set*

$$\{x \in E \mid \mathfrak{L}_G(x) \neq \emptyset\} \cap \mathfrak{U}_G \tag{4.2}$$

is of the second category (S. B. Stečkin [237], theorem 4 and corollary of theorem 5); however, it is not known whether the second result remains valid also for locally uniformly convex spaces.

R. R. Phelps [174], [175] has established some properties of the sets $\pi_G^{-1}(g_0)$ for arbitrary sets $G \subset E$ and elements $g_0 \in G$. Thus, *the set $\pi_G^{-1}(g_0)$ is always closed* (R. R. Phelps [175], p. 867). Also, *if E is an inner product space, then for every $G \subset E$ and every $g_0 \in G$ the set $\pi_G^{-1}(g_0)$ is convex* (R. R. Phelps [174], lemma 4.1). Conversely, *if a normed linear space E has the property that for every $G \subset E$ and every $g_0 \in G$ the set $\pi_G^{-1}(g_0)$ is convex, then*

---

*) We recall that, following A. R. Lovaglia [141], a normed linear space E is said to be *locally uniformly convex*, if for every ε with $0 < \varepsilon \leqslant 2$ and every $x \in E$ with $\|x\| = 1$ there exists a $\delta(\varepsilon, x) > 0$ such that $\left\| \frac{x+y}{2} \right\| \leqslant 1 - \delta(\varepsilon, x)$ $(y \in E, \|y\| \leqslant 1)$. Every locally uniformly convex space is obviously uniformly convex (for the definition of uniformly convex spaces see § 2).

E is equivalent to an inner product space (T. S. Motzkin [154] and R. R. Phelps [174], theorem 4.2); in the case when dim $E \geqslant 3$, the same conclusion remains valid assuming only that for every *convex* $G \subset E$ and every $g_0 \in G$ the set $\pi_G^{-1}(g_0)$ is convex (R. R. Phelps [174], lemma 4.4).

We shall say, following R. R. Phelps [175], that a set A in a normed linear space E is a *nearest-point set* if there exist a set $G \subset E$ and an element $g_0 \in G$ such that $A = \pi_G^{-1}(g_0)$. As has been observed above, all nearest point sets $A \subset E$ are closed and convex if and only if E is equivalent to an inner product space. There arises naturally *the problem of characterizing the normed linear spaces E in which every closed convex set is a nearest-point set.* As has shown R. R. Phelps ([175], theorems 2 and 3), *every complete inner product space has this property, and conversely, every normed linear space E of dimension $\geqslant 3$ with this property is equivalent to a complete inner product space.*

Finally, we mention that some properties of the sets $\{x \in E \mid \dim \mathfrak{Q}_G(x) \geqslant k\}$, for arbitrary sets G in n-dimensional Euclidean spaces, have been given by P. Erdös [54 a].

APPENDIX **II**

BEST APPROXIMATION IN METRIC SPACES BY ELEMENTS OF ARBITRARY SETS

Problems of best approximation arise in a natural way also in spaces other than normed linear spaces. Thus, R. A. Hirschfeld [90] *has raised the problem of best approximation in linear lattices by elements of arbitrary sets* :

$$|x - g_0| = \inf_{g \in G} |x - g|. \tag{1.1}$$

In particular, e.g. in the space $E = C([0,1])$, this problem amounts to pointwise best approximation :

$$|x(t) - g_0(t)| \leqslant |x(t) - g(t)| \quad (g \in G,\ t \in [0, 1]). \tag{1.2}$$

But, if G is a linear subspace of $E = C([0,1])$ then from (1.2) it follows that we must have $g_0(t) = x(t)$ on the set $\bigcup_{g \in G} \{t \in [0,1] | g(t) \neq 0\}$; consequently, taking linear subspaces G for which this set coincides with the whole segment $[0,1]$ (e.g. linear subspaces G containing the constant functions), the problem will admit only the trivial solution $x \in G$, $g_0 = x$.

In order to avoid this inconvenience, one may consider e.g. the following related problem : We shall say that an element $g \in G$ is *closer* to x than an element $g_0 \in G$, if we have

$$|x(t) - g(t)| < |x(t) - g_0(t)| \quad \text{on} \quad [0,1] \setminus Z(x - g_0), \tag{1.3}$$

$$x(t) - g(t) = x(t) - g_0(t) \text{ on } Z(x - g_0), \tag{1.4}$$

where $Z(x - g_0) = \{t \in [0,1] | x(t) - g_0(t) = 0\}$; we shall call an element $g_0 \in G$ a *juxtaelement* of x if there does not exist an element $g \in G$ closer to x than g_0. In the particular case when the scalars are real and $G = [1, t, \ldots, t^{n-1}]$, the *juxtapolynomials*

g_0 of x have been introduced and studied by J. L. Walsh and T. S. Motzkin [260].

The problem of best approximation in normed linear lattices has been studied by D. E. Wulbert ([263], Chap. II).

As concerns spaces more general than normed linear spaces, a first natural step is the study of *best approximation in metric linear spaces by elements of linear subspaces.* G. Albinus [2] has given an example of a metric linear space which has a non-proximinal linear subspace of dimension one. Furthermore, G. Albinus [3] has shown that if E is a linear space having a countable sufficient system of semi-norms $\{p_n\}$, then for the metric

$$\rho(x, y) = \sum_{j=1}^{\infty} \frac{1}{2^j} \frac{p_j(x-y)}{1+p_j(x-y)} \qquad (x, y \in E) \tag{1.5}$$

every finite-dimensional linear subspace G is approximatively compact, whence proximinal; see also G. Albinus [3a]. In p-normed *) metric linear spaces the proximinality of finite-dimensional linear subspaces has been proved by K. Iseki [93]. For best approximation in Fréchet spaces see V. N. Nikolsky [164], [165].

Another class of spaces containing the normed linear spaces is that of locally convex topological linear spaces. R. A. Hirschfeld [90] has raised *the problem of best approximation in a locally convex space E*, the topology of which is defined by a family Γ of semi-norms:

$$p(x-g_0) = \inf_{g \in G} p(x-g) \qquad (p \in \Gamma); \tag{1.6}$$

this problem has been studied by A. M. Flomin [56 a].

In the present Appendix we shall give some results **) on *the problem of best approximation in metric spaces E by elements of arbitrary sets $G \subset E$*:

$$\rho(x, g_0) = \inf_{g \in G} \rho(x, g), \tag{1.7}$$

where ρ denotes the metric in the space E. In this case the notations $\mathfrak{L}_G(x)$, $\pi_G(x)$, $e_G(x), \ldots$ as well as the notions of proximinal set, Čebyšev set, etc. are introduced as an obvious extension of the corresponding notations and notions in normed

*) See e.g. G. Köthe [122], p. 164.

**) It would be interesting to study in metric spaces the problem of best approximation by elements of sets G belonging to certain special classes of sets, for instance *convex* sets $G \subset E$ in the sense of K. Menger [148], i.e. having the property that for any distinct $x, z \in E$ there exists a $y \in E$ with $x \neq y \neq z$ such that $\rho(x, z) = \rho(x, y) + \rho(y, z)$.

linear spaces. Specializing the results which follow to the case when E is a normed linear space and G is a linear subspace (or, eventually, a more general set) of E, the reader will obtain, besides some theorems given in the preceding, a number of results which are exposed only in the present Appendix, within the more general framework of metric spaces (see e.g. § 3).

§ 1. PROPERTIES OF THE SETS $\mathfrak{P}_G(x)$. A CHARACTERIZATION OF ELEMENTS OF BEST APPROXIMATION

For $x \in E$ and $r \geqslant 0$ we shall denote by $S(x, r)$ the cell $\{y \in E \mid \rho(y, x) \leqslant r\}$, by *) $\mathrm{Fr}_m S(x, r)$ the set $\{y \in E \mid \rho(y, x) = r\}$, and by $\mathrm{Int}_m S(x, r)$ the set $\{y \in E \mid \rho(y, x) < r\}$.

THEOREM 1.1. *Let E be a metric space, G a set in E and $x \in E$. We have then* **)

$$\mathfrak{P}_G(x) = G \cap S(x, e_G(x)) = G \cap \mathrm{Fr}_m S(x, e_G(x)). \tag{1.8}$$

Proof. The inclusions

$$\mathfrak{P}_G(x) \subset G \cap \mathrm{Fr}_m S(x, e_G(x)) \subset G \cap S(x, e_G(x)) \tag{1.9}$$

are obvious by the definitions of $\mathfrak{P}_G(x)$ and $e_G(x)$.

Conversely, let $g_0 \in G \cap S(x, e_G(x))$. Then we have

$$e_G(x) = \inf_{g \in G} \rho(x, g) \leqslant \rho(x, g_0) \leqslant e_G(x),$$

whence $g_0 \in \mathfrak{P}_G(x)$. Thus, $G \cap S(x, e_G(x)) \subset \mathfrak{P}_G(x)$, whence, by (1.9), we have (1.8), which completes the proof.

The last equality in (1.8) is equivalent to the equality

$$G \cap \mathrm{Int}_m S(x, e_G(x)) = \varnothing, \tag{1.10}$$

which may be seen also directly, as follows: from $g_0 \in G \cap \cap \mathrm{Int}_m S(x, e_G(x))$ there would result $e_G(x) \leqslant \rho(x, g_0) < e_G(x)$, which is absurd.

---

*) Simple examples show that the "metric boundary" $\mathrm{Fr}_m S(x, r)$ and the "metric interior" $\mathrm{Int}_m S(x, r)$ do not coincide, in general, with the (topological) boundary $\mathrm{Fr}\, S(x, r)$, and the (topological) interior $\mathrm{Int}\, S(x, r)$ respectively.

**) We recall that by definition

$$e_G(x) = \rho(x, G) = \inf_{g \in G} \rho(x, g),$$

$$\mathfrak{P}_G(x) = \{g_0 \in G \mid \rho(x, g_0) = e_G(x)\}.$$

COROLLARY 1.1. *Let E be a metric space and G a set in E. Then*

a) *The set $\mathfrak{L}_G(x)$ is bounded.*
b) *If G is closed, then $\mathfrak{L}_G(x)$ is closed.*
c) *If G is boundedly compact, $\mathfrak{L}_G(x)$ is conditionally compact.*
d) *If G is boundedly compact and A a bounded set in E, then the set* *) $\bigcup_{x \in A} \mathfrak{L}_G(x)$ *is conditionally compact.*

Proof. a), b) and c) are immediate consequences of theorem 1.1. On the other hand, under the hypotheses of d) we have, taking an arbitrary $g_0 \in G$,

$$\delta(A, G) = \sup_{x \in A} e_G(x) \leqslant \sup_{x \in A} \rho(x, g_0) < \infty,$$

and thus, A being bounded, the set $\bigcup_{x \in A} S(x, \delta(A, G))$ is bounded. But by virtue of theorem 1.1 we have

$$\bigcup_{x \in A} \mathfrak{L}_G(x) = \bigcup_{x \in A} [G \cap S(x, e_G(x))] = G \cap \bigcup_{x \in A} S(x, e_G(x)) \subset$$
$$\subset G \cap \bigcup_{x \in A} S(x, \delta(A, G)),$$

hence $\bigcup_{x \in A} \mathfrak{L}_G(x)$ is a bounded subset of G. Consequently, G being boundedly compact, it follows that $\bigcup_{x \in A} \mathfrak{L}_G(x)$ is conditionally compact, which completes the proof.

COROLLARY 1.2. *Let E be a metric space, G a set in E, $x \in E$ with $\mathfrak{L}_G(x) \neq \emptyset$ and $r > 0$ such that*

$$\emptyset \neq G \cap S(x, r) \subset \mathrm{Fr}_m S(x, r). \tag{1.11}$$

Then we have

$$r = e_G(x),$$

and consequently $G \cap S(x, r) = \mathfrak{L}_G(x)$.

Proof. If $r < e_G(x)$, then by the definition of $e_G(x)$ we have $G \cap S(x,r) = \emptyset$, which contradicts (1.11). If $r > e_G(x)$, then by $\mathfrak{L}_G(x) \neq \emptyset$ and (1.8) we have

$$\emptyset \neq \mathfrak{L}_G(x) = G \cap S(x, e_G(x)) \subset G \cap \mathrm{Int}_m S(x, r),$$

contradicting (1.11), which completes the proof.

---

*) Some authors (e.g. C. Berge [13]) use for the set $\bigcup_{x \in A} \mathfrak{L}_G(x)$ the notation $\mathfrak{L}_G(A)$. We have not used this notation in order to avoid a contradiction with the usual meaning of the notation $\mathfrak{L}_G(A)$, namely: the value, in A, of the extension of the mapping $\mathfrak{L}_G : E \to 2^G$ to the set of all subsets (N. Bourbaki [24], § 2), i.e. $\mathfrak{L}_G(A) = \{\mathfrak{L}_G(x) \mid x \in A\}$.

In the particular case when E is a normed linear space and G is a finite-dimensional linear subspace of E, corollary 1.2 has been given by K. Tatarkiewicz ([243], theorem 1.8).

For an arbitrary set $G \subset E$ we shall denote by $\operatorname{Fr} G$ the boundary of G, by $\operatorname{Int} G$ the interior of G (hence $\operatorname{Fr} G = \overline{G} \setminus \operatorname{Int} G$), and by $\mathfrak{M}_G$ the set of all elements of best approximation of the elements $x \in E$ by elements of the set G, i.e.

$$\mathfrak{M}_G = \bigcup_{x \in E} \mathfrak{L}_G(x). \tag{1.12}$$

Obviously $\mathfrak{M}_G \subset G$, but in general $\mathfrak{M}_G \not\subset \operatorname{Fr} G$, as shown by the following example : Let E be the set $\{x = \{\xi_1, \xi_2\} \mid |\xi_1| \geqslant 1\}$ in the Euclidean plane, endowed with the metric induced by the Euclidean metric, let $G = \{x = \{\xi_1, \xi_2\} \in E \mid |\xi_1 + 2| \leqslant 1, |\xi_2| \leqslant 1\}$ and let $x = \{2, 0\} \in E$; then $\mathfrak{L}_G(x)$ consists of the single point $g_0 = \{-1, 0\}$, which $\in \operatorname{Int} G$ (since the "cell" $S\left(g_0, \frac{1}{2}\right) \subset E$ is contained in G)*).

On the other hand, we have

THEOREM 1.2 (M. Nicolescu [161], § 6). *Let E be a metric space and G a proximinal set in E. Then*

$$\operatorname{Fr} G \subset \overline{\mathfrak{M}}_G. \tag{1.13}$$

Proof. If $\operatorname{Fr} G = \varnothing$, the statement is obvious. If $\operatorname{Fr} G \neq \varnothing$, let $g_0 \in \operatorname{Fr} G$ and $\varepsilon > 0$ be arbitrary. Then the cell $S\left(g_0, \frac{\varepsilon}{2}\right)$ contains at least one element $x \in E \setminus G$. Let $\pi_G(x) \in \mathfrak{L}_G(x)$ (it exists, since by hypothesis G is proximinal). We have then

$$\rho(g_0, \pi_G(x)) \leqslant \rho(g_0, x) + \rho(x, \pi_G(x)) \leqslant 2\,\rho(x, g_0) \leqslant \varepsilon,$$

whence, since $\varepsilon > 0$ was arbitrary, we obtain $g_0 \in \overline{\mathfrak{M}}_G$, which completes the proof.

We shall say that a set $A \subset E$ *supports* the cell $S(x, r)$, or that A is a support set of the cell $S(x, r)$, if we have $\rho(A, S(x,r)) = 0$ and $A \cap \operatorname{Int}_m S(x, r) = \varnothing$. We have the following characterization of elements of best approximation :

THEOREM 1.3. *Let E be a metric space, G a non-void set in E, $x \in E \setminus G$ and $g_0 \in G$. We have $g_0 \in \mathfrak{L}_G(x)$ if and only if the set G supports the cell $S = S(x, \rho(x, g_0))$.*

Proof. Assume that $g_0 \in \mathfrak{L}_G(x)$, hence $\rho(x, g_0) = e_G(x)$. Then by (1.10) we have $G \cap \operatorname{Int}_m S = \varnothing$. On the other hand, by

*) Since we have $\rho(x, G) = 3 < \sqrt{10} = \rho(x, \operatorname{Fr} G)$, the same example shows that for $x \in E \setminus G$ in general $\rho(x, G) \neq \rho(x, \operatorname{Fr} G)$.

$g_0 \in G \cap S$ we have $\rho(G, S) = 0$. Consequently, the set G supports the cell S.

On the other hand, assume now that $g_0 \notin \mathfrak{P}_G(x)$, hence $\rho(x, g_0) > e_G(x)$, and let $\varepsilon > 0$ be such that $\rho(x, g_0) > e_G(x) + \varepsilon$. Then there exists a $g \in G$ such that $\rho(x, g) \leqslant e_G(x) + \varepsilon < \rho(x, g_0)$, hence $g \in G \cap \operatorname{Int}_m S$. Consequently, G does not support the cell S, which completes the proof.

§ 2. PROXIMINAL SETS

LEMMA 2.1. *Let E be a metric space and let $x \in E$. Then the functional*

$$\Phi(y) = \rho(x, y) \qquad (y \in E) \tag{2.1}$$

is uniformly continuous on E.

Proof. We have obviously

$$|\Phi(y_1) - \Phi(y_2)| \leqslant \rho(y_1, y_2) \qquad (y_1, y_2 \in E), \tag{2.2}$$

whence the statement follows.

A set G in a metric space E is said to be *approximatively compact* (N. V. Efimov and S. B. Stečkin [53]) if for every $x \in E$ and every sequence $\{g_n\} \subset G$ with $\lim_{n \to \infty} \rho(x, g_n) = \rho(x, G)$ there exists a subsequence $\{g_{n_k}\}$ converging to an element of G.

THEOREM 2.1 (N. V. Efimov and S. B. Stečkin [53]). *Let E be a metric space. Then every non-void approximatively compact set G in E is proximinal.*

Proof. Let $x \in E$ be arbitrary. By the definition of $\rho(x, G)$, from the set of numbers $\{\rho(x, g) \mid g \in G\}$ we can extract a sequence $\{\rho(x, g_n)\}$ such that $\lim_{n \to \infty} \rho(x, g_n) = \rho(x, G)$. Since G is approximatively compact, we can extract from $\{g_n\}$ a subsequence $\{g_{n_k}\}$ converging to a $g_0 \in G$. We then have, by lemma 2.1,

$$\rho(x, g_0) = \rho(x, \lim_{k \to \infty} g_{n_k}) = \lim_{k \to \infty} \rho(x, g_{n_k}) = \rho(x, G),$$

whence $g_0 \in \mathfrak{P}_G(x)$, which completes the proof.

Since every proximinal set is closed *) (by Chap. I, formula (1.1), which obviously remains valid also in general metric spaces), it follows that *every approximatively compact set is closed* (N. V. Efimov and S. B. Stečkin [53]).

---

*) Conversely, V. Klee ([109 a], theorem 6) has proved that *for every closed set $G \neq E$ there exists an equivalent metric on E in which G is proximinal.*

LEMMA 2.2 (N. V. Efimov and S. B. Stečkin [53]). *Let E be a metric space and G a boundedly compact closed set in E. Then G is approximatively compact.*

Proof. Let $x \in E$ and let $\{g_n\} \subset G$ be a sequence such that $\lim_{n\to\infty} \rho(x, g_n) = \rho(x, G)$. Since the sequence $\{\rho(x, g_n)\}$ is then bounded, the set $\{g_n\} \subset G$ is bounded. Consequently, since G is boundedly compact and closed, we can extract from $\{g_n\}$ a subsequence $\{g_{n_k}\}$ converging to an element of G, which completes the proof.

The converse of lemma 2.2 is not true, as shown by the following example : Let $E = l^2$ and $G = \{x = \{\xi_n\} \in E \mid \xi_1 = 0\}$; then G is weakly closed, hence approximatively compact (see Appendix I, § 2), but not boundedly compact, since $\dim G = \infty$.

From theorem 2.1 and lemma 2.2 it follows *)

COROLLARY 2.1. *Let E be a metric space. Then every non-void boundedly compact closed set G in E is proximinal.* **)

In the particular case when E is a normed linear space and G is a finite-dimensional linear subspace of E, from this corollary we find again corollary 2.2 of Chap. I.

We recall that a set G in a topological space T is said to be countably compact, if every infinite subset of G has at least one limit point (i.e. a point $x \in E$ with the property that in every neighbourhood $V(x)$ of x there exists a $g \in G \setminus \{x\}$), or, what is equivalent, if for every decreasing sequence $A_1 \supset A_2 \supset \dots$ of non-void closed subsets of G we have $\bigcap_{n=1}^{\infty} A_n \neq \emptyset$. One can give the following generalization of corollary 2.1 :

THEOREM 2.2. *Let E be a metric space and let τ be an arbitrary topology* ***) *on E. If G is a non-void set of E, such that for every $x \in E$ and $r > 0$ the intersection $G \cap S(x, r)$ is τ-closed and τ-countably compact, then G is proximinal.*

Proof. Put

$$A_n = G \cap S\left(x, e_G(x) + \frac{1}{n}\right) \qquad (n = 1, 2, \dots). \tag{2.3}$$

We have obviously $A_1 \supset A_2 \supset \dots$ and by the definition of e_G each A_n is non-void. Since each A_n is τ-countably compact

---

*) From theorem 2.1 and lemma 2.2 we see also that the notion of approximatively compact set is intermediate between boundedly compact closed sets and proximinal sets.

**) Conversely, V. Klee ([109 a], theorem 5) has proved that *if a set $G \neq E$ is proximinal in every equivalent metric on E, then G is compact.*

***) It is not required that τ be comparable with the initial topology of the space E.

and τ-closed, it follows that there exists a $g_0 \in \bigcap_{n=1}^{\infty} A_n$. We have then

$$e_G(x) \leqslant \rho(x, g_0) \leqslant e_G(x) + \frac{1}{n} \qquad (n = 1, 2, \ldots),$$

whence $g_0 \in \mathfrak{P}_G(x)$, which completes the proof.

Since every τ-compact *) set is τ-closed and τ-countably compact, we have

COROLLARY 2.2. *Let E be a metric space and let τ be an arbitrary topology on E. If G is a non-void set in E, such that for every $x \in E$ and $r > 0$ the intersection $G \cap S(x, r)$ is τ-compact, then G is proximinal.*

In the particular case of normed linear spaces, from theorem 2.2 and corollary 2.2 one obtains again corollaries 2.1, 2.1′ and theorem 2.2 of Chap. I.

A similar result is valid for sets G with the intersections $G \cap S(x, r)$ τ-closed and τ-strictly sequentially compact (since every τ-strictly sequentially compact set is τ-countably compact), and also for sets G with τ-countably compact intersections $G \cap S(x, r)$ and with the property that the functional e_G is τ-lower semi-continuous.

LEMMA 2.3. *Let E be a metric space and let the Cartesian product $E \times E$ be endowed with the metric*

$$\rho^*(\{x_1, y_1\}, \{x_2, y_2\}) = \rho(x_1, x_2) + \rho(y_1, y_2), \tag{2.4}$$

or with an equivalent metric. Then the functional

$$\varphi(\{x, y\}) = \rho(x, y) \qquad (\{x, y\} \in E \times E) \tag{2.5}$$

is uniformly continuous on $E \times E$.

Proof. We have obviously

$$|\rho(x_1 y_1) - \rho(x_2, y_2)| \leqslant \rho(x_1, x_2) + \rho(y_1, y_2)$$
$$(\{x_i, y_i\} \in E \times E, i = 1,2), \tag{2.6}$$

whence the statement follows.

We recall that for two non-void sets $G_1, G_2 \subset E$ one denotes by**) $\rho(G_1, G_2)$ the number $\inf_{\substack{g_1 \in G_1 \\ g_2 \in G_2}} \rho(g_1, g_2)$. We shall now prove the following generalization of corollary 2.1:

---

*) We recall that we use the term "compact" in the sense of N. Bourbaki [25] i.e.: bicompact Hausdorff.

**) Although the number $\rho(G_1, G_2)$ is called the "distance" between G_1 and G_2, this "distance" is not a metric in the set 2^E of all non-void bounded closed subsets of E (it is not transitive and $\rho(G_1, G_2) = 0$ does not imply $G_1 = G_2$). For a metric in the set 2^E see Chap. I, §6, formula (6.79).

THEOREM 2.3. *Let E be a metric space and G_1, G_2 two non-void boundedly compact closed sets in E. Then there exist elements $g_1 \in G_1$ and $g_2 \in G_2$ such that*

$$\rho(g_1, g_2) = \rho(G_1, G_2). \tag{2.7}$$

Proof. By the definition of $\rho(G_1, G_2)$, from the set of numbers $\{\rho(g_1, g_2) | g_1 \in G_1, g_2 \in G_2\}$ we can extract a sequence $\{\rho(g_1^{(n)}, g_2^{(n)})\}$ such that $\lim_{n\to\infty} \rho(g_1^{(n)}, g_2^{(n)}) = \rho(G_1, G_2)$. Since this sequence is bounded, the sets $\{g_1^{(n)}\} \subset G_1$, $\{g_2^{(n)}\} \subset G_2$ are bounded. Consequently, since G_1, G_2 are boundedly compact and closed, we can extract from $\{g_1^{(n)}\}$ and $\{g_2^{(n)}\}$ subsequences $\{g_i^{(n_k)}\}$ $(i = 1,2)$ such that $\lim_{k\to\infty} g_i^{(n_k)} = g_i \in G$ $(i = 1,2)$. By lemma 2.3 we have then

$$\rho(g_1, g_2) = \rho(\lim_{k\to\infty} g_1^{(n_k)}, \lim_{k\to\infty} g_2^{(n_k)}) = \lim_{k\to\infty} \rho(g_1^{(n_k)}, g_2^{(n_k)}) = \rho(G_1, G_2),$$

i.e. (2.7), which completes the proof.

In the particular case when the sets G_1, G_2 are compact, theorem 2.3 has been given by M. Nicolescu ([161], § 3).

We shall make now some remarks in connection with the foregoing results.

In a metric space E, the condition that G be a non-void locally compact closed subspace of E is not sufficient in order that G be proximinal, as shown by the following example: Let E = the real line from which it has been deleted one point p, with the usual metric, and let G = the open half-line $(p, +\infty)$; then G is a non-void locally compact closed subspace E, but it is not proximinal.

The condition of corollary 2.1 and the condition of being a non-void locally compact closed subspace are not necessary in order that G be proximinal, as shown by the example given after lemma 2.2 (the set G in that example is proximinal by virtue of theorem 2.1).

Neither is the condition of theorem 2.1 necessary in order that $G \subset E$ be proximinal, as shown by the following example: Let $E = C([0,1])$, $G = \{y \in E | y(0) = 0\}$; then G is proximinal, but it is not approximatively compact.

A non-void closed subset of a proximinal set is not necessarily proximinal, as shown by the following example: Let $E = l^1$, $G = \{x = \{\xi_n\} \in E | \xi_1 = 0\}$ and $G_0 = \left\{x = \{\xi_n\} \in E \mid \xi_1 = 0, \sum_{n=1}^{\infty} \frac{n}{n+1} \xi_n = 1\right\}$; then G is proximinal, and G_0 is a subset of

G which is not proximinal, since $\mathfrak{P}_{G_0}(0) = \emptyset$. At the same time, this example shows that a non-void closed subset of an approximatively compact set is not necessarily approximatively compact (the above G is approximatively compact, since for $x = \{1, 0, 0, \ldots\}$ and any sequence $\{g_n\} \subset G$ with $\|x - g_n\| \to \to \rho(x, G) = 1$ we have even $g_n \to 0$ and since G is a hyperplane).

Similarly, taking e.g. in $E = l^1$ a finite-dimensional linear subspace and a non-proximinal hyperplane containing it, we see that a closed superset of a proximinal (respectively approximatively compact) set is not necessarily proximinal (respectively approximatively compact).

Finally, let us mention the following result of C. Kuratowsky [132] (see also W. Nitka [171 a]) : *In order that a closed set G in a compact space E be a retract of E (i.e. that there exist a continuous projection of E onto G) it is necessary and sufficient that there exist on E a metric ρ equivalent to the initial topology, such that in this metric G be a Čebyšev set.*

§ 3. PROPERTIES OF THE MAPPINGS $\mathfrak{P}_G$

We recall that for a metric space F one denotes by 2^F the collection of all bounded closed subsets of F. A mapping $\mathfrak{U} : E \to 2^F$ is called *) *upper semi-continuous* (respectively lower semi-continuous) if the set

$$\{x \in E \mid \mathfrak{U}(x) \subset M\} \tag{3.1}$$

is open for every open $M \subset F$ (respectively closed for every closed $M \subset F$), or, what is equivalent, if the set

$$\{x \in E \mid \mathfrak{U}(x) \cap N \neq \emptyset\} \tag{3.2}$$

is closed for every closed $N \subset F$ (respectively open for every open $N \subset F$).

THEOREM 3.1 ([231], theorem 1). *Let E be a metric space and G a non-void approximatively compact set in E. Then $\mathfrak{P}_G$ maps E into 2^G and is upper semi-continuous.*

Proof. By virtue of theorem 2.1 the set G is proximinal, hence $\mathfrak{P}_G(x) \neq \emptyset$ $(x \in E)$. By the remark made after theorem 2.1 and by corollary 1.1, $\mathfrak{P}_G(x)$ is closed and bounded. Thus $\mathfrak{P}_G$ maps E into 2^G.

---

*) See e.g. E. Michael [149].

Let now N be an arbitrary closed subset of G. We shall show that the set

$$B = \{x \in E \,|\, \mathfrak{L}_G(x) \cap N \neq \emptyset\} \tag{3.3}$$

is closed, which will complete the proof. Let $\{x_n\}$ be a sequence in B, converging to an element $x \in E$. Since $\{x_n\} \subset B$, there exists a sequence $\{g_n\} \subset G$ such that $g_n \in \mathfrak{L}_G(x_n) \cap N$ $(n = 1, 2, \ldots)$. Then, by $g_n \in \mathfrak{L}_G(x_n)$ $(n = 1, 2, \ldots)$ we have

$$\rho(x, G) \leqslant \rho(x, g_n) \leqslant \rho(x, x_n) + \rho(x_n, g_n) = \rho(x, x_n) + \\ + \rho(x_n, G) \leqslant 2\rho(x, x_n) + \rho(x, G) \qquad (n = 1, 2, \ldots), \tag{3.4}$$

whence, taking into account $\lim\limits_{n\to\infty} \rho(x, x_n) = 0$, it follows that

$$\lim_{n\to\infty} \rho(x, g_n) = \rho(x, G).$$

Consequently, G being approximatively compact, there exists a subsequence $\{g_{n_k}\}$ of $\{g_n\}$, converging to an element $g_0 \in G$. By (3.4) we then have

$$\rho(x, G) \leqslant \rho(x, g_0) \leqslant \rho(x, g_{n_k}) + \rho(g_{n_k}, g_0) \leqslant \\ \leqslant 2\rho(x, x_{n_k}) + \rho(x, G) + \rho(g_{n_k}, g_0) \qquad (k = 1, 2, \ldots),$$

whence for $k \to \infty$,

$$\rho(x, g_0) = \rho(x, G),$$

i.e. $g_0 \in \mathfrak{L}_G(x)$. On the other hand, since N is closed and $\{g_{n_k}\} \subset N$, $\lim\limits_{k\to\infty} g_{n_k} = g_0$, we have $g_0 \in N$. Consequently, $g_0 \in \mathfrak{L}_G(x) \cap N$, whence $x \in B$, which completes the proof.

THEOREM 3.2 ([231], theorem 2). *Let E be a metric space and G a non-void approximatively compact set in E. Then*

a) $\lim\limits_{n\to\infty} x_n = x$ *implies* $\lim\limits_{n\to\infty} \rho(\mathfrak{L}_G(x_n), \mathfrak{L}_G(x)) = 0$.

b) $\lim\limits_{n\to\infty} x_n = x$ *implies the existence of two sequences* $\{g_n\}$, $\{g_n'\} \subset G$ *with* $g_n \in \mathfrak{L}_G(x_n)$, $g_n' \in \mathfrak{L}_G(x)$ $(n = 1, 2, \ldots)$, *such that* $\lim\limits_{n\to\infty} \rho(g_n, g_n') = 0$.

c) $\lim\limits_{n\to\infty} x_n = x$ *implies the existence, for every sequence* $\{g_n\} \subset G$ *with* $g_n \in \mathfrak{L}_G(x_n)$ $(n = 1, 2, \ldots)$, *of a sequence* $\{g_n'\} \subset \mathfrak{L}_G(x)$ *such that* $\lim\limits_{n\to\infty} \rho(g_n, g_n') = 0$.

d) $\lim\limits_{n\to\infty} x_n = x$ *implies the existence, for every sequence* $\{g_n\} \subset G$ *with* $g_n \in \mathfrak{L}_G(x_n)$ $(n = 1, 2, \ldots)$, *of a subsequence* $\{g_{n_k}\} \subset \{g_n\}$ *and of an element* $g_0 \in \mathfrak{L}_G(x)$ *such that* $\lim\limits_{k\to\infty} g_{n_k} = g_0$.

Proof. The statement d) is implicitly contained in the above proof of theorem 3.1. On the other hand, obviously c) $\Rightarrow$ b) $\Rightarrow$ a). Thus, there remains to prove c).

Assume the contrary, that for a sequence $\{g_n\} \subset G$ with $g_n \in \mathfrak{P}_G(x_n)$ $(n = 1, 2, \ldots)$ there exists no sequence $\{g'_n\} \subset \mathfrak{P}_G(x)$ satisfying $\lim_{n\to\infty} \rho(g_n, g'_n) = 0$. Then, taking $\{g'_n\} \subset \mathfrak{P}_G(x)$ with the property

$$\rho(g_n, g'_n) \leqslant \rho(g_n, \mathfrak{P}_G(x)) + \frac{1}{n} \qquad (n = 1, 2, \ldots),$$

it follows that $\overline{\lim_{n\to\infty}}\, \rho(g_n, \mathfrak{P}_G(x)) \neq 0$, i.e. there exist an infinite subsequence $\{g_{n_k}\}$ of $\{g_n\}$ and an $\varepsilon_0 > 0$ such that

$$\rho(g_{n_k}, \mathfrak{P}_G(x)) \geqslant \varepsilon_0 \qquad (k = 1, 2, \ldots). \tag{3.5}$$

On the other hand, applying d) to $\{x_{n_k}\}$, $\{g_{n_k}\}$, it follows that there exist a subsequence $\{g_{n_{k_m}}\}$ of $\{g_{n_k}\}$ and an element $g_0 \in \mathfrak{P}_G(x)$ such that $\lim_{m\to\infty} g_{n_{k_m}} = g_0$. We have then

$$0 = \lim_{m\to\infty} \rho(g_{n_{k_m}}, g_0) \geqslant \overline{\lim_{m\to\infty}}\, \rho(g_{n_{k_m}}, \mathfrak{P}_G(x)),$$

contradicting (3.5), which completes the proof.

In the particular case when G is compact, theorem 3.2 b) has been proved by M. Nicolescu ([161], theorem 2).

Let us mention separately the following particular case of theorem 3.2 c):

COROLLARY 3.1 ([231], corollary 1). *Let E be a metric space, G a non-void approximatively compact set in E and $x \in E$ such that $\mathfrak{P}_G(x)$ consists of a single element, say g_0. Then $\lim_{n\to\infty} x_n = x$ implies $\lim_{n\to\infty} g_n = g_0$ for every sequence $\{g_n\} \subset G$ with $g_n \in \mathfrak{P}_G(x_n)$ $(n = 1, 2, \ldots)$.*

Sometimes it is also used a different notion of semi-continuity, which we shall call *) (K)-semi-continuity, since it was studied by C. Kuratowski [131]. A mapping $\mathfrak{U}: E \to 2^F$ is called *upper (K)-semi-continuous* (respectively *lower (K)-semi-continuous*) if the relations $\lim_{n\to\infty} x_n = x$, $y_n \in \mathfrak{U}(x_n)$ $(n = 1, 2, \ldots)$, $\lim_{n\to\infty} y_n = y$ imply $y \in \mathfrak{U}(x)$ (respectively, if the relations $\lim_{n\to\infty} x_n = x$, $y \in \mathfrak{U}(x)$ imply the existence of a sequence $\{y_n\}$ with $y_n \in \mathfrak{U}(x_n)$

*) We mention that there are also other denominations for the notions of semi-continuity considered in the present paragraph. There are also other notions of semi-continuity for the mappings $\mathfrak{U}: E \to 2^F$, but those of the present paragraph are more frequently used.

$(n = 1, 2, \ldots)$ such that $\lim_{n\to\infty} y_n = y$). As is well known *), *every upper (lower) semi-continuous mapping is upper (lower) (K)-semi-continuous and, if F is compact, then the converse is also valid*; however, this latter statement is no longer true if F is only approximatively compact.

Let us consider now the mapping $\mathfrak{P}_G$. From the above proof of theorem 3.1 it follows immediately that *if G is an arbitrary proximinal set in a metric space E, then the mapping $\mathfrak{P}_G : E \to 2^G$ is upper (K)-semi-continuous.* In the particular case when E is a Banach space and G is a finite-dimensional linear subspace of E, this result has been proved by K. Tatarkiewicz ([243], theorem 6.4); the proof of K. Tatarkiewicz remains valid also when G is a proximinal set in a metric space E.

We conclude this paragraph by some remarks showing that in certain respects the above results cannot be improved.

If G is a proximinal set, but not approximatively compact, then the conclusions of theorems 3.1, 3.2 and corollary 3.1 may be no longer valid, as shown by the following example ([231], example 1): Let $E = l^2$ and let G be the sequence

$$g_1 = 0, g_n = \Big\{1, \frac{1}{n}, \underbrace{0, \ldots, 0}_{n-1}, 1, 0, \ldots\Big\} \quad (n = 2, 3, \ldots). \tag{3.6}$$

Then G is a proximinal set (since for every $x \in E$ the sequence of non-negative numbers $\{\|x - g_n\|\}$ is convergent, whence $e_G(x) = \inf_{1 \leqslant n < \infty} \|x - g_n\|$ is attained), but it is not approximatively compact (since for $x_1 = \{1, 0, 0, \ldots\} \in E$ we have $\lim_{n\to\infty} \|x_1 - g_n\| = 1 = \rho(x_1, G)$, but $\{g_n\}$ has no convergent subsequence, by virtue of the relations $\|g_i - g_j\| \geqslant 1$ for $i \neq j$). Let N be the closed set $\Big\{x \in E \,|\, \|x\| \geqslant \frac{1}{2}\Big\}$. Then for the sequence $\{x_n\} \subset E$ defined by

$$x_1 = \{1, 0, 0, \ldots\}, x_n = \Big\{1, \frac{1}{n}, 0, 0, \ldots\Big\} \, (n = 2, 3, \ldots) \tag{3.7}$$

we have $\lim_{n\to\infty} x_n = x_1$, $\mathfrak{P}_G(x_n) \cap N \ni g_n$ $(n = 2, 3, \ldots)$ and $\mathfrak{P}_G(x_1) \cap N = \emptyset$, which shows that $\mathfrak{P}_G$ *is not upper semi-continuous.* The same example invalidates also the conclusions of theorem 3.2 and corollary 3.1 for the above non-approximatively compact set G.

The properties of "continuity" given in theorem 3.2 and corollary 3.1 are not uniform, as shown by the following example

*) See e.g. C. Kuratowski [131].

([231], example 2): Let E be the subset $\{x = \{\xi_1, \xi_2\} \mid 0 < |\xi_1| \leqslant 1, \ \xi_2 = 0\} \cup \{1, 1\} \cup \{-1, 1\}$ of the real Euclidean plane, endowed with the induced metric and let $G = \{1, 1\} \cup \cup \{-1, 1\}$; then G is a compact Čebyšev set, but for $x_n = \left\{\frac{1}{n}, 0\right\} \in E$, $y_n = \left\{-\frac{1}{n}, 0\right\} \in E$ $(n = 1, 2, \ldots)$ we have $\lim_{n\to\infty} \rho(x_n, y_n) = 0$ and $\rho(\mathfrak{L}_G(x_n), \mathfrak{L}_G(y_n)) = 2$ $(n = 1, 2, \ldots)$.

Finally, let us observe that the mapping $\mathfrak{L}_G$ is not lower (K)-semi-continuous, hence not lower semi-continuous, as shown by the following example: Let E be the real Euclidean plane and let $G = \{x = \{\xi_1, \xi_2\} \in E \mid \xi_1^2 + \xi_2^2 = 1, \ \xi_1 \geqslant 0, \ \xi_2 \geqslant 0\}$; then $\mathfrak{L}_G$ is not lower (K)-semi-continuous, since for $x_n = \left\{0, \frac{1}{n}\right\} \in E$ we have $\mathfrak{L}_G(x_n) =$ the point $\{0,1\} \in E$ $(n = 1, 2, \ldots)$ and $\lim_{n\to\infty} x_n = 0$, but $\mathfrak{L}_G(0) = G$.

§ 4. PROPERTIES OF THE MAPPINGS π_G AND OF THE FUNCTIONALS e_G

THEOREM 4.1. *Let E be a metric space and G a set in E. Then*

a) *We have*

$$\rho(\pi_G(x), \pi_G(g)) \leqslant 2\,\rho(x, g) \qquad (x \in E, \ g \in G). \tag{4.1}$$

b) *If G is a Čebyšev set, then π_G is continuous at every point $g \in G$.*

c) *If G is a Čebyšev set and approximatively compact, π_G is continuous on E.*

Proof. a) Let $x \in E$ and $g \in G$ be arbitrary. Then, by $\pi_G(g) = g$, we have

$$\rho(\pi_G(x), \pi_G(g)) = \rho(\pi_G(x), g) \leqslant \rho(\pi_G(x), x) + \rho(x, g) \leqslant 2\,\rho(x, g).$$

b) is an immediate consequence of a).

c) follows immediately from corollary 3.1, which completes the proof.

In the particular case when G is a boundedly compact Čebyšev set, theorem 4.1 c) has been given by V. Klee [113], and in other particular cases by R. Fortet [58], Ky Fan and I. Glicksberg [55]; in the general case it has been given in the paper [231], corollary 2, together with some new consequences in Banach spaces ([231], corollaries 4 and 5). For the problem of continuity of the mappings π_G, see also Appendix I, §2.

Finally, let us mention

THEOREM 4.2. *Let E be a metric space and G a set in E. Then we have*

$$|e_G(x) - e_G(y)| \leqslant \rho(x, y) \qquad (x, y \in E), \tag{4.2}$$

hence e_G is uniformly continuous on E.

The proof coincides with that of formula (6.33) of Chap. I.

REFERENCES

1. Н. И. Ахиезер, *Лекции по теории аппроксимации*, изд. 2-е, Москва, 1965.
2. G. Albinus, *Einige Beiträge zur Approximationstheorie in metrischen Vektorräumen*, Wiss. Z. d. Tech. Univ. Dresden, **15**, 1—4 (1966).
3. — *Ein Beitrag zur approximativen Kompaktheit*, Rev. roum. math. pures et appl., **11**, 793—797 (1966).

3 a. *Über Bestapproximationen in metrischen Vektorräumen*, Dissertation, Dresden, 1966.

4. П. С. Александров и П. С. Урысон, *О компактных топологических пространствах*, Труды матем. ин-та им. В. А. Стеклова, 31 (1950).
5. R. F. Arens and J. L. Kelley, *Characterizations of the space of continuous functions over a compact Hausdorff space*, Trans. Amer. Math. Soc., **62**, 499—508 (1947).
6. N. Aronszajn and K. T. Smith, *Invariant subspaces of completely continuous operators*, Ann. of Math., **60**, 345—350 (1954).
7. G. Ascoli, *Sugli spazi lineari metrici e le loro varietà lineari*, Ann. Mat. Pura Appl., (*4*) **10**, 33—81, 203—232 (1932).
8. S. Banach, *Théorie des opérations linéaires*, Monografie Matematyczne, Warszawa, 1932.
9. U. Barbuti, *Sulla teoria della migliore approssimazione nel senso di Tchebychev. I.*, Rend. Sem. Mat. Univ. Padova, **3**, 82—96 (1960).
10. H. Bauer, *Supermartingale und Choquet Rand*, Arch. der Math., **12**, 210—223 (1961).
11. П. К. Белобров, *К вопросу о чебышевском центре множества*, Изв. высш. уч. зав., матем., *1* (*38*), 3—9 (1964).
12. — *О чебышевском центре множества в банаховом пространстве*, Изв. высш.уч. зав., матем., *2* (*39*), 25—30 (1964).
13. C. Berge, *Théorie générale des jeux à n personnes*, Gauthier-Villars, Paris, 1957.
14. E. Berkson, *Some metrics on the subspaces of a Banach space*, Pacific J. Math., **13**, 7—22 (1963).
15. S. Bernstein, *Leçons sur les propriétés extrémales et la meilleure approximation des fonctions analytiques d'une variable réelle*, Gauthier-Villars, Paris, 1926.
16. — *Sur le problème inverse de la théorie de la meilleure approximation des fonctions continues*, Comptes rendus Acad. Sci. Paris, **206**, 1520—1523 (1938).

16 a. A. S. Besicovitch, *On the set of directions of linear segments of a convex surface*, p. 24—25 in *Convexity* (ed. by V. Klee), Proc. Sympos. Pure Math., 7, Amer. Math. Soc., 1963.

17. G. Birkhoff, *Orthogonality in linear metric spaces*, Duke Math. J., **1**, 169—172 (1935).

18. E. Bishop and K. de Leeuw, *The representation of linear functionals by measures on sets of extreme points*, Ann. Inst. Fourier, **9**, 305–331 (1959).
19. В. Г. Болтянский, С. С. Рышков и Ю. А. Шашкин, *О k-регулярных вложениях и их применении к теории приближения функций*, Успехи матем. наук, **15**, 6 (96), 125—132 (1960).
20. F. F. Bonsall, *Dual extremum problems in the theory of functions*, J. London Math. Soc., **31**, 105–110 (1956).
21. K. Borsuk, *Drei Sätze über die n-dimensionale euklidische Sphäre*, Fund. Math., **20**, 177–191 (1933).
22. N. Bourbaki, *Intégration*, Ch. I–IV, Hermann et Cie, Paris, 1952; Ch. V, Hermann et Cie, Paris, 1956.
23. — *Espaces vectoriels topologiques*, Ch. I–II, Hermann et Cie, Paris, 1953; Ch. III–V, Hermann et Cie, Paris, 1955.
24. — *Théorie des ensembles. Fascicule de résultats* (3-ème éd.), Hermann et Cie, Paris, 1958.
25. — *Topologie générale*, Ch. I–II (3-ème éd.), Hermann et Cie, Paris, 1961.
26. J. Bram, *Chebychev approximation in locally compact spaces*, Proc. Amer. Math. Soc., **9**, 133–136 (1958).
26 a. A. Brøndsted, *Convex sets and Chebyshev sets*, Math. Scand., **17**, 5–16 (1965).
26 b. — *Convex sets and Chebyshev sets.* II, Math. Scand., **18**, 5–15 (1966).
27. A. L. Brown, *Best n-dimensional approximation to sets of functions*, Proc. London Math. Soc., **14**, 577–594 (1964).
28. Ю. А. Брудный и А. Ф. Тиман, *Конструктивные характеристики компактных множеств в пространствах Банаха и ε-энтропия*, Доклады Акад. наук СССР, **126**, 927—930 (1959).
29. R. C. Buck, *Applications of duality in approximation theory*, p. 27–42 in *Approximation of functions* (Ed. by H. L. Garabedian), Elsevier, Amsterdam-London-New York, 1965.
30. L. N. H. Bunt, *Bijdrage tot de theorie der konvekse puntverzamelingen*, Thesis, Univ. of Groningen, publ. Amsterdam, 1934.
31. В. Н. Буров, *Аппроксимация со связями в линейных нормированных пространствах*, Украинский мат. ж., **15**, 3—12, 135—144 (1963).
32. П. Л. Чебышев, *Теория механизмов, известных под названием параллелограммов* (1853), p. 23—51 in *Полное собрание сочинений*, Москва-Ленинград, 1947.
33. E. W. Cheney, *An elementary proof of Jackson's theorem on mean-approximation*, Math. Mag., **38**, 189–191. (1965).
34. E. W. Cheney and A. A. Goldstein, *Proximity maps for convex sets*, Proc. Amer. Math. Soc., **10**, 448–450 (1959).
35. — *Tchebycheff approximation in locally convex spaces*, Bull. Amer. Math. Soc., **68** 449–450 (1962).
36. — *Tchebycheff approximation and related extremal problems*, J. Math. Mech., **14**, 87–98 (1965).
37. G. Choquet, *Sur la meilleure approximation dans les espaces vectoriels normés*, Rev. math. pures et appl., **8**, 541–542 (1963).
38. J. A. Clarkson, *Uniformly convex spaces*, Trans. Amer. Math. Soc., **40**, 396–414 (1936).
39. L. Collatz, *Approximation von Funktionen bei einer und bei mehreren Veränderlichen*, Zeitschr. Angew. Math. Mech., **36**, 198–211 (1956).
40. D. F. Cudia, *Rotundity*, p. 73–97 in *Convexity* (ed. by V. Klee), Proc. Sympos. Pure Math., 7, Amer. Math. Soc., 1963.
41. P. C. Curtis, *n-Parameter families and best approximation*, Pacific J. Math., **9**, 1013–1027 (1959).
42. M. M. Day, *Reflexive Banach spaces not isomorphic to uniformly convex spaces*, Bull. Amer. Math. Soc., **47**, 313–317 (1941).
43. — *Normed linear spaces*, Springer Verlag, Berlin-Göttingen-Heidelberg, 1958.

44. J. Dieudonné, *La dualité dans les espaces vectoriels topologiques*, Ann. Sci. Ecole Norm. Sup., **59**, 107–139 (1942).
45. J. Dieudonné et L. Schwartz, *La dualité dans les espaces* ($\mathfrak{F}$) *et* ($\mathfrak{L}\mathfrak{F}$), Ann. Inst. Fourier, **1**, 61–101 (1950).
46. N. Dinculeanu, *Teoria măsurii şi funcţii reale*, Ed. didactică şi pedagogică, Bucureşti, 1964.
47. — *Integrarea pe spaţii local compacte*, Ed. Academiei R.P.R., Bucureşti, 1965.
48. — *Vector measures*, Deutscher Verlag der Wissenschaften, Berlin, 1966.
49. N. Dunford and J. Schwartz, *Linear operators. Part I: General theory*, Interscience Publ., New York, 1958.
50. Н. В. Ефимов и С. Б. Стечкин, *Некоторые свойства чебышевских множеств*, Доклады Акад. наук СССР, **118**, 17—19 (1958).
51. — *Чебышевские множества в банаховых пространствах*, Доклады Акад. наук СССР, **121**, 582–585 (1958).
52. — *Опорные свойства множеств в банаховых пространствах*, Доклады Акад. наук СССР, **127**, 254—257 (1959).
53. — *Аппроксимативная компактность и чебышевские множества*, Доклады Акад. наук СССР, **140**, 522—524 (1961).
54. S. Eilenberg, *Banach space methods in topology*, Ann. of Math., **43**, 568–579 (1942).
54 a. P. Erdös, *On the Hausdorff dimension of some sets in euclidean space*, Bull Amer. Math. Soc., **52**, 107–109 (1946).
54 b. G. Ewald, *Über die Schattengrenzen konvexer Körper*, Abhandl. Math. Sem. Univ. Hamburg, **27**, 167–170 (1964).
55. Ky Fan and I. Glicksberg, *Some geometric properties of the spheres in a normed linear space*, Duke Math. J., **52**, 553–568 (1958).
56. L. Fejér, *Über die Lage der Nullstellen von Polynomen, die aus Minimumforderungen gewisser Art entspringen*, Math. Ann., **85**, 41–48 (1922).
56 a. А. М. Фломин, *Некоторые вопросы теории приближений в произвольных топологических локально-выпуклых пространствах*, Автореферат диссертации канд. физ. -мат. наук, Москва, 1966.
57. S. R. Foguel, *On a theorem by A. E. Taylor*, Proc. Amer. Math. Soc., **9**, 325 (1958).
58. R. Fortet, *Remarques sur les espaces uniformément convexes*, Bull. Soc. Math France, **69**, 23–46 (1941).
59. M. Fréchet, *Les espaces abstraits*, Gauthier-Villars, Paris, 1928.
60. А. Л. Гаркави, *О размерности многогранников наилучшего приближения для дифференцируемых функций*, Изв. Акад. наук СССР, сер. матем., **23**, 93—114 (1959).
61. — *О существовании наилучшей сети и наилучшего поперечника множества в банаховом пространстве*, Успехи матем. наук, **15**, *2* (*92*), 210—211 (1960).
62. — *Общие теоремы об очистке*, Rev. math. pures et appl., **6**, 293–303 (1961).
63. — *О чебышевском центре множества в нормированном пространстве*, р. 328—331 in *Исследования по современным проблемам конструктивной теории функций* (ред. В. И. Смирнова), Москва, 1961.
64. — *Теоремы двойственности для приближений посредством элементов выпуклых множеств*. Успехи матем. наук, **16**, *4* (*100*), 141—145 (1961).
65. — *О наилучшей сети и наилучшем сечении множеств в нормированном пространстве*, Изв. Акад. наук СССР, сер. матем., **26**, 87—106 (1962).
66. — *О наилучшем приближении элементами бесконечномерных подпространств одного класса*, Матем. сб., *62* (*104*), 104—120 (1963).

67. — *О чебышевских и почтичебышевских подпространствах*, Доклады Акад. наук СССР, **149**, 1250—1252 (1963).
68. — *О единственности решения L-проблемы моментов*, Изв. Акад. наук СССР, сер. матем., **28**, 553—570 (1964).
69. — *О чебышевских и почтичебышевских подпространствах*, Изв. Акад. наук СССР, сер. матем., **28**, 799—818 (1964).
70. — *О чебышевском центре и выпуклой оболочке множества*, Успехи матем. наук, **19**, *6(120)*, 139—145 (1964).
71. — *О критерии элемента наилучшего приближения*, Сибирск. мат. ж., **5**, 472—476 (1964).
72. — *Аппроксимативные свойства подпространств конечного дефекта в пространстве непрерывных функций*, Доклады Акад. наук СССР, **155**, 513—516 (1964).
73. — *Чебышевские подпространства конечного дефекта в пространстве суммируемых функций*, Уч. зап. Орехово-Зуевск. пед. ин-та, **22**, 5—11 (1964).
74. — *Почтичебышевские системы непрерывных функций*, Изв. высш. уч. завед., матем., 2 (*45*), 36—44 (1965).
75. И. Ц. Гохберг и М. Г. Крейн, *Основные положения о дефектных числах и индексах линейных операторов*, Успехи матем. наук, **12**, *2* (*74*), 43—118 (1957).
76. И. Ц. Гохберг, и А. С. Маркус, *Две теоремы о растворе подпространств банахова пространства*, Успехи матем. наук, **14**, *5* (*89*), 135—140 (1959).
77. M. Gowurin, *Über die Stieltjessche Integration abstrakter Funktionen*, Fund. Math., **27**, 255—268 (1958).
78. A. Grothendieck, *Critères de compacité dans les espaces vectoriels généraux*, Amer. J. Math., **74**, 168—186 (1952).
79. — *Espaces vectoriels topologiques*, 2-ème éd., Soc. Mat. de São Paulo, São Paulo, 1958.
80. A. Haar, *Die Minkowskische Geometrie und die Annäherung an stetige Funktionen*, Math. Ann., **78**, 294—311 (1918).
81. P. R. Halmos, *The range of a vector measure*, Bull. Amer. Math. Soc., **54**, 416—421 (1948).
82. — *Measure theory*, D. Van Nostrand, New York, 1950.
83. I. Halperin, *The product of projection operators*, Acta Sci. Math., **23**, 96—99 (1962).
84. F. Hausdorff, *Grundzüge der Mengenlehre*, Verlag von Veit, Leipzig, 1914.
85. С. Я. Хавинсон, *О некоторых экстремальных задачах теории аналитических функций*, Уч. зап. Моск. гос. ун-та, **4**, 133—143 (1951).
86. — *О размерности многогранника наилучших приближений в метрике L_1*, Сб. тр. каф. высш. мат. и теор. мех., Моск. инж.-строит. инст., **19**, 18—29 (1957).
87. — *О единственности функции наилучшего приближения в метрике пространства L_1*, Изв. Акад. наук СССР, сер. матем., **22**, 243—270 (1958).
88. R. A. Hirschfeld, *On best approximations in normed vector spaces*, Nieuw Arch. voor Wisk., **6**, 41—51 (1958).
89. — *On best approximations in normed vector spaces*. II, Nieuw Arch. voor Wisk., **6**, 99—107 (1958).
90. — p. 184 in *On approximation theory* (Ed. by P. L. Butzer and J. Korevaar) Birkhauser Verlag, Basel und Stuttgart, 1964.
91. C. R. Hobby and J. R. Rice, *A moment problem in L_1 approximation*, Proc. Amer. Math. Soc., **16**, 665—670 (1965).
92. И. И. Ибрагимов, *Экстремальные свойства целых функций конечной степени*, Изд. Акад. наук Аз. ССР, Баку, 1962.

93. K. Iseki, *An approximation problem in quasi-normed spaces*, Proc. Japan Acad., **35**, 465—466 (1959).
94. В. К. Иванов, *Задача о минимаксе системы линейных функций*, Матем. сб., *28* (*70*), 685—706 (1951).
95. — *О равномерном приближении непрерывных функций*, Матем. сб., *30* (*72*), 543—558 (1952).
96. — *Письмо в редакцию*, Матем. сб., *33* (*75*), 676 (1953).
97. D. Jackson, *A general class of problems in approximation*, Amer. J. Math., **46**, 215—234 (1924).
98. R. C. James, *Orthogonality in normed linear spaces*, Duke Math. J., **12**, 291—302 (1945).
99. — *Orthogonality and linear functionals in normed linear spaces*, Trans. Amer. Math. Soc., **61**, 265—292 (1947).
100. — *Inner products in normed linear spaces*, Bull. Amer. Math. Soc., **53**, 559—566 (1947).
101. — *Characterizations of reflexivity*, Studia Math., **23**, 205—216 (1964).
102. М. И. Кадец, *О гомеоморфизме некоторых пространств Банаха*, Доклады Акад. наук СССР, **92**, 465—468 (1953).
103. — *Про зв'язок між слабою та сильною збіжністю*, Доповіді Акад. наук УРСР, **9**, 949—952 (1959).
104. — *О системах Лозинского-Харшиладзе*, Успехи матем. наук, **18**, *5* (*113*), 167—169 (1963).
105. — *Топологическая эквивалентность всех сепарабельных пространств Банаха*, Доклады Акад. наук СССР, **167**, 23—25 (1966).
106. Л. В. Канторович и Г. П. Акилов, *Функциональный анализ в нормированных пространствах*, Москва, 1959.
107. P. Kirchberger, *Über Tchebyschefsche Annäherungsmethoden*, Math. Ann., **57**, 509—540 (1903).
108. V. Klee, *The support property of a convex set in a linear normed space*, Duke Math. J., **15**, 767—772 (1948).
109. — *A characterization of convex sets*, Amer. Math. Monthly, **56**, 247—249 (1949).
109 a. — *Some characterizations of compactness*, Amer. Math. Monthly, **58**, 389—393 (1951).
110. — *Convex bodies and periodic homeomorphisms in Hilbert space*, Trans. Amer. Math. Soc., **74**, 10—43 (1953).
110 a. — *Research problem No. 5*, Bull. Amer. Math. Soc., **63**, 419 (1957).
111. — *Extremal structure of convex sets*. II, Math. Zeitschr., **69**, 90—104 (1958).
112. — *Mappings into normed linear spaces*, Fund. Math., **49**, 25—34 (1960).
113. — *Convexity of Chebyshev sets*, Math. Ann., **142**, 292—304 (1961).
114. — *On a problem of Hirschfeld*, Nieuw Arch. voor Wisk., **11**, 22—26 (1963).
115. — *Remarks on nearest points in normed linear spaces*, p. 168—176 in Proceedings of a colloquium on Convexity (Copenhagen, 1965), Univ. of Copenhagen, 1966.
116. V. Klee and R. G. Long, *On a method of mapping due to Kadec and Bernstein*, Arch. der Math., **8**, 280—285 (1957).
117. A. N. Kolmogorov, *Über die beste Annäherung von Funktionen einer gegebenen Funktionenklasse*, Ann. of Math., **37**, 107—110 (1936).
118. — *Замечание по поводу многочленов П. Л. Чебышева, наименее уклоняющихся от заданной функции*, Успехи матем. наук, **3**, *1* (*23*), 216—221 (1948).
119. — *О некоторых асимптотических характеристиках вполне ограниченных метрических пространств*, Доклады Акад. наук СССР, **108**, 385—388 (1956).
120. А. Н. Колмогоров и В. М. Тихомиров, *ε-энтропия и ε-емкость множеств в функциональных пространствах*, Успехи матем. наук, **14**, *2* (*86*), 3—86 (1959).
121. П. П. Коровкин, *Линейные операторы и теория приближений*, Москва, 1959.

122. G. Köthe, *Topologische lineare Räume*. I, Springer Verlag, Berlin-Göttingen-Heidelberg, 1960.
123. М. А. Красносельский, *Топологические методы в теории нелинейных уравнений*, Москва, 1956.
124. M. G. Krein, *Sur quelques questions de la géometrie des ensembles convexes situés dans un espace linéaire normé et complet*, Доклады Акад. наук СССР. **14**, 5 — 7 (1937).
125. — *L-проблема в абстрактном линейном нормированном пространстве*, p. 171—199 in Н. И. Ахиезер и М. Г. Крейн, *О некоторых вопросах теории моментов*, ДНТВУ, Харьков, 1938.
126. М. Г. Крейн, М. А. Красносельский и Д. П. Мильман, *О дефектных числах линейных операторов в банаховом пространстве и о некоторых геометрических вопросах*, Сб. трудов ин-та матем. Акад. наук УССР, **11**, 97—112 (1948).
127. М. Г. Крейн, Д. П. Мильман и М. А. Рутман, *Об одном свойстве базиса в пространстве Банаха*, Зап. мат. об-ва (Харьков), **16**, 106—108 (1940).
128. B. R. Kripke and T. J. Rivlin, *Approximation in the metric of* $L^1(X, \mu)$, IBM Research Paper RC—524, 1961.
129. — *Approximation in the metric of* $L^1(X, \mu)$, Trans. Amer. Math. Soc., **115**, 101—122 (1965).
130. B. R. Kripke and R. T. Rockafellar, *A necessary condition for the existence of best approximations*, J. Math. Mech., **13**, 1037—1038 (1964).
131. C. Kuratowski, *Les fonctions semi-continues dans l'espace des ensembles fermés*, Fund. Math., **18**, 148—160 (1932).
132. — *Une condition métrique pour la rétraction des ensembles*, Comptes rendus Soc. Sci. lettres Varsovie, Cl. III, **28**, 156—158 (1936).
133. — *Topologie*, vol. I (4-ème éd.), Monografie Matematyczne, Warszawa, 1958; vol. II (3-ème éd.); Monografie Matematyczne, Warszawa, 1961.
134. А. Ю. Левин, *К задаче о существовании ортогонального элемента к подпространству*. Тр. семин. функц. анал., Воронеж, **6**, 91—92 (1958).
135. А. А. Ляпунов, *О вполне аддитивных векторных функциях*, Изв. Акад. наук СССР, **4**, 465—478 (1940).
136. J. Lindenstrauss, *On nonlinear projections in Banach spaces*, Michigan Math. J., **11**, 263—287 (1964).
137. R. G. Long, *A T-system which is not a Bernstein system*, Proc. Amer. Math. Soc., **8**, 925—927 (1957).
138. E. R. Lorch, *On a calculus of operators in reflexive vector spaces*, Trans. Amer. Math. Soc., **45**, 217—234 (1939).
139. G. G. Lorentz, *Lower bounds for the degree of approximation*, Trans. Amer. Math. Soc. **97**, 25—34 (1960).
140. — *Russian literature on approximation in 1958—1964*, p. 191—215 in *Approximation of functions* (Ed. by H. L. Garabedian), Elsevier, Amsterdam-London-New York, 1965.
141. A. R. Lovaglia, *Locally uniformly convex Banach spaces*, Trans. Amer. Math. Soc., **78**, 225—238 (1955).
142. J. C. Mairhuber, *On Haar's theorem concerning Chebychev approximation problems having unique solution*, Proc. Amer. Math. Soc., **7**, 609—615 (1956).
143. Ю. И. Маковоз, *О чебышевских подпространствах пространства* C, Успехи матем. наук, **19**, *4* (*118*), 185—188.
144. A. I. Markouchevitch, *Sur la meilleure approximation*, Доклады Акад. наук СССР, 44, 262—264 (1944).
145. S. Mazur, *Über konvexe Mengen in linearen normierten Räumen*, Studia Math., **4**, 70—84 (1933).
145 a. T. J. Mc Minn, *On the line segments of a convex surface in* E_3, Pacific J. Math., **10**, 943—946 (1960).
146. E. J. Mc Shane, *Linear functionals on certain Banach spaces*, Proc. Amer. Math. Soc., **1**, 402—408 (1950).

147. D. E. MENCHOFF, *Sur les sommes partielles des séries de Fourier des fonctions continues*, Матем. сб., **15** (*57*), 385—432 (1944).
148. K. MENGER, *Untersuchungen über allgemeine Metrik*, Math. Ann., **100**, 75—163 (1928).
149. E. MICHAEL, *Topologies on spaces of subsets*, Trans. Amer. Math. Soc., **71**, 152—182 (1951).
150. Д. П. МИЛЬМАН, *Достижимые точки функционального кмпакта*, Доклады Акад. наук СССР, 59, 1045—1048 (1948).
151. H. MINKOWSKI, *Theorie der konvexen Körper, insbesondere Begründung ihres Oberflächenbegriffs*, p. 131—229 in *Gesammelte Abhandlungen*, vol. II, Teubner, Berlin, 1911.
152. R. M. MORONEY, *The Haar problem in* L_1, Proc. Amer. Math. Soc., **12**, 793—795 (1961).
153. М. И. МОРОЗОВ, *О некоторых вопросах равномерного приближения непрерывных функций посредством функций интерполяционных классов*, Изв. Акад. наук СССР, сер. матем., **16**, 75—100 (1952).
154. T. S. MOTZKIN, *Sur quelques propriétés caractéristiques des ensembles convexes*, Rend. Ac. Lincei, Cl. VI, **21**, 562—567 (1935).
155. — *Approximation by curves of a unisolvent family*, Bull. Amer. Math. Soc., **55**, 789—793 (1949).
156. T. S. MOTZKIN, E. G. STRAUS and F. A. VALENTINE, *The number of farthest points*, Pacific J. Math., **3**, 221—232 (1953).
157. М. А. НАЙМАРК, *Нормированные кольца*, Москва, 1956.
158. И. П. НАТАНСОН, *Конструктивная теория функций*, Москва-Ленинград, 1949.
159. J. VON NEUMANN, *On rings of operators. Reduction theory*, Ann. of Math., **50**, 401—485 (1949).
160. — *Functional operators. Vol. II: The geometry of orthogonal spaces*, Annals of Math. Studies, no. 21, Princeton Univ. Press, Princeton, 1950.
161. M. NICOLESCU, *Sur la meilleure approximation d'une fonction donnée par les fonctions d'une famille donnée*, Bul. Fac. şti. Cernăuţi, **12**, 120—128 (1938).
162. С. М. НИКОЛЬСКИЙ, *Приближение функций тригонометрическими полиномами в среднем*. Изв. Акад. наук СССР, сер. матем., **10**, 295—332 (1946).
162 a. — *Приближение многочленами функций действительного переменного*, p. 288—318 in *Математика в СССР за тридцать лет, 1917—1947*, *Москва-Ленинград*, 1948.
163. — *Einige Fragen der Approximation von Funktionen durch Polynome*, Proc. Int. Congr. Math., Vol. I, Amsterdam, 1954.
164. В. Н. НИКОЛЬСКИЙ, *Наилучшее приближение и базис в пространстве Фреше*, Доклады Акад. наук СССР, **59**, 639—642 (1948).
165. — *Некоторые вопросы наилучшего приближения в функциональном пространстве*, Уч. зап. Калининск. гос. пед. ин-та, **16**, 119—160 (1954).
166. — *Операторные свойства полиномов наилучшего приближения*, Успехи матем. наук, **12**, *3* (*75*), 353—358 (1957).
167. — *О свойствах операторов наилучшего приближения*, Уч. зап. Калининск. гос. пед. ин-та, **26**, 143—146 (1958).
168. — *Распространение теоремы А. Н. Колмогорова на банаховы пространства*, p. 335—337 in *Исследования по современным проблемам конструктивной теории функций* (ред. В. И. СМИРНОВА), Москва, 1961.
169. — *Наилучшее приближение элементами выпуклых множеств в линейных нормированных пространствах*, Уч. зап. Калининск. гос. пед. ин-та, 29, 85—119 (1963).
170. — *О некоторых свойствах рефлексивных пространств*, Уч. зап. Калининск. гос. пед. ин-та, **29**, 121—125 (1963).
171. — *Некоторые замечания о пространствах, обладающих* (*B*)-*свойством*, Уч. Зап. Калининск. гос. пед. ин-та, **39**, 48—52 (1964).

171 a. W. Nitka, *Une généralisation du théorème de Kuratowski sur la caractérisation métrique de la rétraction.* Colloq. Math. **8**, 35—37 (1961).

172. D. del Pasqua, *Su una nozione di varietà lineari disgiunte di uno spazio di Banach,* Rend. Math. Pura Appl., **13**, 406—422 (1955).

173. S. Paszkowski, *Sur l'approximation uniforme avec des nœuds,* Ann. Pol. Math., **2**, 118—135 (1955).

174. R. R. Phelps, *Convex sets and nearest points,* Proc. Amer. Math. Soc., **8**, 790—797 (1957).

175. — *Convex sets and nearest points.* II, Proc. Amer. Math. Soc., **9**, 867—873 (1958).

176. — *Uniqueness of Hahn-Banach extensions and unique best approximation,* Trans. Amer. Math. Soc., **95**, 238—255 (1960).

177. — *Čebyšev subspaces of finite codimension in $C(X)$,* Pacific J. Math., **13**, 647—655 (1963).

177 a. — *Čebyšev subspaces of finite dimension in L_1,* Proc. Amer. Math. Soc., **17**, 646—652 (1966).

178. D. Pompeiu, *Sur la continuité des fonctions de variables complexes,* Ann. de Toulouse (2) **7**, 264—315 (1905).

179. Л. С. Понтрягин, *Основы комбинаторной топологии,* Москва-Ленинград, 1947.

180. V. Pták, *A remark on approximation of continuous functions,* Czechoslovak Math. J., **8**, 251—256 (1958).

181. — *On approximation of continuous functions in the metric* $\int_a^b |x(t)|\,dt$, Czechoslovak Math. J., **8**, 267—273 (1958).

182. — *Supplement to the article "On approximation of continuous functions in the metric* $\int_a^b |x(t)|\,dt$*",* Czechoslovak Math. J., **8**, 464 (1958).

183. H. Rademacher and I. J. Schoenberg, *Helly's theorems on convex domains and Tchebycheff's approximation problem,* Canad. J. Math., **2**, 245—256 (1950).

184. Е. Я. Ремез, *Про методи найкращого, в розумінні Чебишева, наближеного представлення функцій,* Вид.-во Укр. Акад. наук, Київ, 1935.

185. — *О чебышевских приближениях в комплексной области,* Доклады Акад. наук СССР, **77**, 965—968 (1951).

186. — *Некоторые вопросы чебышевского приближения в комплексной области.* Укр. матем. ж., **5**, 3—48 (1953).

187. — *Общие вычислительные методы чебышевского приближения,* Изд. Акад. наук Укр. ССР, Киев, 1957.

188. J. R. Rice, *Best approximations and interpolating functions,* Trans. Amer. Math. Soc. **101**, 477—498 (1961).

189. — *Tchebycheff approximation in several variables,* Trans. Amer. Math. Soc., **109**, 444—466 (1963).

190. — *Approximation with convex constraints,* J. Soc. Indust. Appl. Math., **11**, 15—32 (1963).

191. — *Nonlinear approximation,* p. 111—133 in *Approximation of functions* (Ed. by H. L. Garabedian), Elsevier, Amsterdam-London-New York, 1965.

192. — *The approximation of functions. Vol. I: Linear theory.* Addison-Wesley, Reading, Mass., 1964.

193. T. J. Rivlin and H. S. Shapiro, *Some uniqueness problems in approximation theory,* Comm. Pure Appl. Math., **13**, 35—47 (1960).

194. — *A unified approach to certain problems of approximation and minimization,* J. Soc. Indust. Appl. Math. **9**, 670—699 (1961).

195. B. D. Roberts, *On the geometry of abstract vector spaces,* Tôhoku Math. J., **39**, 42—59 (1934).

196. W. W. Rogosinski, *Extremum problems for polynomials and trigonometric polynomials*, J. Lond. Math. Soc., **29**, 259—275 (1954).
197. — *Continuous linear functionals on subspaces of $\mathfrak{L}^p$ and $\mathfrak{C}$*, Proc. Lond. Math. Soc., **6**, *22*, 175—190 (1956).
198. З. С. Романова, *О размерности многогранников наилучших приближений в пространстве непрерывных функций*. Литовский матем. сб., **2**, 181—191 (1963).
199. Г. Ш. Рубинштейн, *Об одном методе исследования выпуклых множеств*, Доклады Акад. наук СССР, **103**, 451—454 (1955).
200. — *Об одной экстремальной задаче в линейном нормированном пространстве*, Сибирск. матем. ж., **6**, 711—714 (1965).
201. W. Rudin and K. T. Smith, *Linearity of best approximation: A characterization of ellipsoids*, Proc. Nederl. Akad. Wet., Ser. A., **64**, 97—103 (1961).
202. A. F. Ruston, *Conjugate Banach spaces*, Proc. Cambridge Phil. Soc., **45**, 576—580 (1957).
203. — *Auerbach's theorem and tensor products of Banach spaces*, Proc. Cambr. Phil. Soc., **58**, 476—480 (1964).
204. I. J. Schoenberg, *On the question of unicity in the theory of best approximation*, Ann. of the New York Acad. Sci., **86**, 682—692 (1960).
205. I. J. Schoenberg and C. T. Yang, *On the unicity of solutions of problems of best approximation*, Ann. Mat. Pura Appl., **54**, 1—12 (1961).
206. J. Schwartz, *A note on the space L_p^**, Proc. Amer. Math. Soc., **2**, 270—275(1951).
207. H. S. Shapiro, *Applications of normed linear spaces to function theoretic extremal problems*, in *Lectures on functions of a complex variable* (ed. by W. Kaplan), Univ. of Michigan Press, Ann Arbor, 1955.
208. K. Sieklucki, *Topological properties of sets admitting the Tschebycheff systems*, Bull. Ac. Pol. Sci., Sér. Sci. math. astr. phys., **6**, 603—606 (1958).
209. I. Singer, *Sur l'extension des fonctionnelles linéaires*, Rev. math. pures et appl., **1**, *2*, 99—106 (1956).
210. — *Asupra reprezentării concrete a spaţiilor Banach*, Bul. şti. Acad. R.P.R., secţ. şti. mat. fiz., **8**, 31—37 (1956).
211. — *Proprietăţi ale suprafeţei sferei unitate şi aplicaţii la rezolvarea problemei unicităţii polinomului de cea mai bună aproximaţie în spaţii Banach oarecare*, Studii şi cercet. mat., **7**, 95—145 (1956).
212. — *Caractérisation des éléments de meilleure approximation dans un espace de Banach quelconque*, Acta. Sci. Math., **17**, 181—189 (1956).
213. — *Asupra L-problemei teoriei momentelor în spaţii Banach*, Bul. şti. Acad. R.P.R., secţ. şti. mat. fiz., **9**, 19—28 (1957).
214. — *Unghiuri abstracte şi funcţii trigonometrice în spaţii Banach*, Bul. şti. Acad. R.P.R., secţ. şti. mat. fiz., **9**, 29—42 (1957).
215. — *Asupra unicităţii elementului de cea mai bună aproximaţie în spaţii Banach oarecare*, Studii şi cercet. mat., **8**, 234—244 (1957).
216. — *Линейные функционалы на пространстве непрерывных отображений бикомпактного хаусдорфого пространства в пространство Банаха*, Rev. Math. pures et appl., **3**, 201—215 (1957).
217. — *Sur la meilleure approximation des fonctions abstraites continues à valeurs dans un espace de Banach*, Rev. math. pures et appl., **2**, 245—262 (1957).
218. — *Sur quelques théorèmes de W. W. Rogosinski et S. I. Zoukhovitzky*, Rev. math. pures et appl., **3**, 117—130 (1958).
219. — *Les points extrémaux de la boule unité du dual d'un produit tensoriel normé inductif d'espaces de Banach*, Bull. Sci. Math., **82**, 73—80 (1958).
220. — *Quelques applications d'un dual du théorème de Hahn-Banach*, Comptes rendus Acad. Sci. (Paris), **247**, 846—849 (1958).
221. — *Sur un dual du théorème de Hahn-Banach et sur un théorème de Banach*, Rendic. dei Lincei, **25**, 443—446 (1958).
222. — *La meilleure approximation interpolative dans les espaces de Banach*, Rev. math. pures et appl., **4**, 95—113 (1959).
223. — *Sur un théorème de Ch. J. de la Vallée Poussin*, Rev. math. pures et appl., **4**, 317—324 (1959).

224. — *Sur les applications linéaires intégrales des espaces de fonctions continues.* I., Rev. math. pures et appl., **4**, 391—401 (1959).
225. — *On the set of the best approximations of an element in a normed linear space,* Rev. math. pures et appl., **5**, 383—402 (1960).
226. — *On best approximation of continuous functions,* Math. Ann., **140**, 165—168 (1960).
227. — *On a theorem of V. Pták concerning best approximation of continuous functions in the metric* $\int_a^b |x(t)|\,dt$, Czechoslovak Math. J., **10** (85), 425—431 (1960).
228. — *Some remarks on best approximation in normed linear spaces.* I, Rev. math. pures et appl., **6**, 357—362 (1961).
229. — *On best approximation of continuous functions.* II., Rev. math. pures et appl., **6**, 507—511 (1961).
230. — *Choquet spaces and best approximation,* Math. Ann., **148**, 330—340 (1962).
231. — *Some remarks on approximative compactness,* Rev. roum. math. pures et appl., **9**, 167—177 (1964).
232. — *On the extension of continuous linear functionals and best approximation in normed linear spaces.* Math. Ann., **159**, 344—355 (1965).
233. — *Some remarks on best approximation in normed linear spaces.* II, Rev. roum. math. pures et appl., **11**, 799—807 (1966).
234. — *Extremal points, Choquet boundary and best approximation,* Rev. roum. math. pures et appl., **11**, 1173—1185 (1966).
235. В. И. Смирнов и Н. А. Лебедев, *Конструктивная теория функций комплексного переменного*, Москва-Ленинград, 1964.
236. С. Б. Стечкин, *О приближении абстрактных функций*, Rev. math. pures et appl., **1**, *3*, 79—84 (1956).
237. — *Аппроксимативные свойства множеств в линейных нормированных пространствах*, Rev. math. pures et appl., **8**, 5—18 (1963).
238. W. J. Stiles, *Closest-point maps and their products,* Nieuw Arch. voor Wisk., **13**, 19—29 (1965).
238 a. — *A solution to Hirschfeld's problem,* Nieuw Arch. voor Wisk., **13**, 116—119 (1965).
238 b. — *Closest-point maps and their products.* II, Nieuw Arch. voor Wisk., **13**, 212—225 (1965).
239. S. Straszewicz, *Über exponierte Punkte abgeschlossener Punktmengen,* Fund. Math., **24**, 139—143 (1935).
240. Ю. А. Шашкин, *Топологические свойства множеств, связанные с теорией приближения функций*, Изв. Акад. наук СССР, сер. матем., **29**, 1085—1094 (1965).
241. Л. Г. Шнирелман, *О равномерных приближениях*, Изв. Акад. наук СССР, сер. матем., **2**, 53—60 (1938).
242. K. Tatarkiewicz, *Quelques remarques sur la convexité des sphères,* Ann. Univ. Mariae Curie-Sklodowska, **6**, 19—30 (1952).
243. — *Une théorie généralisée de la meilleure approximation,* Ann. Univ. Mariae Curie-Sklodowska, **6**, 31—46 (1952).
244. A. E. Taylor, *The extension of linear functionals,* Duke Math. J., **5**, 538—547 (1939).
245. — *A geometric theorem and its applications to biorthogonal systems,* Bull. Amer. Math. Soc., **53**, 614—616 (1947).
246. С. А. Теляковский, *О нормах линейных полиноминальных операторов*, Матем. сб., **68** (*110*), 561—569 (1965).
247. В. М. Тихомиров, *Поперечники множеств в функциональных пространствах и теория наилучших приближений*, Успехи матем. наук, **15**, *3*(*93*), 81—120 (1960).
248. — *Одно замечание об n-мерных поперечниках множеств в банаховых пространствах*, Успехи матем. наук, **20**, *1* (*121*), 227—230 (1965).
249. — *Некоторые вопросы теории приближений*, Доклады Акад. наук СССР, **160**, 774—777 (1965).

250. А. Ф. ТИМАН, *Теория приближения функций действительного переменного*, Москва, 1960.
251. И. С. ТЮРЕМСКИХ, *(B)-свойство гильбертовых пространств*, Уч. зап. Калининск. гос. пед. ин-та, **39**, 53—64 (1964).
252. — *Чебышевские свойства последовательностей подпространств пространства Банаха*, Уч. зап. Калининск. гос. пед. ин-та, **39**, 73—75 (1964).
253. L. TONELLI, *I polinomi d'approssimazione di Tchebychev*, Ann. di Mat., **15**, 47—119 (1908).
254. L. TORNHEIM, *On n-parameter families of functions and associated convex functions*, Trans. Amer. Math. Soc., **69**, 457—467 (1950).
255. CH. J. de la VALLÉE POUSIN, *Leçons sur l'approximation des fonctions d'une variable réelle*, Paris, 1919.
256. В. С. ВИДЕНСКИЙ, *О равномерном приближении в комплексной плоскости*, Успехи матем, наук, **11**, 5 (71), 169—175 (1956).
257. — *Качественные вопросы теории наилучшего приближения функций комплексного переменного*, p. 258—272 in *Исследования по современным проблемам теории функций комплексного переменного* (ред. А. И. МАРКУШЕВИЧ), Москва, 1960.
258. А. Г. ВИТУШКИН, *Оценка сложности задачи табулирования*, Москва, 1959.
259. Л. П. ВЛАСОВ, *О чебышевских множествах в банаховых пространствах*, Доклады Акад. наук СССР, **141**, 19—20 (1961).
259 a. — *Аппроксимативно выпуклые множества в банаховых пространствах*, Доклады Акад. наук СССР, 163, 18—21 (1965).
260. J. L. WALSH and T. S. MOTZKIN, *Polynomials of best approximation on an interval*, Proc. Nat. Acad. Sci. of the U.S.A., **45**, 1523—1528 (1959).
261. N. WIENER, *On the factorization of matrices*, Comment. Math. Helv., **29**, 97—111 (1955).
262. N. WIENER and P. MASANI, *The prediction theory of multivariate stochastic processes*. II, Acta Math., **99**, 93—137 (1958).
263. D. E. WULBERT, *Continuity of metric projections—approximation theory in a normed linear lattice*, Thesis, Univ. of Texas, 1966.
264. J. W. YOUNG, *General theory of approximation by functions involving a given number of arbitrary parameters*, Trans. Amer. Math. Soc., **8**, 331—344 (1907).
265. С. И. ЗУХОВИЦКИЙ, *Про точку, що найменш відхиляється (в розумінні П. Л. Чебишова) від даної системи точок*, Наукови зап. Луцьк. пед. ін-ту, сер. физ. матем., **1**, 3—24 (1953).
266. — *Некоторые теоремы теории чебышевских приближений в пространстве Гильберта*, Матем. сб., *37* (*79*), 3—20 (1955).
267. — *О приближении действительных функций в смысле П. Л. Чебышева*, Успехи матем. наук, **11**, *2* (*68*), 125—159 (1956).
268. — *О минимальных расширениях линейных функционалов в пространстве непрерывных функций*, Изв. Акад. наук СССР, сер. матем., **21**, 409—422 (1957).
269. С. И. ЗУХОВИЦКИЙ и Г. И. ЭСКИН, *О приближении абстрактных непрерывных функций неограниченными оператор-функциями*, Доклады Акад. наук СССР, **116**, 731—734 (1957).
270. — *Некоторые теоремы о наилучшем приближении неограниченными оператор-функциями*, Изв. Акад. наук СССР, сер. матем., **24**, 93—102 (1960).
271. С. И. ЗУХОВИЦКИЙ и М. Г. КРЕЙН, *Замечание об одном возможном обобщении теорем А. Хаара и А. Н. Колмогорова*, Успехи матем. наук, **5**, *1* (*35*), 217—229 (1950).
272. С. И. ЗУХОВИЦКИЙ и С. Б. СТЕЧКИН, *О приближении абстрактных функций со значениями в гильбертовом пространстве*, Доклады Акад. наук СССР, **106**, 385—388 (1956).
273. — *О приближении абстрактных функций со значениями в банаховом пространстве*, Доклады Акад. наук СССР, **106**, 773—776 (1956).

Notation index

a.e. = almost everywhere, p. 30

$A_q(E)$, p. 76

(B), p. 151

(B_f), p. 153

$C(Q)$, p. 29

$C_E(Q)$, p. 191

$C_R(Q)$, p. 33

$C^1(Q, \nu)$, p. 55

$C^1_R(Q, \nu)$, p. 55

$\mathfrak{C}_f$, p. 128

$d_n(A) = d_n(A, E)$, p. 268

$d^n(A)$, p. 357

E^*, p. 18

$E_{(r)}$, p. 26

e_G, p. 139

$(\mathfrak{E}_*)$, p. 94

$\mathfrak{E}(A)$, p. 58

Fr_m, p. 379

$G^\perp$, p. 20

$G_{n,m}$, p. 372

$\widehat{G_1, G}$, p. 161

Int_l, p. 128

Int_m, p. 379

$L^p(T, \nu)$, p. 45

$L^1_R(T, \nu)$, p. 121

$\mathfrak{L}^p$, p. 45

$\mathfrak{M}_f$, p. 128

$\mathfrak{M}_q(E)$, p. 76

$\mathfrak{L}_G(A)$, p. 380

$\mathfrak{L}_G(x)$, p. 15

$r(G)$, p. 127

S_E^*, p. 58

$S(x, r)$, p. 25

$S(\mu)$, p. 29

sign α, p. 19

(U), p. 107

$(\mathfrak{U}_*)$, p. 105

$\mathfrak{U}_G$, p. 375

$\underset{e \cup A}{\mathrm{Var}} f_e$, p. 192

$\|x\|_\Gamma$, p. 20

$x \perp y$, p. 91

$Z(x')$, p. 46

$\gamma(E)$, p. 76

$\Gamma_\perp$, p. 20

$\delta(A, G)$, p. 156

$\delta''(A, G)$, p. 160

$\tilde{\delta}(A, G)$, p. 357

$\Delta(A, B)$, p. 161

$\theta(G_1, G)$, p. 161

$\tilde{\theta}(G_1, G_2)$, p. 162

π_G, p. 139

$\pi_G^{-1}(g_0)$, p. 143

$\rho(x, y)$, p. 14, 15

$\tau(x, y)$, p. 88

Author Index

Ahiezer N. I., 57, 98, 110, 111, 151, 191, 218, 237, 266, 359, 369.
Akilov G. P., 366
Albinus G., 378
Alexandrov P. S., 323
Auerbach H., 273
Arens R. F., 73, 200
Aronszajn N., 142
Ascoli G., 24, 89

Banach S., 21, 22, 47, 247, 252, 267, 273, 343, 349
Barbuti U., 69
Bauer H., 76, 83
Belobrov P. K., 288
Berge C., 380
Berkson K., 162
Bernstein S. N., 150, 151, 182, 184, 189, 245, 246, 262, 264
Besicovitch S. N., 215
Birkhoff G., 91
Bishop E., 61, 76, 83
Boltiansky V. G., 242
Bonsall F. F., 22
Borsuk K., 270
Botts T., 47
Bourbaki N., 24, 29, 32, 43, 45, 57, 76, 80, 81, 91, 114, 150, 217, 248, 287, 306, 322, 340, 365, 380, 384
Bram J., 182
Brondsted A., 364
Brown A. L., 159, 268, 275, 276, 279, 282, 284, 287
Brudnyĭ Yu. A., 282
Buck R. C., 22, 162
Bunt L. N. H., 364
Burov V. N., 362

Čebyšev P. L., 13, 14, 184, 359, 369
Cheney E. W., 44, 45, 181, 237, 363
Choquet G., 59, 61
Clarkson J. A., 109, 368
Collatz L., 224
Cudia D. F., 111, 113
Curtis P. C., 219

Day M. M., 27, 58, 97, 100, 110, 111, 222, 339
Dieudonné J., 22, 102, 103
Dinculeanu N., 29
Dunford N., 18, 19, 23, 29, 45, 47, 58, 60, 69, 72, 76, 88, 89, 100, 118, 168, 222, 223, 257, 260, 292, 317, 322, 329, 334, 360, 369

Eberlein W. F., 99, 100
Efimov N. V., 23, 93, 103, 364, 365, 368, 369, 373, 374, 382, 383
Eilenberg S., 73, 74, 75
Erdös P., 376
Eskin G. I., 226
Ewald G., 215

Fan Ky, 390
Fejér L., 289
Ficken F. A., 370
Foguel S. R., 113
Fortet R., 390
Fréchet M., 267

Garkavi A. L., 16, 67, 97, 114, 115, 175, 224, 225, 233, 235, 242, 268, 281, 284, 286, 288, 289, 292, 293, 294, 299, 300, 302, 310, 314, 315, 316, 317, 319, 321, 323, 324, 325, 330, 331, 332, 334, 336, 337, 339, 344, 347, 360, 361
Gelfand I. M., 357
Gohberg I. Ts., 162, 269

Goldstein A. A., 44, 45, 181, 363
Grothendieck A.,103, 339, 345

Haar A., 215, 218, 224
Halmos P. R., 29, 39, 45, 50, 230
Halperin I., 147
Haršiladze F., 257, 261
Hausdorff F., 161
Havinson S. Ya., 22, 51, 124, 135, 237
Helly E., 292, 309
Hirschfeld R. A., 101, 111, 142, 146, 249, 250, 377, 378
Hobby C. R., 233

Ibragimov I. I., 55, 57
Iseki K., 378
Ivanov V. K., 181, 184, 187

Jackson D., 236
James R. C., 89, 92, 99, 100, 250
Jerison M., 223
Jordan D., 248, 249, 351
Jung H. W. E., 288

Kadec M. I., 257, 266, 267, 356
Kantorovič L. V., 366
Kakutani S., 248, 339, 352
Kelley J. L., 73, 200
Kirchberger P., 69
Klee V. L., 23, 59, 93, 97, 98, 101, 102, 113, 114, 116, 144, 147, 153, 266, 267, 356, 368, 369, 370, 371, 382, 383, 390
Kolmogorov A. N., 69, 70, 72, 75, 215, 241, 268, 282, 290,
Korovkin P. P., 258
Köthe G., 23, 29, 89, 110, 150, 153, 295, 367, 368, 378
Krasnoselsky M. A., 269, 270, 272, 273
Krein M. G., 16, 22, 23, 51, 54, 110, 111, 226, 229, 231, 233, 236, 237, 261, 269, 272, 273, 371, 375
Kripke B. R., 45, 51, 121, 232, 237
Kuratowsky C., 370, 386, 388

Lebedev N. A., 181, 184, 187
Lebesgue H., 288
Leeuw K. de, 76, 83
Levin A. Yu., 270
Liapunov A. A., 230
Lindenstrauss J., 114, 146
Lorentz G. G., 282
Long R. G., 266, 267
Lorch E. R., 161
Lovaglia A. R., 375
Lozinsky S. M., 257, 261

Mairhuber J. C., 219
Makovoz Yu. I., 324
Markus A. S., 162
Markuševič A. I., 264
Masani P., 146
Mazur S., 28, 29, 59, 89, 219, 367
Menger K., 378
Menšov D. E., 264
Michael E., 386
Milman D. P., 114, 116, 261, 269, 272, 273, 334
Minkowsky H., 166
Minn Mc. T. J., 215
Moroney R. M., 232
Morozov M. I., 272
Motzkin T. S., 160, 237, 364, 372, 373, 376, 378

Naimark M. A., 223
Natanson I. P., 257, 262
Neumann von J., 146, 147, 248, 249
Nicolescu M., 16, 142, 147, 364, 381, 385, 388
Nikolsky, S. M., 22, 160
Nikolsky V. N., 51, 57, 61, 70, 99, 105, 120, 151, 153, 237, 253, 255, 258, 261, 353, 354, 355, 378
Nitka W., 386

Pasqua del D., 161
Paszkowski S., 139
Phelps R. R., 16, 93, 100, 102, 103, 107, 108, 109, 110, 113, 120, 144, 218, 223, 228, 229, 231, 233, 249, 250, 251, 289, 293, 295, 296 299, 309, 315, 316, 317, 320, 323, 324, 332, 334, 336, 337, 339
Pompeiu D., 161
Pontryagin L. S., 125
Pták V., 124, 218, 237

Rademacher H., 288
Remez E. Ya., 125, 166, 180, 181, 184, 187, 191
Rice J. R., 233, 363, 372, 373, 374
Rivlin T. J., 51, 121, 181, 218, 224, 232, 237
Roberts B. D., 92

Rockafellar R. T., 45
Rogosinski W. W., 22, 37
Romanova Z. S., 32, 119, 133
Rubinstein G. Š., 241, 362
Rudin W., 250
Ruston A. F., 273, 287
Rutman M. A., 261
Ryškov S. S., 242

Saškin Yu. A., 242
Schauder J., 366
Schnirelman L. G., 187
Schoenberg I. J., 218, 219, 220, 222, 288
Schwartz J., 18, 19, 23, 29, 45, 47, 58, 60, 69, 72, 76, 88, 89, 100, 118, 168, 203, 222, 223, 257, 260, 292, 317, 322, 329, 344, 360, 369
Shapiro H. S., 22, 181, 218, 224
Sieklucki K., 219
Smirnov V. I., 181, 184, 187
Smith K. T., 142, 250
Smolyak S., 357
Stečkin S. B., 23, 93, 103, 203, 226, 364, 368, 369, 373, 375, 382, 383
Stiles W. J., 147, 249
Straszewicz E., 114
Straus E. G., 160

Tatarkiewicz K., 73, 111, 212, 266, 381, 389
Taylor A. E., 113, 293
Teliakovsky S. A., 257
Timan A. F., 55, 57, 262, 273, 282
Tihomirov V. M., 160, 268, 275, 277, 278, 281, 282, 290, 357, 358
Tonelli L., 69
Tornheim L., 372
Tukey J. W., 360
Tyuremskih I. S., 151, 153, 155

Uryson P. S., 323

Vallée-Poussin Ch. J., 187, 189
Valentine F. A., 160
Vidensky V. S., 181, 184, 187
Vituškin A. G., 282
Vlasov L. P., 368

Walsh J. L., 237, 378
Wiener N., 146, 147
Wulbert D. E., 370, 378

Yang C. T., 219, 220, 222
Young J. W., 372

Zuhovitzky S. I., 28, 91, 180, 184, 187, 188, 203, 226, 288

Subject Index

Almost Čebyšev subspace, p. 116
Alternating method, p. 146
Ascoli-Mazur theorem, p. 89
Atom, p. 229

Banach limit, p. 343
Banach space,
uniformly convex, p. 368
uniformly convex in every direction, p. 288
Banach-Stone theorem, p. 222
Baricentrically independent, p. 125
Bernstein system, p. 267
Best n-covering, p. 290
Best n-dimensional secant, p. 268
Best n-dimensional $\mathfrak{V}$-secant, p. 287
Best n-net, p. 289
Boundary, p. 82

C_1-set, p. 370
C_2-set, p. 370
Čebyšev center, p. 288
Čebyšev linear manifold, p. 243
Čebyšev point, p. 288
Čebyšev rank,
of a subspace, p. 127
of a system of functions, p. 242
Čebyšev set, p. 360, 378
Čebyšev subspace = Haar subspace, p. 103
Čebyšev system, p. 182
Choquet boundary, p. 76
Choquet space, p. 59
Closest point, p. 288
Codimension, p. 99
Compact, p. 29
approximatively, p. 368
boundedly, p. 365
conditionally = relatively, p. 365
strictly sequentially, p. 97
Cone, p. 128
Conical point, p. 23
Convex combination, p. 172
Convex set, p. 58

Deviation,
of two functions, p. 14
of a set from a linear subspace, p. 156
Diameter, p. 290
n-dimensional, p. 268
n-dimensional linear, p. 282
of order n, p. 357
Distance set, p. 93

Equivalent (= linearly isometric) spaces, p. 47
E-equivalent points, p. 75
Element of best approximation, p. 15
Element of ε-approximation, p. 162
Existence set, p. 93
Extremal element, p. 159
Extremal n-dimensional linear subspace, p. 268
Extremal point, p. 58
Extremal subset, p. 58

Face of a cell, p. 73
Farthest point, p. 160

Gowurin integral, p. 192

Haar subspace = Čebyšev subspace, p. 103
Hahn-Banach theorem, p. 18, 23, 71, 78, 99, 101, 150, 168
Hilbert space = complete inner product space, p. 146
Hyperplane, p. 24
extremal, p. 67
real, p. 27
support, p. 25
tangent, p. 28

Inclination, p. 161
Index of disjunction, p. 161
Inner product space = separated prehilbertian space, p. 57

Inverse problem of the theory of best approximation, p. 264

Juxtaelement, p. 377
Juxtapolynomial, p. 377

k-Čebyšev subspace, p. 126
k-semi-Čebyšev subspace, p. 126
(K)-semi-continuous, p. 388
  lower, p. 388
  upper, p. 388
Kernel, p. 37
Krein-Milman theorem, p. 60, 73, 74, 83, 169

Lebesgue theorem, p. 195
Linear manifold, p. 24
  real, p. 26
Linear subspace, p. 17
Locally solvent, p. 372
Locally unisolvent, p. 372
Lozinsky-Haršiladze system = Λ-system, p. 257

Maximal element, p. 19, 37
Mazur theorem, p. 29, 59, 367
Metric boundary, p. 379
Metric interior, p. 379
Metric projection = normal projection, p. 146
Monoplane point, p. 23

N-dimensional surface, p. 371
n-net, p. 289
N-parametric set, p. 371
Non-linear set, p. 359
Normal element, p. 23
Normed linear space,
  k-strictly convex, p. 127
  smooth, p. 111
  strictly convex = strictly normed = rotund, p. 110

Opening, p. 161
Orthogonal, p. 91
  in the isosceles sense, p. 92
  in the pythagorean sense, p. 92
  in the sense of B. D. Roberts, p. 92
  in the sense of [214], p. 92

Polynomial, p. 13, 165
  of best approximation, p. 14, 165
Property (B), p. 151
  absolute, p. 151
  with respect to a family of subspaces, p. 153
Property (B_f), p. 153
Property ($\mathfrak{S}_*$), p. 94
Property (U), p. 107
Property ($\mathfrak{U}_*$), p. 105
Proximinal set, p. 359, 378
Proximinal subspace, p. 93
Proximity point, p. 288

Quasi-analytic,
  element, p. 262
  function, p. 262
Quasi-normal, p. 91

Segment, p. 58
Semi-Čebyšev subspace, p. 103
Semi-continuous, p. 386
  lower, p. 386
  upper, p. 386
Simultaneous characterization, p. 22
Smooth point, p. 23
Support set, p. 25, 381

T-system, p. 266

Unisolvent, p. 372

Variation, p. 192
Varisolvent, p. 374
Vector-valued function,
  continuous (strongly), p. 191
  simple, p. 192
Vector-valued set function,
  completely additive, p. 192
  of bounded variation, p. 192
  weakly* completely additive, p. 192
  weakly* regular, p. 192
Very non-Čebyšev subspace, p. 116
Very non-proximinal subspace, p. 100

105. Rinow: Die innere Geometrie der metrischen Räume. DM 83,— . . . . US $ 22.90
106. Scholz/Hasenjaeger: Grundzüge der mathematischen Logik. DM 98,— . . . . US $ 27.00
107. Köthe: Topologische lineare Räume I. DM 78,— . . . . . US $ 21.50
108. Dynkin: Die Grundlagen der Theorie der Markoffschen Prozesse. DM 33,80 . . . . US $ 9.30
110. Dinghas: Vorlesungen über Funktionentheorie. DM 69,— . . US $ 19.00
111. Lions: Equations différentielles opérationnelles et problèmes aux limites. DM 64,— . . . . US $ 17.60
112. Morgenstern/Szabó: Vorlesungen über theoretische Mechanik. DM 69,— . . . . US $ 19.00
113. Meschkowski: Hilbertsche Räume mit Kernfunktion. DM 58,— US $ 16.00
114. MacLane: Homology. DM 62,— . . . . US $ 15.50
115. Hewitt/Ross: Abstract Harmonic Analysis Vol. 1: Structure of Topological Groups, Integration Theory, Group Representations. DM 76,— . . . . US $ 20.90
116. Hörmander: Linear Partial Differential Operators. DM 42.— US $ 10.50
117. O'Meara: Introduction to Quadratic Forms. DM 48,— . . . US $ 13.20
118. Schäfke: Einführung in der Theorie der Speziellen Funktionen der mathematischen Physik. DM 49,40 . . . . US $ 13.60
119. Harris: The Theory of Branching Processes. DM 36,— . . US $ 9.90
120. Collatz: Funktionalanalysis und numerische Mathematik. DM 58,— . . . . US $ 16.00
121.
122. Dynkin: Markov Processes. DM 96,— . . . . US $ 26.40
123. Yosida: Functional Analysis. DM 66,— . . . . US $ 16.50
124. Morgenstern: Einführung in die Wahrscheinlichkeitsrechnung und mathematische Statistik. DM 38,— . . . . US $ 10.50
125. Itô/McKean: Diffusion Processes and Their Sample Paths. DM 58,— . . . . US $ 16.00
126. Lehto/Virtanen: Quasikonforme Abbildungen. DM 38,— . . US $ 10.50
127. Hermes: Enumerability, Decidability, Computability DM 39,— . . . . US $ 10.80
128. Braun/Koecher: Jordan-Algebren. DM 48.— . . . . US $ 13.20
129. Nikodým: The Mathematical Apparatus for Quantum-Theori s. DM 144,— . . . . US $ 36.00
130. Morrey: Multiple Integrals in the Calculus of Variations. DM 78,— . . . . US $ 19.50
131. Hirzebruch: Topological Methods in Algebraic Geometry. DM 38,— . . . . US $ 9.50
132. Kato: Perturbation Theory for Linear Operators. DM 79,20 . US $ 19.80
133. Haupt/Künneth: Geometrische Ordnungen. DM 68,— . . . US $ 18.70
134. Huppert: Endliche Gruppen I. DM 156,— . . . . US $ 42.90
135. Handbook for Automatic Computation. Vol. 1/Part a: Rutishauser: Description of ALGOL 60. DM 58,— . . . . . . US $ 14.50
136. Greub: Multilinear Algebra. DM 32,— . . . . US $ 8.00

137. Handbook for Automatic Computation. Vol. 1/Part b: Grau/Hill/Langmaack: Translation of ALGOL 60. DM 64,— . . US $ 16.00
138. Hahn: Stability of Motion. DM 72,— . . . . US $ 19.80
139. Mathematische Hilfsmittel des Ingenieurs, Herausgeber: Sauer/Szabó. 1. Teil. DM 88,— . . . . US $ 24.20
140. Mathematische Hilfsmittel des Ingenieurs. Herausgeber: Sauer/Szabó. 2. Teil. DM 136,— . . . . US $ 37.40
141. Mathematische Hilfsmittel des Ingenierus. Herausgeber: Sauer/Szabó. 3. Teil. DM 98,— . . . . US $ 27.00
142. Mathematische Hilfsmittel des Ingenieurs. Herausgeber: Sauer/Szabó. 4. Teil. DM 124,— . . . . US $ 34.10
143. Shur/Grunsky: Vorlesungen über Invariantentheorie. DM 32,— US $ 8.80
144. Weil: Basic Number Theory. DM 48,— . . . . US $ 12.00
145. Butzer/Berens: Semi-Groups of Operators and Approximation. DM 56,— . . . . US $ 14 00
146 Treves: Locally Convex Spaces and Linear Partial Differential Equations. DM 36,— . . . . US $ 9.90
147. Lamotke: Semisimpliziale algebraische Topologie. DM 48,— US $ 13.20
148. Chandrasekharan: Introduction to Analytic Number Theory. DM 28,— . . . . US $ 7.00
149. Sario/Oikawa: Capacity Functions. DM 96,— . . . . US $ 24.00
150. Iosifescu/Theodorescu: Random Processes and Learning. DM 68,— . . . . US $ 18.70
151. Mandl: Analytical Treatment of One-dimensional Markov Processes. DM 36,— . . . . US $ 9.80
152. Hewitt/Ross: Abstract Harmonic Analysis. Vol. II. Structure and Analysis for Compact Groups. Analysis on Locally Compact Abelian Groups. DM 140,— . . . . US $ 38.50
153. Federer: Geometric Measure Theory. DM 118,— . . . . . US $ 29.50
154. Singer: Bases in Banach Spaces I. DM 112,— . . . . . . US $ 30.80
155. Müller: Foundations of the Mathematical Theory of Electromagnetic Waves. DM 58,— . . . . US $ 16.00
156. van der Waerden: Mathematical Statistics. DM 68,— . . . . US $ 18.70
157. Prohorov/Rozanov: Probability Theory. DM 68,— . . . . . US $ 18.70
159. Köthe: Topological Vector Spaces I. DM 78,— . . . . . US $ 21.50
160. Agrest/Maksimov: Theory of Incomplete Cylindrical Functions and their Applications. In preparation
161. Bhatia/Szegö: Stability Theory of Dynamical Systems. In preparation
162. Nevanlinna: Analytic Functions. DM 76,— . . . . US $ 20.90
163. Stoer/Witzgall: Convexity and Optimization in Finite Dimensions I. DM 54,— . . . . US $ 14.90
164. Sario/Nakai: Classification Theory of Riemann Surfaces. DM 98,— . . . . US $ 27.00
165. Mitrinovič: Analytic Inequalities. DM 88,— . . . . US $ 24.20
166. Grothendieck/Dieudonné: Eléments de Géometrie Algébrique I. En préparation

167. Chandrasekharan: Arithmetical Function. DM 58,— . . . . . US $ 16.00
168. Palamodov: Linear Differential Operators with Constant Coefficients. DM 98,— . . . . US $ 27.00
169. Rademacher: Topics in Analytic Number Theory. In preparation.
170. Lions: Optimal Control Systems Governed by Partial Differential Equations. DM 78,— . . . . US $ 21.50
171. Singer: Best Approximation in Normed Linear Spaces by Elements of Linear Subspaces. DM 78,— . . . . US $ 21.50

PRINTED IN ROMANIA